# Statics and Strength of Materials

By Alfred Jensen and Harry H. Chenoweth

Applied Engineering Mechanics, Fourth Edition
Applied Strength of Materials, Fourth Edition
Statics and Strength of Materials, Fourth Edition

# Statics and Strength of Materials

**Fourth Edition**

**Alfred Jensen**
Late Professor, Engineering
University of Washington, Seattle

**Harry H. Chenoweth**
Associate Professor, Civil Engineering
University of Washington, Seattle

**Gregg Division**
**McGraw-Hill Book Company**
New York  Atlanta  Dallas  St. Louis  San Francisco  Auckland  Bogotá
Guatemala  Hamburg  Johannesburg  Lisbon  London  Madrid
Mexico  Montreal  New Delhi  Panama  Paris  San Juan  São Paulo
Singapore  Sydney  Tokyo  Toronto

Sponsoring Editor: Cary F. Baker
Editing Supervisor: Evelyn Belov
Design and Art Supervisor: Nancy Axelrod
Production Supervisor: Priscilla Taguer

Cover Designer: Ampersand Studio
Cover Photo: Susan Berkowitz
Technical Studio: Fine Line Inc.


**Library of Congress Cataloging in Publication Data**

Jensen, Alfred.
    Statics and strength of materials.

    Includes index.
    1. Statics.   2. Strength of materials.
I. Chenoweth, Harry H.   II. Title.
TA351.J4   1982    620.1'12    82-12642
ISBN 0-07-032494-8

## Statics and Strength of Materials, Fourth Edition

   2 3 4 5 6 7 8 9 0 DOCDOC 8 9 8 7 6 5 4 3

ISBN   0-07-032494-8

# Contents

**Authors' Note:** The material contained in Chapter 10, Centroids and
Centers of Gravity, and Chapter 11, Moments of Inertia of Areas, is
also found in Chapter 5 of Part 2. It was deleted here to avoid
duplication. See Preface, third paragraph.

# Preface

The subjects of statics and strength of materials have been arranged under one cover in this text in response to many requests from teachers at schools where the study of dynamics is not included in certain specialized curricula.

This material has been taken, unchanged, from two of the authors' previously published books. "Statics," contained in Part 1 of this volume, was taken from Part One of *Applied Engineering Mechanics*. "Strength of Materials," contained in Part 2, was taken in its entirety from *Applied Strength of Materials*.

To avoid duplication of material, Chapters 10 and 11 of *Statics* were deleted from Part 1 because the same material appears in Chapter 5 of Part 2 in slightly rearranged form. The authors recommend, however, that the material in Chapter 5, Centroids and Moments of Inertia of Areas, be taught in the statics course, as is usual. By so doing, a better balance of quantity of material in each of the two courses will be achieved.

Virtually all the subject matter usually taught in college courses in statics and strength of materials has been included in this book. However, formulas and mathematical relationships have been derived using only algebra and elementary trigonometry. The text material is based on easily and commonly understood physical concepts and principles rather than on the more abstract mathematical relationships which often are not so well understood. Therefore, a working knowledge of high school mathematics is the only prerequisite. For those who would like to see a more advanced treatment of some of the derivations, using the calculus, a number of such derivations are shown in Appendixes A to D.

The authors' primary aim has been to develop and to present the material in an easily understood manner in order to make the text suitable in content and approach both for college students in engineering and architecture, where a rigorous mathematical treatment of the subjects is not required, and for those in junior and community colleges, technical institutes, and in many industrial training and Armed Services programs. Further, the authors believe that this text will be of continuing value to the many practicing engineers, architects, and engineering aides who have found the earlier separate editions useful as reference texts.

During many years of teaching these subjects to community college and technical institute students, and in industrial training programs, the authors' experiences suggest that students learn more easily by studying the two subjects separately rather than in integrated form. Statics is essentially force analysis; that is, the determination of the *total* internal forces produced in members of a structure by externally applied loads. Statics in itself is not design of any part of a structure but is only a first step leading to design. In strength of materials these total internal forces, together with predetermined allowable unit stresses are then used to determine the size of a structural member or part required to safely resist the external loads. This is design. In practice, these two steps definitely require separate solutions.

The content of Parts 1 and 2 is divided into 22 chapters, each composed of several articles. Each article presents additional theory, a new concept, or a different aspect and contains one or more completely solved problems illustrating the application of some theory or concept. A number of problems then follow, carefully arranged in order of difficulty, more than one-third of which are supplied with answers. Classroom instructors may, upon request to the publisher, obtain solutions to the problems. Each chapter closes with a summary of the important points and formulas covered in the chapter. This summary affords students a quick review of the essential parts. The summary too is followed by a series of review problems and by a number of review questions carefully arranged to test the student's grasp of the subject matter.

In arranging and developing the subject matter, the authors have proceeded very gradually from the elementary problems to those more difficult, in the honest belief that, by such gradation, students learn more quickly and easily. Their experience has been that students learn more efficiently by thoroughly mastering one step at a time and later integrating the various steps and concepts into a completed whole.

The more than 1,200 problems in this text, of which 130 are solved in complete detail, together with more than 300 review questions, give students ample opportunity to test their understanding of the subject matter. Chapter 12, Part 1, contains some interesting articles and many problems explaining and dealing with such varied, yet commonly encountered, engineering problems as hydrostatic pressure, buoyancy, hydrostatic loads, stability of retaining walls, flexible cables, and arches. Practical problems are used whenever possible, because they stimulate the students' interest and often lie within their personal experience or observation.

In the solution of analytical problems, mainly in statics, great emphasis is placed on the *complete* free-body diagram. Students are encouraged to develop this diagram gradually as the solution progresses until, upon completion of the problem, it shows all forces or their components. Simple arithmetical summations then prove the correctness of the solution without further computation. Analytical and graphical problems are presented side by side in order to show their close relationship and to encourage students to use them concurrently, one as a check on the other. Experience has shown that graphical solutions enable a student to *visualize* force analysis and definitely better the student's understanding of corresponding processes in analytical solutions.

In strength of materials, strong emphasis has been placed on the actual design of structural members, parts, and connections. This emphasis is of particular advantage to the many students who do not later take additional courses in design. It is also very helpful to those already employed in industry who have need of such a book for reference and self-study.

Because relatively few students have the opportunity to study separately the methods of manufacture and the properties of the various so-called engineering materials, a brief chapter discussing these items has been included in Part 2. The subject matter of this chapter is informative rather than technical and is intended to give students only an elementary understanding of some of the most commonly used engineering materials.

Deflections of beams are determined (1) by the beam-diagram method and (2) by the moment-area method. In the beam diagram method, two additional diagrams—the slope diagram and the deflection diagram—have been added to the load, shear, and moment diagrams to make a total of five consecutive diagrams, all of which are easily constructed by use of the two laws of beam diagrams. While this method is applicable only to a limited number of types of beams and loadings, it nevertheless has great teaching value because students understand it better than other methods and can easily construct the diagrams. More difficult problems are solved by the more complex, but excellent, moment-area method.

Special subjects, such as (1) axial stresses in members of two materials, (2) eccentrically loaded riveted joints, (3) design of laterally unsupported steel beams, (4) statically indeterminate beams, and (5) eccentrically loaded columns, have been included in the firm belief that students should be familiar with them. However, students may skip what does not seem relevant; such omissions will not be detrimental.

The authors appreciate the response of many readers who offered constructive suggestions. Appreciation is also extended to the American Institute of Steel Construction for permission to reproduce from their handbook many of the tables in the Appendix.

Alfred Jensen
Harry H. Chenoweth

# Statics and Strength of Materials

# Part 1
# Statics

# Chapter 1
# Introduction

## 1-1 DEFINITION OF MECHANICS

The science of **mechanics** treats of motion, forces, and the effects of forces on the bodies on which they act.

**Applied engineering mechanics** concerns itself mainly with applications of the principles of mechanics to the solution of problems commonly met within the field of engineering practice.

**Mechanics** is generally divided into two main branches of study: (1) statics and (2) dynamics. **Statics** is that branch which deals with forces and with the effects of forces acting on rigid bodies at rest. The subject of statics, therefore, is essentially one of *force analysis,* that is, a study of force systems and of their solutions. **Dynamics** deals with motion and with the effects of forces acting on rigid bodies in motion. Dynamics is divided into two branches: (1) **kinematics,** the study of motion without consideration of the forces causing the motion, and (2) **kinetics,** the study of forces acting on rigid bodies in motion and of their effect in changing such motion.

## 1-2 PROBLEMS IN APPLIED MECHANICS

Engineers conceive, plan, design, and construct buildings, machines, airplanes, and countless other objects for the comfort and use of the human race. Each of these objects serves a definite and useful purpose; behind each lies an absorbingly interesting story of engineering skill and achievement.

Having in mind a definite purpose for the object to be constructed, the engineer then conceives its appropriate form, either entirely new, or an improvement on one already in existence. The next step is to analyze and determine the forces, known and unknown, acting on the object, and the motions, if any, of its various related parts. To do this successfully, *the engineer must have a thorough knowledge and understanding of the principles of mechanics and of their applications to the particular problem.*

Having thus determined the forces and the motions involved, the engineer may proceed with the design of the object, using available materials of suitable strength and other requisite properties. The final size and shape of the object, and of each of its separate parts, may then be expressed in blueprints, after which the object is ready for production.

From this brief analysis of the work of the engineer, we see clearly that *a knowledge of mechanics is as fundamental to success in the field of engineering as is an understanding of the alphabet to those who would learn to read and write their own language.* The extent of this knowledge of mechanics may have an important bearing on the opportunities that will open to the student in this great field of work. Certainly, without some knowledge and training in mechanics, the student would have little or no chance of entering the engineering profession.

## 1-3 PROCEDURES IN THE SOLUTION OF MECHANICS PROBLEMS

Successful and efficient solution of any engineering problem calls for a well-organized and logical method of attack, involving a number of steps, each of which must first be well understood and then carefully executed. Among these steps, the following five include in a general way the entire process of solving any problem:

1. Analyze carefully the given data and ascertain the known quantities and the unknown quantities to be determined.
2. Recognize all the acting forces, known and unknown.
3. Decide on a suitable type of solution to use to determine the unknown quantities.
4. Formulate the steps to be taken to complete this solution.
5. Execute these steps, using available methods of checking the results.

The necessity for checking intermediate as well as final results as the solution progresses cannot be overemphasized, and yet it is most difficult to impress this fact on students, especially during the early part of their training; too many insist on dashing on to some answer, often finding it to be wrong and then, on a recheck, discovering some foolish or careless mistake which a second glance at the proper time would have quickly revealed.

## 1-4 STANDARDS OF QUALITY IN PROBLEM SOLUTION

Because of the great responsibility that attaches to the practice of engineering, high standards of work are demanded by the profession. These standards call for clear and neat figures and letters, and for uncrowded and logical arrangement of all computations and diagrams. Squared paper, 8½ by 11 inches (in.), and a straightedge for drawing diagrams are recommended. All diagrams must be complete; that is, they must show all forces, dimensions, and other items that are parts of the problem.

In order that computations may readily be checked, as they must be, all work except the simplest additions, subtractions, multiplications, and divisions must be shown. Of course, if the slide rule is used, no multiplications or divisions need be shown, but the processes must be indicated. One of the best ways of checking data and computations is to glance over them *as soon as they are completed,* to see that no mistake has been made and that the result obtained so far is *reasonable.* Often an absurd answer or a misplaced decimal point is thus quickly detected. The use of scratch paper encourages sloppy work and should, therefore, not be allowed.

In most engineering computations, a *degree of accuracy* to three significant figures is considered satisfactory. (The numbers 64,800 and 0.0648 both contain three significant figures, 6, 4, and 8.) The process of learning the subject matter and of attaining the required standards of quality is progressive. Students are encouraged, therefore, to file completed problems in a loose-leaf notebook which should always be available for ready reference.

# Chapter 2
# Basic Principles
# of Statics

## 2-1 FORCE

In his *first law of motion,* Sir Isaac Newton states that a body will continue in its state of rest or of uniform motion unless acted on by a force that changes or tends to change its state. Therefore, we may state that **force is an action that changes or tends to change the state of motion of the body on which it acts.** In statics we are interested only in bodies at rest. When applied to bodies at rest, as in statics, this definition more appropriately is that **force is an action that changes the shape of the body on which it acts.**

In his *third law of motion,* Newton states that **to every action there is an equal and opposed reaction.** A force, then, being an action, is always opposed by an equal reaction. Therefore, *forces exist not singly, but always in pairs, equal and opposite.* In analyzing forces and their effects, however, we may consider a force singly in order to study and evaluate its effect.

**A body** is any object, or any part of an object, which may be considered separately. When two objects are in contact, equal and opposite forces are produced at the contacting surfaces. A ball resting on a person's hand presents an example of two equal and opposite forces: the weight of the ball pressing down on the hand, and the hand pushing upward with a force equal to this weight, with force exerted on the two surfaces of contact.

When acted on by forces, a body is necessarily somewhat deformed; thus the relative positions of the points at which the forces are applied are changed slightly. In solid bodies, such as are normally encountered in engineering practice, these changes are so insignificant that, for the purpose of force analysis, the bodies may be considered to be "rigid" and are therefore referred to as *rigid bodies.*

6

## 2-2 TYPES OF FORCES

Force can be exerted only through the action of one physical body on another, either in contact or at a distance. Accordingly, forces may be classified under two general headings: (1) contacting or **applied forces,** such as a *push* or a *pull* which might be produced by muscular or mechanical effort; for example, a pull on a rope, the push of steam on the piston of a steam engine, or the retarding force produced by applying the brakes on a moving automobile, and (2) noncontacting or **nonapplied forces,** such as the gravitational pull of the earth on all physical bodies, magnetic force, or inertia force such as manifests itself within a body when its state of motion is changed.

A distinction is made between the **external forces** acting on a body and the resulting **internal forces,** also called **stresses,** produced in its various parts. In Fig. 2-1, for example, the external forces acting on the truss produce internal forces in its various members. The existence of these internal forces is easily visualized and can readily be shown, for should a single member be cut the truss would collapse.

A further illustration of external and internal forces is Fig. 2-2, in which an external force $P$ must be applied to the loose end of the rope passing over the sheave, in order to counteract the downward pull of the weight $W$, which is also an external force. Quite obviously, if friction on the sheave pin is disregarded, an internal force equal to the pull $P$ is produced in the rope.

External forces are further divided into **acting and reacting forces.** Gravity forces, forces caused by wind or water pressures, and all applied forces are examples of *acting forces.* Resisting forces at the supports of a structure are called *reacting forces.* Therefore, of the forces on the truss in Fig. 2-1, $P$, $T$, and $F$ are acting forces, and $R_A$, $B_V$, and $B_H$ are reacting forces. A short diagonal bar across the shank of the arrow denoting a force indicates that it is a *reacting force.*

Since mechanical forces must be resisted by material of limited strength, they must necessarily be distributed over some area of material. However, to

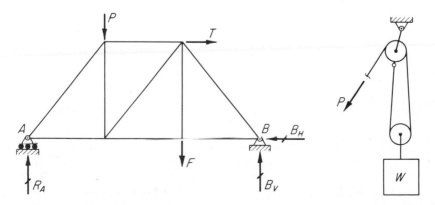

Fig. 2-1  Acting and reacting forces.                    Fig. 2-2  Internal force.

simplify the solutions of many force systems, we regard as a **concentrated force** one which is resisted by an area of material relatively so small that, for practical purposes and with inappreciable error, it may be considered as a point. The pull in a rope or a cable, the pressure of the leg of a table or a chair on the floor, and the forces exerted on the rails by the wheels of a locomotive are examples of concentrated forces.

A **distributed force** is one which acts on an area relatively too large to be considered as a point without introducing an appreciable error. The pressure on the floor of a warehouse caused by piles of goods, the wind pressure against the side of a building, and water pressure on the bottom and against the side of a tank are examples of distributed forces.

## 2-3 CHARACTERISTICS AND UNITS OF A FORCE

A force is completely described through statement of its (1) magnitude, (2) direction,[1] (3) sense, and (4) point of application. These items are called the *characteristics of the force.*

The *units* most commonly used for expressing the **magnitude** of a force are as follows:

| *English Units* | | *SI Units* | |
|---|---|---|---|
| pound | (lb) | newton | (N) |
| kilopound (1000 lb) | (kip) | kilonewton (1000 N) | (kN) |
| ton (2000 lb) | (ton) | | |

The **direction** of a force is the direction of the line along which it acts, and may be expressed as vertical, horizontal, or at some angle with the vertical or horizontal. Since an applied force is transmitted from one body to another by contact, the **point of application** is the point of contact between the two bodies. A straight line extending through the point of application in the direction of the force is called its *line of action.* The **sense** of a force is indicated by the *way* it acts along its line of action, upward or downward, to the right or to the left, and so on, and is generally denoted by an arrowhead. In Fig. 2-3, for

[1]Some authors of texts on mechanics prefer to include the *sense* of a force within the meaning of the term *direction.* However, in most problems involving the solution of an unknown force, its direction is known but its magnitude and sense are unknown. Preferably, therefore, a distinction is drawn between direction and sense.

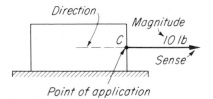

Fig. 2-3   Characteristics of a force.

example, the force applied to the block may be described thus: a 10-lb (magnitude) force acting horizontally (direction) to the right (sense) through point $C$ (point of application).

## 2 4 VECTOR AND SCALAR QUANTITIES

All quantities, such as forces, velocities, accelerations, and moments, that possess direction and sense as well as magnitude are called *vector quantities.* The graphical representation of a vector quantity in its proper magnitude, direction, and sense is called a **vector.** Vectors may be added graphically (geometrically). For example, in Fig. 2-4 the vector $AC$ in each of the three illustrations is the vector sum, or geometric sum, of vectors $AB$ and $BC$.

Quantities that have magnitude only, such as pounds, feet, and dollars, are referred to as *scalar quantities;* these may be added algebraically, and the algebraic sum, or scalar sum, has magnitude only. Figure 2-4$a$ shows that, when a number of vectors lie along the same straight line, they may be added algebraically as scalar quantities, since the direction of the vector sum remains unchanged.

## 2-5 TRANSMISSIBILITY OF FORCE

The horizontal beam shown in Fig. 2-5 supports two vertical loads, $P$ and $T$, having a common line of action and applied midway between the end reactions. Each end reaction $R$ then is $\frac{1}{2}(P + T)$. These end reactions will remain unchanged if $T$ is moved upward along the common line of action and is supported on the top of the beam, or if $P$ is moved downward and is supported by the rope.

The principle illustrated, known as the *principle of transmissibility,* is that *the point of application of an external force acting on a body may be transmitted anywhere along its line of action without changing other external forces also acting on the body.*

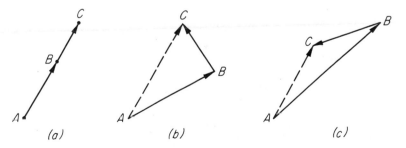

Fig. 2-4  **Graphical vector diagram.**

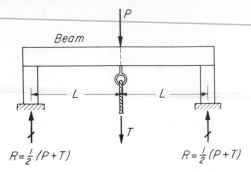

**Fig. 2-5    Principle of transmissibility.**

## 2-6 TYPES OF FORCE SYSTEMS

To simplify their solutions, force systems are conveniently classified into six groups, the solutions of which differ according to their differing characteristics. Before these groups are named, some definitions are necessary. When all forces of a system lie in one plane, the system is said to be **coplanar.** Conversely, when they do not all lie in one plane, the system is **noncoplanar.** When the action lines of all the forces of a system intersect at a common point, the system is said to be **concurrent.** Conversely, when the action lines do not intersect at a common point, the system is **nonconcurrent.** In many force systems, all forces are **parallel.** When all forces in a parallel system act along a single line of action, illustrated by a tug of war in which two or more persons pull opposite ways on a straight rope, the system is said to be **collinear.** This system is a special case of coplanar, concurrent forces.

**The six force systems,** then, are (1) *coplanar, parallel,* (2) *coplanar, concurrent,* (3) *coplanar, nonconcurrent,* (4) *noncoplanar, parallel,* (5) *noncoplanar, concurrent,* and (6) *noncoplanar, nonconcurrent.* Each of these six systems is more fully discussed in following chapters.

## 2-7 COMPONENTS OF A FORCE

**A force may be replaced by two or more components that will produce the same effect as the force they replace.** The solution of many engineering problems is greatly simplified by reducing the force system involved to one containing only vertical and horizontal forces. To obtain such a system, forces that are neither vertical nor horizontal are replaced by their vertical and horizontal components. This principle applies to all vector quantities, forces, velocities, and accelerations. Its validity may be illustrated as follows:

Consider a person walking due north from $O$ in Fig. 2-6, at the rate of 3 kilometers per hour (km/h) across the deck of a ship which is traveling due east at the rate of 4 km/h. Let these velocities be represented by the vectors $OA$ and $OB$, respectively. Evidently, the *actual* path traveled is directly from $O$ toward $C$. The vector $OC$ then represents the person's *re-*

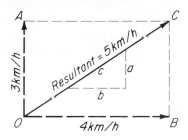

Fig. 2-6  Components of a vector.

*sultant velocity,* while *OA* and *OB* represent her or his *component ve-locities,* north and east. When the desired components are *rectangular* (at right angles to each other), they may be evaluated analytically as follows:

If the angle $\theta$ defines the direction of *OC*, then, by trigonometry,

$$OA = OC \sin \theta \quad \text{and} \quad OB = OC \cos \theta \tag{2-1}$$

If the direction of *OC* is defined by the **slope triangle** *abc,* and vector *OA* is placed in the position of side *BC,* we have *two similar triangles* in which, by geometry, corresponding sides are proportional. That is,

$$\frac{a}{OA} = \frac{b}{OB} = \frac{c}{OC}$$

By cross multiplication, we obtain

$$OA = \frac{a}{c} \cdot OC \quad \text{and} \quad OB = \frac{b}{c} \cdot OC \tag{2-1a}$$

from which it is clearly seen that $a/c = \sin \theta$ and $b/c = \cos \theta$.

Graphically, these components may readily be evaluated by laying out the diagram in Fig. 2-6 to some convenient scale.

## PROBLEMS _____

**2-1.** Determine the vertical and horizontal components of each of the forces *P* and *Q* in Fig. 2-7. Check graphically (scale: 1 in. = 500 lb).

Ans. $P_V = P_H = 0.707$ kip; $Q_V = 0.866$ kip, $Q_H = 0.5$ kip

**2-2.** In Fig. 2-8, compute the vertical and horizontal components of each of forces *P* and *Q*. Check graphically.

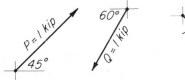

Fig. 2-7  Prob. 2-1.          Fig. 2-8  Prob. 2-2.

**2-3.** Calculate the vertical and horizontal components of each of forces $P$ and $Q$ in Fig. 2-9. Check graphically.

*Ans.* $P_V = 300$ N, $P_H = 160$ N; $Q_V = 240$ N, $Q_H = 100$ N

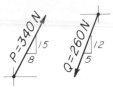

**Fig. 2-9  Prob. 2-3.**

## 2-8  RESULTANT OF TWO CONCURRENT FORCES

**Two or more forces may be replaced by a single resultant force which will produce the same result as the forces it replaces.** Clearly, then, in Fig. 2-10, if forces $OA$ and $OB$ are the components of $OC$, then $OC$ is the *resultant* of forces $OA$ and $OB$.

If the two forces are at right angles, we find by geometry that

**Resultant force:**

$$OC = \sqrt{\overline{OA}^2 + \overline{OB}^2} \qquad (2\text{-}2)$$

When the forces are not at right angles to each other, as in Fig. 2-11a, they may be laid out *graphically* to scale, each with its proper magnitude, direction, and sense. The diagonal $OC$ of the completed *parallelogram* then is the resultant force. The angle $\theta$, defining the direction of the resultant force, may be scaled by a protractor, or its tangent may be scaled as follows: From $O$, along $OB$, lay off one unit of distance (or 10) to any convenient scale. The vertical distance $t$ is then the tangent of $\theta$, the value of which may be obtained from a table of trigonometric functions.

If, in Fig. 2-11a, force $OA$ is placed in the position of line $BC$, a *triangle* $OBC$ is obtained whose diagonal $OC$ is the resultant force, as shown in Fig. 2-11b. Thus, a resultant force is obtained by the principle of vector addition, described in Art. 2-4.

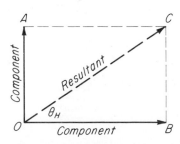

**Fig. 2-10  Resultant force.**

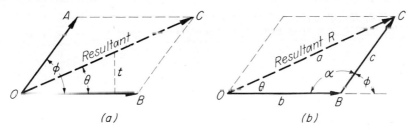

**Fig. 2-11** **Resultant of two concurrent forces.** *(a)* **Parallelogram method.** *(b)* **Triangle method.**

The resultant of two nonrectangular forces may also be found *trigonometrically* by means of the cosine law, as in Prob. 2-4, and *algebraically* by means of summations of rectangular components, as in Prob. 2-5.

A force exactly equal and opposite to a resultant force is called an **equilibrant force.**

### ILLUSTRATIVE PROBLEMS

**2-4.** Determine by trigonometry the resultant of forces $OA$ and $OB$ in Fig. 2-11 when $OA = 8$ N, $OB = 10$ N, and $\phi = 60°$. Determine also the angle $\theta$.

*Solution:* By the law of cosines, $a^2 = b^2 + c^2 - 2\,bc\,\cos\alpha$. (The cosine of $\alpha$ is positive for values of $\alpha$ up to 90° and is negative for values of $\alpha$ between 90 and 180°.) Since $\alpha = 180° - 60° = 120°$, and $\cos 120° = -\cos(180° - 120°) = -0.5$, we have

$$R^2 = (10)^2 + (8)^2 - (2)(10)(8)(-0.5) = 244 \qquad \text{and} \qquad R = 15.6 \text{ N } \textit{Ans.}$$

To find $\theta$ by the law of sines,

$$\frac{\sin\theta}{c} = \frac{\sin\alpha}{R} \qquad \text{or} \qquad \sin\theta = \frac{c\sin\alpha}{R}$$

Since $c = 8$ N, $R = 15.6$ N, and $\sin\alpha = \sin(180° - 120°) = 0.866$, we obtain

$$\sin\theta = \frac{(8)(0.866)}{15.6} = 0.4441 \qquad \text{and} \qquad \theta = 26°22' \qquad\qquad \textit{Ans.}$$

**2-5.** Determine the resultant $R$ of the two forces shown in Fig. 2-12a, and its angle of inclination $\theta_H$ with the horizontal.

*Solution:* The vertical and horizontal components of forces $P$ and $Q$ are computed in accordance with the methods described in Art. 2-7 and are then recorded in the component diagram. The components are

$$P_V = P \sin 60° = (200)(0.866) = 173.2 \text{ lb}$$
$$P_H = P \cos 60° = (200)(0.5) = 100 \text{ lb}$$
$$Q_V = \frac{5}{13} Q = \frac{5}{13}(260) = 100 \text{ lb}$$
$$Q_H = \frac{12}{13} Q = \frac{12}{13}(260) = 240 \text{ lb}$$

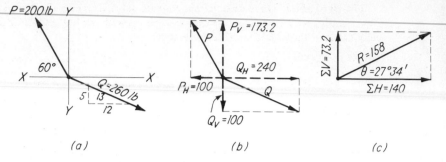

Fig. 2-12   Resultant force. Algebraic method. *(a)* Space diagram. *(b)* Component diagram. *(c)* Summation diagram.

For convenience, let upward-acting forces and forces acting to the right be positive. Summing up the vertical and horizontal components then gives

$$\Sigma V = 173.2 - 100 = 73.2 \text{ lb}$$
$$\Sigma H = 240 - 100 = 140 \text{ lb}$$

from which

$$R = \sqrt{(73.2)^2 + (140)^2} = \sqrt{5358 + 19{,}600} \quad \text{or} \quad R = 158 \text{ lb} \quad \textit{Ans.}$$

and

$$\tan \theta_H = \frac{\Sigma V}{\Sigma H} = \frac{73.2}{140} = 0.522 \quad \text{or} \quad \theta_H = 27°34' \quad \textit{Ans.}$$

**PROBLEMS**

**2-6.** Determine the resultant $R$ of the two forces shown in Fig. 2-13, and the angle $\theta_H$ it makes with the horizontal. Check graphically.

> *Ans.* $R = 4.36$ kN; $\theta_H = 36.6°$

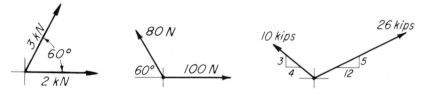

Fig. 2-13   Prob. 2-6.     Fig. 2-14   Probs. 2-7     Fig. 2-15   Prob. 2-9.
                                  and 2-10.

**2-7.** Using the cosine law, solve for the resultant $R$ of the forces shown in Fig. 2-14 and the angle $\theta_H$ it makes with the horizontal. Check graphically. (See Prob. 2-4.)

> *Ans.* $R = 91.7$ N; $\theta_H = 49°10'$

**2-8.** Using the cosine law, compute the resultant $R$ of upward-acting forces of 4 kips to the left at 45° to the horizontal and 7 kips to the right at 30° to the horizontal. Compute also the angle $\theta_H$ that the force $R$ makes with the horizontal. Check graphically (scale: 1 in. = 2 kips).

**2-9.** By algebraic summations of components, solve for the resultant $R$ of the forces shown in Fig. 2-15. Find also its direction angle $\theta_H$ with the horizontal. Check graphically (scale: 1 in. = 6 kips). (See Prob. 2-5.)

*Ans.* $R$ = 22.6 kips; $\theta_H$ = 45°

**2-10.** Solve Prob. 2-7 by the component method as shown in Prob. 2-5.

**2-11.** Calculate the resultant $R$ of forces $A$ and $B$ shown in Fig. 2-16. Use the *component method* of Prob. 2-5. Calculate also the angle $\theta_H$.

*Ans.* $R$ = 431 lb; $\theta_H$ = 21°50′

**2-12.** Determine the resultant $R$ of forces $A$, $B$, and $C$, shown in Fig. 2-17, and the angle $\theta_H$ it makes with the horizontal. (*a*) Solve analytically using the summations of $V$ and $H$ components. (*b*) Solve graphically, using the method of triangles.

**2-13.** Using an algebraic summation of components, calculate the resultant $R$ of forces $A$, $B$, and $C$ shown in Fig. 2-18. Find also the angle $\theta_H$ it makes with the horizontal.

*Ans.* $R$ = 100 lb; $\theta_H$ = 36°52′

**2-14.** Solve Prob. 2-13 if $A$ = 400 N, $B$ = 340 N, and $C$ = 160 N.

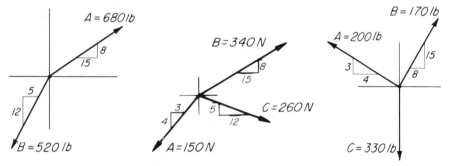

Fig. 2-16   Prob. 2-11.        Fig. 2-17   Prob. 2-12.        Fig. 2-18   Probs 2-13 and 2-14.

## 2-9 MOMENT OF A FORCE

The moment of a force is *its tendency to produce rotation of the body on which it acts,* about some axis. **The measure of a moment is the product of the force and the perpendicular distance between the axis of rotation and the line of action of the force.** This distance is called the **moment arm.** The intersection of the axis of rotation with the plane of the force and its moment arm is called the **center of moments.** Since *the center of moments is a point,* it is often referred to as such.

These definitions are illustrated in Fig. 2-19, in which the force $F$, applied to the end of a wrench in an effort to tighten a nut, rotates the wrench and the nut about an axis through the center of the nut. When $F$ is vertical,

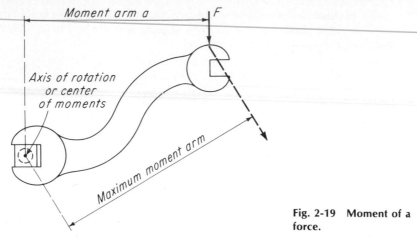

Fig. 2-19   Moment of a force.

its moment arm is *a,* and the moment of *F* then is $F \cdot a$ and is clockwise. To obtain a maximum turning effort, *F* must be so directed that its moment arm is maximum.

**Units of Moments.**  Since a moment is the product of force and distance, its unit is likewise the product of the respective units of force and distance. The basic unit of moment is the pound-foot (lb·ft) and newton-meter (N·m).[2] Other units are the pound-inch (lb·in.), the kip-foot (kip·ft), the kip-inch (kip ·in.), and the kilonewton-meter (kN·m).

The moment of a force has *magnitude* and has *direction* clockwise or counterclockwise with respect to a given moment axis. To aid in visualizing its direction at a glance, the following is recommended:

In Fig. 2-20, the magnitudes and directions of the moments of each of the four forces acting on bar *AB* about the moment center *A* are to be determined. Imagine the bar to be *hinged* at the moment center. Then, placing a finger at the moment center, consider *all* supports and forces removed except that force

[2]No agreement has yet been reached among authors as to whether the unit of moment should be the pound-foot or the foot-pound. Many authors prefer to use the foot-pound as the unit of moment, work, energy, torque, and bending moment, each of which is the product of force and distance, or of their equivalents. Others prefer the pound-foot as the unit of moment and of torque to distinguish these quantities from those of work and energy.

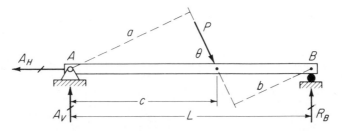

Fig. 2-20   Magnitude and direction of the moment of a force.

whose moment is desired. *The direction in which the given force will rotate the bar about the moment center is the direction of the moment* and should be obvious at a glance.

About $A$, then, the moment of $P$ is $P \cdot a$ clockwise, and the moment of $R_B$ is $R_B \cdot L$ counterclockwise. The moments of $A_V$ and $A_H$ about $A$ must both be zero, because their moment arms are zero. Similarly, about $B$ as the moment center, the moment of $P$ is $P \cdot b$ counterclockwise, that of $A_V$ is $A_V \cdot L$ clockwise, and that of $A_H$ is zero, since its moment arm with respect to $B$ is zero.

We may now state, as an important fact, that **the moment of a force about an axis passing through its line of action is zero.**

## PROBLEMS

In the following problems let counterclockwise moments be positive.

**2-15.** Compute the moment of force $F$, in Fig. 2-21, about each of points $A$, $B$, and $C$.

$$\textit{Ans.} \quad M_A = +60 \text{ lb·ft} \curvearrowleft \ ; \ M_B = +60 \text{ lb·ft} \curvearrowleft \ ; \ M_C = 0$$

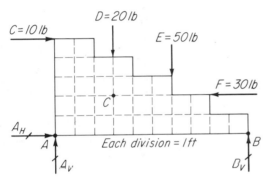

Fig. 2-21    Probs. 2-15 through 2-18.

**2-16.** Calculate the moment of force $D$, in Fig. 2-21, about each of points $A$, $B$, and $C$.

**2-17.** Find the algebraic sum of the moments of forces $C$ and $D$, in Fig. 2-21, about each of points $A$, $B$, and $C$.

$$\textit{Ans.} \quad M_A = 110 \text{ lb·ft} \curvearrowleft \ ; \ M_B = 90 \text{ lb·ft} \curvearrowleft \ ; \ M_C = 30 \text{ lb·ft} \curvearrowleft$$

**2-18.** If, in Fig. 2-21, the algebraic sum of the moments of forces $C$, $D$, $E$, $F$, and $B_V$ is zero about point $A$, what must be the value of force $B_V$?

## 2-10 THE PRINCIPLE OF MOMENTS. VARIGNON'S THEOREM

A force may be replaced by its components without changing its total effect, as is stated in Art. 2-7. Therefore, **about any point, the algebraic sum of the moments of the components of a force equals the moment of the force.** This principle is known as *Varignon's theorem.* To prove the theorem, consider the force $R$ in Fig. 2-22 and its two components $T$ and $P$, obtained by the parallelogram method. Let $\alpha$ (alpha), $\beta$ (beta), and $\gamma$ (gamma) be the angles

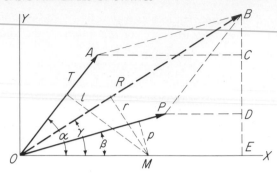

Fig. 2-22  Varignon's theorem.

between the axis $OX$ and the forces $T$, $P$, and $R$, respectively. If $t$, $p$, and $r$ are the moment arms of forces $T$, $P$, and $R$, respectively, from the center of moments $M$, we must prove that $R \cdot r$, the moment of $R$, equals $T \cdot t + P \cdot p$, the algebraic sum of the moments of $T$ and $P$, the components of $R$.

Line $AB$ is parallel to $P$ and is equal in length to $P$. Therefore, $BC$ equals $DE$. Also, we see that

$$BE = CE + BC \tag{a}$$

or $$R \sin \gamma = T \sin \alpha + P \sin \beta \tag{b}$$

Without changing the total result, each term in this equation may be multiplied by the distance $OM$, which gives

$$R(OM \sin \gamma) = T(OM \sin \alpha) + P(OM \sin \beta) \tag{c}$$

However, from Fig. 2-22, $r = OM \sin \gamma$, $t = OM \sin \alpha$, and $p = OM \sin \beta$. Substituting these values of $r$, $t$, and $p$ in Eq. ($c$), we obtain

$$R \cdot r = T \cdot t + P \cdot p \tag{2-3}$$

which is the required proof.

The forces of a system may be regarded as the components of their resultant force. Hence, **about any point, the moment of the resultant of a system of forces equals the algebraic sum of the moments of the separate forces.**

**ILLUSTRATIVE PROBLEMS**

**2-19.** An application of Varignon's theorem is shown in Fig. 2-23. For the purpose of determining $R_C$, we wish to obtain the moment of force $P$ about $A$ as the moment center. This moment may be determined in three ways:

1. The moment is $P \cdot a$. Triangles $AEB$ and $DFB$ are similar. Hence

$$\frac{a}{6} = \frac{4}{5} \quad \text{or} \quad a = \frac{(6)(4)}{5} = 4.8 \text{ m} \quad \text{and} \quad M_A = (10)(4.8) = 48 \text{ kN} \cdot \text{m}$$

2. Replace $P$ with its $V$ and $H$ components, and apply these at $D$. The components are

$$P_V = \frac{4}{5}(10) = 8 \text{ kN} \qquad \text{and} \qquad P_H = \frac{3}{5}(10) = 6 \text{ kN}$$

The algebraic sum of their moments about point $A$ then is

$$\Sigma M_A = (8)(3) + (6)(4) = 48 \text{ kN} \cdot \text{m}$$

In general, method 2 is preferable to method 1.

3. According to the principle of transmissibility (Art. 2-5), we may transmit the point of application of $P$ to any other point along its line of action without changing its effect on $R_C$. If we apply $P$ at $B$ and there replace it with its components, again we obtain

$$\Sigma M_A = (8)(6) + (6)(0) = 48 \text{ kN} \cdot \text{m}$$

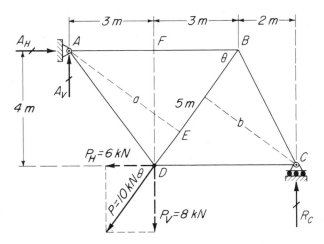

**Fig. 2-23** The moment of a force equals the algebraic sum of the moments of its components.

**2-20.** Compute the moment of the force $P_1$ in Fig. 2-24 about point $A$. $P_1 = 26$ kN, $l = 4$ m, and $h = 3$ m.

**2-21.** Calculate the moment of force $P_2$ in Fig. 2-24 about point $A$. $P_2 = 50$ kips, $l = 8$ ft., and $h = 6$ ft.

*Ans. $M_A = 240$ kip·ft*

**2-22.** Determine the resultant moment of forces $P_3$ and $P_4$ in Fig. 2-24 about $A$. $P_3 = P_4 = 170$ kN, $l = 12$ m, and $h = 5$ m.

**2-23.** In Fig. 2-24 the algebraic sum of the moments of forces at $B$, $C$, $D$, $E$, and $F$ about point $A$ is zero. Compute the value of $R_E$. Assume that $P_1 = 26$ kips, $P_2 = 10$ kips, $P_3 = 15$ kips, $P_4 = 17$ kips, $l = 16$ ft, and $h = 12$ ft.

*Ans. $R_E = 45.7$ kips*

**2-24.** The algebraic sum of the moments about point $E$ of all the forces acting on the truss shown in Fig. 2-24 is zero. Calculate $A_V$. Find $A_H$ by $\Sigma H = 0$. $P_1 = 26$ kN, $P_2 = 10$ kN, $P_3 = 15$ kN, $P_4 = 17$ kN, $l = 8$ m, and $h = 6$ m.

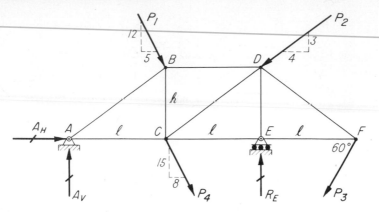

**Fig. 2-24    Probs. 2-20 through 2-24.**

## 2-11 COUPLES

Many special combinations of forces are encountered in engineering practice. Among them is a **couple,** which consists of **two equal, opposite, and parallel forces having separate lines of action.** The plane in which the forces lie is called the **plane of the couple,** and the perpendicular distance between their lines of action is the **arm of the couple.**

Such a couple is shown in Fig. 2-25. In this figure the two forces $F$ are equal and opposite and might represent the push and pull on the opposite ends of a die holder cutting threads on a steel bar. Other examples of couples are the two forces applied by the hands on opposite sides of the steering wheel of an automobile and the forces applied by thumb and forefinger in turning a nut on a bolt.

A couple has the following characteristics:

1. *The resultant force of a couple is zero.*
2. *The moment of a couple is the product of one of the forces and the perpendicular distance between their lines of action.*
3. *The moment of a couple is the same for all points in the plane of the couple.*

(To prove this fact, we will select three different moment centers in Fig. 2-25: *A, B,* and a point *C* chosen at random. Then,

$M_A = F \cdot a$ clockwise; $M_B = F \cdot a$ clockwise
$M_C = F(x + a) - F \cdot x = F \cdot x + F \cdot a - F \cdot x$
$\quad = F \cdot a$ clockwise

thus proving the statement.)

4. *A couple can be balanced only by an equal and opposite couple in the same, or in a parallel plane.* (That is, a couple cannot be balanced by a single force, since the moment of a couple is constant for all points in its plane, while the moment of a force is dependent on the position of its moment center.)

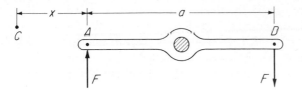

**Fig. 2-25   Moment of a couple.**

## 2-12  RESULTANT OF TWO PARALLEL FORCES

In Fig. 2-26, the magnitude of the resultant $R$ of the parallel forces $A$ and $B$ equals their algebraic sum, $A + B$. **The distances $a$ and $b$ from the action line of $R$ to the action lines of forces $A$ and $B$, respectively, are inversely proportional to the magnitudes of these forces.** That is,

$$\frac{A}{b} = \frac{B}{a} \tag{2-4}$$

Proof is obtained by the principle of moments (Art. 2-10), according to which the moment of $R$ about any point equals the algebraic sum of the moments of $A$ and $B$ about the same point. With point $c$ as the moment center, letting the counterclockwise moment be positive, we have

$$R \cdot 0 = A \cdot a - B \cdot b \qquad \text{or} \qquad Aa = Bb$$

from which

$$\frac{A}{b} = \frac{B}{a} \tag{2-4a}$$

which is the required proof.

The action line of $R$ may be located *graphically* by the **inverse-proportion method,** as shown in Fig. 2-26. On one side of any reference line, $xx$, and on the action line of $B$, is laid off to scale the magnitude of force $A$; on the opposite side of $xx$, and on the action line of $A$, is laid off the magnitude of $B$. Point $c$, where a diagonal line crosses $xx$, then lies on the action line of $R$. That this is true is seen by the fact that the two triangles thus formed are similar and that, therefore, $A/b = B/a$, as was stated above.

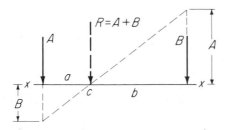

**Fig. 2-26   Resultant of two parallel forces.**

If $A$ and $B$ are oppositely directed, the resultant $R$ equals the difference between $A$ and $B$, and both are laid off *on the same side of xx.* The intersection $c$ then lies to the right of $B$ when $A$ is less than $B$, and to the left of $A$ when $A$ is greater than $B$. The following convenient rules should be remembered: *When both forces have similar senses,* their resultant lies between them; *when they have different senses,* the resultant lies outside the space between them. It always lies *near the greater* of the two forces.

## 2-13 RESOLUTION OF A FORCE INTO TWO PARALLEL COMPONENTS

Suppose that we wish to resolve the force $F$ shown in Fig. 2-27 into two parallel components, one component to be located a distance $a$ to the left of $F$ and the other a distance $b$ to the right of $F$. The magnitudes of the components will be inversely proportional to their distances from the force $F$. That is,

$$\frac{A}{b} = \frac{B}{a} \qquad \qquad (2\text{-}4b)$$

Proof is obtained by the principle of moments. Taking moments about point $c$ in Fig. 2-27a gives

$$Aa = Bb \qquad \text{from which} \qquad \frac{A}{b} = \frac{B}{a}$$

which is the required proof. The sum of components $A$ and $B$ must equal $F$.

The components $A$ and $B$ may be found graphically by the **inverse-proportion method** as shown in Fig. 2-27b. The force $F$ is drawn to scale at $c$. A line $dc$ is then drawn parallel to $xx$. The line $de$ is drawn next. This line will divide the force $F$ into the two required components $A$ and $B$.

If both components are to lie on the same side of the force, the graphical procedure shown in Fig. 2-28 should be followed. Draw the force $F$ to scale. Draw $cd$ parallel to $xx$. Next draw the line $de$ and extend it to $f$. Then draw $fg$ parallel to $xx$ and $ch$ parallel to $fe$. The line $gi$ will give the magnitude of component $A$, and $eh$ will give the magnitude of component $B$. It will be noted

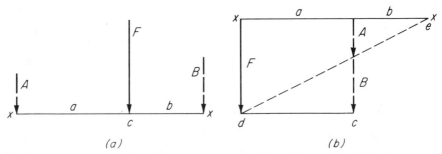

Fig. 2-27   Resolution of a force into two parallel components. *(a)* Force $F$ to be resolved into two components. *(b)* Method of resolution.

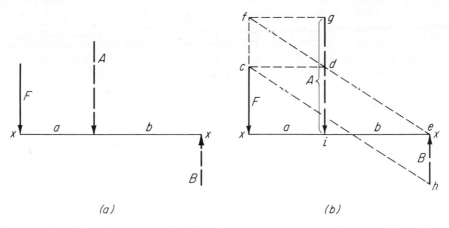

*(a)*                                     *(b)*

**Fig. 2-28  Resolution of a force into two parallel components on the same side of the force. (a) Force and location of components. (b) Method of resolution.**

that the component nearest the force $F$ will be *larger* than $F$ and that the component farthest from $F$ will be reversed in *sense* from that of $F$.

## PROBLEMS

In each of the following problems determine *graphically by the inverse-proportion method* the magnitude of the resultant force $R$ and the horizontal distance $x$ from $A$ to its line of action. Check analytically by moments.

**2-25.** See Fig. 2-29.

*Ans. $R = 14$ kips↓; $x = 3$ ft right*

**2-26.** See Fig. 2-30.

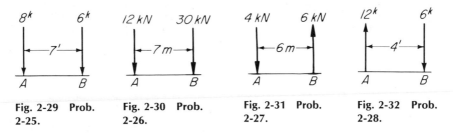

**Fig. 2-29  Prob. 2-25.**   **Fig. 2-30  Prob. 2-26.**   **Fig. 2-31  Prob. 2-27.**   **Fig. 2-32  Prob. 2-28.**

**2-27.** See Fig. 2-31.

*Ans. $R = 2$ kN↑; $x = 18$ m right*

**2-28.** See Fig. 2-32.

In each of the following problems determine *graphically by the inverse-proportion method* the magnitudes of the parallel components at $A$ and $B$ of the force shown. Check analytically by moments.

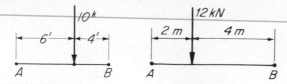

Fig. 2-33  Prob. 2-29.    Fig. 2-34  Prob. 2-30.

**2-29.** See Fig. 2-33.

$Ans.\ A = 4$ kips; $B = 6$ kips

**2-30.** See Fig. 2-34.
**2-31.** See Fig. 2-35.

$Ans.\ A = 90$ N↓; $B = 30$ N↑

**2-32.** See Fig. 2-36.

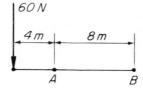

Fig. 2-35  Prob. 2-31.          Fig. 2-36  Prob. 2-32.

## 2-14 EQUILIBRIUM OF FORCE SYSTEMS

Equilibrium is essentially a state of balance. Equilibrium in a force system is, then, a state of balance between opposing forces within the system, and a balance about any point of the moments of all the forces in the system. If a body at rest is acted on by a system of forces and remains at rest, the system is *balanced* and the body is said to be in **static equilibrium.** Within the subject of statics all bodies are presumed to remain at rest.

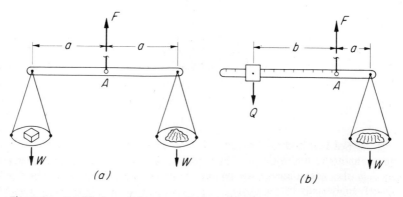

Fig. 2-37  Equilibrium of forces and of moments.

The meaning of equilibrium may be simply illustrated as in Fig. 2-37. The simple balance shown in Fig. 2-37*a* is symmetrical about the fulcrum *A,* and the weights of its various parts therefore balance about *A.* The magnitude of an unknown weight *W,* placed on one scale pan, is ascertained when a known and equal weight, placed on the opposite pan, balances the weighing arm in a horizontal position.

To have *complete equilibrium,* the single upward-acting force *F* must equal the sum of all the downward-acting forces, which are the two weights *W* and the weight of the entire balance, and the tendencies of the forces to the right of *A* to cause clockwise rotation about *A* must be equaled by the tendencies of the forces on the left to cause counterclockwise rotation.

Figure 2-37*b* shows especially well a balance of moments of unequal forces. The scale balances with *W* and *Q* removed. When the unknown weight *W* is placed on the weighing pan, the position of *Q* is adjusted until the arm balances horizontally. Then, the clockwise moment $W \cdot a$ about *A* is equaled by the counterclockwise moment $Q \cdot b$, and a state of equilibrium has been reached.

**A body at rest is in static equilibrium under the combined action of all forces acting on it.** When *a* forces acting on a body are considered, we have a **complete force system.** Hence, **a complete force system is always in equilibrium.**

**Force Law of Equilibrium.** Equilibrium implies a *balance of opposing forces* within a system. Therefore, *in any direction,* **the algebraic sum of all forces acting on a body in static equilibrium is zero.** This statement is known as the *force law of equilibrium.* Stating the law in symbols, we have

$$\Sigma F = 0 \tag{2-5}$$

The Greek letter $\Sigma$ (sigma) means "the algebraic sum of." To be assured of *complete* force equilibrium, the algebraic sum of the forces must be shown to be zero in at least two directions, usually vertical and horizontal (*V* and *H*). We may then substitute for Eq. (2-5) the following *two equations of force equilibrium:*

$$\Sigma V = 0 \tag{2-6}$$
$$\Sigma H = 0 \tag{2-7}$$

If the directions of the summations are not vertical and horizontal, the summations may be indicated thus: $\Sigma F_x = 0$ and $\Sigma F_y = 0$, where *x* and *y* are any two rectangular axes. The principle embodied in this law is especially useful in determining the unknown forces of a *concurrent* system in equilibrium.

**Moment Law of Equilibrium.** Equilibrium also implies a *balance of opposing moments* of the forces within a system. Therefore, about *any* moment axis, **the algebraic sum of the moments of all forces acting on a body in static equilibrium is zero.** This statement is known as the *moment law of equilibrium.* Stating this law in symbols, we obtain *the equation of moment equilibrium:*

$$\Sigma M = 0 \qquad\qquad (2\text{-}8)$$

About *any* axis, then, the sum of the clockwise moments must be equal to the sum of the counterclockwise moments. The principle embodied in this law is especially useful in determining the unknown forces of a *nonconcurrent* system in equilibrium.

**Signs of Forces and Moments.** The preceding discussions have indicated that no sign convention is necessary in the summations of forces and of moments of force in equilibrium. That is, *in any one direction,* the sum of the forces acting *one way* is simply equated to the sum of the forces acting the *opposite way.* Similarly, the sum of the clockwise moments is equated to the sum of the counterclockwise moments.

Occasionally, when dealing with groups of forces *not* in equilibrium, and in some instances of force systems *in* equilibrium, signs are conveniently given to forces and moments. When signs are so given, *forces acting upward or to the right are positive, and forces acting downward or to the left are negative.* Similarly, *counterclockwise moments are usually positive, and clockwise moments are negative.* [3]

## 2-15 PRINCIPLES OF FORCE EQUILIBRIUM

The following principles are extremely important, for they are very useful in the solution of many engineering problems:

**The Two-force Principle. When two forces are in equilibrium, they must be equal, opposite, and collinear.** Proof of this principle is embodied in Newton's third law, which states: To every action there is an equal and opposed reaction.

**The Three-force Principle. When three nonparallel forces are in equilibrium, their lines of action must intersect at a common point.** The proof of this principle may be deduced from the two-force principle and the principle of resultant force. Let the three forces $P$, $Q$, and $S$, acting on the body shown in Fig. 2-38, be in equilibrium. Then let $Q$ and $S$ be replaced by their resultant $R$. Now the two forces $R$ and $P$ are in equilibrium and hence are collinear. The intersection $O$ of the action lines of $Q$ and $S$ lies on the action line of $P$. Hence the three forces are concurrent; that is, their lines of action intersect at a common point.

Also, *when the three concurrent forces are in equilibrium, they must be coplanar.* This relation we deduce from the fact that the intersecting action

---

[3]The convention of designating forces acting upward and to the right as being positive is perhaps universal. In the matter of moments, however, no agreement exists among authors. But in mechanics, most authors designate the counterclockwise moment as being positive.

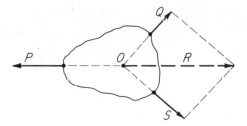

**Fig. 2-38   The three-force principle.**

lines of two concurrent forces form a *plane.* The action line of the third force must also lie in that plane; otherwise, that force would have an *unbalanced component* perpendicular to the plane, and the system would not be in equilibrium.

**The Four-force Principle.  When four forces are in equilibrium, the resultant of any two of the forces must be equal, opposite, and collinear with the resultant of the other two.** Again, proof of this principle may be deduced from the two-force principle and the principle of resultant force. The four forces divide into two pairs. When each pair is replaced with a resultant, the two resultants are then in equilibrium and hence must be equal, opposite, and collinear. This principle of two equal and opposite resultant forces applies, of course, to any number of forces in equilibrium.

## 2-16  SUPPORTS AND SUPPORT REACTIONS

When a body exerts a force against a point of support, the support will *react* with a force of equal magnitude but with opposite sense. Here, then, action and reaction are equal, opposite, and collinear. To be solved by the laws of statics only, a single, rigid body may be supported at not more than two points in one plane, or at three points in space.

The truss shown in Fig. 2-39*a* is held to its support at *A* by a pin, presumed to be frictionless, thus allowing *rotation* of the truss in its plane but preventing

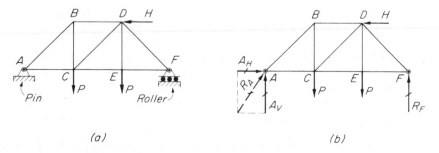

*(a)*            *(b)*

**Fig. 2-39  Supports and support reactions. *(a)* Bridge truss and supports. *(b)* Supports replaced by reactions.**

*displacement.* ~~To allow for small expansions or contractions of the truss due~~
to changing loads or to changes in temperature, some freedom of displacement
is provided at the other support *B* by placing that end of the truss on **rollers.**
No resistance (force) is then offered to displacement parallel to the surface
supporting the rollers.

For lighter structures or single members, such as the beam shown in Fig.
2-40, an equivalent condition is provided by assuming one end to rest on an
ideally *smooth surface* which likewise offers no resistance (force) to displacement parallel to itself.

Hence **the reacting force at a roller support, or at a smooth surface, is
always perpendicular to the supporting surface.** Only one reacting force exists
at each support. The direction of the reaction at a pinned support is generally
unknown. When so unknown, the reaction is usually replaced with its rectangular components acting in known directions, as at support *A* in Fig. 2-39*b,*
thus simplifying analytical solution.

## 2-17 FREE-BODY DIAGRAMS

Two common engineering problems are (1) to determine the external forces,
called *support reactions,* at the supports of a structure and caused by the loads
it carries and (2) to determine the internal forces, called *stresses,* which the
external forces produce in the various members or parts of the structure. The
first essential step in determining such unknown forces is to recognize *where*
and *in which direction* they act. The following paragraphs outline an important
aid in *visualizing* all these unknown factors.

For example, to determine the reactions at supports *A* and *F* of the truss
in Fig. 2-39, the supports are removed and are replaced with the support
reactions $R_A$ and $R_F$ which hold it in equilibrium. The vertical and horizontal
components $A_V$ and $A_H$ at support *A* in turn replace the reaction $R_A$, since the

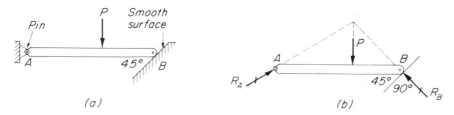

**Fig. 2-40   Support reactions.** *(a)* **Beam with supports.** *(b)* **Supports replaced by
reactions.**

direction of $R_A$ is unknown. When so isolated from its supports, a body is said to be *free* and is called a **free body.** A diagram of the body showing *all* forces acting on it is called a **free-body diagram.** The unknown forces may now be determined because *the complete force system thus shown is in equilibrium.* A similar situation is shown in Fig. 2-40.

When all external forces acting on a body are known, the *internal forces,* or stresses, in its various members or parts may be determined by several methods. In one method of determining stresses in members of trusses, the members are cut near a joint, as at joints $A$ and $B$ in Fig. 2-41a. The joint is then isolated as a free body, and a *free-body diagram* is drawn. The external forces acting at the joint, together with the internal forces in the cut members, then form a *complete* force system which is in equilibrium, and the unknown forces may thus be determined.

In another method of determining stresses in members of trusses, a section is cut through the entire truss, as in Fig. 2-41b. Either of the two parts is then isolated as a free body, and a *free-body diagram* is drawn. Again a *complete* force system is formed by the external forces acting on the part isolated and the internal forces in the cut members; this system then is in equilibrium and the unknown forces may thus be determined.

From the foregoing discussion, we conclude that **a body isolated from its supports, or any part of such a body isolated from the remainder, is in static equilibrium under the influence of all forces acting on it.**

## 2-18 PROBLEMS IN EQUILIBRIUM OF COPLANAR FORCE SYSTEMS

By use of the three equations of equilibrium, $\Sigma V = 0$, $\Sigma H = 0$, and $\Sigma M = 0$, we may determine three unknown quantities of any coplanar, nonparallel

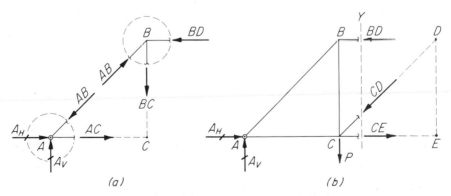

**Fig. 2-41   Free-body diagram.** *(a)* A joint is isolated. *(b)* A section is isolated.

force system in equilibrium, or two unknown quantities of a parallel system in equilibrium. An unknown quantity may be (1) the magnitude and sense of a force, (2) the direction of the line of action of a force, or (3) the loction of the line of action of a force or of its point of application. The most commonly occurring problems in force analysis involve the determination of either two or three unknown magnitudes of forces together with their unknown senses. Problem 2-33 illustrates the determination, in a single problem, of all these three quantities. To emphasize methods and to simplify solutions, the weight of a structure mentioned in a problem is generally disregarded.

The true sense of an unknown force can usually be visualized at a glance. If not, a probable sense is assumed. If the assumption is correct, the numerical value of the force, when solved for, will be positive; if incorrect, it will be negative.

## ILLUSTRATIVE PROBLEMS

**2-33.** The horizontal plank shown in Fig. 2-42a is acted on by forces $P$ and $Q$ and is supported by a frictionless pin at $C$. We need to determine (a) the magnitude of the reacting force $R$ at support $C$, (b) the direction of its line of action, as indicated by the direction angle $\theta$, and (c) the location of its point of application $C$ in order that equilibrium may be maintained. All forces are in pounds. The weight of the plank is disregarded.

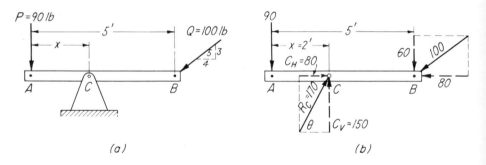

**Fig. 2-42   Static equilibrium. (a) Space diagram. (b) Free-body diagram.**

*Solution:* Our first step is to remove the support, thus isolating the plank as a free body, and then to draw the free-body diagram shown in Fig. 2-42b. To simplify the solution, force $Q$ is replaced by its $V$ and $H$ components. By similar triangles,

$$\frac{Q_V}{3} = \frac{Q_H}{4} = \frac{100}{5} \quad \text{or} \quad Q_V = \frac{3}{5}(100) = 60 \text{ lb and } Q_H = \frac{4}{5}(100) = 80 \text{ lb}$$

Since the direction of $R_C$ is unknown, we replace it with its components $C_V$ and $C_H$. We now have a *complete* force system in equilibrium. The component $C_V$ clearly acts upward, and $C_H$ acts to the right. Applying the two equations of force equilibrium, we have

$[\Sigma V = 0]$    $C_V = 90 + 60 = 150$ lb
$[\Sigma H = 0]$    $C_H = 80$ lb
Then        $R_C = \sqrt{C_V^2 + C_H^2} = \sqrt{(150)^2 + (80)^2} = 170$ lb        *Ans.*

We next obtain the direction angle $\theta$,

$$\tan \theta = \frac{150}{80} = 1.875 \quad \text{and} \quad \theta = 61°56' \qquad Ans.$$

All forces are now known. A moment equation about $A$ (or $B$) will give the location of support $C$ required for equilibrium. When $A$ is chosen as the center of moments, the moments of each of the two 80-lb forces become zero and are therefore eliminated from consideration. Thus

$$[\Sigma M_A = 0] \quad 150x = (60)(5) = 300 \quad \text{and} \quad x = 2 \text{ ft} \qquad Ans.$$

We note with interest here that this simple problem contains in its solution nearly all the basic principles of statics, namely: (1) force, (2) components of a force, (3) resultant force, (4) moment of a force, (5) the principle of moments, (6) force equilibrium, and (7) moment equilibrium.

**2-34.** The beam shown in Fig. 2-43a is resting freely on supports $A$ and $B$. Determine the reacting forces $R_A$ and $R_B$ necessary at these supports for equilibrium.

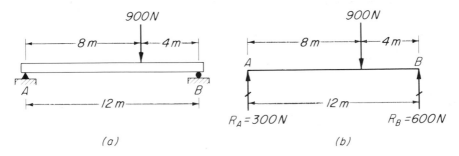

**Fig. 2-43   Equilibrium of a parallel force system.** *(a)* **Space diagram.** *(b)* **Free-body diagram.**

*Solution:* The supports are removed and are replaced by the two reacting forces, $R_A$ and $R_B$, as shown in the free-body diagram (Fig. 2-43b). The roller at $B$ rests on a horizontal surface. Hence, $R_B$ is vertical. The force $R_A$ is also vertical, because, when two of three forces in equilibrium are parallel, all must be parallel. Both reactions are clearly upward-acting forces. Then only two unknown magnitudes need be determined. A moment equation with $B$ as the moment center will eliminate $R_B$ from consideration since its moment about $B$ is zero, and $R_A$ may be solved for. A second moment equation with $A$ as the moment center will give $R_B$ *independently.* A summation of vertical forces will furnish the required check. The equations are

$[\Sigma M_B = 0]$   $12R_A = (900)(4) = 3600$    and    $R_A = 300$ N          *Ans.*

$[\Sigma M_A = 0]$   $12R_B = (900)(8) = 7200$    and    $R_B = 600$ N          *Ans.*

$[\Sigma V = 0]$      $900 = 300 + 600 = 900$                                  *Check*

**2-35.** Figure 2-44 shows a ladder supporting a person weighing 200 lb. The ladder rests on the floor at $A$ and against a smooth wall at $B$. The negligible friction at $B$ may be disregarded. Compute the horizontal reaction $R_B$ exerted by the wall at $B$, the vertical reaction component $A_V$ exerted by the floor at $A$, and the horizontal reaction component $A_H$ (friction or applied) necessary to prevent the bottom of the ladder from slipping to the left. Neglect the weight of the ladder.

*Solution:* Three unknown magnitudes of forces are to be determined. Their correct senses are readily deduced, and are as shown in the free-body diagram, Fig. 2-44*b*. Three simple steps solve this problem: (*a*) a moment equation about $A$ eliminates the moments of $A_V$ and $A_H$, and $R_B$ may be solved for; (*b*) $A_V$ is then found by a summation of vertical forces; and (*c*) $A_H$ by a summation of horizontal forces. A moment equation with $C$ as moment center will then *check* the results obtained. The equations are

$[\Sigma M_A = 0]$     $20R_B = (200)(4) = 800$    and    $R_B = 40$ lb          *Ans.*
$[\Sigma V = 0]$        $A_V = 200$ lb                                              *Ans.*
$[\Sigma H = 0]$        $A_H = 40$ lb                                               *Ans.*
$[\Sigma M_C = 0]$   $(200)(4) = (40)(10) + (40)(10)$    or    $800 = 800$          *Check*

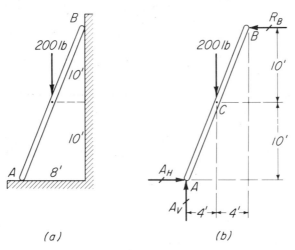

(a)                    (b)

**Fig. 2-44   Equilibrium of a nonconcurrent force system.** *(a)* **Space diagram.** *(b)* **Free-body diagram.**

PROBLEMS ——————————————————————————————————

**2-36.** Compute the reacting forces $R_A$ and $R_B$ at the ends of the beam shown in Fig. 2-45, required to maintain equilibrium.

*Ans.* $R_A = 70$ lb; $R_B = 30$ lb

**2-37.** The beam shown in Fig. 2-46 is resting freely on supports at $A$ and $B$. Draw a free-body diagram of the beam and compute the vertical reacting forces $R_A$ and $R_B$.

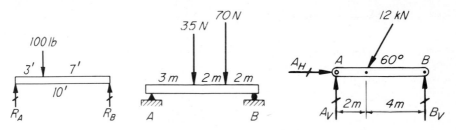

Fig. 2-45  Prob. 2-36.      Fig. 2-46  Prob. 2-37.      Fig. 2-47  Prob. 2-39.

**2-38.** Calculate the reaction components shown in Fig. 2-23.

**2-39.** In Fig. 2-47 is shown a beam supported at $A$ and $B$. Compute the reacting forces $R_B$, $A_V$, $A_H$, and $R_A$ necessary for equilibrium.

$$Ans. \ R_B = 3.46 \ kN; \ A_V = 6.93 \ kN; \ A_H = 6 \ kN; \ R_A = 9.17 \ kN$$

**2-40.** The frame shown in Fig. 2-48 carries a vertical load at $C$ and is supported at $A$ and $E$. Compute the reaction $R_A$ and the reaction components $E_V$ and $E_H$ required for equilibrium.

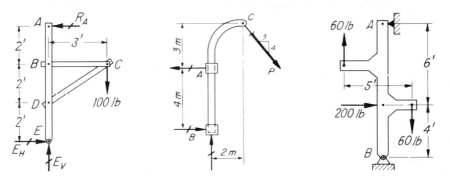

Fig. 2-48  Prob. 2-40.          Fig. 2-49  Prob. 2-41.          Fig. 2-50   Prob. 2-42.

**2-41.** Figure 2-49 is shows a mast acted on by a 10-kN force at $C$ and supported at $A$ and $B$. Compute the reacting forces.

**2-42.** The frame shown in Fig. 2-50 is acted on by a force and a couple as shown and is supported at $A$ and $B$. No vertical force can be resisted at $A$. Draw a free-body diagram of the frame, and solve for the vertical and horizontal reacting forces, at $A$ and $B$, required for equilibrium.

**2-43.** The materials hoist shown in Fig. 2-51 is held against motion by the cable tension $T$. The load $P$ is 4 kN. Find the horizontal reactions on the wheels at $A$ and $B$. Neglect the weight of the hoist. The hoist dimensions are $a = 0.5$ m, $b = 0.5$ m, and $c = 2$ m.

$$Ans. \ A_H = 1 \ kN \leftarrow; \ B_H = 1 \ kN \rightarrow$$

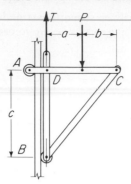

**Fig. 2-51    Probs. 2-43 and 2-44.**

**2-44.** Solve Prob. 2-43 if $P = 5$ kN, $a = 0.75$ m, $b = 0.5$ m, and $c = 3.75$ m.

### SAMPLE PROBLEM SOLUTION SHEET ⎯⎯⎯⎯⎯⎯⎯⎯⎯⎯⎯⎯⎯⎯

**Paper:** 8½ by 11 in. squared.

**Heading:** Left to right, course number, date, name, problem number.

**Problem Number:** The number 18 represents the student's consecutive problem number; 2.35 is the number of the problem in the book; and the third number is the page number in the book.

**Upper Right Corner:** *Upper,* student's consecutive sheet number; *lower,* total number of sheets in the solution.

**2-45.** The frame shown in Fig. 2-52 is subjected to the horizontal load of 6 kips as shown. Find the vertical reaction at $B$ and the vertical and horizontal reactions at $A$

$Ans.$ $B_V = 4$ kips↑; $A_V = 4$ kips↓; $A_H = 6$ kips←

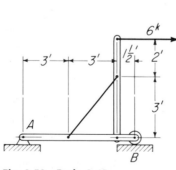

**Fig. 2-52    Prob. 2-45.**

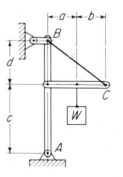

**Fig. 2-53    Probs. 2-46 and 2-47.**

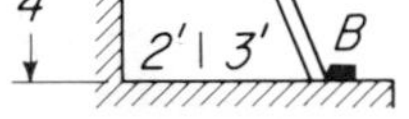

**Fig. 2-54    Prob. 2-48.**

**2-46.** The frame shown in Fig. 2-53 supports a weight $W = 1200$ lb. The link at $B$ can supply only a horizontal reaction at $B$. Find the horizontal reaction at $B$ and the vertical and horizontal reacting forces at $A$. The dimensions of the frame are $a = 2$ ft., $b = 2$ ft, $c = 5$ ft, and $d = 3$ ft.

$Partial$ $Ans.$ $A_H = 300$ lb→

| C.E. 112 | 1-16-80 | Jones, R. C. | 18-2.35 (Page No.) | I |
|---|---|---|---|---|

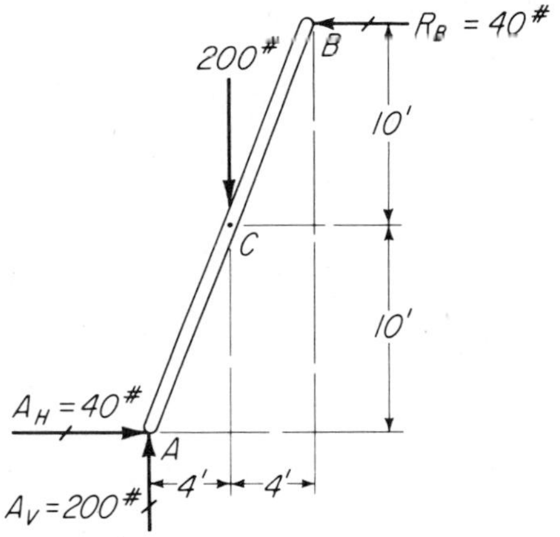

Free-body diagram

Find: $R_B$, $A_H$, $A_V$

| | |
|---|---|
| To find $R_B$, $\Sigma M_A = 0$:<br>$20 R_B = (200)(4) = 800$ | $R_B = 40^{\#} \leftarrow$    $R_B$ |
| To find $A_H$, $\Sigma H = 0$:<br>$A_H = R_B = 40^{\#}$ | $A_H = 40^{\#} \leftarrow$    $A_H$ |
| To find $A_V$, $\Sigma V = 0$:<br>$A_V = 200$ | $A_V = 200^{\#} \leftarrow$    $A_V$ |
| Check, $\Sigma M_C = 0$:<br>$(200)(4) = (40)(10) + (40)(10)$<br>$800 = 400 + 400$ | Check |

**2-47.** Solve Prob. 2-46 if $W = 5$ kN; $a = 1$ m, $b = 1$ m, $c = 3$ m, and $d = 2$ m.

**2-48.** A ladder $AB$ is leaning against a smooth, vertical wall at $A$ in Fig. 2-54 and is resting on a horizontal floor at $B$. A person weighing 200 lb exerts a vertical force at $C$; another person applies a horizontal force of 60 lb at $D$ to help reduce the horizontal pressure against the block at $B$. Draw a free-body diagram of this ladder; compute the horizontal reacting force at $A$, and the vertical and horizontal reaction components at $B$, necessary to maintain equilibrium. Check by moments about $E$.

*Ans.* $R_A = 70$ lb→; $B_V = 200$ lb↑; $B_H = 10$ lb←

**2-49.** The horizontal beam in Fig. 2-55 is acted upon by a single force $P = 3400$ N at $B$ and is supported at $A$ and $D$ as shown. Draw a free-body diagram of the beam, and compute the vertical and horizontal reacting forces at pins $A$ and $C$.

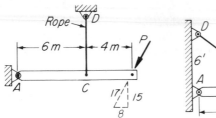

**Fig. 2-55   Prob. 2-49.**          **Fig. 2-56   Prob. 2-50.**

**2-50.** Draw a free-body diagram of the horizontal member $AB$, shown in Fig. 2-56, and compute the vertical and horizontal reacting forces at pins $A$ and $C$. Compute also the tension $T$ in the rope $CD$.

$$Ans. \ A_V = 500 \ \text{lb}\downarrow; \ A_H = 2000 \ \text{lb}\rightarrow;$$
$$C_V = 1500 \ \text{lb}\uparrow; \ C_H = 2000 \ \text{lb}\leftarrow; \ T = 2500 \ \text{lb}$$

**SUMMARY** (By Article Number) ————— ——————————————————

**2-1. Force** is an action which changes or tends to change the state of motion of the body upon which it acts. A force is completely defined when its (1) magnitude, (2) direction, (3) sense, and (4) point of application are known. Force is a *vector quantity.*

**2-4. A vector** is a graphical scale representation of a vector quantity. When vectors are added graphically, the result is called their **vector sum.**

**2-5. Transmissibility** of a force means that its point of application may be transmitted to any other point along its line of action, without changing other *external* forces also acting on the body.

**2-6. Coplanar forces** all lie in one plane; **noncoplanar forces** do not all lie in one plane. The lines of action of **concurrent forces** all pass through a common point; those of **nonconcurrent forces** do not. All force systems are combinations of these four types. When all forces of a system have a common line of action they are said to be **collinear.**

**2-7. The components of a force** acting through its points of application may replace it without change in effect.

**2-8. A resultant force** will produce the same *external* effect as the forces it replaces. An **equilibrant force** is exactly equal and opposite to a resultant force.

**2-9. The moment** of a force is its tendency to produce rotation of the body on which it acts and is the product of the force and the perpendicular distance between the axis of rotation and the line of action of the force. This distance is called the **moment arm.** In coplanar systems, the moment axis becomes a point, called the **center of moments.**

**2-10. The Principle of Moments. Varignon's Theorem.** The algebraic sum of the moments of the components of a force about any point in their plane equals the moment of the force about the same point.

**2-11. A couple** consists of two equal, opposite, and parallel forces with separate lines of action. The plane in which the forces lie is called the **plane of the couple.** The **moment of a couple,** *about any point in this plane,* is the product of one force and the perpendicular distance between the action lines of the two forces. This perpendicular distance is called the **arm of the couple.**

**2-14. Equilibrium** is a balance of opposing forces and opposing moments. For all systems of forces in equilibrium, **the algebraic sum of the acting and reacting forces is zero,** or, in symbols, $\Sigma F = 0$. Likewise, **the algebraic sum of the acting and reacting moments is zero,** or $\Sigma M = 0$.

**2-15. Principles of Force Equilibrium.** Three principles of force equilibrium are:

**The Two-force Principle.** When *two forces* are in equilibrium, they must be equal, opposite, and collinear.

**The Three-force Principle.** When *three nonparallel forces* are in equilibrium, their lines of action intersect at a common point, and they are coplanar.

**The Four-force Principle.** When *four forces* are in equilibrium, the resultant of any two of the forces must be equal, opposite, and collinear with the resultant of the other two.

**2-16.** The reaction of a **roller support,** or at a **smooth surface,** is always perpendicular to the supporting surface. The direction of the reaction at a pinned support is generally unknown.

## REVIEW PROBLEMS _____

**2-51.** Determine the $V$ and $H$ *components* of a 4-kip force acting upward and to the right at an angle of 70° with the horizontal. Check graphically (scale: 1 in. = 1 kip).

*Ans.* $V = 3.76$ kips↑; $H = 1.37$ kips→

**2-52.** An 18-N force acts upward and to the right with a slope of two units vertical to three units horizontal. Compute the $V$ and $H$ *components.* Check graphically (scale: 1 cm = 10 N).

**2-53.** Using the *sine law,* resolve a 100-lb force acting upward and to the right at 45° with the horizontal, into two *components P* and *Q* also acting upward and to the right at angles of 20° and 80° with the horizontal, respectively. Check graphically (scale: 1 in. = 20 lb).

*Ans.* $P = 66.3$ lb; $Q = 48.8$ lb

**2-54.** Using the *cosine law,* solve for the *resultant R* of two 10-kip forces, one acting upward and to the right at 30° with the horizontal and the other upward and to the left at 70° with the horizontal. Also find the angle $\theta_H$ the resultant makes with the horizontal. Check graphically (scale: 1 in. = 4 kips).

*Ans.* $R = 15.3$ kips; $\theta_H = 70°$

**2-55.** The right end of the beam shown in Fig. 2-57 rests on a smooth, frictionless plane. Calculate the vertical and horizontal components of the reactions at supports *A* and *B*.

**2-56.** The truss shown in Fig. 2-58 is loaded as shown. Solve for the vertical and horizontal components of the reaction at *A* and the vertical reaction at *B*. $T = 10$ kN, $P = 25$ kN.

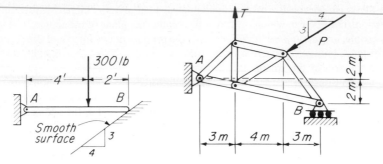

**Fig. 2-57   Prob. 2-55.**          **Fig. 2-58   Prob. 2-56.**

**2-57.** The pole $AB$ shown in Fig. 2-59 leans against a smooth, frictionless wall at $B$. Calculate the vertical and horizontal components of the reactions at $A$ and $B$.

Ans. $A_V = 300$ lb↑; $A_H = 150$ lb→; $B_H = 50$ lb←

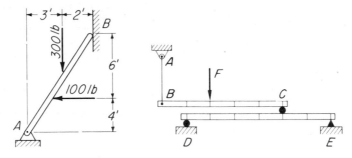

**Fig. 2-59   Prob. 2-57.**     **Fig. 2-60   Prob. 2-59.**

**2-58.** Solve Prob. 2-54 by the algebraic-component method. Draw three diagrams similar to those shown in Fig. 2-12.

**2-59.** Beam $BC$ in Fig. 2-60 is supported at its left end by the rope $AB$, and its right end is resting on another beam $DE$. Compute the tension $T$ in the rope and the *reactions* $R_D$ and $R_E$ required to maintain *equilibrium*. Draw separate free-body diagrams of $BC$ and $DE$. The force $F$ is 250 N.

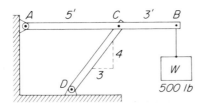

**Fig. 2-61   Prob. 2-60.**

**2-60.** Draw a free-body diagram of bar $AB$ in Fig. 2-61. Solve for the $V$ and $H$ components of the *reacting forces* at $A$ and $C$, necessary for *static equilibrium*, and compute the reactions $R_A$ and $R_C$.

Ans. $R_A = 671$ lb↙; $R_C = 1000$ lb↗

**2-61.** The weight $W$ shown in Fig. 2-62 is 400 lb. Calculate the reactions at $A$ and $C$.

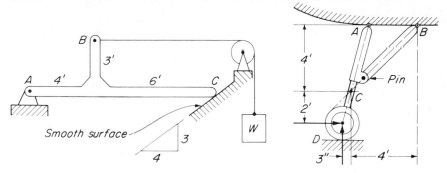

Fig. 2-62   Prob. 2-61.

Fig. 2-63   Probs. 2-62 and 2-63.

**2-62.** The nose landing gear shown in Fig. 2-63 is subjected to a horizontal force $D_H = 1400$ lb and a vertical force $D_V = 7000$ lb. Calculate the magnitude and directions of the reaction at $A$ and $B$. Note that the resultant reaction at $B$ must point along the line $BC$.

**2-63.** Solve Prob. 2-62 if $D_H = 0$ and $D_V = 10,000$ lb.

$$Ans. \ A = 12,320 \text{ lb}, \ \theta_H = 80°; \ B = 3020 \text{ lb}, \ \theta_H = 45°$$

**2-64.** Calculate the crushing force exerted on the pipe shown in Fig. 2-64 if the piston force is 300 N. Neglect all friction.

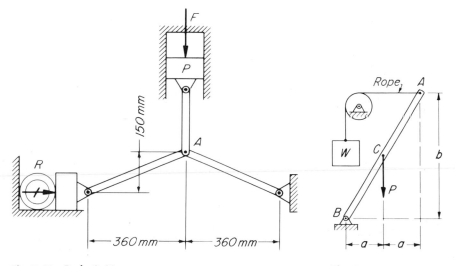

Fig. 2-64   Prob. 2-64.

Fig. 2-65   Prob. 2-65.

**2-65.** The bar $AB$ in Fig. 2-65 is supported at $B$ as shown and is held at $A$ by a horizontal rope passing over a sheave. Draw a free-body diagram of bar $AB$ and compute the weight $W$ required for *static equilibrium*, if the load $P$ is 200 N, $a = 1$ m, and $b = 4$ m. Compute also reaction components $B_V$ and $B_H$, and the *reaction* $R_B$. Neglect sheave friction and weight of bar.

*Ans.* $W = 50$ N, $R_B = 216.2$ N

## REVIEW QUESTIONS

**2-1.** Define briefly the following terms: (*a*) mechanics, (*b*) statics, and (*c*) force.

**2-2.** Give the four characteristics of a force.

**2-3.** What is meant by (*a*) a vector quantity, and (*b*) a scalar quantity? Give a few examples of each.

**2-4.** Explain briefly the principle of transmissibility.

**2-5.** Name the six types of force systems.

**2-6.** State briefly the principle of components.

**2-7.** Give briefly the principle of resultants.

**2-8.** Define the moment of a force. By what product is the moment of a force measured?

**2-9.** State the principle of moments known as Varignon's theorem.

**2-10.** What is a couple? What is the algebraic sum of the forces of a couple? What is the moment of a couple?

**2-11.** State, or illustrate by a diagram, how to locate the line of action of the resultant of two parallel forces.

**2-12.** Explain the essential meaning of equilibrium.

**2-13.** What is meant by a complete force system? Is such a system always in equilibrium?

**2-14.** State the force law of equilibrium.

**2-15.** Give the moment law of equilibrium.

**2-16.** When dealing with force systems in equilibrium, is a sign convention necessary? If it is used, which forces and moments are generally positive?

**2-17.** State the two-force principle.

**2-18.** Define the three-force principle.

**2-19.** Explain the four-force principle.

**2-20.** Give the specific direction of the reaction at a roller support or at a smooth supporting surface.

**2-21.** What is meant by a free body? What is the main purpose of a free-body diagram?

# Chapter 3
# Coplanar, Parallel
# Force Systems

## 3-1 INTRODUCTION

Because of the attraction of the gravitational force of the earth on all outside bodies, vertical forces and systems of vertical forces (parallel) are commonly encountered in engineering practice.[1] The usual problem in this, as in all other types of force systems, is (1) to determine the magnitude of a resultant force and the location of its line of action or (2) to determine the unknown reacting forces of a system in equilibrium. Both analytical and graphical solutions are available and are described in the following articles.

## 3-2 RESULTANT OF COPLANAR, PARALLEL FORCES

The solutions of problems involving parallel force systems are often simplified by use of the resultant force. The *analytical method of moments* is simple, as is shown in the example below, and is generally used. Graphical methods are more involved but are useful at times when all-graphical solutions are desired, or for checking results obtained analytically. The graphical method of inverse proportion, described in Art. 2-12, may be applied to three or more parallel forces by combining the resultant of any two forces, obtained as in Fig. 2-26, with a third force, and by similarly combining the resultant of those three with

---

[1]The *gravity forces* of a system are not truly parallel; actually, they are *concurrent* at the center of the earth's attraction. In such problems as normally arise in engineering practice, we do, of course, consider them to be parallel. Theoretically, a system of applied forces can be made truly parallel. The *inertia forces* acting on the various parts of a body in rectilinear motion are truly parallel. The *centrifugal forces* acting on the various parts of an automobile traveling in a circular path on a horizontal road appear *parallel* when projected onto a vertical plane and appear *concurrent* through the axis of rotation when projected onto a horizontal plane.

41

a fourth force, and so on. The method of determining a resultant force by the *graphical string-polygon method* is described in Chap. 5. Although this description refers to the determination of the resultant of a system of nonparallel forces, it applies as well, of course, to a system of parallel forces.

*The magnitude of the resultant* of a system of parallel forces is their algebraic sum. *Its line of action* is parallel to those of the forces of the system. *The center of a system of parallel forces* is a point through which the resultant passes. Specifically, it is the point where the action line of the resultant pierces any chosen perpendicular plane.

## ILLUSTRATIVE PROBLEM

**3-1.** Determine the magnitude of the resultant $R$ of the four forces shown in Fig. 3-1 and the distance $x$ from $A$ to its line of action.

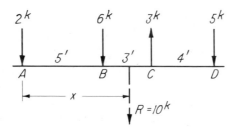

**Fig. 3-1  Resultant of a coplanar, parallel force system.**

*Solution:* The resultant of a system of parallel forces must be equal in magnitude to the *algebraic sum* of the forces of the system it replaces. Also, its moment about any point must equal the *algebraic sum* of the moments of the separate forces about that point, and its line of action is parallel to those of the forces. Then, if upward-acting forces are positive,

$$R = -2 - 6 + 3 - 5 = -10 \text{ kips}$$

Let $A$ be the moment center to determine $x$. Then

$$-10x = -(6)(5) + (3)(8) - (5)(12) = -66 \quad \text{and} \quad x = 6.6 \text{ ft} \quad Ans.$$

The negative sign of $R$ shows it to be downward-acting. The positive sign of $x$ indicates that the action line of $R$ lies to the right of $A$, as was assumed.

## PROBLEMS

**3-2.** Compute the resultant $R$ of the forces shown in Fig. 3-2 and the distance $x$ from $A$ to its line of action.

*Ans. $R = 9$ kN; $x = 7$ m*

**3-3.** Calculate the resultant $R$ of the forces shown in Fig. 3-3 and the distance $x$ from $A$ to its line of action.

*Ans. $R = -12$ kips; $x = 4.25$ ft*

**3-4.** In Fig. 3-2, change the force at $D$ to 3 kN up; solve for $R$ and for distance $x$ from $A$.

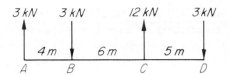

Fig. 3-2   Probs. 3-2 and 3-4.

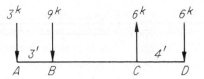

Fig. 3-3   Probs. 3-3 and 3-5.

**3-5.** Change force $B$ in Fig. 3-3 from 9 kips downward-acting to 6 kips upward-acting. Then solve for $R$ and for distance $x$ from $A$.

*Ans.* $R = 3$ kips; $x = -2$ ft (to the left of $A$)

## 3-3 RESULTANTS OF DISTRIBUTED LOADS

Distributed forces were defined and discussed in Art. 2-2. They are usually referred to as **distributed loads.** The pressure of warehouse contents on the floor and of the floor planking on the supporting joists or beams are examples of distributed loads. Unless clearly otherwise distributed, such loads on beams are usually considered to be uniformly distributed along the length of the beam. In order that the beam reactions may be computed, *a distributed load must be replaced by its equivalent concentrated resultant load,* which always acts through the center of gravity of the load, or through the centroid of the load area.

By geometric construction, the centroid of a rectangular area is at the intersection $C$ of its diagonals, as in Fig. 3-4; that of a triangular area is at the intersection $C$ of its medians, which is always one-third of any altitude above its base, as shown in Fig. 3-5.

Nonuniformly distributed loads may usually be reduced to either the triangular form, shown in Fig. 3-6, or the trapezoidal, as in Fig. 3-7, without appreciable error. A trapezoidal load is usually replaced with its equivalent rectangular and triangular loads, whose two resultants then have the same effect as the single resultant of the trapezoidal load.

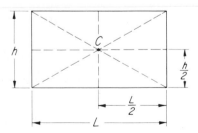

Fig. 3-4   Centroid of rectangle.

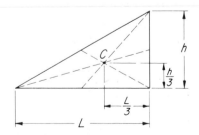

Fig. 3-5   Centroid of triangle.

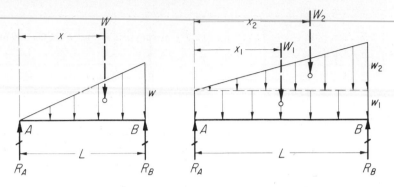

**Fig. 3-6   Triangular load.**          **Fig. 3-7   Trapezoidal load.**

## 3-4 EQUILIBRIUM OF COPLANAR, PARALLEL FORCE SYSTEMS

The usual problem is that of determining two unknown support reactions. Analytical solutions by moments are simplest; a graphical solution using the string and force polygons is shown in Chap. 5.

### ILLUSTRATIVE PROBLEM

**3-6.** Solve by moments for the reactions $R_A$ and $R_B$ at the supports of the beam shown in Fig. 3-8.

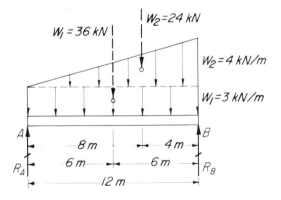

**Fig. 3-8   Support reactions.**

*Solution:* The trapezoidal load area is divided into a rectangle and a triangle. The resultant $W_1$ is (12)(3) or 36 kN, and $W_2$ is (½)(12)(4) or 24 kN. When these resultants are applied at their respective centroids, we find $R_A$ by moments about $B$, and $R_B$ by moments about $A$. A summation of vertical forces will then provide a sufficient check. The equations are

$[\Sigma M_B = 0]$    $12R_A = (4)(24) + (6)(36) = 312$    and    $R_A = 26$ kN    *Ans.*
$[\Sigma M_A = 0]$    $12R_B = (6)(36) + (8)(24) = 408$    and    $R_B = 34$ kN    *Ans.*
$[\Sigma V = 0]$    $36 + 24 = 26 + 34$    or    $60 = 60$    *Check.*

**PROBLEMS**

**3-7.** Compute the reactions $R_A$ and $R_B$ at supports $A$ and $B$ of the beam shown in Fig. 3-9.

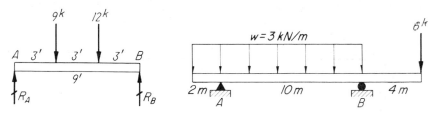

Fig. 3-9   Prob. 3-7.                Fig. 3-10   Prob. 3-8.

**3-8.** Calculate the reacting forces at supports $A$ and $B$ caused by the loads on the beam shown in Fig. 3-10.

**3-9.** Find the reactions at support $A$ and $B$ of the beam shown in Fig. 3-11.

Ans. $R_A = 47.83$ kips; $R_B = 27.17$ kips

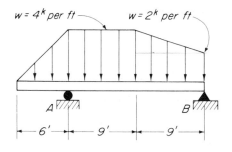

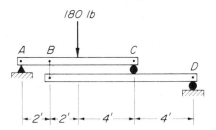

Fig. 3-11   Prob. 3-9.                Fig. 3-12   Prob. 3-10.

**3-10.** In Fig. 3-12, compute the vertical forces acting at $A$, $B$, $C$, and $D$. (Draw three free-body diagrams.)

Ans. $A = 120$ lb; $B = 40$ lb; $C = 100$ lb; $D = 60$ lb

**3-11.** Figure 3-13 shows a typical arrangement of platform-scale levers, the two identical levers $A$ and levers $B$ and $C$. If, in all these levers, the ratio of distances $a$

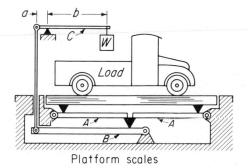

Platform scales

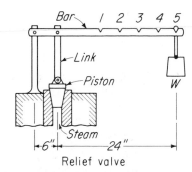

Relief valve

Fig. 3-13   Probs. 3-11 and 3-12.        Fig. 3-14   Probs. 3-13 and 3-14.

and $b$ is 1 to 9, determine the load $L$ on the platform which will be balanced by a 1-N weight $W$ on the right end of weighing arm $C$. Does it matter whether or not the load is centered on the platform?

*Ans. $L = 900$ N; no!*

**3-12.** Solve Prob. 3-11 when the ratios of distances $a$ and $b$ on levers $A$, $B$, and $C$ are 1 to 9, 1 to 14, and 1 to 8, respectively.

**3-13.** Figure 3-14 shows a simple form of relief valve. The weight of the link and the piston is 10 lb. The horizontal bar weighs 20 lb, which may be considered to be concentrated at its midpoint. If the circular surface of the bottom of the piston is 2 in. in diameter, compute the least weight $W$ required to maintain a steam pressure of 30 psi.

*Ans. $W = 6.85$ lb*

**3-14.** If, in Prob. 3-13, the notches in the horizontal bar are 4 in. apart and a weight of 30 lb is placed in notch 3, compute the maximum steam pressure $p$, in psi, under which the valve will remain closed.

**3-15.** A steel beam 6 m long and weighing 90 N/m is supported at its ends and carries a triangular load, as in Fig. 3-6. If the rate of load $w$ at $B$ is 1500 N/m per ft, compute the reactions $R_A$ and $R_B$.

*Ans. $R_A = 1770$ N; $R_B = 3270$ N*

**3-16.** In Fig. 3-15 block $A$ has two sheaves and block $B$ has one. Determine ($a$) the pull $P$ required to lift a weight $W$, and ($b$) the tension $T$ in the rope $C$. Neglect friction. The ropes may be assumed to be coplanar and parallel with inappreciable error. (HINT: Cut ropes just above block $B$, and draw a free-body diagram.)

*Ans. $P = (W/3)$; $T = (4W/3)$*

**3-17.** Determine ($a$) the pull $P$ required to lift a weight $W$, using sheaves as arranged in Fig. 3-16, and ($b$) the tension $T$ in each of ropes $A$, $B$, and $C$. Neglect friction.

**3-18.** Figure 3-17 shows a differential chain hoist such as is commonly used in shops and garages. Two sprocket sheaves $A$ and $B$ of radii $R$ and $r$ are fastened together. A third sprocket sheave $C$ of diameter $r + R$ supports the weight to be lifted. A

**Fig. 3-15   Prob. 3-16.**

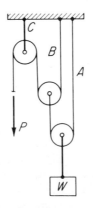

**Fig. 3-16   Prob. 3-17.**

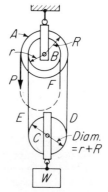

**Fig. 3-17   Probs. 3-18 and 3-19.**

continuous chain passes over sheave $A$, then under sheave $C$, and up over sheave $B$, returning to sheave $A$. A pull $P$ on the chain raises side $D$ and lowers side $E$ in the ratio of $R$ to $r$. There is no stress in side $F$ of the chain. Assuming $P$, $D$, and $E$ to be parallel, cut chains $D$ and $E$, and show that, for equilibrium,

$$P - \frac{W(R - r)}{2R}$$

**3-19.** In the differential chain hoist shown in Fig. 3-17, let $R = 4$ in. and $r = 3.5$ in. Neglect friction. ($a$) What pull $P$ is required to lift a weight $W$ of 1000 lb? ($b$) If $R$ is 4 in., what should $r$ be in order that the pull $P$ will be equal to 5 percent of $W$?

*Ans.* ($a$) $P = 62.5$ lb; ($b$) $r = 3.6$ in.

**3-20.** The load $P$ in Fig. 3-18 is 5 kips. Calculate the vertical forces at $A$, $B$, $C$, and $D$ if $a$ equals 10 ft.

**3-21.** The load $P$ in Fig. 3-18 is 7 kips. Calculate the vertical forces at $A$, $B$, $C$, and $D$ if $a$ equals 14 ft.

*Ans.* $A = 1$ kip, $B = 9$ kips, $C = 15$ kips, and $D = 6$ kips

**3-22.** In Fig. 3-19, compute the reactions at $A$ and $B$.

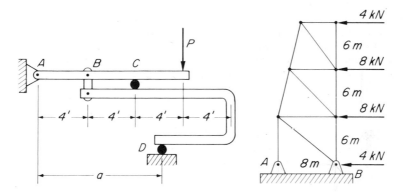

**Fig. 3-18   Probs. 3-20 and 3-21.**          **Fig. 3-19   Prob. 3-22.**

**SUMMARY** (By Article Number) _____

**3-1. Parallel force systems** are common, since probably most forces encountered in nature and in engineering practice arise from gravitational attraction of bodies, thus creating parallel vertical forces.

**3-2. The resultant of a parallel force system** is most easily found analytically by moments. Graphical solutions may be obtained either by the inverse-proportion method or by the string-polygon method outlined in Chap. 5.

**3-3. Resultants of distributed loads** are always applied at the **center of gravity** of the load, which is the centroid of the load area.

**3-4.** The unknown quantities of a **parallel force system in equilibrium** are generally the unknown magnitudes of two *support reactions*. These are most easily found by moments.

**REVIEW PROBLEMS**

**3-23.** Find the *resultant R* of the forces shown in Fig. 3-20 and the horizontal distance x from A to its line of action. All forces are in newtons.

*Ans. R = 100 N; x = 13.5 m*

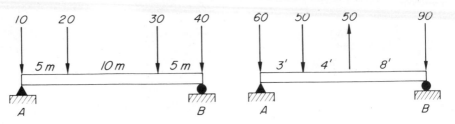

**Fig. 3-20   Prob. 3-23.**                **Fig. 3-21   Prob. 3-24.**

**3-24.** In Fig. 3-21, determine the *resultant force R*, and the horizontal distance x from A to its line of action. All forces are in pounds.

**3-25.** Compute by moments the *reactions $R_A$ and $R_B$* of the beam shown in Fig. 3-22. (HINT: See Art. 2-11.)

*Ans. $R_A$ = 6 kN↑; $R_B$ = 9 kN↑*

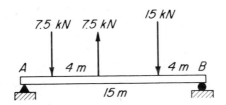

**Fig. 3-22   Prob. 3-25.**                **Fig. 3-23   Prob. 3-26.**

**3-26.** Compute by moments the *reactions $R_A$ and $R_B$* of the beam shown in Fig. 3-23.

**3-27.** Compute by moments the *reactions $R_A$ and $R_B$* of the beam shown in Fig. 3-24.

*Ans. $R_A$ = 3 kN↑; $R_B$ = 6 kN↑*

**3-28.** Compute by moments the *reactions $R_A$ and $R_B$* of the beam shown in Fig. 3-25.

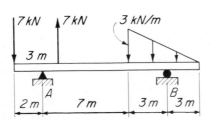
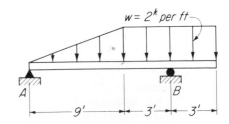

**Fig. 3-24   Prob. 3-27.**                **Fig. 3-25   Prob. 3-28.**

**3-29.** Compute by moments the *reactions* $R_A$ and $R_B$ of the beam shown in Fig. 3-26.

Ans. $R_A = 4$ kips; $R_B = 14$ kips

**3-30.** The beam in Fig. 3-27 is subjected to a load as shown. Compute *reactions* $R_A$ and $R_B$.

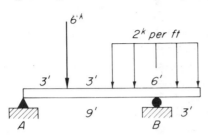

Fig. 3-26    Prob. 3-29.                    Fig. 3-27    Prob. 3-30.

**3-31.** Draw separate *free-body diagrams* of beams $AB$ and $CD$ in Fig. 3-28; then compute and show all forces acting on each beam.

Ans. $R_C = 8$ kN

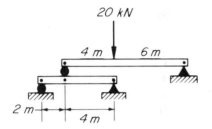

Fig. 3-28    Prob. 3-31.                    Fig. 3-29    Prob. 3-32.

**3-32.** In Fig. 3-29, draw separate *free-body diagrams* of the two horizontal levers; then compute and show values of all forces acting on each lever.

**3-33.** The solid, rectangular stone block shown in Fig. 3-30 weighs 4000 N. Compute the vertical force $P$ required barely to lift the right end off the horizontal supporting surface by using the bar as shown.

Ans. $P = 60$ N

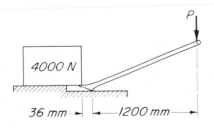

Fig. 3-30    Prob. 3-33.                    Fig. 3-31    Probs. 3-34 through 3-36.

**3-34.** Two 6-ft planks are held together by a single bolt at $B$, as shown in Fig. 3-31. Compute the tensile force $T$ in the bolt. Spacers are used at $B$ and $C$. Show complete free-body diagrams.

**3-35.** If the bolt at $B$ in Fig. 3-31 will safely stand 420 lb in tension, what maximum additional downward-acting vertical load $P$ may be placed at $C$?

**3-36.** Add an 80-lb load to the beam shown in Fig. 3-31 midway between $A$ and $B$. Calculate the forces acting at $A$, $B$, $C$, and $D$ due to the two loads.

**3-37.** The dead weight $W_1$ of the revolving crane shown in Fig. 3-32, but not including the counterweight $W_2$, is 6 kips. (*a*) If the reaction at $C$ is zero when the load $P$ is zero, what is the value of $W_2$? (*b*) If $a = 50$ ft, the reaction at $B$ is zero and the counterweight $W_2$ is as calculated in (*a*), what is the value of $P$? (*c*) Repeat part (*b*) if $a$ is limited to 20 ft.

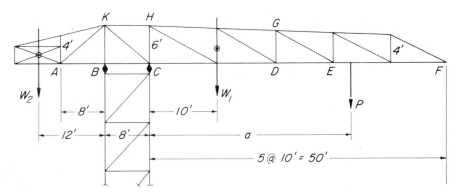

**Fig. 3-32    Prob. 3-37.**

## REVIEW QUESTIONS

**3-1.** What is the magnitude of the resultant of a system of parallel forces? What is known about its line of action?

**3-2.** What is meant by the center of a system of parallel forces?

**3-3.** By what principle of moments may the line of action of the resultant of a system of parallel forces be located?

**3-4.** At what point is the resultant of a distributed load assumed to be concentrated?

**3-5.** Define the locations of the centroids of (*a*) a rectangle and (*b*) a triangle.

# Chapter 4
# Coplanar,
# Concurrent
# Force Systems

## 4-1 INTRODUCTION

In a coplanar, concurrent force system the lines of action of all forces lie in one plane and intersect at a common point, the *point of concurrence*. This type of force system is most commonly encountered in problems involving the determination of stresses in members of trusses and also, as is shown later in this chapter, in certain problems involving the determination of support reactions.

## 4-2 RESULTANTS OF COPLANAR, CONCURRENT FORCE SYSTEMS

Two or more coplanar, concurrent forces may be combined into a single resultant force by three methods: (1) algebraically by summations of rectangular components, as in Prob. 2-5, and graphically (2) by the parallelogram method or (3) by the triangle method, both of which are discussed in Art. 2-8 and are illustrated in Fig. 2-11 *a* and *b*, respectively. *The resultant of a concurrent force system is fully defined when its magnitude, sense, and the direction of its line of action are known.* It always acts through the point of concurrence.

### ILLUSTRATIVE PROBLEM

**4-1.** Determine the resultant force $R$ of the forces $P$, $Q$, and $S$, shown in Fig. 4-1 *a*, when $P = 6$ lb, $Q = 7$ lb, and $S = 10$ lb. Find also the angle $\theta_H$ it makes with the horizontal. (The algebraic solution by summations of rectangular components is similar to that shown in Prob. 2-5; hence only the graphical solutions are shown here.)

*a. Solution by the Parallelogram Method:* The three forces are drawn to scale, each in its proper direction and sense, as shown in Fig. 4-1b. We obtain the resultant $R_{PQ}$ of $P$ and $Q$ by completing the parallelogram 0-1-3-2-0. This resultant is then combined with force $S$ by completing the second parallelogram 0-3-5-4-0. The diagonal 0-5 is then the resultant $R_{PQS}$ of forces $P$, $Q$, and $S$. The angle $\theta_H$ may be scaled with a protractor, or its tangent may be scaled by the method explained in Art. 2-8.

*b. Solution by the Triangle Method:* If the three forces $P$, $Q$, and $S$ are laid out to scale *end to end,* as in Fig. 4-1c, the diagonal 0-3 is the resultant of forces $P$ and $Q$, and 0-5 is the resultant $R_{PQS}$ of $R_{PQ}$ and $S$, or of $P$, $Q$, and $S$. Intermediate resultants such as $R_{PQ}$ are generally omitted in this method. Note that in both methods, the polygons 0-1-3-5-0 are identical.

*Ans.* $R = 18.5$ lb $\nearrow$ ; $\theta_H = 36°13'$

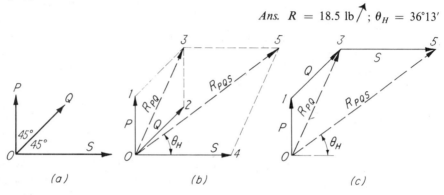

**Fig. 4-1   Graphical determination of resultant force.** *(a)* **Space diagram.** *(b)* **Parallelogram method.** *(c)* **Triangle method.**

**PROBLEMS** _____

**4-2.** Using the parallelogram method determine the resultant $R$ of the forces shown in Fig. 4-2 and its direction angle $\theta_H$ with the horizontal. Check by the triangle method. (Scale on 8½ by 11: 1 cm = 10 N.)

**4-3.** Solve for the resultant $R$ of the three forces shown in Fig. 4-3 by the algebraic summations of components. (See Prob. 2-5.) Find also its direction angle $\theta_H$ with the

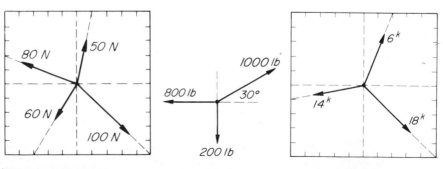

**Fig. 4-2   Prob. 4-2.**          **Fig. 4-3   Prob 4-3.**          **Fig. 4-4   Prob 4-4.**

horizontal. Check graphically by the triangle method. (Scale on 8½ by 11: 1 in. = 200 lb.)

Ans. $R = 307$ lb $\nearrow$; $\theta_H = 77°36'$

**4-4.** Determine the resultant $R$ of the forces shown in Fig. 4-4 by the parallelogram method. Find also its direction angle $\theta_{II}$ with the horizontal. Check by the triangle method. (Scale on 8½ by 11: 1 in. = 5 kips.)

## 4-3 EQUILIBRIUM OF COPLANAR, CONCURRENT FORCE SYSTEMS

When a system of forces is in equilibrium, the algebraic sums of their vertical and horizontal components must, respectively, equal zero. Since all forces of a concurrent system acting on a body at rest act through the same point, they cannot cause rotation of the body. Therefore, the unknown elements of a concurrent force system may be determined entirely by the following equations of force equilibrium: $\Sigma V = 0$ and $\Sigma H = 0$.

Figure 4-5 shows the graphical equivalent of the algebraic summations obtained by $\Sigma V = 0$, and $\Sigma H = 0$. The four forces $P$, $Q$, $S$, and $T$ are in equilibrium. Their resultant, therefore, is zero. Consequently, when the forces are laid out to scale end to end, *the resulting force polygon must close.* Further proof of this fact is indicated by the vertical and horizontal components, shown above and to the right of the force polygon. A glance indicates that their algebraic sums are zero. Thus, **for a system of forces in equilibrium, the force polygon must be closed.** This principle is known as the **polygon law.** In such a polygon *the arrowheads will always follow each other,* regardless of the order in which the forces are laid out.

A closed force polygon, however, denotes only free equilibrium and not necessarily moment equilibrium. If, for example, force $T$ is displaced, as shown in Fig. 4-5c, force equilibrium still exists, and the force polygon remains unchanged, but the forces are no longer in moment equilibrium.

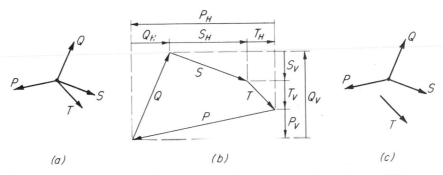

(a)    (b)    (c)

**Fig. 4-5   The graphical force polygon.** *(a)* System of forces in complete equilibrium. *(b)* Graphical force polygon showing equilibrium of forces. (c) System in force equilibrium only.

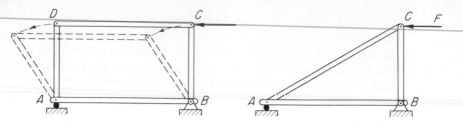

**Fig. 4-6   Unstable frame.**          **Fig. 4-7   Stable frame.**

## 4-4 TRUSSES

The primary function of a structure is to support loads and to transmit these loads from their points of application through the various parts or members of the structure to its supports. This job can be done efficiently only if the members are carefully arranged.

In Fig. 4-6, four members are fastened together at their ends by bolts. This frame, however, is *unstable,* because tightening the bolts sufficiently to prevent the indicated collapse would be difficult. A diagonal member from $A$ to $C$ would obviously overcome the instability. Should the members be so arranged as to form a *triangle,* as is shown in Fig. 4-7, a *stable* structure is obtained, called a **truss.** When several loads are to be carried, several triangular units are connected together, as in Fig. 4-8. **A truss** is a structural unit in which the bars are arranged to form one or more connected triangles. **A joint** is the connection between two or more bars. **A member** is a bar extending from joint to joint.

## 4-5 STRESSES IN MEMBERS OF TRUSSES

The forces (loads and reactions) acting on a truss are generally applied at the joints and cause **internal forces** or **stresses** to be produced in the various

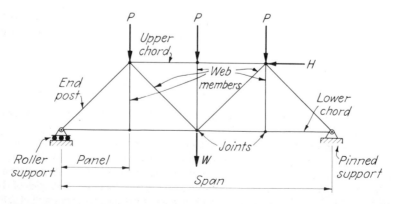

**Fig. 4-8   Typical pin-connected truss.**

members. Unless the members are rather rigidly connected at the joints, for the purpose of stress analysis they are considered to be connected with smooth, frictionless pins. Such trusses are said to be **pin-connected.** Both ends of each member are presumed to be free to rotate about the connecting pins, and bending in any one member therefore cannot be transmitted to other members. For simplicity of analysis and with no appreciable error, common types of roof and bridge trusses made of timbers and steel rods, and relatively light roof trusses and transmission towers made of steel angles riveted together at the joints, are considered as being pin-connected.

A **two-force member** is one on which *forces act at two points only,* usually at its ends. When isolated as a free body, a two-force member is then in equilibrium under the action of two forces, one at each end; these forces therefore must be equal, opposite, and collinear. Their common line of action clearly passes through the centers of the two pins and, in a straight member, coincides with the axis of the member. Hence, *when a two-force member is cut, the force (stress) within it is known to act along its axis.* When all forces on a truss are applied at its joints, all members are two-force members.

**Type of Stress.** If the stress *pulls* on the member, or stretches it, it is called a **tensile stress** and the member is said to be in **tension.** Conversely, if the stress *pushes* on the member, or compresses it, it is called a **compressive stress** and the member is in **compression.** Tension and compression are referred to as *types of stresses* and are usually denoted by T and C, respectively. Stresses in two-force members are also called **axial** or **direct stresses,** as distinguished from *bending stresses.*

Axial stresses may be determined by various methods, such as (1) the *geometric,* based on the geometric similarity between the force and space triangles in certain problems, as in Prob. 4-5, and (2) the *trigonometric,* using the sine law, as in Prob. 4-9. The practical applications of these two methods are limited to special cases involving only three forces at a joint. The more general and practical methods, such as the *analytical method of joints* and the *graphical method of joints,* are fully outlined in later articles.

**Notation.** In analytical solutions, supports and joints of trusses are generally designated by capital letters. A member, as well as the stress within it, is then identified by the two letters appearing at its ends.

## ILLUSTRATIVE PROBLEM

**4-5.** Determine the stresses in members $AC$ and $BC$ of the simple truss shown in Fig. 4-9a, caused by a load $W$ of 60 lb.

*Solution by Geometry:* The two members are cut near joint $C$, and that joint is isolated as shown in the free-body diagram in Fig. 4-9b. The known force $W$ and the unknown stresses $AC$ and $BC$ form a concurrent system in equilibrium whose force polygon (or diagram), shown in Fig. 4-9c, then must close. The triangles of the space and force diagrams are clearly similar, since corresponding sides are parallel. Therefore, by proportion,

$$\frac{60}{3} = \frac{BC}{4} = \frac{AC}{5}$$

or  $AC = \dfrac{5}{3}(60) = 100$ N T  and  $BC = \dfrac{4}{3}(60) = 80$ N C  *Ans.*

In the force diagram, the sense of $W$ is known, and, because *the arrows must follow each other around the diagram,* $BC$ is seen to be comprehensive and $AC$ to be tensile. (NOTE: If the space and force diagrams are drawn to scale, a complete graphical solution is obtained, thus providing a convenient check.)

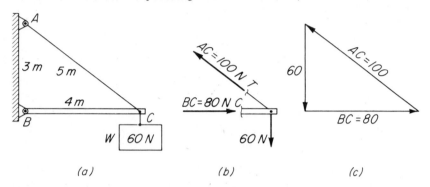

(a)    (b)    (c)

**Fig. 4-9   Stresses in truss members by the geometric method. *(a)* Space diagram. *(b)* Free-body diagram. *(c)* Force diagram.**

**PROBLEMS** _____

**4-6.** The truss of Fig. 4-7 has dimensions $AB = 7.5$ m, $BC = 4$ m, and $AC = 8.5$ m. The force $F$ is 1500 N. Solve by geometry for the stresses $AC$ and $AB$.

**4-7.** Determine by geometry the stresses in members $AC$ and $BC$ of the truss shown in Fig. 4-10.

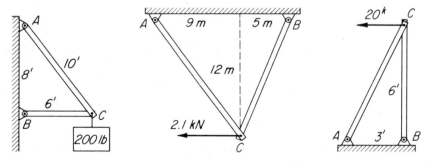

**Fig. 4-10   Prob. 4-7.**    **Fig. 4-11   Prob. 4-8.**    **Fig. 4-12   Prob. 4-9.**

**4-8.** Solve by geometry for the stresses $AC$ and $BC$ in the truss shown in Fig. 4-11.

**4-9.** Find the stresses in $AC$ and $BC$ in Fig. 4-12, by geometry.

*Ans.*  $AC = 44.7$ kips C; $BC = 40$ kips T

## ILLUSTRATIVE PROBLEM

**4-10.** In members $AC$ and $BC$ of the truss in Fig. 4-13a, solve by trigonometry for the stresses caused by the applied force $P$.

*Solution by Trigonometry:* When the two members are cut, the joint may be isolated as a free body, as in Fig. 4-13b. A force diagram is then constructed (Fig. 4-13c) in which the forces $P$, $AC$, and $BC$ are drawn parallel, respectively, to force $P$ and members $AC$ and $BC$, as shown in the space diagram. From the given data, we determine the angles of the force triangle (with great care). Now, with one side and all angles known, by the law of sines, we have

$$\frac{AC}{\sin 70°} = \frac{BC}{\sin 35°} = \frac{100}{\sin 75°} \quad \text{or} \quad \frac{AC}{0.9397} = \frac{BC}{0.5736} = \frac{100}{0.9659}$$

Hence
$$AC = \frac{(100)(0.9397)}{0.9659} = 97.2 \text{ lb T} \qquad \qquad Ans.$$

and
$$BC = \frac{(100)(0.5736)}{0.9659} = 59.4 \text{ lb C} \qquad \qquad Ans.$$

The sense of force $P$ is known, and the senses of $AC$ and $BC$, as found in the force diagram where the arrowheads always follow each other, indicate, respectively, tension and compression.

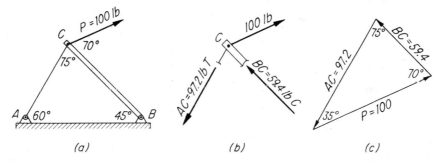

**Fig. 4-13   Stresses in truss member by trigonometry.** *(a)* **Space diagram.** *(b)* **Free-body diagram.** *(c)* **Force diagram.**

## PROBLEMS

**4-11.** Solve by trigonometry for the stresses in members $AC$ and $BC$ of the frame shown in Fig. 4-14.

*Ans.* $AC = 89.7$ lb T; $BC = 73.2$ lb C

**4-12.** Calculate by trigonometry the force in the spring and the link in Fig. 4-15.

**4-13.** Find the stresses $AC$ and $BC$ in Fig. 4-16, by trigonometry.

*Ans.* $AC = 612.84$ N C; $BC = 538.92$ N T

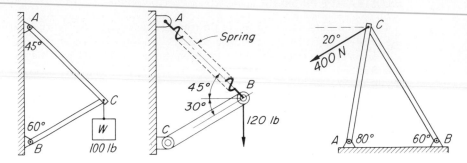

Fig. 4-14   Prob. 4-11.    Fig. 4-15   Prob. 4-12.      Fig. 4-16   Prob. 4-13.

## 4-6 ROPES OVER SHEAVES AND PULLEYS

Sheaves are used to change the direction of a force applied at one end of a rope or flexible cable. They are widely used in hoisting equipment. When a load is applied to a joint of a frame or truss through a sheave, as in Fig. 4-17a, **the sheave may be removed and the forces may be applied directly at the joint without change in total effect,** as is indicated in Fig. 4-17b. The proof is shown in Fig. 4-17c.

The relatively small pin friction is usually neglected. Then $T$ and $W$ are equal, and their resultant $R$ passes through the pin $O$, where it may again be resolved into the forces $T$ and $W$. For convenience, $T$ may be shown as pushing on the pin, without change in its point of application.

## 4-7 STRESSES IN TRUSSES; ANALYTICAL METHOD OF JOINTS

Although applicable to all trusses, regardless of size, the analytical method of joints is most conveniently applied to trusses in which most members are at

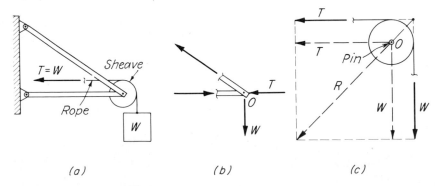

(a)                    (b)                    (c)

Fig. 4-17   Forces $T$ and $W$ may be applied at pin. (a) Rope over sheave. (b) Free-body diagram. (c) Forces applied at pin.

right angles to each other, as in trusses having parallel chords. The joints must be isolated in such order that no more than two unknown stresses need be found at any joint. A stress, or its component, is then found by application of the two equations of force equilibrium, $\Sigma V = 0$ and $\Sigma H = 0$.

## ILLUSTRATIVE PROBLEMS

**4-14.** Compute the stresses in members $AC$ and $BC$ of the pin-connected truss shown in Fig. 4-18$a$.

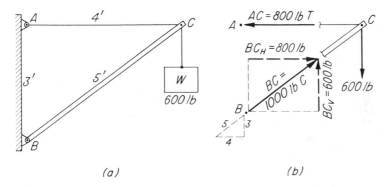

*(a)*          *(b)*

**Fig. 4-18** **Stresses by the analytical method of joints.** *(a)* **Space diagram.** *(b)* **Free-body diagram.**

*Solution:* First, we cut members $AC$ and $BC$ near joint $C$ and isolate that joint. To facilitate summations of vertical and of horizontal forces, $BC$ is replaced with its $V$ and $H$ components, as is shown in the free-body diagram. A *slope triangle* indicating the slope of its line of action is also drawn.

We begin the solution with a summation of forces in the direction having only one unknown force, here the vertical direction. Then, assuming $BC_V$ to be upward-acting, as clearly it is, we have

$$[\Sigma V = 0] \qquad\qquad BC_V = 600 \text{ lb}$$

If we now imagine $BC_H$ to be placed in the position of the bottom side of the parallelogram of forces, we find, by similarity of the force and slope triangles, that

$$\frac{BC_V}{3} = \frac{BC_H}{4} = \frac{BC}{5} = \frac{600}{3} = 200$$

Hence                          $BC_H = (4)(200) = 800 \text{ lb}$

and                            $BC = (5)(200) = 1000 \text{ lb C}$          *Ans.*

Then, if $AC$ is assumed to be a tensile stress,

$$[\Sigma H = 0] \qquad\qquad AC = 800 \text{ lb T} \qquad\qquad\qquad \textit{Ans.}$$

**4-15.** Solve for the stresses in the four members of the pin-connected derrick shown in Fig. 4-19$a$ by the analytical method of joints.

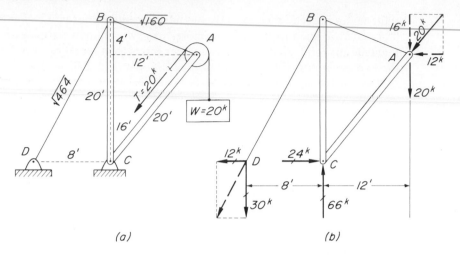

*(a)*                                   *(b)*

**Fig. 4-19 Stresses in derrick by analytical method of joints.** *(a)* **Space diagram.** *(b)* **Free-body diagram.**

*Solution:* A free-body diagram of the derrick is shown in Fig. 4-19*b*. The sheave is removed, and the two 20-kip forces are applied directly at the pin (see Art. 4-6). We may take moments about *C* to find the vertical component of the reaction at *D*.

$$[\Sigma M_C] = 0 \qquad 8(D_V) = 12(20) \qquad \text{or} \qquad D_V = 30 \text{ kips}$$

By geometry, $\qquad \dfrac{D_H}{8} = \dfrac{30}{20} \qquad$ or $\qquad D_H = 12$ kips

The reaction components at *C* can now be found from equilibrium.

$$[\Sigma V = 0] \qquad\qquad C_V = 30 + 20 + 16 = 66 \text{ kips}$$
$$[\Sigma H = 0] \qquad\qquad C_H = 12 + 12 = 24 \text{ kips}$$

The stress *BD* can be found from geometry,

$$\frac{BD}{\sqrt{464}} = \frac{12}{8} \qquad \text{or} \qquad BD = 32.4 \text{ kips (T)} \qquad\qquad Ans.$$

Joint *B* may be isolated as shown in Fig. 4-20*a*. The horizontal component of the stress in member *AB* can be found from equilibrium (*AB* is shown as a tensile force, which it obviously is).

$$[\Sigma H = 0] \qquad\qquad AB_H = 12 \text{ kips}$$

By geometry, $\qquad \dfrac{AB_V}{4} = \dfrac{12}{12} \qquad$ or $\qquad AB_V = 4$ kips

and $\qquad \dfrac{AB}{160} = \dfrac{4}{4} \qquad$ or $\qquad AB = 12.65$ kips (T) $\qquad\qquad Ans.$

The force in the vertical member can now be found from

$$[\Sigma V = 0] \qquad\qquad BC = 30 + 4 = 34 \text{ kips (C)} \qquad\qquad Ans.$$

The force components in member *AC* may now be found from the free-body diagram of joint *A* (see Fig. 4-20*b*); *AC* was assumed to be compressive.

$[\Sigma V = 0]$ $\qquad\qquad AC_V = 20 + 16 - 4 = 32$ kips

$[\Sigma H = 0]$ $\qquad\qquad AC_H = 12 + 12 = 24$ kips

By geometry $\qquad\qquad \dfrac{AC}{20} = \dfrac{32}{16}$ $\quad$ or $\quad AC = 40$ kips (C) $\qquad\qquad$ *Ans.*

When all forces are shown on the free-body diagrams, a glance will show that equilibrium exists. The true senses of the forces were correctly assumed, since all the computed stresses are positive (see Art. 2-18).

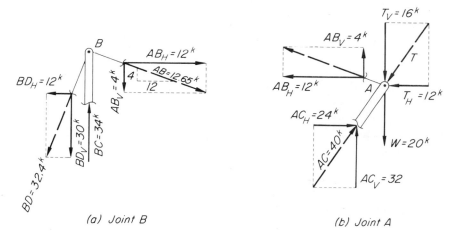

(a) Joint B $\qquad\qquad\qquad\qquad\qquad\qquad$ (b) Joint A

**Fig. 4-20   Joint free-body diagrams.**

**4-16.** Determine the stresses in all members of the Howe roof truss shown in Fig. 4-21.

*Solution:* The general procedure is similar to that outlined in Probs. 4-14 and 4-15. The joints are isolated in such order that no more than two unknown stresses are to be determined at any joint. Most of the stresses to be determined are dependent on other stresses previously found, and no positive check on results obtained is available until the last joint is solved. Hence, to minimize cumulative inaccuracies and chances of error, the joints should be solved from each end to the middle, or in the order *A, B, C, E,* and then *H, F, G, D.* Then we obtain a check from a

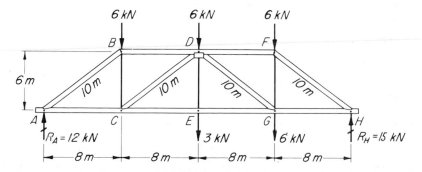

**Fig. 4-21   Space diagram. Stresses by the analytical method of joints.**

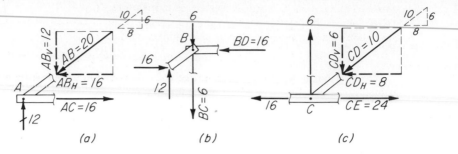

**Fig. 4-22  Free-body diagrams. (a) Joint A. (b) Joint B. (c) Joint C.**

complete free-body diagram of the last joint $D$ showing the forces there to be in equilibrium.

The free-body diagram of joint $A$ is shown in Fig. 4-22a. From $\Sigma V = 0$, $AB_V$ is found. By similarity of slope and force triangles, $AB$ and $AB_H$ are obtained, and $AC$ is then found by $\Sigma H = 0$. At joint $B$, shown in Fig. 4-22b, stresses $BC$ and $BD$ are unknown. $BC$ is found by $\Sigma V = 0$, and $BD$ by $\Sigma H = 0$. At joint $C$, $CD_V$ is found by $\Sigma V = 0$. Forces $CD$ and $CD_H$ are obtained by similarity of triangles, and $CE$ by $\Sigma H = 0$. When more than one or two joints are to be solved, we find it helpful to record the computed stresses on a *stress-summation diagram* as in Fig. 4-23. (Students are urged to complete this problem.)

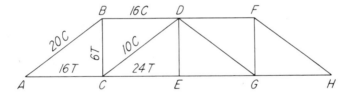

**Fig. 4-23  Stress-summation diagram.**

**PROBLEMS**

Stresses in trusses: the analytical method of joints.

**4-17.** Solve for the stresses in members $AC$ and $BC$ of the truss shown in Fig. 4-24.

*Ans.* $AC = 340$ lb T; $BC = 160$ lb C

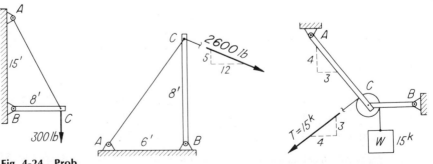

**Fig. 4-24  Prob. 4-17.**

**Fig. 4-25  Prob. 4-18.**

**Fig. 4-26  Prob. 4-19.**

**4-18.** In Fig. 4-25, determine the stresses in members $AC$ and $BC$.

**4-19.** A flexible hoisting rope passes over a sheave at joint $C$ of the truss shown in Fig. 4-26. Find the resulting stresses in members $AC$ and $BC$.

*Ans. $AC$ = 30 kips T; $BC$ = 30 kips T*

**4-20.** Determine the stresses in members $AC$, $AD$, and $CD$ of the roof truss shown in Fig. 4-27.

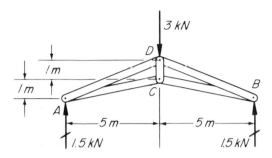

Fig. 4-27    Prob. 4-20.

**4-21.** Compute the stresses in all members of the truss shown in Fig. 4-28. Assume pinned joints at $B$ and $C$. Draw a stress-summation diagram.

*Ans. $AB$ = $DE$ = 5kN T; $BC$ = $CD$ = 5 kN C; $AC$ = $CE$ = 4 kN C; $BD$ = 8 kN T*

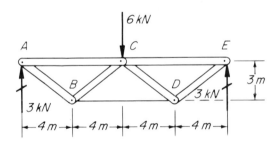

Fig. 4-28    Prob. 4-21.

**4-22.** Find the value of $W$ in Fig. 4-29 required to produce a stress of 13 kN in member $AB$. Determine also the stresses in $BC$, $BE$, and $CD$.

**4-23.** In Fig. 4-30, a power-hoist cable, passing over sheaves at $C$ and $D$, is supported by four coplanar members as shown. Find the stresses in these four members due to the 36-kip load. Neglect all cable and sheave friction.

*Ans. $AC$ = 18 kips C; $BC$ = 20 kips C; $DE$ = 52 kips C; $DF$ = 30 kips C*

**4-24.** The frame shown in Fig. 4-31 supports two 4-kip loads as shown. Find the vertical and horizontal components of the reactions at $A$ and $E$. Then find the stresses in the members by the method of joints. Solve joints in order $A$, $E$, $B$, and $D$. Check answers by use of joint $C$.

*Partial Ans. $A_H$ = 8 kips←; $A_V$ = 4 kips↑; $BD$ = 8 kips C; $CD$ = 20 kips C*

**4-25.** Find the stresses in the truss shown in Fig. 4-31. Assume that the 4-kip load at $D$ acts toward $F$ along the line $DF$.

*Ans. $AB$ = 11.5 kips T; $BC$ = 17.9 kips T; $CD$ = 20 kips C; $BD$ = 5.76 kips C; $BE$ = 13.12 kips T; $DE$ = 16.6 kips C*

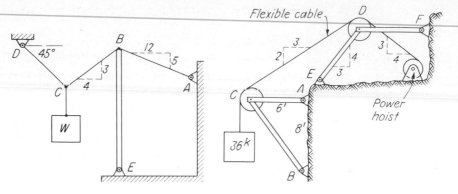

**Fig. 4-29   Prob. 4-22.**          **Fig. 4-30   Prob. 4-23.**

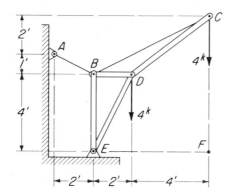

**Fig. 4-31   Probs. 4-24 and 4-25.**          **Fig. 4-32   Prob. 4-26.**

**4-26.** The truss shown in Fig. 4-32 supports the two vertical loads shown. Find the vertical and horizontal components of the reactions at $A$ and $E$. Then find the stresses in all members by the method of joints.

**4-27.** The truss shown in Fig. 4-33 supports the two vertical loads shown. Find the stresses in all truss members by the analytical method of joints procedure.

*Partial Ans. BC* = 30 kips C; *BD* = 24 kips C; *CE* = 40 kips T

**4-28.** Find the stresses in all members of the truss shown in Fig. 4-34. Use the analytical method of joints.

**4-29.** Determine the stresses in all members of the water tower truss shown in Fig. 4-35. Draw a stress-summation diagram. (The reacting forces at supports $E$ and $F$ are $E_H = 9$ kips $\leftarrow$, $E_V = 4$ kips $\uparrow$, and $F_V = 44$ kips $\uparrow$.)

*Ans. AB* = 6 kips C; *AC* = 24 kips C; *BC* = 10 kips T; *BD* = 32 kips C; *CE* = 16 kips C; *CD* = 9 kips C; *DE* = 15 kips T; *DF* = 44 kips C

**4-30.** The loading-platform truss shown in Fig. 4-36 is subjected to vertical dead loads and to a horizontal wind load, as indicated. Compute the stresses in all members. Draw a stress-summation diagram. (The reacting forces at supports $F$ and $G$ are $F_V = 37$ kN $\uparrow$, $F_H = 16$ kN $\rightarrow$, and $G_V = 3$ kN $\uparrow$.)

## 4-8 STRESSES IN TRUSSES; THE GRAPHICAL METHOD OF JOINTS

This method is the graphical equivalent of the analytical method of joints described in Art 4-7. In this method we use *the closed graphical force polygon,* which *constitutes a geometric summation of forces in equilibrium,* as was described and illustrated in Art. 4-3 and Fig. 4-5, respectively. In addition to the space and force diagrams, carefully drawn to scale, a free-body diagram of the isolated joint is often helpful, since it shows clearly all forces acting at the joint. Also, this diagram assists greatly in determining tension and compression. One distinct advantage of this graphical method over the analytical method is that geometric summations of all coplanar, concurrent force systems, *regardless of the direction of the forces,* are made with equal facility.

**Bow's Notation.** In solutions using the graphical force polygon, a special system of notation, referred to as *Bow's notation* after its originator, is universally used and is, indeed, indispensable. In this system, *letters are placed in the*

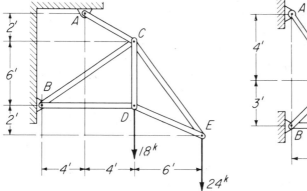

Fig. 4-33   Prob. 4-27.

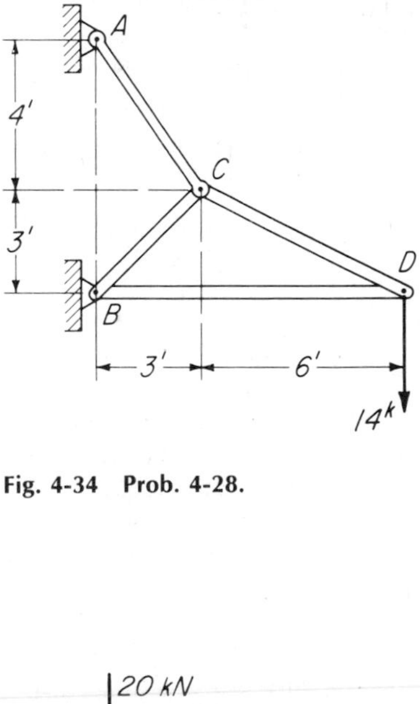

Fig. 4-34   Prob. 4-28.

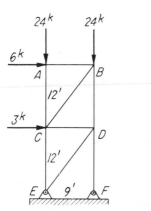

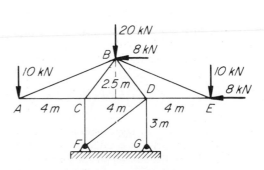

Fig. 4-35   Probs. 4-29 and 5-72.

Fig. 4-36   Probs. 4-30 and 5-118.

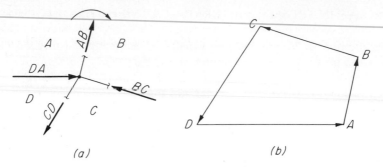

**Fig. 4-37  Bow's notation.** *(a)* **Free-body diagram.** *(b)* **Force diagram.**

*spaces between external forces and in each triangular panel of a truss,* generally in a clockwise order around a truss or a joint. Each force and member, and the force within each member, will then lie between two letters and is identified by those letters.

For example, Fig. 4-37*a* shows the free-body diagram of an isolated joint. The letters *A, B, C,* and *D* are placed in the spaces *between* the forces, *usually in clockwise order around the joint,* as is indicated by the curved arrow. The forces are now designated *AB, BC, CD,* and *DA*. In the force diagram, however, the letters are placed *at the ends* of the lines representing the forces.

## ILLUSTRATIVE PROBLEMS

**4-31.** Determine graphically the stresses in members *CD* and *DA* of the truss shown in Fig. 4-38*a*.

*Solution:* The space diagram is carefully drawn to scale *with the members shown as single lines,* as in Fig. 4-38*a*. We then cut the two members, isolate the joint as a free

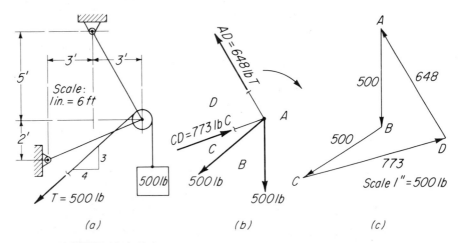

**Fig. 4-38  Stresses by the graphical method of joints.** *(a)* **Space diagram.** *(b)* **Free-body diagram.** *(c)* **Force diagram.**

body, as in Fig. 4-38*b* and insert the letters *A, B, C,* and *D* in the spaces between the four forces. The curved arrow indicates that the forces will be considered in *clockwise order* around the joint; that is, first the known forces *AB* and *BC*, then the unknown forces *CD* and *DA*.

In the force diagram, to some convenient scale, *AB* and *BC* are laid off *end to end* in their true directions and magnitudes and with correct senses shown. Here the two letters designating a force now appear at its ends. Because these four forces are in equilibrium, the force diagram must close. It is closed with a line from *C parallel to member CD* and a line from *A parallel to member DA,* thus locating the unknown intersection *D*. The magnitude of *CD* is scaled from *C* to *D*, that of *DA* from *D* to *A*. The sense of each force is readily determined, because the *arrows must follow each other around the polygon.* These values and senses are then recorded on the free-body diagram. Member *CD* scales 773 lb and is seen to be comprehensive; *DA* scales 648 lb and is tensile.

**4-32.** Determine graphically the stresses in the boom and the boom cable and in the two structural members of the caterpillar derrick, when all are in the position shown in Fig. 4-39. (The weight of the tackle and of all members has been neglected.)

*Solution:* The upper right joint is isolated first, since at that joint the only unknown stresses are those in the boom and the boom cable. Because of the arrangement of the tackle at this joint, the tension in the haulback cable is half of the load, or $T = 1.5$ kN. The free-body diagram of that joint (Fig. 4-40*a*) is then drawn. In order to have all known forces in succession, as is preferable, the force *T* is transmitted back along its line of action until it pushes on the joint; this transmission, of course, does not affect the ultimate results. (See Transmissibility of Force, Art. 2-5. When forces are transmitted in this manner, Bow's notation is most conveniently applied to the free-body diagrams only.) The stresses in the boom and the boom cable are readily found in the force diagram (Fig. 4-40*b*). Lines *AB* and *BC* are drawn to scale, with proper direction and sense. The diagram is then closed by lines *CD* and *DA*, drawn parallel to those members.

Next the free-body diagram is constructed for the second joint (Fig. 4-40*c*). Here the force *F* equals the stress in the boom cable. Now *F* may be transmitted back along its line of action, simply in order not to interfere with the stress *DE*. With the known forces

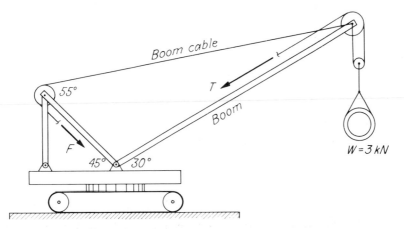

**Fig. 4-39   Caterpillar derrick. Stresses by the graphical method of joints.**

*FA* and *AD* drawn to scale (Fig. 4-40*d*), the stresses *DE* and *EF* are found by closing the diagram; *DE* scales 18.17 kN and is seen to be compressive, and *EF* scales 8.80 kN and is tensile.

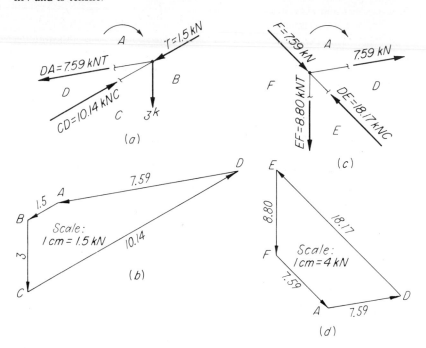

Fig. 4-40   Free-body and force diagrams.

## PROBLEMS

Stresses by the graphical method of joints.

**4-33.** Determine the stresses in members *BC* and *CA* of the truss shown in Fig. 4-41. (Scales on 8½ by 11: 1 in. = 2 ft, 1 in. = 5 kips.)

*Ans. BC* = 14.4 kips C; *CA* = 20 kips T

**4-34.** Solve for stresses *CD* and *DA* in the truss shown in Fig. 4-42. (Scale: 1 cm = 0.5 kN.) *P = W* = 3.2 kN.

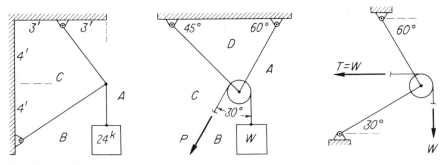

Fig. 4-41   Prob. 4-33.          Fig. 4-42   Prob. 4-34.          Fig. 4-43   Prob. 4-35.

**4-35.** Find the stresses in the two members of the frame shown in Fig. 4-43. $W = 2000$ N. When forces are transposed, apply Bow's notation to free-body and force diagrams only.

*Ans.*  Upper = 732 N T; lower = 2732 N C

**4-36.** The cable spool in Fig. 4-44 is supported by a sling as shown. Determine the stresses in the crossbar $AD$ and in the sling rope. (Scale on 8½ by 11: 1 in. = 500 lb.)

**4-37.** Find the stresses in the four members of the truss shown in Fig. 4-45. (Scales on 8½ by 11: 1 in. = 2 ft, 1 in. = 300 lb.)

*Ans.*  $BE = 721$ lb T; $EA = 400$ lb C; $CD = 541$ lb T; $DE = 180$ lb C

**4-38.** In the construction derrick shown in Fig. 4-46, the stress in the guy wire is maximum when the wire, the boom, and the mast all lie in one plane. Solve for the forces $T$, $F$, and $P$, and for the stresses in the boom and the mast, when $W$ is 10 kips. Neglect sheave friction. (Scales on 8½ by 11: 1 in. = 20 ft., 1 in. = 5 kips.)

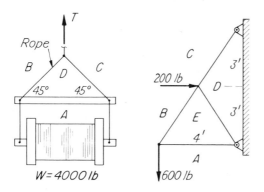

**Fig. 4-44   Prob. 4-36.**        **Fig. 4-45   Prob. 4-37.**

**4-39.** Figure 4-47 illustrates a typical yard crane. Find the stresses in all members caused by a load $W$ of 4 kips. Neglect sheave friction. (SUGGESTIONS: Apply Bow's

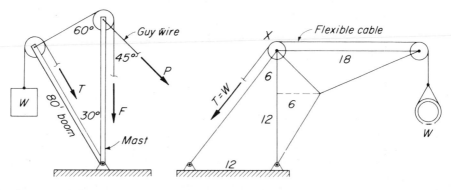

**Fig. 4-46   Prob. 4-38.**          **Fig. 4-47   Prob. 4-39.**

notation to each separate joint only. Analyze the three upper joints only. Scales on 11 by 17: 1 in. = 4 ft, 1 in. = 2 kips.) Lengths shown are in feet.

*Ans.* Horizontal member, 4 kips T; vertical member, 2.0 kips C

## SAMPLE PROBLEM SOLUTION SHEET

**Paper:** 8½ by 11 in. squared.

**Heading:** Left to right, course number, date, name, problem number.

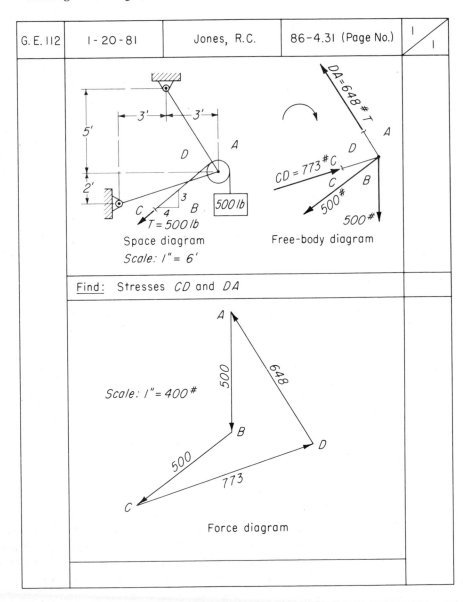

| G. E. 112 | 1-20-81 | Jones, R.C. | 86-4.31 (Page No.) |
|---|---|---|---|

Space diagram
Scale: 1" = 6'

Free-body diagram

Find: Stresses *CD* and *DA*

Scale: 1" = 400 #

Force diagram

**Problem Number:** The number 86 represents the student's consecutive problem number; 4-31 is the number of the problem in the book; and the third number is the page number in the book.

**Upper Right Corner:** *Upper,* student's consecutive sheet number; *lower,* total number of sheets in the solution.

**4-40.** Solve Prob. 4-39 if $W = 24$ kN and if the pull $T$ is vertically downward from the pulley at *x.* Lengths shown are in meters.

**4-41.** Steam exerts a 500-lb force on the piston of Fig. 4-48. What upward vertical force must be applied at joint $B$ to hold the piston stationary?

*Ans.* $P = 1500$ lb

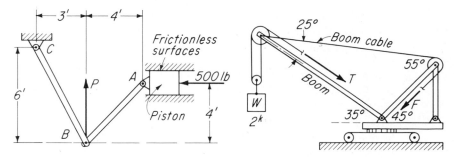

**Fig. 4-48   Prob. 4-41.**                     **Fig. 4-49   Prob. 4-42.**

**4-42.** Figure 4-49 shows a typical caterpillar construction derrick. Find the stresses in the boom and the boom cable, and in the two structural members when they are in the position shown. (Scale on 11 by 17: 1 in. = 1 kip.)

## 4-9 STRESSES IN TRUSSES; THE GRAPHICAL METHOD OF COMBINED DIAGRAMS

When several joints of a truss need to be solved, as in many commonly used roof and bridge trusses, all the separate force diagrams may be superimposed upon one another until they form a single **combined diagram.** This is possible because the force diagrams of the forces at adjacent joints usually have one or two or three sides in common. Hence, less work is involved than in the joint method, and greater precision is obtainable. This method, therefore, is most commonly used in practice. The general procedure is as follows:

### PROCEDURE _____

1. Draw to scale a good-sized space diagram using *single lines* for members and showing *all* external forces (loads and reactions) and all usual dimensions. Be certain that the external forces are in force and moment equilibrium. Use Bow's notation *clockwise* around the truss.

2. Draw a force diagram of *all* external forces (loads and reactions) considering them *clockwise* around the truss. Since these forces necessarily are in equilibrium, this diagram *must* close. Show magnitude and sense of each known force. This is known as the **load diagram.**

3. Beginning at a joint where only two forces are unknown, locate in the load diagram the known forces and solve for the unknown forces by closing the diagram of the forces at the joint. *Do not show senses of stresses solved for on the force diagram.* Continue this process until all forces are found. All stress lines in the force diagram must be drawn parallel to the corresponding members in the space diagram. The distance by which the *closing line* fails to close on the *final point* is called the **error of closure.** Scale this distance perpendicular from point to line, and note it as a dimension.

*Sources of Error.* The error of closure may have two main causes: (1) If it is relatively small, the error may be due to lack of precision in (*a*) the space diagram or (*b*) the load diagram, or to failure in drawing the stress lines *parallel* to the space lines. (2) If the error is relatively large, it may be due to *mistakes* in either the space or the load diagram. Also, the external forces may be in *force* equilibrium; then the load diagram will close; but if the forces are not also in *moment* equilibrium, the stress diagram will not close.

*Permissible Error of Closure.* In general, the error of closure should not exceed *2 percent of the average of all stresses in the truss.*

4. Scale the magnitudes of all stresses and record them on the force diagram. Determine the *type* of each stress solved for, as indicated by its sense, by considering all forces at a joint in clockwise order, noting whether the stress pushes (compression) or pulls (tension) on the joint. If desired, this may be done at each joint as soon as its stresses are determined.

5. Record magnitude and type of each stress on the space diagram or on a separate stress-summation diagram.

**ILLUSTRATIVE PROBLEMS** _____

**4-43.** Find the stresses in all members of the Fink roof truss shown in Fig. 4-50*a* by the graphical method of combined diagrams.

*Solution:* The space diagram is first drawn to scale in accordance with item 1 of the procedure. The load diagram is next constructed (item 2). Since all known forces are parallel, the reactions *DE* and *EA* overlie the loads in the load diagram in Fig. 4-50*b*. Although the load diagram is a straight line, it is nevertheless a closed polygon, since force *EA* closes on the point of beginning *A*. The separate force diagrams are then completed (item 3) in the following order, the forces being considered clockwise around each joint: (1) *EAFE*, (2) *FABGF*, (3) *EFGHE*, (4) *DEJD*, (5) *CDJIC*, and (6) *IJEHI*. The uppermost joint need not be solved, since all stresses are now known. Failure of the last stress line *HI* to close on point *I* would disclose the *error of closure*, which then

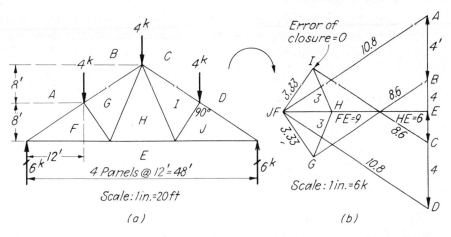

Fig. 4-50   Stresses by the graphical method of combined diagrams. *(a)* Space diagram. *(b)* Force diagram.

should be scaled and noted as a dimension (item 3). Arrowheads are shown on all external forces but are omitted from all internal stress lines.

The stresses are now scaled (item 4), and their magnitudes are noted on the force diagram. The magnitudes of stresses whose stress lines overlie each other, as do *FE* and *HE*, should be indicated separately as shown. The *kind of stress* is now to be determined. Consider the forces acting at the joint on the left. Clockwise, they are *EA, AF,* and *FE.* Their stress lines are retraced in the force diagram: *EA* is upward acting; *AF* acts downward and to the left, *pushing on the joint, and is therefore compressive;* and *FE* acts from left to right, *pulling on the joint, and is therefore tensile.* Finally (item 5), the magnitude and type of each stress are recorded on the corresponding member on the space diagram or on a separate stress-summation diagram.

**4-44.** Find the stresses in all the members of the tower shown in Fig. 4-52 by the graphical method of combined diagrams.

*Solution:* The space diagram is drawn to scale first. The load diagram is constructed next. The loads are laid off in the order *AB, BC, CD, DE, EF,* and *FA.* The load diagram is closed. The separate force diagrams are drawn in the following order, the forces being taken consecutively in a clockwise order around each joint: (1) *ABGA,* (2) *AGHA,* (3) *HGBCIH,* (4) *ICJI,* and (5) *AHILEA.* The stresses are scaled and are recorded on the

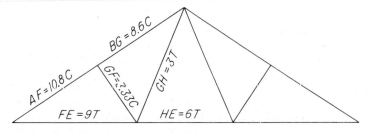

Fig. 4-51   Stress-summation diagram.

force diagram (Fig. 4-53) and also on the respective members of the space diagram (Fig. 4-52), or on a separate stress-summation diagram similar to those shown in Figs. 4-51 and 4-55.

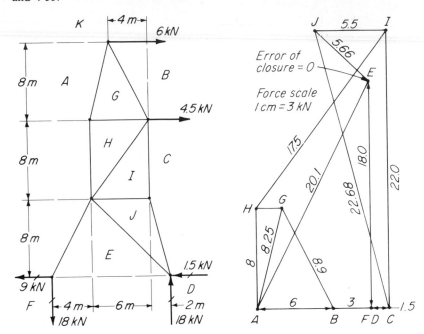

**Fig. 4-52   Space diagram.**          **Fig. 4-53   Force diagram.**

**4-45.** Using the graphical method of combined diagrams, find the stresses in all members of the truss shown in Fig. 4-54.

*Solution:* Having drawn the space diagram to scale, as in Fig. 4-54a, we discover that there are a minimum of three unknown stresses to be found at each joint, one more than may be determined by the usual graphical method. This difficulty may be overcome by the temporary use of a **substitute member** as follows.

Temporarily remove members *FG, GH,* and *HF* and substitute for them the single member *XH* indicated by the dashed line. We should note here that in all cases the substitute member must be so placed that the truss remains stable and that all stresses can then be solved for by the usual procedure.

If now we cover the left two-thirds of the space diagram, it becomes apparent that the stresses in the members of the right half of the truss do not depend in any way on the particular arrangement of the members in the left half. That is, with a section cut through members *AHID* and the part of the truss on the right then isolated (see Figs. 2-41b and 5-62), it becomes apparent that stresses *AH, HI,* and *ID* may be found as follows: *AH* by a summation of moments about the center joint of the top chord, the vertical component of *HI* by a summation of vertical forces, and *ID* by a summation of horizontal forces. Consequently, *the stress in member AH,* which member is common to both halves of the truss, *will be the true stress,* whether found by solving the lower right joint with the original arrangement of members, or by solving the lower left joint using the substitute member.

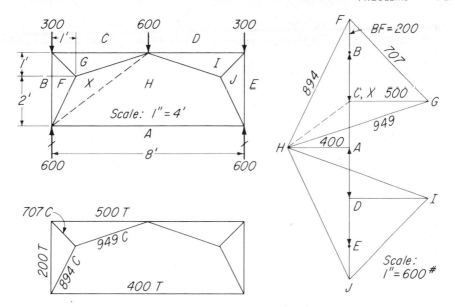

**Fig. 4-54  Stresses by the graphical method of combined diagrams using substitute member.** *(a)* **Space diagram.** *(b)* **Stress-summation diagram.** *(c)* **Force diagram.**

The procedure now is as follows. In the force diagram (Fig. 4-54c), lay off to scale all external forces, taking them in order clockwise around the truss. Then remove members *FG, GH,* and *HF* and substitute for them the single member *XH.* Next solve the upper left joint for stresses *CX* and *XB,* and then the lower left joint for stresses *XH* and *HA.* Of these stresses, only *HA* will be a true stress. Now remove the substitute member *XH* and replace members *FG, GH,* and *HF.* A second solution of the lower left joint, superimposed on the first, and using the true stress *HA,* will give also the true stresses in *BF* and *FH.* From here the force diagram is completed as usual.

It is interesting to note here that this problem may be solved quite easily by the analytical method of joints, using the substitute member as in the graphical solution. The steps are as follows: (1) remove members *FG, GH,* and *HF* and substitute for them member *XH;* (2) solve the upper left joint; (3) solve the lower left joint (*HA* is true stress); (4) remove *XH,* replace *FG, GH,* and *HF,* and resolve the lower left joint using the true stress *HA;* (5) proceed with the balance of the solution in the normal manner.

## PROBLEMS

Stresses by the graphical method of combined diagrams.

**4-46.** Determine the stresses in all members of the truss shown in Fig. 4-55. (Scale: 1 cm = 100 N.)

**4-47.** Find the stresses in all members of the truss shown in Fig. 4-56. (Scales on 8½ by 11: 1 in. = 2 ft, 1 in. = 2 kips.)

*Ans.* *BF* = 8.44 kips C; *FA* = 2.67 kips T; *FE* = 6.01 kips T; *EC* = 3 kips T;
*BE* = 6.71 kips C

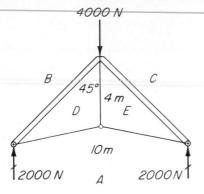

Fig. 4-55   Prob. 4-46.

**4-48.** Solve Prob. 4-44 if a vertical downward load $AK$ of 9 kN is added at the top of the tower.

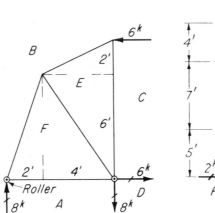

Fig. 4-56   Prob. 4-47.          Fig. 4-57   Prob. 4-49.

**4-49.** The signboard truss in Fig. 4-57 is subjected to horizontal wind loads as shown. Determine the stresses in all members caused by these loads. (Scales on 8½ by 11: 1 in. = 4 ft, 1 in. = 2 kips.)

*Ans.* $BG$ = 1.60 kips T; $GA$ = 2.56 kips C; $GH$ = 1.71 kips T; $HA$ = 2.74 kips C; $CI$ = 5.05 kips T; $IH$ = 5.85 kips C; $EI$ = 4.98 kips T; $AE$ = 8.25 kips C

**4-50.** The truss shown in Fig. 4-58 is one of two similar trusses supporting a water tank. It is also subjected to a horizontal wind load. Determine the stresses in all members due to these loads. (Scales on 8½ by 11: 1 in. = 4 ft, 1 in. = 4 kips.)

**4-51.** A punch frame is shown in Fig. 4-59. Note that the reaction forces are zero under the loading shown. Solve for the stresses in the members.

*Ans.* $AC$ = $AH$ = 24 kN T; $BC$ = $HI$ = 26 kN C; $AF$ = $CD$ = $GH$ = 28 kN T; $BD$ = $GJ$ = 30 kN C; $BE$ = 18 kN C; $EF$ = $BJ$ = 0

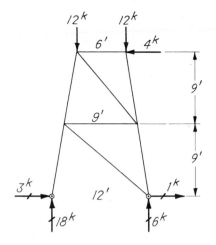

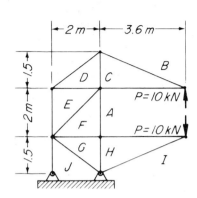

**Fig. 4-58   Prob. 4-50.**          **Fig. 4-59   Prob. 4-51.**

**4-52.** Determine the stresses in all members of the scissors roof truss shown in Fig. 4-60. (Scale: 1 cm = 1 kN)

*Partial Ans.* Stress in vertical member is 12 kN T

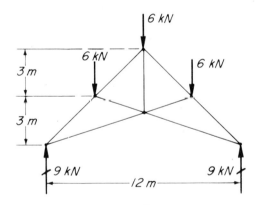

**Fig. 4-60   Prob. 4-52. Scissors truss.**

**4-53.** In Fig. 4-61 the wind loads act perpendicular to the sloping chord of the Howe roof truss shown. Assume the reactions at the left and right supports to be parallel to the wind loads. Find the reactions by the inverse-proportion method described in Art. 2-13. Then solve for the stresses in all members. (Scales on 8½ by 11: 1 in. = 10 ft, 1 in. = 2 kips.)

*Partial Ans.* $R_L$ = 4.13 kips; $R_R$ = 1.87 kips

**4-54.** In Fig. 4-62 the wind loads act perpendicular to the sloping chord of the Fink roof truss shown. Determine the stresses in all members. (Scales on 11 by 17: 1 in. = 4 ft, 1 in. = 1 kip.)

**4-55.** Solve for the stresses in all members of the truss shown in Fig. 5-66. (Scales on 8½ by 11: 1 in. = 5 ft, 1 in. = 5 kips. Use Bow's notation.)

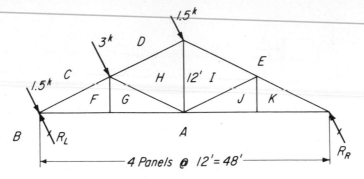

**Fig. 4-61  Prob. 4-53. Howe roof truss, wind loads.**

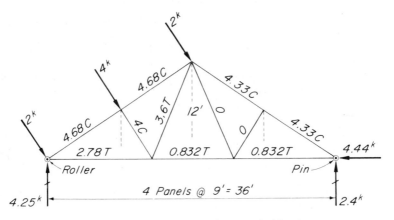

**Fig. 4-62  Probs. 4-54 and 4-93. Fink roof truss, wind loads.**

**4-56.** Solve for the stresses in all members of the truss shown in Fig. 5-61. (Scales on 8½ by 11: 1 in. = 10 ft, 1 in. = 5 kips. Use Bow's notation.)

**4-57.** Find the stresses in all members of the roof truss shown in Fig. 4-63. (Scales on 11 by 17: 1 in. = 3 ft, 1 in. = 3 kips. Answers are shown on right half of truss.)

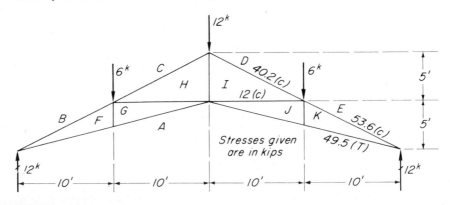

**Fig. 4-63  Prob. 4-57.**

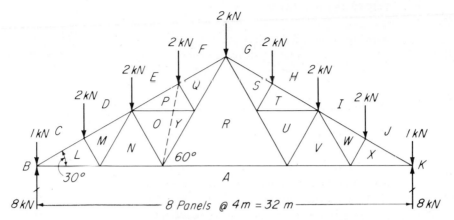

**Fig. 4-64   Prob. 4-59. Fink roof truss to be solved by the substitute-member method.**

**4-58.** Solve for the stresses in all members of the truss shown in Fig. 4-36. (Scale: 1 cm = 4 kN. Use Bow's notation.)

**4-59.** Using the substitute-member method, solve for the stresses in all members of the compound Fink roof truss shown in Fig. 4-64. (HINT: Remove members *OP* and *PQ* and substitute for them the member *OY* indicated by the dashed line. The stress in member *RA* is true stress. (Scale: 1 cm = 2 kN.)

**4-60.** Using the substitute-member method, do the *alternate graphical solution* suggested in Prob. 5-65. (Space scale: 1 in. = 10 in. Force scale: *AD* by four-force principle, 1 in. = 200 lb. Force scale for combined diagram: 1 in. = 100 lb. Use Bow's notation for combined diagram.)

**4-61.** Using the substitute-member method, solve for the stresses in all members of the truss shown in Fig. 4-65. (NOTE: In this problem, try substituting the member *AH* indicated by the dashed line in Fig. 4-66. Begin the solution at the lower left joint. Stress *AI* will be true stress. Scales on 8½ by 11: 1 in. = 10 ft, 1 in. = 4 kips. See explanation of the *all-graphical solution* of Prob. 5-65.)

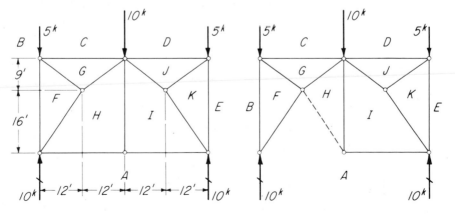

**Fig. 4-65   Prob. 4-61.**

**Fig. 4-66   Substitute member. Prob 4-61.**

## 4-10 THREE-FORCE MEMBERS

A member on which forces act at three (or more) points is called a *three-force member* (or a multiple-force member). These forces necessarily act transversely to its axis or have transverse components, thus causing it to bend, as in the case of member $AD$ in Fig. 4-67. The stresses in three-force members are complex, being usually a combination of axial, bending, and shearing stresses. Such stresses are considered under the subject of strength of materials. *In statics, therefore, a three-force member should never be cut.*

In Fig. 4-67, $BC$ is clearly a two-force member, since forces act on it at two points only, and the reacting force $R_B$ at $B$ therefore lies parallel to $BC$. Because the force $W$ acts transversely to member $AD$, the reacting forces $R_A$ and $R_C$ also act transversely to $AD$, or must have transverse components. This is true of all three-force members. Hence **the reacting force at any support of a three-force member never lies parallel to the axis of that member.** These facts are of importance in the determination of support reactions, outlined in the following article.

## 4-11 GRAPHICAL DETERMINATION OF REACTIONS USING THE THREE-FORCE PRINCIPLE

Supports and support reactions were discussed in Art. 2-16. As is stated therein, bodies subjected to coplanar force systems are generally supported at two points, at one of which the direction of the reacting force will often be known, while at both the magnitudes will be unknown. The usual graphical problem then is to determine the unknown direction of one support reaction and magnitudes of both. When only three nonparallel forces act on the body, such as one load and two reactions, this unknown direction may be established by the three-force principle, whereafter the two unknown magnitudes are found by a closed graphical force diagram, as shown in Prob. 4-55. When two

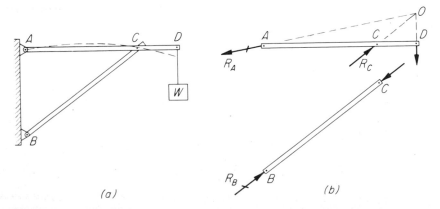

(a)

(b)

**Fig. 4-67  Two- and three-force members.** *(a)* Space diagram. *(b)* Free-body diagrams.

or more *known* forces act on the body, they must be combined into their single resultant force (Art. 2-18) before the three-force principle can be applied, as shown in Prob. 4-63.

## ILLUSTRATIVE PROBLEMS

**4-62.** Determine graphically the reactions at supports $A$ and $B$ of the frame shown in Fig. 4-68*a*.

*Solution:* From the space diagram (Fig. 4-68*a*), we recognize that $BC$ is a two-force member (Art. 4-5). The reaction $R_B$ then lies parallel to $BC$. The extended lines of action of $W$ and $R_B$ then establish $O$, their point of intersection, through which the action line of reaction $R_A$ also must pass, according to the three-force principle. From the force diagram, we find $R_A$ and $R_B$ to be 206 N and 250 N, respectively. Students should record these values on the space diagram.

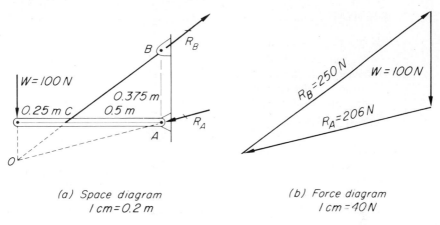

(a) Space diagram
1 cm = 0.2 m

(b) Force diagram
1 cm = 40 N

**Fig. 4-68   Graphical determination of reactions using the three-force principle.**

**4-63.** Determine graphically the support reactions at $A$ and $B$ of the truss shown in Fig. 4-69*a*.

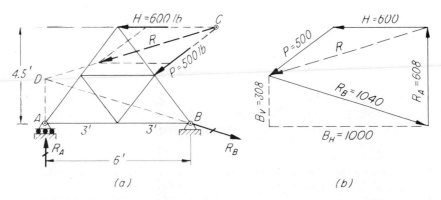

(a)

(b)

**Fig. 4-69   Graphical determination of reactions using the three-force principle. *(a)* Space diagram. 1 in. = 4 ft. *(b)* Force diagram. 1 in. = 600 lb.**

*Solution:* Four forces act on this truss, the known forces $H$ and $P$ and the unknown reactions $R_A$ and $R_B$. We must therefore combine $H$ and $P$ into their resultant $R$. This is done by first extending their action lines backward to their intersection $C$. From $C$ along the respective action lines we lay off 600 and 500 lb. The diagonal of the completed parallelogram is the action line of $R$. Now only three forces remain whose action lines must intersect at a common point.

Because the rollers at $A$ rest on a horizontal surface, the action line of $R_A$ is vertical and intersects that of $R$ at $D$, the intersection through which the action line of $R_B$ must then also pass. A force diagram of the four forces, here taken clockwise around the truss, then gives $R_A$ and $R_B$.. Greater accuracy will be obtained by using the original forces $H$ and $P$ rather than their resultant. As a check against an analytical solution, $B_V$ and $B_H$ may also be scaled.

## PROBLEMS

Graphical determination of reactions.

**4-64.** The cylinder shown in Fig. 4-70 is 1 m in diameter and weighs 300 N. Determine the horizontal push $P$ required to hold it in the position shown, when block $B$ is removed. Neglect friction. (Scale on 8½ by 11: 1 cm = 30 N.)

**4-65.** Solve Prob. 4-64 if the push $P$ is parallel to incline.

*Ans.* $P = 111.4$ N

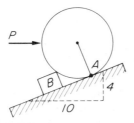

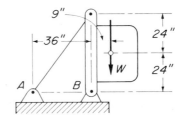

**Fig. 4-70   Probs. 4-64 and 4-65.**          **Fig. 4-71   Prob. 4-66.**

**4-66.** The weight $W$ in Fig. 4-71 is 480 lb. Determine the support reactions at $A$ and $B$. (Scale on 8½ × 11: 1 in. = 1 ft; 1 in. = 100 lb.)

**4-67.** Determine the force $F$ required to start to tip the block shown in Fig. 4-72 about corner $C$ and the reacting force $R_C$ at $C$. (Scales on 8½ by 11: 1 cm = 0.2 m; 1 cm = 50 N.)

**4-68.** The Fink roof truss shown in Fig. 4-73 is subjected to a resultant wind load at the left midjoint. Determine the support reactions $R_A$ and $R_B$.. (Scales on 8½ by 11: 1 in. = 10 ft, 1 in. = 600 lb.)

**4-69.** In Fig. 4-74 the 900-lb weight is suspended from a flexible cable passing over a sheave. Find the tension in the rope $AB$ and the reaction at $C$. (Scales on 8½ by 11: 1 in. = 2 ft, 1 in. = 200 lb.)

*Ans.* $T = 300$ lb↑; $R_C = 1,082$ lb↗

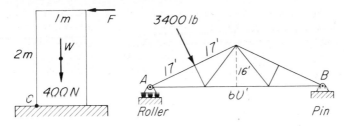

**Fig. 4-72   Prob. 4-67.**    **Fig. 4-73   Prob. 4-68.**

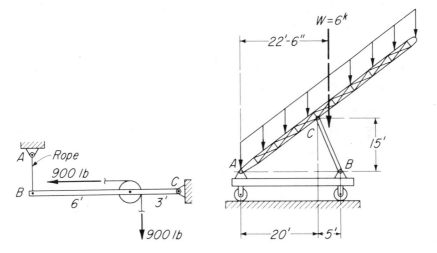

**Fig. 4-74   Prob. 4-69.**          **Fig. 4-75   Prob. 4-70.**

**4-70.** A portable conveyor is shown diagrammatically in Fig. 4-75. If the conveyor and its contents weigh 6 kips, determine graphically the reactions at $A$ and $B$.

**4-71.** A crane is equipped with a fixed jib as shown in Fig. 4-76. Make a complete graphical solution and determine the forces acting on pins $A$ and $B$. The weight $W$ is 3000 lb.

**4-72.** The three-hinged arch shown in Fig. 4-77 is acted on by two loads which are concurrent through point $D$. Determine the support reactions $R_A$ and $R_B$ (Scales on 8½ by 11: 1 in. = 20 ft, 1 in. = 10 kips.)

*Ans.* $R_A$ = 33.5 kips ↗; $R_B$ = 19.2 kips ↖

**4-73.** Figure 4-78 shows an end view of an air-compressor tank 1.5 m in diameter and weighing 2 kN, half of which (1 kN) is supported by each of two identical frames $ACB$. Determine (a) the force $P$ exerted by the tank at $E$ and (b) the support reactions at $A$ and $B$. Neglect friction between tank and supports. [HINT: (a) Isolate the tank as a free body. (b) Only force $P$ produces reactions at $A$ and $B$.]

*Ans.* $P$ = 1.16 kN ↙; $R_A$ = 0.43 kN ; $R_B$ = 0.81 kN ↗

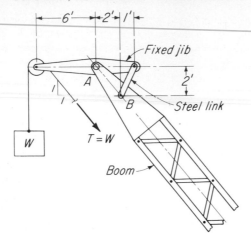

Fig. 4-76   Prob. 4-71.

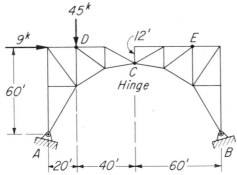

Fig. 4-77   Probs. 4-72 and 4-100.

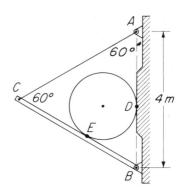

Fig. 4-78   Prob. 4-73.

---

**SUMMARY** (By Article Number) _____

**4-1. A coplanar, concurrent force system** is one in which all forces lie in one plane, and their lines of action all intersect at a common point. The internal forces acting in the members meeting at the joints of trusses illustrate this type of system.

**4-2. The resultant of a coplanar, concurrent force system** passes through the point of concurrency and may be determined analytically (*a*) by algebraic summation of forces or components parallel to two rectangular axes or (*b*) graphically by the parallelogram or the triangle method.

**4-3. In a system of coplanar, concurrent forces in equilibrium,** $\Sigma V = 0$ and $\Sigma H = 0$. Hence the resultant of such a system is zero. When laid off graphically to scale, *the forces of a system in equilibrium will always form a closed polygon, in which the arrows always follow each other.*

**4-4. A truss** is a structural unit in which the bars are arranged to form one or more connected triangles. **A joint** is the connection between two or more bars. **A member** is a bar extending from joint to joint.

**4-5. Stresses in members of trusses** which are pin-connected and are loaded only at the joints always lie parallel to the principal axes of the members. Such members are **two-force members.** When the stress pushes on the member, it is compressive; when it pulls on the member, it is tensile.

**4-6.** When a load is applied at a joint by means of **a rope passing over a sheave or a pulley,** the sheave or pulley may be removed and the forces may be applied directly at the joint without changing their effect.

**4-7.** To determine *stresses in trusses* by the **analytical method of joints,** the joints are successively isolated, beginning with one at which only two stresses are unknown, and proceeding in such order that not more than two unknown stresses need be determined at any joint. Sloping forces are replaced by vertical and horizontal components. The unknown stresses are then found by solving the two equations of force equilibrium, $\Sigma V = 0$ and $\Sigma H = 0$.

**4-8.** In the **graphical method of joints,** stresses are found by drawing to scale a separate closed force diagram of the forces at each joint. The arrowheads always follow each other around the diagram, thus determining whether stresses are tensile or compressive.

**4-9.** In the **graphical method of combined diagrams,** all external forces are first laid off to scale, thus forming a closed diagram. Thereafter the stresses are found as in the graphical method of joints. The separate force diagrams are joined into a single *combined diagram.*

**4-10. A three-force member** is one on which forces act at three or more points. Internal stresses in a three-force member are complex, and the total resultant stress does not lie parallel to the axis of the member. Consequently, in statics, a three-force member should never be cut.

**4-11. The three-force principle** states that, when three nonparallel forces are in equilibrium, they must be coplanar and their lines of action must intersect at a common point. This principle is very useful in *graphical determinations of support reactions,* when only three forces are involved or when they can conveniently be reduced to three. The reaction at the end of a two-force member lies parallel to the member; the reaction at the end of a three-force member does not.

## REVIEW PROBLEMS

**4-74.** Using algebraic summations of rectangular components, solve for the *resultant force R* of the three forces shown in Fig. 4-79 and the angle $\theta_H$ it makes with the horizontal. Check graphically by the triangle method. (Scale: 1 in. = 50 lb.)

*Ans.* $R = 170$ lb $\swarrow$; $\theta_H = 43°25'$

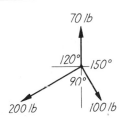

**Fig. 4-79   Prob. 4-74.**

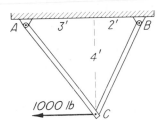

**Fig. 4-80   Prob. 4-75.**

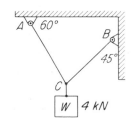

**Fig. 4-81   Prob. 4-76.**

**4-75.** Solve by *geometry* for the *stresses* in members *AC* and *BC* of the truss shown in Fig. 4-80. If space and force diagrams are drawn to scale (1 in. = 2 ft, 1 in. = 400 lb), a graphical check is obtained.

*Ans. AC* = 1000 lb C; *BC* = 894 lb T

**4-76.** Solve by *trigonometry* for the *stresses* in members *AC* and *BC* of the truss in Fig. 4-81. Check graphically.

**4-77.** The weight *W* shown in Fig. 4-82 is 3.6 kN. Find the *stresses* in members *AB* and *BC* by *trigonometry.*

*Ans. AB* = 2.35 kN T; *BC* = 3.17 kN C

**4-78.** The truss shown in Fig. 4-83 supports a weight *W* of 6300 lb. Using the *geometric method* (similar triangles), solve for the *stresses* in members *AB* and *BC*.

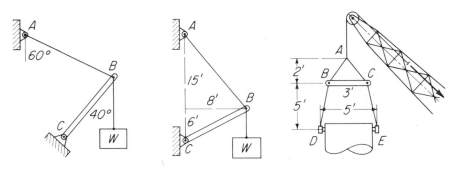

Fig. 4-82   Prob. 4-77.          Fig. 4-83   Prob.          Fig. 4-84   Prob. 4-79.
4-78.

**4-79.** The crane sling shown in Fig. 4-84 supports a concrete bucket weighing 16 kips. Determine the *stress* in the horizontal bar *BC* using the *analytical method of joints.* The bar *BC* is 3 ft long.

*Ans. BC* = 4.4 kips C

**4-80.** Using the *analytical method of joints,* solve for *stresses AB* and *AC* in Fig. 4-85.

**4-81.** Calculate the load *W* and pull *P* if the cable tension must not exceed 4000 N when the carriage is in the position shown in Fig. 4-86.

*Ans. W* = 4400 N; *P* = 264 N

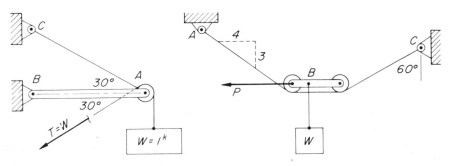

Fig. 4-85   Prob. 4-80.          Fig. 4-86   Prob. 4-81.

**4-82.** In Fig. 4-87, solve analytically for the *stress* in the link and for the angle $\theta_H$ it makes with the horizontal. Assume static conditions, and neglect sheave friction.

*Ans.* $T = 22.36$ kips; $\theta_H = 63°27'$

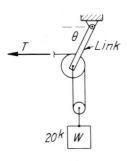

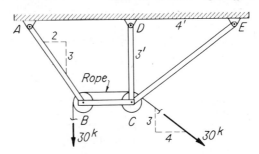

**Fig. 4-87   Prob. 4-82.**          **Fig. 4-88   Prob. 4-83.**

**4-83.** Using the *analytical method of joints,* solve for the *stresses* in all members of the frame shown in Fig. 4-88. Neglect friction.

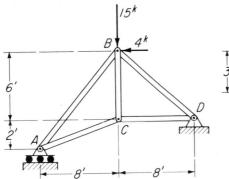

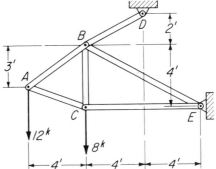

**Fig. 4-89   Prob. 4-84.**              **Fig. 4-90   Prob. 4-85.**

**4-84.** Using the *analytical method of joints,* solve for the *stresses* in all members of the truss shown in Fig. 4-89.

**4-85.** Determine the *stresses* in all members of the truss shown in Fig. 4-90 using the *analytical method of joints.*

*Ans.* $AB = 15$ kips T; $AC = 12.4$ kips C; $BC = 11$ kips T; $BD = 29.1$ kips T; $BE = 15.7$ kips C; $CE = 12$ kips C

**4-86.** Using the *analytical method of joints,* solve for the *stresses* in all members of the roof truss shown in Fig. 4-91.

*Ans.* $AB = 38$ kips C; $AC = 34$ kips T; $BC = 1$ kip T; $BD = 45$ kips C; $CD = 3.6$ kips T; $CE = 31$ kips T; $DF = 47$ kips C; $EF = 31$ kips T; $DE = 3$ kips T

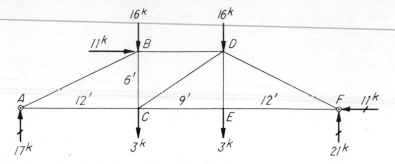

**Fig. 4-91    Probs. 4-86, 4-91, 5-69, and 5-70.**

**4-87.** Using the *graphical method of joints*, determine the *stresses* in the two members of the frame shown in Fig. 4-92. (Scales on 8½ by 11: 1 in. = 2 ft, 1 in. = 100 lb.)

*Ans.* Left = 189 lb T; right = 596 lb C

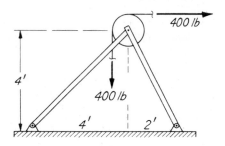

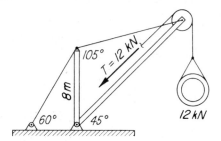

**Fig. 4-92    Prob. 4-87.**                **Fig. 4-93    Prob. 4-88.**

**4-88.** Find the *stresses* in all members of the derrick shown in Fig. 4-93, by the *graphical method of joints.*

**4-89.** Using the combined-diagram method, determine graphically the stresses in all members of the truss shown in Fig. 4-94. Record these stresses on a space diagram and indicate whether *T* or *C.* (Force scale: 1 in. = 5 kips.)

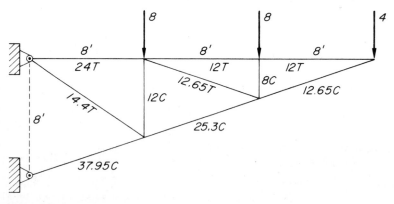

**Fig. 4-94    Prob. 4-89.**

**4-90.** The cambered Fink roof truss in Fig. 4-95 is subjected to wind loads as shown. Using the *three-force principle,* determine the *support reactions.* Then find all *stresses* by the *graphical combined-diagram method.* (Scales on 11 by 17: 1 in. = 6 ft, 1 in. = 2 kips.)

*Ans.* $R_A = 2.41$ kips↑; $R_D = 6.14$ kips

**4-91.** Using the *graphical combined-diagram method,* solve for all stresses in the roof truss shown in Fig. 4-91. (Scales on 11 by 17: 1 in. = 4 ft, 1 in. = 4 kips.)

*Ans.* See Prob. 4-86.

**4-92.** Using the *three-force principle,* check the *support reactions* given in Prob. 5-25

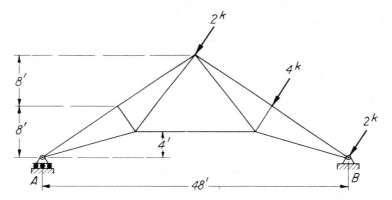

**Fig. 4-95    Prob. 4-90. Cambered Fink roof truss, wind loads.**

**4-93.** Check the *support reactions* in Fig. 4-62 by the *three-force principle.* (Scales on 8½ by 11: 1 in. = 10 ft, 1 in. = 2 kips.)

**4-94.** Using the *three-force principle,* determine the *reactions AD* and *DC* for the truss of Fig. 4-96. Then find the *stresses* in all members by the graphical *combined-diagram method.* Superimpose the combined-diagram solution on the three-force diagram. (Scales on 8½ by 11: 1 in. = 1 ft, 1 in. = 5 kips.)

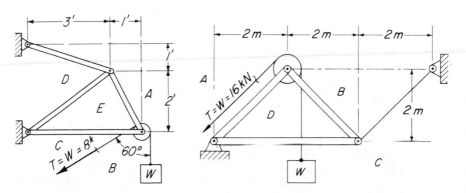

**Fig. 4-96    Prob. 4-94.**          **Fig. 4-97    Prob. 4-95.**

**4-95.** Using the *three-force principle*, determine graphically the *reactions* at the supports of the truss shown in Fig. 4-97. Continue the graphical solution to determine the *stresses* in all members.

**4-96.** Using the *three-force principle*, determine graphically the *reactions* at the supports of the truss shown in Fig. 4-98. Continue the graphical solution to find the *stresses* in all truss members. (Scales on 11 by 17: 1 in. = 4 ft, 1 in. = 3 kips.)

**4-97.** Using the *three-force principle*, determine graphically the belt tension $T$ and the reaction at $A$ of the assembly shown in Fig. 4-99 if $W = 100$ lb. (Scales on 8½ by 11: 1 in. = 2 in., 1 in. = 20 lb.)

*Ans.* $T = 13.4$ lb, $A_V = 75$ lb, $A_H = 6.7$ lb

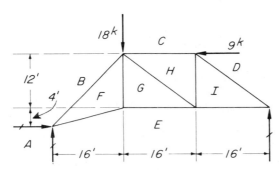

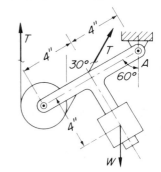

**Fig. 4-98    Prob. 4-96.**                    **Fig. 4-99    Prob. 4-97.**

**4-98.** Using the *three-force principle*, determine graphically the *reactions* at supports $A$ and $B$ of the frame shown in Fig. 4-100.

*Ans.* $R_A = 6.4$ kN $\uparrow$; $R_B = 5.66$ kN $\nwarrow$

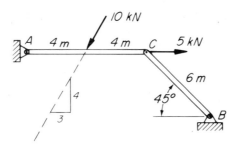

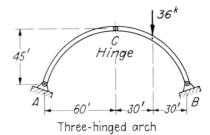

**Fig. 4-100    Prob. 4-98.**                    **Fig. 4-101    Prob. 4-99.**

**4-99.** Determine graphically the *reactions* at $A$ and $B$ of the three-hinged arch shown in Fig. 4-101. Use the *three-force principle*. (Scales on 8½ by 11: 1 in. = 30 ft, 1 in. = 10 kips.)

*Ans.* $R_A = 15$ kips $\nearrow$; $R_B = 29.5$ kips $\nwarrow$

**4-100.** In Fig. 4-77, determine the *reactions* at $A$ and $B$ if an additional vertical load of 30 kips acts downward at $E$. Use the *three-force principle*. (HINT: Determine separately the reactions due to loads on each half of the arch; then find the resultant reactions. Scales on 11 by 17: 1 in. = 20 ft, 1 in. = 10 kips.)

**REVIEW QUESTIONS** _____

**4-1.** Define a coplanar, concurrent force system.

**4-2.** What must be known in order to fully define the resultant of a coplanar, concurrent force system?

**4-3.** By what three methods may the resultant of a coplanar, concurrent force system be determined?

**4-4.** Why must the graphical force polygon of a system of forces in equilibrium always close?

**4-5.** In the language of structures, what is a truss, a joint, and a member?

**4-6.** State the principal feature of a pin-connected truss.

**4-7.** What is a two-force member?

**4-8.** What is true about the forces acting on a two-force member?

**4-9.** What two types of stresses are commonly encountered in pin-connected trusses?

**4-10.** Name some of the methods by which stresses in members of trusses are determined.

**4-11.** Outline briefly how stresses in trusses are determined by the analytical method of joints.

**4-12.** State briefly how stresses in trusses are determined by the graphical method of joints.

**4-13.** Explain briefly how stresses in trusses are obtained by the graphical combined-diagram method.

**4-14.** What is a three-force member?

**4-15.** Explain briefly the types of stresses found in three-force members.

**4-16.** Can the direction of the reaction at the support of a three-force member be parallel to the member?

**4-17.** By means of what simple principle of statics may support reactions be determined graphically? State the principle itself.

**4-18.** In problems to determine support reactions, just what does the three-force principle establish?

# Chapter 5 _____
# Coplanar, _____
# Nonconcurrent _____
# Force Systems _____

## 5-1 INTRODUCTION

In a nonconcurrent force system, the action lines of the forces do not meet at a common point. Each force in such a system will then tend to rotate the body, on which it acts, about some axis. Therefore, *the principle of moments must be used in solving these systems.*

## 5-2 RESULTANT OF A COPLANAR, NONCONCURRENT FORCE SYSTEM

The resultant of a coplanar, nonconcurrent force system is fully known only when the following items are determined: (1) its magnitude and sense, (2) the angle of inclination $\theta$ of its action line with some reference line (usually horizontal), and (3) the location of some point on its line of action. The resultant may be determined analytically, or graphically as in Prob. 5-1. A graphical solution by the string-polygon method is outlined in Prob. 5-12.

### ILLUSTRATIVE PROBLEM _____

**5-1.** Determine fully the resultant $R$ of the three forces, $A$, $B$, and $C$, acting on bar $DE$ shown in Fig. 5-1a and the angle $\theta_H$ which the action line of $R$ makes with the horizontal. All forces are in pounds.

*Algebraic Solution:* The method followed here is similar to that shown in Prob. 2-5. Sloping forces are replaced by their vertical and horizontal components. Summations of vertical forces and components, giving $\Sigma V$, and of horizontal forces and components, giving $\Sigma H$, are then made. From these we determine (*a*) the magnitude of $R$ and (*b*) the inclination $\theta_H$ of its line of action with the horizontal.

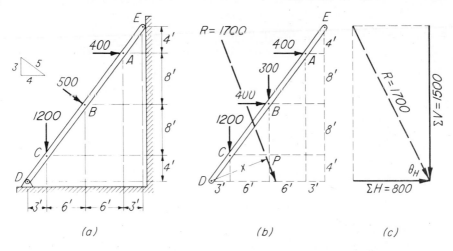

**Fig. 5-1** **Resultant of coplanar, nonconcurrent forces. Analytical solution.** *(a)* **Loaded member.** *(b)* **V and H components.** *(c)* **Components of R.**

The $V$ and $H$ components of force $B$ are found to be 300 and 400 lb, respectively, as shown in Fig. 5-1b. Then $\Sigma V$ is $300 + 1200$ or 1500 lb and $\Sigma H$ is $400 + 400$ or 800 lb (Fig. 5-1c). Hence

$$[R = \sqrt{(\Sigma V)^2 + (\Sigma H)^2}] \qquad R = \sqrt{(1500)^2 + (800)^2} = 1700 \text{ lb} \qquad Ans.$$

Also $\qquad \tan \theta_H = \dfrac{\Sigma V}{\Sigma H} = \dfrac{1500}{800} \qquad$ or $\qquad \theta_H = 61°56' \qquad Ans.$

A point $P$ on the action line of $R$ may now be determined as follows: According to the principle of moments, the moment of $R$ about any point equals the algebraic sum

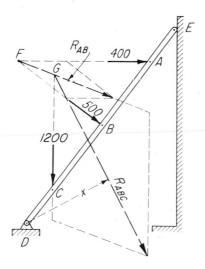

**Fig. 5-2** **Resultant force; parallelogram method.**

of the moments of the separate forces about the same point. Let $D$ be the moment center and $x$ the moment arm of $R$, perpendicular to its action line. Then

$$1700x = (1200)(3) + (400)(12) + (300)(9) + (400)(20) = 19,100$$

and                                             $x = 11.2 \text{ ft}$                                             *Ans.*

*Graphical Solution:* When the bar has been drawn to scale, the three forces are combined into their single resultant by the parallelogram method, as shown in Fig. 5-2. The action lines of forces $A$ and $B$ are extended backward to their intersection $F$ from which $A$ and $B$ are laid off. The diagonal $R_{AB}$ of the completed parallelogram is the resultant of $A$ and $B$. Next we combine $R_{AB}$ and $C$ by laying off their magnitudes from $G$, the intersection of their action lines. The diagonal $R_{ABC}$ of the second parallelogram is then the resultant of $A$, $B$, and $C$. It scales 1700 lb. The angle $\theta_H$ scales 62°, and $x$ scales 11.2 ft.

## PROBLEMS

In the following problems let $R$ be the resultant force, $\theta_H$ the angle which the action line of $R$ makes with the horizontal, and $x$ the moment arm from $O$ (or other designated moment center) to the action line of $R$.

**5-2.** Determine algebraically $R$, $\theta_H$, and $x$ (from $O$) of the four forces shown in Fig. 5-3.

**5-3.** Find algebraically $R$, $\theta_H$, and $x$ (from $O$) of the four forces shown in Fig. 5-4. Check graphically using the parallelogram method. (SUGGESTION: Combine $A$ and $B$, then $C$ and $D$, and then $R_{AB}$ and $R_{CD}$.

*Ans.* $R = 12.2 \text{ N}$, $\theta_H = 18°$, $x = 3.74 \text{ m}$

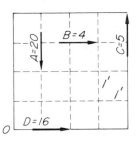

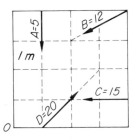

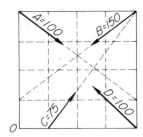

**Fig. 5-3    Prob. 5-2.**          **Fig. 5-4    Prob. 5-3.**          **Fig. 5-5    Prob. 5-4.**

**5-4.** Using the graphical parallelogram method, determine $R$, $\theta_H$, and $x$ (from $O$) of the four forces shown in Fig. 5-5. (SUGGESTION: Combine $A$ and $B$, then $C$ and $D$, and then $R_{AB}$ and $R_{CD}$. Scale on 8½ by 11: 1 in. = 1 ft, 1 in. = 30 lb.)

**5-5.** Determine algebraically $R$, $\theta_H$, and $x$ (from $E$) of the three forces acting on the bar shown in Fig. 5-6.

*Ans.* $R = 1063 \text{ lb}$; $\theta_H = 41°11'$; $x = 5.93 \text{ ft}$

**5-6.** Solve graphically for $R$, $\theta_H$, and $x$ (from $E$) in Fig. 5-6. Then, using the three-force principle, determine the reactions at supports $D$ and $E$. (Scales on 8½ by 11: 1 in. = 3 ft, 1 in. = 300 lb.)

*Ans.* $R_D = 731 \text{ lb} \nearrow$; $R_E = 590 \text{ lb} \rightarrow$

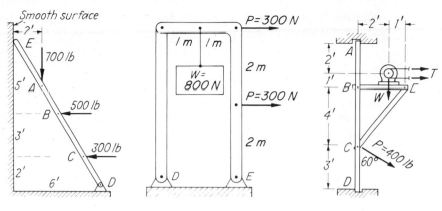

Fig. 5-6  Probs. 5-5
and 5-6.

Fig. 5-7  Probs. 5-7
and 5-8.

Fig. 5-8  Probs. 5-9
and 5-10.

**5-7.** The frame shown in Fig. 5-7 is acted on by three forces. Find algebraically $R$, $\theta_H$, and $x$ (from point $D$).

*Ans.* $R = 1000$ N; $\theta_H = 53.1°$; $x = 2.6$ m

**5-8.** In Fig. 5-7, determine graphically $R$, $\theta_H$, and $x$ (from $D$). Then, using the three-force principle, determine the reactions at supports $D$ and $E$.

**5-9.** The frame shown in Fig. 5-8 supports a motor weighing 400 lb and is also subjected to a force of 400 lb at $C$. The total belt tension $T$ (representing both sides of the belt) is 300 lb and is assumed to act through the center of the shaft. Find algebraically $R$, $\theta_H$, and $x$ (from point $C$).

*Ans.* $R = 882$ lb; $\theta_H = 42°54'$; $x = 2.61$ ft

**5-10.** In Prob. 5-9, determine graphically $R$, $\theta_H$, and $x$ (from $C$). Then, using the three-force principle, determine the support reactions $R_A$ and $R_D$. (Scales on 8½ by 11: 1 in. = 2 ft, 1 in. = 200 lb.)

**5-11.** The concrete bridge pier shown in Fig. 5-9 must support the two vertical loads from the bridge trusses and a horizontal ice load at the water surface as indicated. Satisfactory stability of the pier requires that the resultant $R$ of the four forces pass

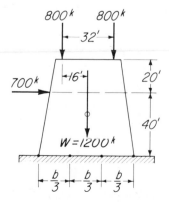

Fig. 5-9  Prob. 5-11.

within the middle third of its base. Determine graphically the minimum permissible width $b$ of the pier. (Scales on 8½ by 11: 1 in. = 10 ft, 1 in. = 500 kips.)

*Ans.  $b$ = 60 ft*

**Resultant Force; Graphical String-polygon Method.** The use of the graphical parallelogram method, described in Prob. 5-1, often becomes impractical, especially when the action lines of a group of forces whose resultant is desired become nearly parallel. The so-called string-polygon method, explained in Prob. 5-12, is readily applied to all problems.

**ILLUSTRATIVE PROBLEMS** —————————————————————————

**5-12.** Using the *graphical string-polygon method,* determine the resultant force $R$ of the system shown in Fig. 5-10a.

*Graphical Solution; String-polygon Method:* To determine $R$ fully, we must find its magnitude and direction, and a point on its line of action must be located. If the forces *A, B, C,* and *D* are laid down to scale to form the *load line,* and in the order shown in the *force diagram* (Fig. 5-10b), $R$ will be the magnitude of their resultant shown in its true direction.

We may now locate a point on the action line of $R$ as follows. The point $O$, called the **pole,** may be located at random. The **rays** *AB, BC, CD, DR,* and *RA* are then drawn from $O$ to the respective *intersections* on the load line of forces *A* and *B, B* and *C, C* and *D, D* and *R,* and *R* and *A. These rays are in reality the components of the various forces;* that is, rays *AB* and *BC* are the components of force *B,* rays *BC* and *CD* are the components of force *C,* and so on. In like manner, rays *DR* and *RA* are the components of the resultant force *R.* When we study this completed force diagram, we recognize that the *AB* component of force *A* is canceled by the *AB* component of force *B,* because they are identical except for opposite senses. Likewise, the *BC* and *CD* components are canceled, leaving only the uncanceled components *DR* and *RA* of the resultant force *R.*

However, the opposing components *AB, BC,* and *CD* can cancel only if they have common lines of action. To establish these, an *action line,* or *string, polygon* is drawn,

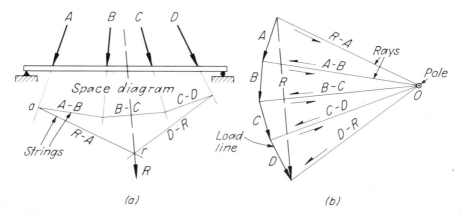

**Fig. 5-10   Resultant force; graphical string-polygon method.** *(a)* Action line or string polygon. *(b)* Force polygon.

in which *the strings are the action lines of the opposing components,* as shown in Fig. 5-10*a*. We first select any suitable point, such as *a,* on the action line of force *A* as the starting point of the string polygon. From this point, on the action line of force *A,* to action line *B,* draw the **string** *AB* parallel to ray *AB.* From action line *B* to action line *C,* draw string *BC* parallel to ray *BC.* From action line *C* to action line *D,* draw string *CD* parallel to ray *CD.* Point *r* on the action line of *R* is then established by drawing *strings RA* and *DR* parallel, respectively, to *rays RA* and *DR.* The resultant *R* in the space diagram is drawn parallel to *R* in the force polygon.

## PROBLEMS

Resultant force; string-polygon method.

**5-13.** Determine the resultant *R* of forces *A, B,* and *C* in Fig. 5-11. (Scales on 8½ by 11: 1 in. = 5 ft, 1 in. = 400 lb.)

*Ans. R* = 1350 lb

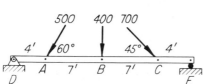

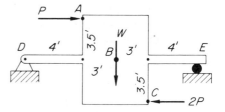

Fig. 5-11   Prob. 5-13.                Fig. 5-12   Prob. 5-14.

**5-14.** In Fig. 5-12 find the resultant *R* of forces *A, B,* and *C.*

**5-15.** Find the resultant *R* of forces *A, B,* and *C* in Fig. 5-13. (Scales on 8½ by 11: 1 in. = 4 ft, 1 in. = 300 lb.)

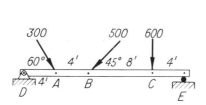

Fig. 5-13   Probs. 5-15 and 5-96.            Fig. 5-14   Probs. 5-16 and 5-31.

**5-16.** Determine the resultant of the forces *P* and *W* in Fig. 5-14 if *P* = 80 lb and *W* = 200 lb. (Force scale on 8½ by 11: 1 in. = 100 lb.)

## 5-3 EQUILIBRIUM OF COPLANAR, NONCONCURRENT FORCE SYSTEMS

Nonconcurrent force systems are most frequently encountered in problems involving the determination of unknown externally reacting forces at the supports of structures, structural members, machines, or machine parts. These force systems are also encountered in problems involving determination of stresses in members of trusses, as shown in Art. 5-7. Analytical solutions to determine reactions are based on the three equations of equilibrium, $\Sigma V = 0$, $\Sigma H = 0$, and $\Sigma M = 0$, as in Art 5-5. Graphical solutions employ the three-force principle as in Art 4-11 or the principles of the string and force polygons as is illustrated in Art. 5-4.

## 5-4 DETERMINATION OF REACTIONS; THE GRAPHICAL STRING-POLYGON METHOD

The basis of this method is similar to that explained in Art. 5-2 for determination of resultant force. When a system of forces is in equilibrium, however, all components of the forces in the force polygon necessarily cancel out; hence *the string polygon must close with a closing string from the last action line to the point of beginning on the first action line.* The application of the method is most easily understood by noting carefully the various steps outlined in the following example.

**ILLUSTRATIVE PROBLEM** _____

**5-17.** Determine graphically by the string-polygon method the reactions at supports *A* and *D* of the beam shown in Fig. 5-15.

*Solution:* The space diagram is first drawn to scale. Next the loads *B* and *C* are laid off to scale to form the beginning of the force polygon, which later will be closed by the unknown reacting forces *D* and *A*. The direction of force *D* is seen to be vertical; that of *A* is unknown. Now we choose a pole and draw rays *AB*, *BC*, and *CD*. The

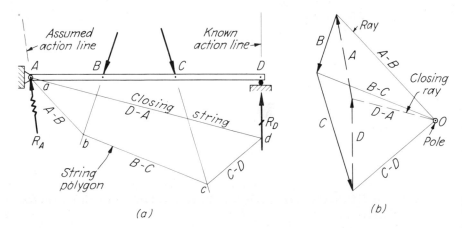

Fig. 5-15    Determination of reactions. String-polygon method. *(a)* Space diagram. *(b)* Force polygon.

*closing ray DA,* the direction of which determines the magnitudes of reactions *D* and *A,* can be located only *after* the completion of the string polygon. The unknown direction of reaction *A* is indicated by the "buckled" arrow.

*The string polygon must always start at the point of application of the reaction whose direction is unknown,* since that is the only *known* point on the action line of reaction *A.* Then, from this point on action line *A* to action line *B,* draw string *AB* parallel to ray *AB;* from action line *B* to action line *C,* draw string *BC* parallel to ray *BC;* and from action line *C* to action line *D,* draw string *CD* parallel to ray *CD.*

Only one ray, *DA,* remains to be located. *This closing ray must be parallel to the closing string* which extends from action line *D* to the point of beginning on line *A.* The string polygon may now be closed by string *DA,* after which ray *DA* locates the intersection of reactions *D* and *A,* thus determining the magnitudes of both and the direction of *A.*

If the directions of all action lines are known, as in parallel force systems, the string polygon may be started at any point on any action line. *The string polygon is so called because it takes the exact form of a (weightless) string, which is acted upon by forces A, B, C, and D at "joints" a, b, c, and d, respectively.* If now each "joint" is solved, we find that the "stress" in each string is given by the correspondingly lettered force component (ray) in the force polygon. Hence we recognize that the force polygon (Fig. 5-15*b*) is in reality the combined stress diagram giving the stresses in strings *AB, BC, CD,* and *DA,* produced by forces *A, B, C,* and *D.*

## PROBLEMS

Reactions; string-polygon method.

**5-18.** Determine the reactions at supports *A* and *E* of the beam shown in Fig. 5-16. Check by moments. (Scales on 8½ by 11: 1 in. = 5 ft, 1 in. = 500 lb.)

*Ans.* $R_A = 1075$ lb↑; $R_E = 1325$ lb↑

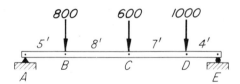 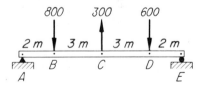

**Fig. 5-16   Prob. 5-18.**             **Fig. 5-17   Prob. 5-19.**

**5-19.** In Fig. 5-17, find the reactions at supports *A* and *E.* Check by moments.

*Ans.* $R_A = 610$ N↑; $R_B = 490$ N↑

**5-20.** Find the reactions produced by loads *B* and *C* at supports *A* and *D* of the beam shown in Fig. 5-18. Check $R_D$ by moments about *A.*

**5-21.** Determine the reactions produced at supports *A* and *D* of the beam shown in Fig. 5-19 by loads *B* and *C.* Check $R_D$ by moments about *A.* (Scales on 8½ by 11: 1 in. = 3 ft, 1 in. = 5 kips.)

*Ans.* $R_A = 14.2$ kips ↗; $R_D = 9.84$ kips ↖

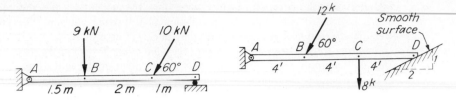

**Fig. 5-18** Prob. 5-20.                         **Fig. 5-19** Prob. 5-21.

**5-22.** The Fink roof truss shown in Fig. 5-20 is acted on by the resultant dead load $B$ and the resultant wind load $C$. Find $R_A$ and $R_D$. (Scales on 8½ by 11: 1 in. = 10 ft, 1 in. = 5 kips.)

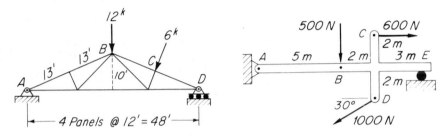

**Fig. 5-20** Prob. 5-22.                    **Fig. 5-21** Probs. 5-23 and 5-28.

**5-23.** Determine the reactions at supports $A$ and $E$ of the beam shown in Fig. 5-21.

*Ans.* $R_A = 287\text{N}$, $R_E = 893\text{N}$

**5-24.** Figure 5-22 shows a roof truss acted on by combined dead and wind loads. Determine first the reactions at supports $A$ and $E$. Then determine the stresses in all members, using the graphical combined-diagram method. (An all-graphical solution is especially desirable for this type of problem. Scales on 11 by 17: 1 in. = 6 ft, 1 in. = 3 kips.)

*Ans.* $R_A = 11.04$ kips ↗; $R_E = 8.55$ kips↑

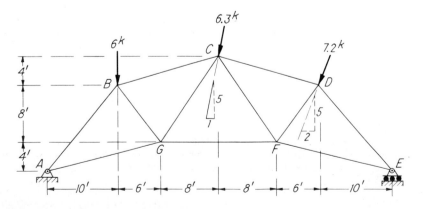

**Fig. 5-22** Prob. 5-24.

## 5-5  DETERMINATION OF REACTIONS; THE ANALYTICAL METHOD

At every support of a structure subjected to loads there is a single reacting force, the direction of which may or may not be known. If the direction as well as the magnitude of a reaction is unknown, it is generally replaced with two components of known directions, such as vertical and horizontal. Two unknowns, a magnitude and a direction, have then been replaced by two unknown magnitudes. The equations of solution are $\Sigma V = 0$, $\Sigma H = 0$, and $\Sigma M = 0$. Suitable graphical solutions have been indicated in the following problems.

### ILLUSTRATIVE PROBLEMS

**5-25.** The roof truss shown in Fig. 5-23a is subjected to wind loads acting perpendicularly to the sloping chord. Solve analytically for the reaction $R_A$ at support $A$, and the vertical and horizontal components $E_V$ and $E_H$ of the reaction at support $E$.

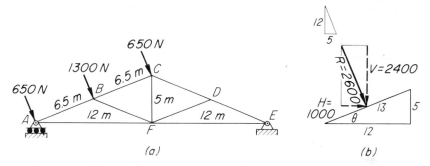

Fig. 5-23   Analytical determination of support reactions. *(a)* Space diagram. *(b)* Components of resultant.

*Solution:* Since at $A$ a roller is supported on a horizontal surface, the reaction $R_A$ is vertical. The direction of the reaction at $E$ is unknown. Hence we replace it with, and solve for, its $V$ and $H$ components. This load system is symmetrical about joint $B$. Therefore, the resultant force ($R = 2600$ N), applied at $B$, may replace the three loads. The solution is further simplified by replacing $R$ with its $V$ and $H$ components. From the diagram in Fig. 5-23b, we have

$$\frac{V}{12} = \frac{H}{5} = \frac{2600}{13}$$

or     $V = \dfrac{(12)(2600)}{13} = 2400$ N     and     $H = \dfrac{(5)(2600)}{13} = 1000$ N

If now a separate diagram of the truss is drawn, as in Fig. 5-24, with these components of $R$ acting at $B$, $R_A$ may be found by moments about $E$, and $E_V$ by moments about $A$. A check on these is found by $\Sigma V = 0$. Finally, $E_H$ is found by $\Sigma H = 0$. The equations are

$[\Sigma M_E = 0]$    $24R_A + (1000)(2.5) = (2400)(18)$    and    $R_A = 1696$ N↑  *Ans.*
$[\Sigma M_A = 0]$    $24E_V = (1000)(2.5) + (2400)(6)$    and    $E_V = 704$ N↑  *Ans.*

$[\Sigma V = 0]$         $1696 + 704 = 2400$         *Check*

$[\Sigma H = 0]$         $E_H = 1000 \text{ N} \leftarrow$         *Ans.*

These answers should then be recorded on the free-body diagram to show *visually* that equilibrium exists.

A *graphical solution* is readily obtained by means of the three-force principle. When the resultant $R$ is applied at $B$, only three forces, $R$, $R_A$, and $R_E$, act on the truss, and their lines of action therefore meet at a common point. Having thus established the action line of $R_E$, a force polygon will give $R_A$ and $R_E$.

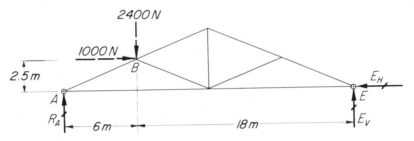

**Fig. 5-24   Free-body diagram.**

**5-26.** Determine the vertical and horizontal reaction components at supports $A$ and $B$ of the frame shown in Fig. 5-25a.

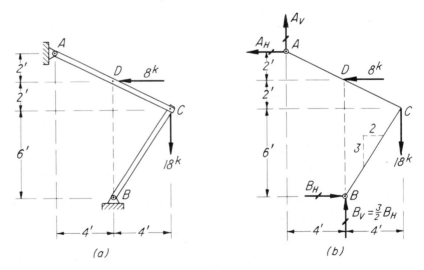

**Fig. 5-25   Analytical determination of support reactions. *(a)* Space diagram. *(b)* Free-body diagram.**

*Solution:* The two supports are removed and are replaced by the $V$ and $H$ components of the support reaction, as in Fig. 5-25b. We shall assume their senses to be as indicated. Because $AC$ is a *three-force member* (Art. 4-10), no known relationship exists between the components $A_V$ and $A_H$. Member $BC$, however, is a *two-force member* with a slope of 3 vertical to 2 horizontal. Therefore, $B_V/3 = B_H/2$, from which $B_V = \tfrac{3}{2}B_H$. We now solve for $B_H$ by taking moments about $A$. If we assume that

$R_B$ acts *against* point B, both of its components will also act *against* B. (Unless this rule is strictly followed, the results obtained will be incorrect, and the mistake cannot be detected until a final check is made by a summation of moments.) Then

$$[\Sigma M_A = 0] \quad 10B_H + (4)\left(\frac{3}{2}B_H\right) = (8)(2) + (18)(8)$$

or $\quad\quad 10B_H + 6B_H = 16 + 144 = 160 \quad$ and $\quad B_H = 10 \text{ kips} \rightarrow \quad Ans.$

Hence $\quad\quad\quad\quad\quad\quad B_V = \frac{3}{2}B_H = 15 \text{ kips}\uparrow \quad\quad\quad\quad\quad Ans.$

The positive sign of $B_H$ indicates a correct assumption as to its sense. The reaction components at A are now found by the force-equilibrium equations. That is,

$$[\Sigma V = 0] \quad\quad A_V + 15 = 18 \quad \text{and} \quad A_V = 3 \text{ kips}\uparrow \quad\quad\quad Ans.$$
$$[\Sigma H = 0] \quad\quad A_H + 8 = 10 \quad \text{and} \quad A_H = 2 \text{ kips}\leftarrow \quad\quad Ans.$$

*The values thus obtained should now be checked by a moment equation in which is included all or most of the computed forces.* Choosing D as a moment center, we have

$$[\Sigma M_D = 0] \quad (18)(4) - (10)(8) - (2)(2) + (3)(4) = 0 \quad \text{or} \quad 84 = 84 \quad Check$$

This problem is easily checked graphically if the two known forces are combined into their resultant. The action line of $R_A$ may then be found by the three-force principle and $R_A$ and $R_B$, by a simple force diagram.

**5-27.** Find the V and H reaction components at supports A and B of the two-member frame shown in Fig. 5-26.

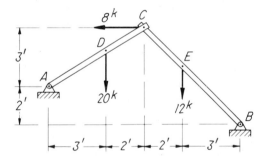

**Fig. 5-26  Reactions.**

*Solution:* In this problem, AC and BC are both three-force members. Therefore, no known relationship exists between the V and H components at either support, such as simplified the solution of Prob. 5-26. A summation of moments about either A or B will give an equation involving two unknown forces, and a second and independent equation must be written involving the same two unknowns. A simultaneous solution of these two equations will then give the unknown forces.

Using the entire frame as a free body, we select point A as the center of moments, and assume that $B_V$ and $B_H$ both act *against* joint B. Then, from Fig. 5-27a, we have

$$[\Sigma M_A = 0] \quad\quad 10B_V + (8)(3) = 2B_H + (20)(3) + (12)(7)$$
or $\quad\quad\quad\quad\quad\quad 10B_V - 2B_H = 120 \quad\quad\quad\quad\quad\quad\quad\quad (a)$

If now member BC is isolated, as in Fig. 5-27b, we may write a second independent equation by a summation of moments about C. The reacting force at C is replaced by its V and H components, as yet unknown. Their moments are zero when C is the center of moments. Then

$$[\Sigma M_C = 0] \quad\quad 5B_V = 5B_H + (12)(2) \quad \text{or} \quad 5B_V - 5B_H = 24 \quad\quad (b)$$

Multiplying (b) by 2 and subtracting it from (a), we have

$$10B_V - 2B_H = 120 \qquad (a)$$
$$\underline{10B_V - 10B_H = 48} \qquad (c)$$
$$8B_H = 72 \quad \text{from which} \quad B_H = 9 \text{ kips} \leftarrow \qquad Ans.$$

Substituting this value of $B_H$ in (a), we have

$$10B_V - (2)(9) = 120 \quad \text{or} \quad 10B_V = 138 \quad \text{and} \quad B_V = 13.8 \text{ kips} \uparrow \quad Ans.$$

By use of the values of $B_H$ and $B_V$ just obtained, the reaction components at $A$ are now readily solved for by applying the equations $\Sigma V = 0$ and $\Sigma H = 0$. If we assume $A_V$ to be upward-acting and $A_H$ to act from left to right, we obtain

$$[\Sigma V = 0] \qquad A_V + 13.8 = 20 + 12 \quad \text{or} \quad A_V = 18.2 \text{ kips} \uparrow \qquad Ans.$$
$$[\Sigma H = 0] \qquad A_H = 9 + 8 \quad \text{or} \quad A_H = 17 \text{ kips} \rightarrow \qquad Ans.$$

The positive signs of all computed components indicate a correct assumption as to their senses. (If an unknown sense is incorrectly assumed, the correct numerical value may be obtained but will be preceded by a negative sign.) The components $A_V$ and $A_H$ are dependent on $B_V$ and $B_H$. Therefore, *a check by summation of moments of all forces acting on the frame, preferably about a point that does not eliminate any computed force, is absolutely essential.* If we select point $C$ as the center of moments, we have

$$[\Sigma M_C = 0] \qquad (18.2)(5) - (17)(3) - (20)(2) + (12)(2) + (9)(5) - (13.8)(5) = 0$$
or
$$160 = 160 \qquad Check$$

This problem is of a type generally referred to as the **three-hinged arch.** A graphical solution is most easily obtained by removing loads $C$ and $E$ and solving for the reactions at $A$ and $B$ due to load $D$ only. Next, remove load $D$ and solve for the reactions at $A$ and $B$ due to loads $C$ and $E$ only. The two reactions at each support are then readily combined into a resultant reaction by constructing parallelograms, from which the true reactions or their components may be scaled.

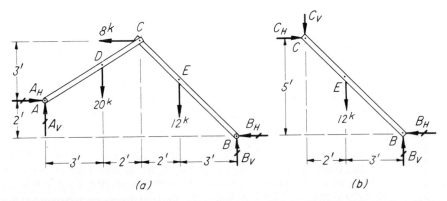

**Fig. 5-27  Analytical determination of support reactions. (a) Entire frame as free body. (b) Member $BC$ as free body.**

**PROBLEMS** _____

Analytical determination of reactions. Graphical check solutions are suggested.

**5-28.** Solve for the components of the reactions at $A$ and $E$ in Fig. 5-21 by the analytical method.

**5-29.** Compute the reaction at support $G$ of the Fink roof truss shown in Fig. 5-28, and the $V$ and $H$ reaction components at support $A$. (See Prob. 5-25.)

Ans. $R_G = 13.75$ kN↑; $A_V = 6.25$ kN↑; $A_H = 10$ kN

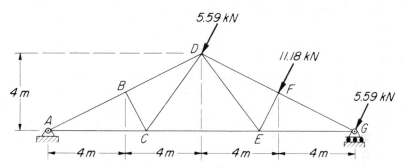

**Fig. 5-28    Prob. 5-29. Fink truss, wind loads.**

**5-30.** Find the reaction at support $A$ of the sawtooth truss shown in Fig. 5-29 and the $V$ and $H$ components of the reaction at support $G$. (See Prob. 5-27.)

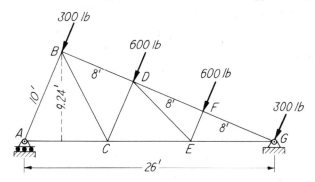

**Fig. 5-29    Prob. 5-30. Sawtooth roof truss, wind loads.**

**5-31.** Solve for the reaction components at $D$ and $E$ in Fig. 5-12.

**5-32.** Determine the $V$ and $H$ reaction components at supports $A$ and $D$ of the truss shown in Fig. 5-30.

Ans. $A_V = 5$ kips↑; $A_H = 20$ kips←; $D_V = 10$ kips↑; $D_H = 20$ kips→

**5-33.** Find the $V$ and $H$ reaction components at supports $A$ and $G$ of the truss shown in Fig. 5-31.

**5-34.** Determine the vertical and horizontal components of the reactions at $A$ and $B$ of the crane shown in Fig. 5-32. ($W = 12$ kips).

Ans. $A_V = 20$ kips↑; $A_H = 8$ kips←; $B_V = 8$ kips↓; $B_H = 8$ kips→

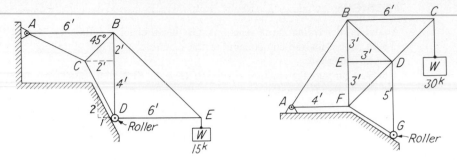

Fig. 5-30  Probs. 5-32 and 5-125.       Fig. 5-31   Prob. 5-33.

**5-35.** Determine the vertical and horizontal components of the reactions at $A$ and $G$ of the truss shown in Fig. 5-33.

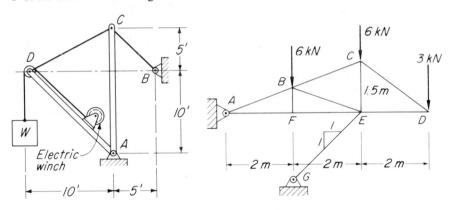

Fig. 5-32   Prob. 5-34.         Fig. 5-33   Prob. 5-35.

**5-36.** Compute the $V$ and $H$ reaction components at supports $A$ and $B$ of the truss shown in Fig. 5-34.

*Ans.* $A_V = 2$ kips↑; $A_H = 3$ kips←; $B_V = 10$ kips↑; $B_H = 15$ kips→

**5-37.** Find the reactions $R_A$ and $R_B$ at the two supports of the truss shown in Fig. 5-35.

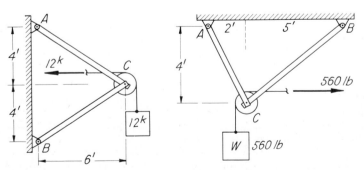

Fig. 5-34   Prob. 5-36.         Fig. 5-35   Prob. 5-37.

**5-38.** Compute the vertical and horizontal reaction components at *A* and *B* for the truss shown in Fig. 5 36.

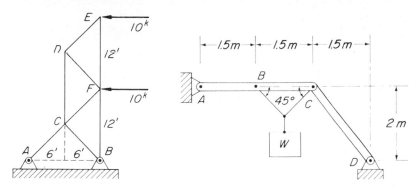

Fig. 5-36   Prob. 5-38.          Fig. 5-37   Prob. 5-39.

**5-39.** Compute the vertical and horizontal reaction components at *A* and *D* in Fig. 5-37. The weight *W* = 7200 N.

**5-40.** Determine the *V* and *H* reaction components at supports at *A* and *B* of the truss shown in Fig. 5-38.

*Ans.* $A_V$ = 3.78 kips↑; $A_H$ = 3.03 kips←; $B_V$ = 10.22 kips↑; $B_H$ = 13.03 kips→

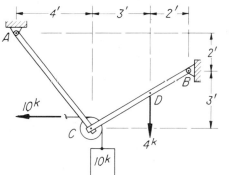 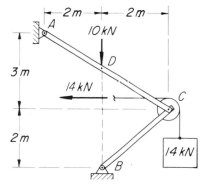

Fig. 5-38   Prob. 5-40.                    Fig. 5-39   Prob. 5-41.

**5-41.** Solve for the reactions $R_A$ and $R_B$ at supports *A* and *B* in Fig. 5-39.

**5-42.** Find the *V* and *H* reaction components at supports *A* and *F* of the truss shown in Fig. 5-40. (SUGGESTION: Detach member *EF* and solve for the forces at *E*.)

*Ans.* $A_V$ = 12 kips↓; $A_H$ = 4 kips→; $F_V$ = 36 kips↑; $F_H$ = 24 kips←

**5-43.** The symmetrical steel radio tower shown in Fig. 5-41 is 5 ft square and 200 ft high and weighs 60 kips. It is supported against horizontal wind loads by four guy wires attached 130 ft above the central base *B*. Its vertical projected area of 1000 ft² is subjected to horizontal wind pressure of 15 pounds per square foot (psf). When the wind blows from right to left, only guy wire *AC* is active. Compute the *V* and *H* reaction components at *A* and *B* and the tension *T* in *AC*.

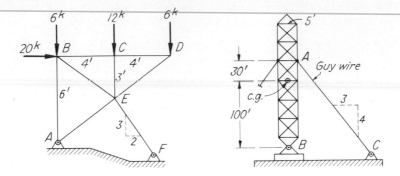

**Fig. 5-40** Probs. 5-42, 5-91, and 5-124. **Fig. 5-41** Prob. 5-43.

**5-44.** Compute the $V$ and $H$ reaction components at supports $A$ and $E$ of the frame shown in Fig. 5-42. (See Prob. 5-27.)

*Ans.* $A_V = 9.35$ kips↑; $A_H = 9.53$ kips←; $E_V = 9.65$ kips↑; $E_H = 9.53$ kips→

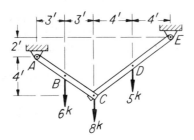

**Fig. 5-42** Prob. 5-44.  **Fig. 5-43** Prob. 5-45.

**5-45.** Solve for the reactions $R_A$ and $R_B$ at supports $A$ and $B$ of the three-hinged arch shown in Fig. 5-43, and for the $V$ and $H$ reaction components at the crown hinge $C$. (See Prob. 5-27.)

**5-46.** Figure 5-44 shows a loading crane with its supporting framework. The weight of the horizontal truss is 5 kips and is concentrated at its center of gravity. Compute

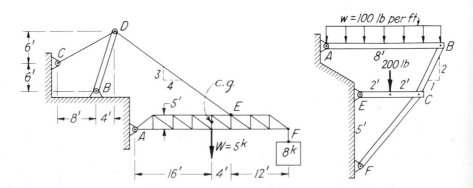

**Fig. 5-44** Prob. 5-46.  **Fig. 5-45** Prob. 5-47.

the $V$ and $H$ reaction components at supports $A$, $B$, and $C$ due to the weight of the truss and the load at $F$ as shown. Neglect weight of other members.

*Ans.* $A_V = 0.4$ kips↑; $A_H = 16.8$ kips→; $B_V = 25.2$ kips↑; $B_H = 8.4$ kips→ $C_V$
$= 12.6$ kips↓; $C_H = 25.2$ kips←

**5-47.** Find the $V$ and $H$ reaction components at supports $A$, $E$, and $F$ of the frame shown in Fig. 5-45. (SUGGESTION: Isolate member $AB$ and consider first the unknown forces at $A$ and $B$. Next isolate $EC$ and $CF$ as a unit.)

*Ans.* $A_V = 400$ lb↑

**5-48.** Figure 5-46 shows the boom of a concrete-road paver with its supporting framework. At $E$ the flexible wire rope passes over a sheave, the small diameter of which may be disregarded. The weight of the boom is 600 lb, which may be concentrated at its center. Compute the tension $T$ in the wire rope, and the $V$ and $H$ reaction component at supports $A$ and $B$ due to the weight of the boom and the load as shown.

*Ans.* $T = 14.42$ kips; $A_V = 9.68$ kips↓; $A_H = 12.9$ kips←; $B_V = 28.7$ kips↑; $B_H$
$= 12.9$ kips→

**5-49.** A cylindrical tank, 2 m in diameter and weighing 7500 N, is supported on an incline by a bracket, as shown in Fig. 5-47. Compute the reacting forces $R_A$ and $R_B$ acting perpendicularly to the supporting surfaces at $A$ and $B$. Disregard possible friction ($EC = 2$ m and $CD = 1.5$ m).

**5-50.** Compute the stress in brace $DE$ of the bracket shown in Fig. 5-47. (See Prob. 5-49.)

*Ans.* 3125N C

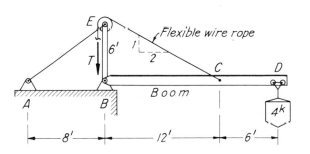

**Fig. 5-46   Prob. 5-48.**

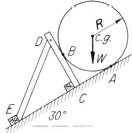

**Fig. 5-47   Probs. 5-49 and 5-50.**

## 5-6 PIN REACTIONS; THE METHOD OF MEMBERS

Trusses are generally so constructed, loaded, and supported that all forces act at the joints, thereby producing simple axial or direct stresses (tensile or compressive) in the members which, then, are two-force members. In the design of structures and equipment, and in machines, three-force members (Art. 4-10) are avoided whenever possible, because the material in such members is less efficiently used than in two-force members. Frames in which three-force members *must* be used are often connected by pins. Forces produced by loads on such frames are transmitted through members to the pins at the joints and by the pins to other connected members, and are referred to as **pin reactions.** In computing these pin reactions, any possible pin friction and friction between members connected at a joint are disregarded.

The usual procedure to determine these pin reactions is as follows. *All determinable support reactions, or their components, are first computed, using the frame as a whole.* Each three-force member is then isolated as a free body, beginning with one on which *known* forces act. When *all* forces acting on any single member are considered, they constitute a system in equilibrium. Hence the unknown forces, or their components, may be found by application of the equations of static equilibrium, $\Sigma V = 0$, $\Sigma H = 0$, and $\Sigma M = 0$. From these components the actual pin reactions are readily computed. Possible difficulty concerning the senses of these forces may be overcome by remembering that *a force acting on any given member represents the action, pushing or pulling, of another member on the given member.* Since each member is thus considered individually, the method is logically referred to as the **method of members.**

**ILLUSTRATIVE PROBLEM** ————————————————————

**5-51.** The frame shown in Fig. 5-48 is held in place by a pin connection at $A$ and rests on a smooth surface at $F$. Determine the reactions at $A$ and $F$ and the pin reactions at $B$ and $C$.

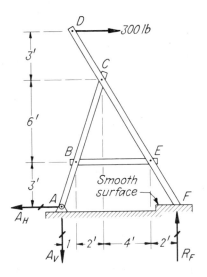

Fig. 5-48   Pin reactions.

*Solution:* The support reactions at $A$ and $F$ are first determined, $R_F$ by moments about $A$, $A_V$ by moments about $F$, and $A_H$ by $\Sigma H = 0$. Then $R_A = \sqrt{A_V{}^2 + A_H{}^2}$. The equations are

$[\Sigma M_A = 0]$    $9R_F = (300)(12) = 3600$    and    $R_F = 400$ lb↑    *Ans.*
$[\Sigma M_F = 0]$    $9A_V = (300)(12) = 3600$    and    $A_V - 400$ lb↓
$[\Sigma H = 0]$                      $A_H = 300$ lb←
Then    $R_A = \sqrt{(400)^2 + (300)^2} = \sqrt{250,000}$    or    $R_A = 500$ lb ↙ *Ans.*

Next we draw free-body diagrams of the three members as in Fig. 5-49. The unknown forces to be determined are the $V$ and $H$ components at $B$ and $C$. Clearly, $BE$ is a two-force member, since forces act on it only at $B$ and $E$, and the pin reactions at $B$ and $E$ are therefore horizontal. Also, $BE$ is apparently in tension, which we shall assume. Using member $AC$, we may determine force $B_H$ by moments about $C$. Thereafter the components at $C$ are found by $\Sigma V$ and $\Sigma H$. Using member $AC$, we have

$[\Sigma M_C = 0]$    $6B_H + (400)(3) = (300)(9)$    and    $B_H = 250$ lb    *Ans.*
$[\Sigma V = 0]$        $C_V = 400$ lb                    $C_V = 400$ lb
$[\Sigma H = 0]$        $C_H + 250 = 300$    and    $C_H = 50$ lb

The pin reaction $R_C$ at $C$ then is

$$R_C = \sqrt{(400)^2 + (50)^2} = \sqrt{162,500} \qquad \text{or} \qquad R_C = 403 \text{ lb} \quad Ans.$$

These forces acting at $B$ and $C$ are now transferred to member $DF$ but *with opposite senses*, since, if they are considered to be *actions* on member $AC$, they must be *reactions* when applied to member $DF$, and *action and reaction are always equal, opposite, and collinear.*

We now obtain the final check by a moment equation about $D$. That is,

$[\Sigma M_D = 0]$  $(50)(3)+(400)(2)+(250)(9) = (400)(8)$ or $3200 = 3200$        *Check*

*Graphical Solution:* The action lines of forces $D$ and $R_F$ are extended to their intersection through which the action line of $R_A$ then must pass, according to the three-force principle. $R_A$ and $F$ are then found by a closed diagram of forces $D$, $R_A$, and $R_F$. Similarly, using a scale-drawn free-body diagram of member $AC$, the action line of $R_C$ passes through the intersection of the known action lines of forces $R_A$ and $B$. Forces $B$ and $R_C$ are then found from a closed diagram of forces $R_A$, $B$, and $R_C$.

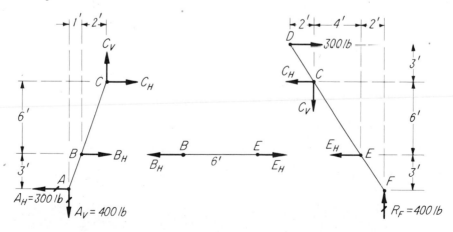

**Fig. 5-49  Free-body diagrams.**

**PROBLEMS**

**5-52.** Determine the reactions at supports $A$ and $F$ of the frame shown in Fig. 5-50, and the pin reactions at $B$ and $C$.

Ans. $R_A = 200$ lb↑; $R_F = 40$ lb↑; $R_B = 150$ lb; $R_C = 250$ lb

**5-53.** The crane shown in Fig. 5-51 is supported at the top and bottom in loose-fitting sockets. Compute the horizontal reactions at $A$, the $V$ and $H$ reaction components at $D$, the pin reaction at $B$, and the stress in member $CF$ if $W = 1500$ lb.

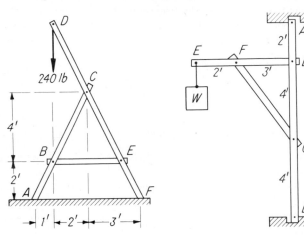

**Fig. 5-50   Prob. 5-52.**           **Fig. 5-51   Prob. 5-53.**

**5-54.** The frame shown in Fig. 5-52 rests on a smooth surface at $A$ and $E$. Find the reactions at $A$ and $E$, and the $V$ and $H$ components of the pin reactions at $B$, $C$, and $D$.

Ans. $R_A = 3$ kN; $R_E = 4.5$ kN; $B_V = 7.5$ kN; $B_H = 12$ kN; $C_V = 12$ kN; $C_H = 12$ kN; $D_V = 15$ kN; $D_H = 12$ kN

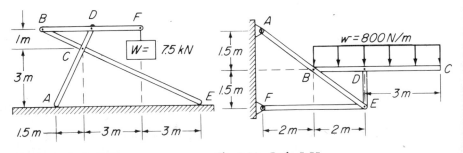

**Fig. 5-52   Prob. 5-54.**           **Fig. 5-53   Prob. 5-55.**

**5-55.** The frame shown in Fig. 5-53 supports a uniformly distributed load of 800 N/m on member $BC$. Determine the $V$ and $H$ components of the pin reactions at $A$, $B$, and $F$.

**5-56.** The horizontal member of the frame of Fig. 5-54 supports a uniformly distributed load of 2 kips per ft. Determine the vertical and horizontal components of the reactions at $A$ and $B$.

**5-57.** The cable shown in Fig. 5-55 passes over the drum $G$ and is fastened to the frame at $C$. Determine the vertical and horizontal components of the reactions at $A$ and $F$. Also determine the vertical and horizontal components of the pin reactions at $B$, $D$, and $E$.

*Ans.* $A_V = 8$ kips; $F_V = 10$ kips; $B_V = 9$ kips; $B_H = 12$ kips; $D_V = 1$ kip; $D_H = 6$ kips; $E_V = 9$ kips; $E_H = 6$ kips

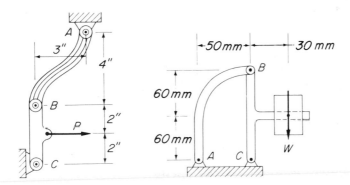

**Fig. 5-54   Prob. 5-56.**      **Fig. 5-55   Prob. 5-57.**

**5-58.** The pull $P$ in Fig. 5-56 is 60 lb. Calculate the reactions at $A$ and $C$.

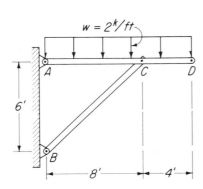

**Fig. 5-56   Prob. 5-58.**      **Fig. 5-57   Prob. 5-59.**

**5-59.** The weight $W$ in Fig. 5-57 is 200 N. Calculate the reactions at $A$ and $C$.

*Ans.* $R_A = 130$ N; $R_C = 324$ N

**5-60.** The load $P$ in Fig. 5-58 is 100 lb. Calculate the tensile force in the spring and the reaction at $C$.

**5-61.** Determine the $V$ and $H$ components of the pin reactions at all pins of the frame shown in Fig. 5-59.

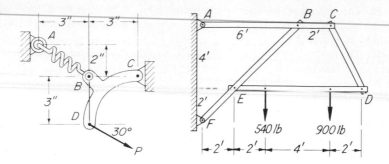

**Fig. 5-58   Prob. 5-60.**          **Fig. 5-59   Prob. 5-61.**

**5-62.** A log is supported by two identical frames of a sawbuck, one of which is shown in Fig. 5-60. Each frame supports a weight of 90 lb and rests on a smooth surface. Determine the reactions at $A$ and $G$, the stress in member $BF$, and the pin reaction at $C$. Neglect all friction.

*Ans. $R_A = R_G = 45$ lb↑; $BF = 104$ lb T; $R_C = 164$ lb*

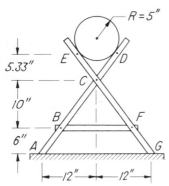

**Fig. 5-60   Prob. 5-62.**

## 5-7 STRESSES IN TRUSSES; THE METHOD OF SECTIONS

In Art. 4-7 was described an analytical method of determination of stresses in trusses—the analytical method of joints—in which the joints were successively isolated as free bodies. The forces acting at each joint then constituted a concurrent system of forces in equilibrium, the unknown quantities of which were found by the equations $\Sigma V = 0$ and $\Sigma H = 0$. In the **method of sections** *a section is cut through the entire truss and one of the two parts is isolated as a free body.* The forces acting on each part, namely, the external forces acting on the part and the internal forces, or stresses, in the cut members, then constitute a *nonconcurrent system of forces in equilibrium,* the unknown quantities of which are found by the equations $\Sigma V = 0$, $\Sigma H = 0$, and $\Sigma M = 0$.

*In the method of sections, each stress may be found independently of all others.* A truss may therefore be cut at any section, and *any single stress may be found independently of all other stresses in the truss.* Not more than three members with *unknown* stresses may be cut at any section, except in the special case where all but one of the cut members intersect at a common point. In general, *a section should cut the least number of members, and the smaller of the two parts should be isolated as a free body.*

Trusses in which the two main chords are parallel, as in Prob. 5-63, are solved by a special adaptation called the **shear method of sections.** Other trusses are most easily solved by the **component method of sections,** as in Prob. 5-64.

In some instances the members of a truss are so arranged that there will be three or more unknown stresses at some or all joints, as, for example, in Fig. 5-67, and the truss cannot, therefore, be solved by the analytical method of joints or by either of the graphical methods. In such problems, one stress in the truss is readily determined by the method of sections, and the solution may then be completed graphically, or analytically by the joint method. This is illustrated in Prob. 5-65.

A graphical solution to obtain three unknown stresses at a section is possible by use of the four-force principle. All known external forces are first combined into their resultant force. The resultant of this force and one of the unknown stresses is then equal, opposite, and collinear with the resultant of the other two unknown stresses. Problem 5-65 may be solved in this manner.

## ILLUSTRATIVE PROBLEMS

**5-63.** Using the *shear method of sections,* determine the stresses in members *BD, CD,* and *CE* of the Howe roof truss shown in Fig. 5-61a. Determine also stresses *AB, AC, BC, DE,* and *EG.* Draw a stress-summation diagram for recording the stresses, letting *T* and *C* denote tension and compression, respectively. (Students should solve for the remainder of the stresses in this truss.)

*Solution:* Vertical section 3 is cut through the second panel of the truss containing members *BD, CD,* and *CE,* whose stresses we must find. The part on the left is then isolated as a free body, as shown in Fig. 5-62. Six forces act on this part, the reaction at *A,* the loads *B* and *C,* and the unknown stresses in the three cut members.

The stress in the diagonal *CD* should first be determined. Its vertical component $CD_V$ is easily found and is seen to be downward-acting. That is,

$$[\Sigma V = 0] \qquad CD_V + 6 + 6 = 18 \qquad \text{and} \qquad CD_V = 6 \text{ kips}$$

Then, by similarity of slope and force triangles,

$$\frac{CD_V}{6} = \frac{CD_H}{8} = \frac{CD}{10} = \frac{6}{6} = 1$$

or $\qquad CD_H = (8)(1) = 8 \text{ kips} \qquad$ and $\qquad CD = (10)(1) = 10 \text{ kips C} \qquad$ *Ans.*

Because the vertical component of *CD* acts *against* member *CD,* the horizontal component and the stress *CD* also act *against* it, thus indicating compression.

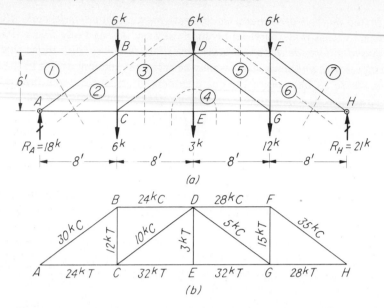

(a)

(b)

**Fig. 5-61. Determination of stresses in trusses; the shear method of sections.** *(a)* **Space diagram.** *(b)* **Stress-summation diagram.**

To find *BD independently* of *CD* and *CE*, we must take moments about *C*, thus eliminating the moments of *CD* and *CE*. Likewise, we find *CE independently* by moments about *D*, thus eliminating the moments of *BD* and *CD*. If we assume *BD* to be compressive and *CE* to be tensile (as they apparently are), we have

$$[\Sigma M_C = 0] \qquad 6BD = (18)(8) = 144 \qquad \text{and} \qquad BD = 24 \text{ kips C} \quad Ans.$$
$$[\Sigma M_D = 0] \quad 6CE + (6)(8) + (6)(8) = (18)(16) \text{ and} \qquad CE = 32 \text{ kips T} \quad Ans.$$

The positive signs of *BD* and *CE* indicate that the *type of stress* in each was correctly assumed. When these forces are shown on the free-body diagram, as was done in Fig.

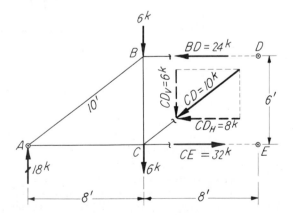

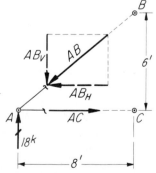

**Fig. 5-62 Free-body diagram, section 3.**

**Fig. 5-63 Free-body diagram, section 1.**

5-62, a summation of horizontal forces clearly indicates force equilibrium, which is a sufficient check. (Students should *always* record computed values on the free-body diagrams to obtain a visual check.) That is,

$$[\Sigma H = 0] \qquad 32_\rightarrow = 24_\leftarrow + 8_\leftarrow \qquad \text{or} \qquad 32_\rightarrow = 32_\leftarrow \qquad \textit{Check}$$

We see clearly now that any stress at any section may be found *independently* of all other stresses in the truss.

If all stresses in this truss are desired, a number of sections are cut such as those indicated by the encircled numbers in Fig. 5-61a. A free-body diagram is then drawn, usually of the smallest part, and the unknown stresses are determined. To find stresses *AB* and *AC*, for example, section 1 is cut and the portion on the left is isolated as in Fig. 5-63. From $\Sigma V = 0$, $AB_V$ is found to be 18 kips. Then, by similarity of slope and force triangles, $AB_H$ and $AB$ are found to be 24 and 30 kips, respectively. The stress *AC* is found independently by moments about *B*. A check of horizontal forces then indicates force equilibrium.

To find *AC, BC,* and *BD,* section 2 is cut, isolating the part shown in Fig. 5-64. Then *BC* is found from $\Sigma V = 0$, *BD* by moments about *C,* and *AC* by moments about *B*. The stress *DE* is most easily found by cutting the circular section 4 (Fig. 5-65) and using $\Sigma V = 0$. Then *EG* is found from $\Sigma H = 0$. *In general, joints should not be isolated* except as in sections 1, 4, and 7 in Fig. 5-61a. Sections are generally cut from left to right in the order of the numbers 1 to 7.

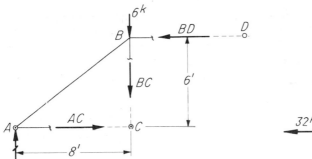

**Fig. 5-64   Free-body diagram, section 2.**

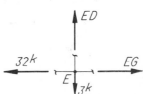

**Fig. 5-65   Free-body diagram, section 4.**

**5-64.** By means of the *component method of sections,* find the stresses in members *AC, BC,* and *BD* of the truss shown in Fig. 5-66.

*Solution:* The most suitable section to cut is one through members *AC, BC,* and *BD.* We then use the part on the right as a free body, as shown in Fig. 5-66b. *By proper selection of moment centers, we may determine each of the three stresses independently of the other two.* That is, we find *AC* by moments about *B, BC* by moments about *E,* and *BD* by moments about *C.* In each of these three moment equations, the moments of the other two stresses are zero about the moment center selected and are thus eliminated. Assuming *AC* to be a tensile stress, as apparently it is, and with *B* as a moment center, thus eliminating *BC* and *BD,* we have

$$[\Sigma M_B = 0] \quad 6AC_H = 6(4) + 6(4) + 9(8) \qquad \text{and} \qquad AC_H = 20 \text{ kN}$$

By similar triangles

$$\frac{AC_H}{4} = \frac{AC}{5} \qquad \text{and} \qquad AC = \frac{5}{4}(20) = 25 \text{ kN T} \qquad Ans.$$

Next, to find $BC$, we select $E$ as the moment center and replace $BC$ with its $V$ and $H$ components, applied at either $B$ or $C$, but most conveniently at $B$. Assuming $BC$ to be a compressive stress, we solve for its *vertical* component $BC_V$. That is,

$$[\Sigma M_E = 0] \qquad 8BC_V = 6(4) + 6(4) \qquad \text{and} \qquad BC_V = 6 \text{ kN}$$

By similar triangles

$$\frac{BC_V}{3} = \frac{BC}{5} \qquad BC = \frac{5}{3}(6) = 10 \text{ kN C} \qquad Ans.$$

Finally, to obtain $BD$, we select $C$ as the moment center. Then, assuming $BD$ is compressive,

$$[\Sigma M_C = 0] \qquad 3BD = 9(4) \qquad \text{and} \qquad BD = 12 \text{ kN C} \qquad Ans.$$

Since all results are positive, the type of stresses were correctly assumed. We can find the other components of $AC$ and $BC$ by similar triangles. This has been done, and the results are shown on the free-body diagram shown in Fig. 5-66b. As a final check,

$$[\Sigma V = 0] \quad 15{\uparrow} + 6{\downarrow} + 9{\downarrow} + 6{\downarrow} + 6{\uparrow} = 0 \quad \text{or} \quad 21{\uparrow} = 21{\downarrow} \qquad Check$$
$$[\Sigma H = 0] \quad 20 + 8 + 12 = 0 \quad \text{or} \quad 20 = 20 \qquad Check$$

Occasionally, the moment center which will eliminate the moments of two of the three stresses to be found lies outside the truss and may be inconvenient to locate. The desired stress may then be found by $\Sigma V$ and $\Sigma H$, provided a final check is made by $\Sigma M$ about a center so chosen that no moment of a computed stress is eliminated. In any event, *a final check is essential.*

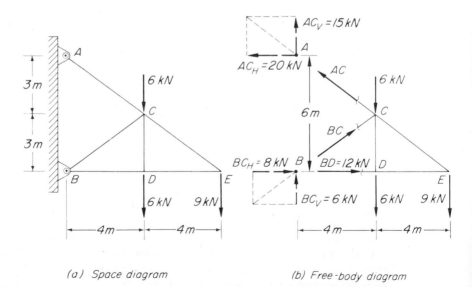

(a) Space diagram                    (b) Free-body diagram

**Fig. 5-66  Determination of stresses in trusses; the component method of sections.**

**5-65.** A graphical solution to determine the stresses in the members of the truss shown in Fig. 5-67a is desired, but a start on it cannot be made, since there are three or more unknown stresses at every joint. Using the *method of sections,* determine analytically the stress in one member in order that the solution may then be completed graphically. All forces are in pounds.

*Solution:* The simplest section to cut is clearly through members *BE, CE,* and *AD.* When the portion on the left is isolated as a free body, as shown in Fig. 5-67b, the stress in *AD* is readily found by a summation of moments about point *E.* That is,

$$[\Sigma M_E = 0] \quad 24AD + (90)(32) = (300)(32) \quad \text{and} \quad AD = 280 \text{ lb T} \quad \textit{Ans.}$$

An *all-graphical solution* of this problem is made possible by the **substitute-member method** (see Prob. 4-45) in which the three members, *AC, BC,* and *CE* are removed and are replaced with a single straight member *AE,* partly indicated by the dashed lines. From study of Fig. 5-67b, we see that *the stress AD is unaffected by this substitution,* since, by moments about *E, AD* is still 280 lb T. Because only two unknown stresses now exist at joint *B,* joints *B* and *A* may be solved graphically. Of the stresses so found, *only AD will be a true stress,* and the others are therefore disregarded. The original members, *AC, BC,* and *CE,* are now replaced, and the graphical solution may then begin at joint *A*—since stress *AD* is now known—and may proceed through joints *B, C, D, E, F, G,* and *H.*

An *alternate graphical solution* is to obtain stress *AD* by use of the four-force principle (Art. 2-15), in which the resultant of stresses *BE* and *CE* is applied at *E* in Fig. 5-67b and its equal and opposite resultant is applied at *A,* is as follows. Remove members *AC, BC,* and *CE* and substitute for them a member *AE.* When joint *B* is now solved, the stress in *AB* is found to be a 90-lb compression. The resultant of this stress *AB* and the 300-lb reaction at *A* is 210 lb upward. When this 210-lb force is laid off upward from *A* as the vertical side of a parallelogram of forces, the horizontal side will be the stress *AD* and the diagonal will be the stress in the substitute member *AE,* which stress, then, is equal and opposite to the resultant of stresses *BE* and *CE,* applied at *E.* When *AD* has thus been determined, the stresses in other members may then be determined by the usual combined-diagram method. (See Prob. 4-60 for scales to use.)

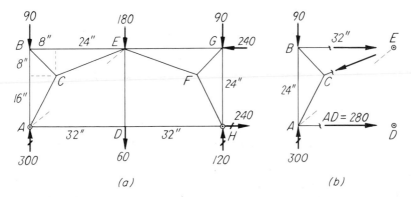

*(a)*                                    *(b)*

**Fig. 5-67   Determination of a single stress by method of sections. (a) Loaded truss. (b) Free-body diagram.**

## PROBLEMS

Solve Probs. 5-66 through 5-74 by the *shear method of sections.*

**5-66.** Determine the stresses in members *BD, CD,* and *CE* of the roof truss shown in Fig. 5-68.

*Ans. BD* = 20 kN C; *CD* = 5 kN C; *CE* = 20 kN T

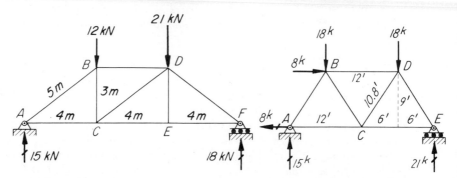

**Fig. 5-68   Probs. 5-66 and 5-67.**          **Fig. 5-69   Probs. 5-68 and 5-117.**

**5-67.** Find the stresses in members *AC, BC,* and *AB* of the truss shown in Fig. 5-68.

**5-68.** Solve for the stresses in members *BD, CD,* and *CE* of the Warren roof truss shown in Fig. 5-69.

*Ans. BD* = 16 kips C; *CD* = 3.6 kips T; *CE* = 14 kips T

**5-69.** In Fig. 4-91, solve for the stresses in members *AC, BC,* and *BD.*

*Ans. AC* = 34 kips T; *BC* = 1 kip T; *BD* = 45 kips C

**5-70.** In Fig. 4-91, solve for the stresses in members *BD, CD,* and *CE.*

**5-71.** Figure 5-70 shows one truss of a traveling crane. Determine the stresses in members *DE, DG,* and *GH.*

*Ans. DE* = 20 kips C; *DG* = 5 kips C; *GH* = 24 kips T

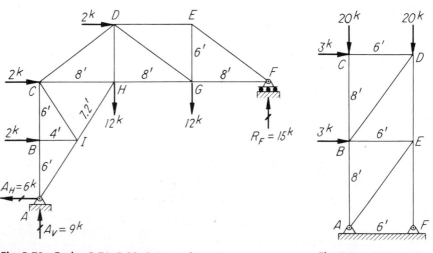

**Fig. 5-70   Probs. 5-71, 5-80, 5-81, and 5-116.**          **Fig. 5-71   Prob. 5-73.**

**5-72.** Find the stresses in members *BD, CD,* and *CE* of the tower truss shown in Fig. 4-35.

**5-73.** The vertical tower truss shown in Fig. 5-71 supports a water tank and is also subjected to horizontal wind loads. Compute the stresses in all members.

*Ans. AB* = 16 kips C; *BD* = 5 kips T; *EF* = 32 kips C

**5-74.** Find the stresses *BD, CD,* and *CE* of the truss shown in Fig. 5-72. Loads are $P_1$ = 18 kN and $P_2$ = 8 kN.

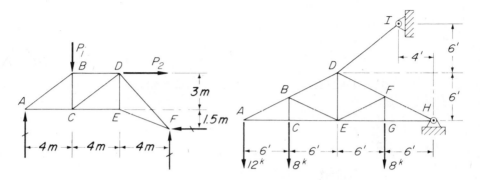

**Fig. 5-72   Prob. 5-74.**                **Fig. 5-73   Probs. 5-75 and 5-101.**

**5-75. Compute the stresses *BD, BE,* and *CE* of the truss shown in Fig. 5-73.**

*Ans. BD* = 35.7 kips T; *BE* = 9.0 kips C; *CE* = 24.0 kips C

**5-76.** The bowstring truss in Fig. 5-74 supports a live load of 72 kN at joint *F.* Compute the support reactions and the stresses in members *AB, BH,* and *GH* produced by this load.

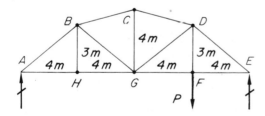

**Fig. 5-74   Probs. 5-76 and 5-77.**

**5-77.** Calculate the support reactions at *A* and *E,* and determine the stresses in members *BC, BG,* and *GH* of the bowstring truss shown in Fig. 5-74 produced by the 72 kN load at joint *F.*

*Ans. BC* = 37.2 kN C; *BG* = 15 kN T; *GH* = 24 kN T

**5-78.** Find the stresses *DF* and *EF* in Fig. 5-75.
**5-79.** Find the stresses *DF, DE,* and *CE* in Fig. 5-76.

*Ans. DF* = 2370 lb T; *DE* = 902 lb C; *CE* = 1500 lb C

**5-80.** Solve for the stresses in members *BC, CI,* and *HI* in the traveling crane truss shown in Fig. 5-70.

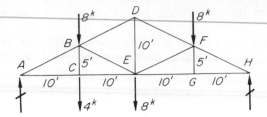

**Fig. 5-75   Prob. 5-78.**

**5-81.** Find the stresses in members *CD, CH,* and *HI* of the traveling crane truss shown in Fig. 5-70.

*Ans. CD* = 27.5 kips C; *CH* = 19 kips T; *HI* = 9 kips T

**5-82.** Find the stresses in members *AB, AD,* and *DE* of the truss shown in Fig. 5-77.

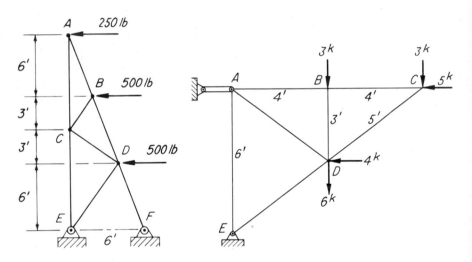

**Fig. 5-76   Prob. 5-79.**            **Fig. 5-77   Prob. 5-82.**

**5-83.** In Fig. 5-78 is shown one vertical truss of a tower supporting a heavy construction crane. The 12-kN horizontal load at *C* represents the total wind pressure against the crane and the truss. Determine the stresses in members *BC, CE,* and *DE* ($P_2$ = 36 kN, *CD* = 2 m, *h* = 3 m, and *AF* = 4 m).

*Ans. BC* = 24.3 kN C; *CE* = 15.6 kN C; *DE* = 36.5 kN C

**5-84.** The bowstring roof truss of Fig. 5-79 is loaded as shown. Find the *stresses* in members *CE, CD,* and *BD.*

**5-85.** Find the stresses in members *EG, EF,* and *DF* of the bowstring truss of Fig. 5-79.

*Ans. EG* = 27.2 kips C; *EF* = 0.6 kips C; *DF* = 27.4 kips T

**5-86.** Compute the support reactions at *A* and *F* in Fig. 5-78 and determine the stresses in members *BC, BE,* and *EF* ($P_1$ = 24 kN, $P_2$ = 72 kN, *CD* = 2 m, *h* = 3 m, and *AF* = 4 m).

## 5-8 COUNTERDIAGONALS IN TRUSSES

In many trusses subjected to moving or other varying loads, some of the diagonals may undergo a reversal of stress from tension to compression, or vice versa. Such reversal of stress might be produced by a train moving across a railroad bridge or by the change in direction of wind pressure against a vertical tower truss.

In trusses such as the one shown in Fig. 5-80, the inside diagonals *BE, DG, GH,* and *IJ* are usually designed to carry tensile stress only and are apt to be so slender that they could easily buckle under a compressive stress. To prevent possible failure of this type, *counterdiagonals EF* and *FI* are inserted in all panels where a reversal of stress might occur, due to either wind loads or moving loads. The assumption then is that *only the tension diagonal is active and carries stress.*

Let the truss in Fig. 5-80 be one of a pair supporting a railroad bridge, and consider the two loads at panel points *I* and *K* to be caused by a load moving across the bridge. The resulting reaction at *A* is 12 kips upward. If we now cut section 4-4 through panel 4 and isolate the left half as a free body, the vertical component of the *active diagonal stress* is 12 kips downward, thus

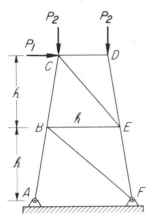

**Fig. 5-78** Probs. 5-83, 5-86, and 5-98.

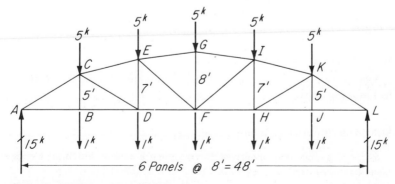

**Fig. 5-79** Probs. 5-84 and 5-85.

indicating that member *FI* is the active tension diagonal. Member *GH* is in compression and is therefore assumed to be inactive.

When the two loads are applied at *G* and *I*, $R_A$ will be 20 kips upward and the vertical component in the active diagonal at section 4 will be 4 kips upward, thus indicating that *GH* is now the active tension diagonal.

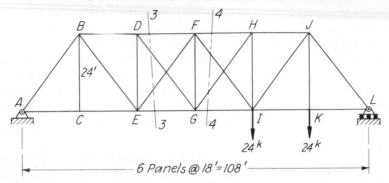

**Fig. 5-80   Probs. 5-87, 5-88, and 5-119. Counterdiagonals in trusses.**

PROBLEMS _____

**5-87.** Determine the stress in the active tension diagonal at section 4-4 of the truss shown in Fig. 5-80, and the stresses in the chords *FH* and *GI*.

*Ans. FI* = 15 kips T; *FH* = 36 kips C; *GI* = 27 kips T

**5-88.** Solve Prob. 5-87, with the two 24-kip loads applied at panel points *G* and *I*.

**5-89.** In Fig. 5-81, find the stress in the active tension diagonal in the third panel from the left and the stresses in chords *DF* and *EG*. (Isolate the left half as a free body.)

*Ans. DG* = 25 kips T; *DF* = 54.2 kips C; *EG* = 38.4 kips T

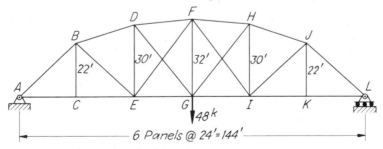

**Fig. 5-81   Probs. 5-89 and 5-90.**

**5-90.** Solve Prob. 5-89 when the 48-kip load is moved from *G* to *I*. (Isolate the left half as a free body.)

SUMMARY(By Article Number) _____

**5-1. A coplanar, nonconcurrent force system** is one in which all forces lie in one plane but whose action lines do not meet at a common point.

**5-2. The resultant force** may be determined analytically by algebraic summations of forces and of moments and graphically by the parallelogram method or by the string-polygon method.

**5-3.** The unknown quantities of **coplanar, nonconcurrent force systems in equilibrium** are generally solved analytically by algebraic summations of forces and of moments. Graphical solutions are also available. Typical problems are (1) determination of support reactions, (2) determination of the forces acting at the pins of certain pin-connected frames by the **method of members,** and (3) determination of stresses in members of trusses by the **method of sections.**

**5-4.** In determinations of **reactions by the graphical string-polygon method,** the string polygon must always start at the point of application of the reaction whose direction is unknown. Problems in which most of the forces are neither vertical nor horizontal are most suitably solved by this method.

**5-5. Analytical determination of reactions** is most suitable for problems in which most of the forces are either vertical or horizontal. The equations of solution are $\Sigma V = 0$, $\Sigma H = 0$, and $\Sigma M = 0$.

**5-6. Pin reactions** at connections of frames containing three-force members are obtained by the **method of members.** First, all determinable support reactions are found. Thereafter, each three-force member is isolated as a free body. Unknown forces are then readily determined, because the forces acting on each member so isolated are in equilibrium.

**5-7. The method of sections** enables us to determine the stress in any single member of a truss *independently* of all other stresses. The truss is cut in two through the least number of members, and either part—usually the smaller—is then isolated as a free body in equilibrium. Each unknown stress is then determined by one of the equations of equilibrium, $\Sigma V = 0$, $\Sigma H = 0$, or $\Sigma M = 0$.

**5-8.** When **counterdiagonals** are used in trusses, *only the tension diagonal is considered to be active.* The compressive diagonal is assumed to be inactive.

## REVIEW PROBLEMS

**5-91.** Determine analytically the *resultant R* of the four forces acting on the truss shown in Fig. 5-40. Find also the direction angle $\theta_H$ that the action line of $R$ makes with the horizontal and the distance $x$ (from $C$) to its line of action. Check by the *graphical parallelogram method.*

*Ans. R* $= 31.2$ kips   ; $\theta_H = 50°12'$; $x = 0$

**5-92.** Using the graphical *string-polygon method,* determine the *resultant R* of forces $B$ and $C$ in Fig. 5-82. (Scales on 8½ by 11: 1 in. $= 2$ ft, 1 in. $= 500$ lb.)

**5-93.** Determine the *reactions A* and $D$ at supports $A$ and $D$ in Fig. 5-82, using the graphical *string-polygon method.* (Scales on 8½ by 11: 1 in. $= 5$ ft, 1 in. $= 500$ lb.)

*Ans. A* $= 1185$ lb $\uparrow$, $D = 1035$ lb $\nwarrow$

**5-94.** The truss shown in Fig. 5-83 is subjected to the roof loads and the hoist load shown. Determine the *reactions $R_A$* and $R_G$, using the graphical *string-polygon method.* (Scales on 11 by 17: 1 in. $= 5$ ft, 1 in. $= 5$ kips.)

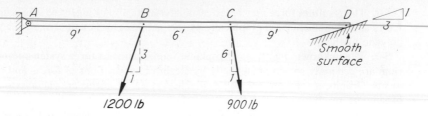

**Fig. 5-82   Probs. 5-92 and 5-93.**

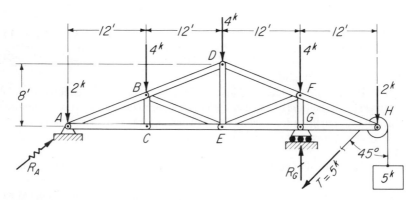

**Fig. 5-83   Prob. 5-94.**

**5-95.** Calculate the reactions at $A$ and $E$ in Fig. 5-84.

*Ans.* $A_V = 240$ lb↑; $A_H = 360$ lb→; $E = 300$ lb

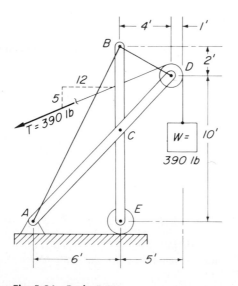

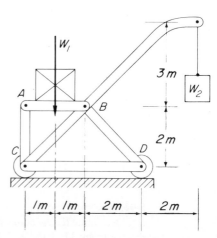

**Fig. 5-84   Prob. 5-95.**              **Fig. 5-85   Prob. 5-97.**

**5-96.** Find graphically the *reactions* at supports *D* and *E* of the beam shown in Fig. 5-13. Use the *string-polygon method.* (Scales on 8½ by 11; 1 in. = 4 ft. 1 in. = 300 lb.)

*Ans. D* = 577 lb ↗; *E* = 673 lb↑

**5-97.** The cart shown in Fig. 5-85 has four wheels, and the loads $W_1 = W_2 = 4800$ N are midway between the near and far set of wheels. Calculate the reaction forces under the near wheels *C* and *D*.

*Ans. C* = 600 N; *D* = 4200 N

**5-98.** Solve analytically for the *V* and *H reaction components* at supports *A* and *F* of the truss shown in Fig. 5-78 ($P_1 = 12$ kN, $P_2 = 36$ kN, $h = 3$ m, $CD = 2$ m, and $AF = 4$ m).

*Ans.* $A_V = 18$ kN↑; $A_H = 3$ kN→; $F_V = 54$ kN↑; $F_H = 15$ kN←

**5-99.** The truss of Fig. 5-86 supports the hoist sheave shown. Solve analytically for the *V* and *H reaction components* at supports *A* and *B*. As a check on $A_V$, select a moment center which will eliminate $B_V$, $B_H$, and $A_H$, and recalculate $A_V$.

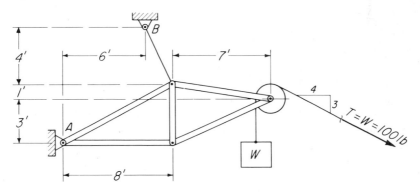

**Fig. 5-86    Prob. 5-99.**

**5-100.** In Fig. 5-87 solve analytically for the *V* and *H reaction components* at supports *A* and *E*.

*Ans.* $A_V = 38$ kN↑; $A_H = 16$ kN→; $E_V = 6$ kN↑; $E_H$ 24 kN←

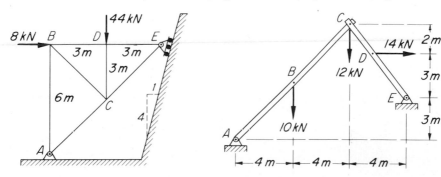

**Fig. 5-87    Probs. 5-100 and 5-123.**        **Fig. 5-88    Prob. 5-102.**

**5-101.** Calculate the stresses *DF, EF,* and *EG* in the truss of Fig. 5-73.

*Ans. DF* $= 0$; *EF* $= 9$ kips C; *EG* $= 24$ kips C

**5-102.** The frame shown in Fig. 5-88 is of the three-hinged-arch type. Determine analytically the *V* and *H reaction components* at supports *A* and *E*.

*Ans. $A_V = 7.89$ kN↑; $A_H = 2.89$ kN→; $E_V = 14.11$ kN↑; $F_{II} = 16.89$ kN←*

**5-103.** The roof hoist shown in Fig. 5-89 is resting on blocks at *A* and *B* and is prevented from sliding and overturning by a small block at *B* and a counterweight *W* at *A*. Compute the minimum value of the counterweight *W* required to prevent overturning and the *V* and *H reaction components* at *B*. Neglect possible friction.

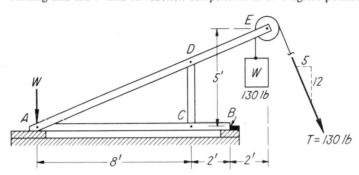

**Fig. 5-89    Prob. 5-103.**

**5-104.** Figure 5-90 shows one of two identical frames supporting an air-compressor tank having a diameter of 4 ft and weighing 1200 lb. The load *W* on each frame then is 600 lb. Compute the *V* and *H* components of the force exerted by the tank on bar *AB* and the *V* and *H reaction components* at supports *A* and at pin *B*. Neglect friction.

*Ans. $F_V = 600$ lb↓; $F_H = 347$ lb→; $A_V = 213$ lb↑; $A_H = 218$ lb←;*
*$B_V = 387$ lb↑; $B_H = 129$ lb←*

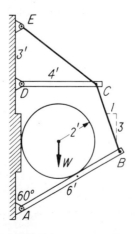

**Fig. 5-90    Probs. 5-104 and 5-105.**

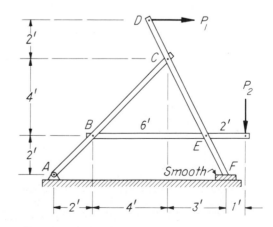

**Fig. 5-91    Prob. 5-109.**

**5-105.** Determine the *reactions* at supports D and E of the frame shown in Fig. 5-90. (See Prob. 5-104 for values of $B_V$ and $B_H$.)

**5-106.** In Fig. 5-93 the rigid bar AB is resting against smooth surfaces at A and B and is held at C by the rope CD. Compute the *reactions* at A and B and the stress in CD.

Ans. $R_A = 620$ lb↑; $R_B = 447$ lb↖; $CD = 500$ lb T

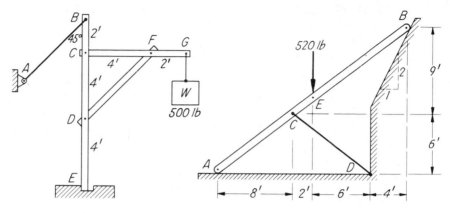

Fig. 5-92    Prob. 5-110.                Fig. 5-93    Probs. 5-106 and 5-108.

**5-107.** Solve Prob. 5-106 graphically, using the *four-force principle*. (HINT: Find first the resultant of E and B, then the resultant of A and CD. Scales on 8½ by 11: 1 in. = 5 ft, 1 in. = 300 lb.)

**5-108.** Solve Prob. 5-106 when an additional horizontal force of 260 lb, acting from left to right, is applied at C.

Ans. $R_A = 395$ lb↑; $R_B = 313$ lb↖; $CD = 25$ lb T

**5-109.** Using the *method of members*, compute the *pin reactions* at A, B, and E in Fig. 5-91. The frame rests on a smooth block at F and the loads are $P_1 = 90$ lb and $P_2 = 270$ lb.

Ans. $R_A = 142.1$ lb; $R_B = 108.2$ lb; $R_C = 36.1$ lb; $R_E = 365.0$ lb

**5-110.** Using the *method of members*, compute the V and H components of the forces acting at B, C, D, and E of the frame shown in Fig. 5-92. Find also the stresses in members AB and DF.

**5-111.** The forces P and W in Fig. 5-94 are 210 lb and 420 lb, respectively. Using the *method of members*, compute the vertical and horizontal components of the pin *reactions* at B, C, and D.

Ans. $B_V = 280$ lb; $B_H = 245$ lb; $C_V = 406$ lb; $C_H = 245$ lb; $D_V = 70$ lb; $D_H = 245$ lb

**5-112.** Solve Prob. 5-111 if $P = 300$ lb and $W = 150$ lb.

**5-113.** Using the *method of members*, compute the vertical and horizontal components of the pin *reactions* at B and C of Fig. 5-95.

**5-114.** The sliding joint at D in Fig. 5-96 is frictionless. Calculate the vertical and horizontal components of the pin reactions at E, C, and D.

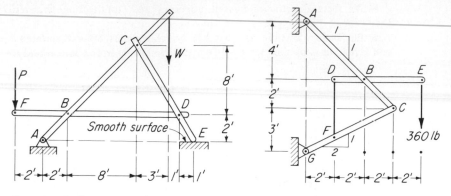

**Fig. 5-94    Probs. 5-111 and 5-112.**        **Fig. 5-95    Prob. 5-113.**

**5-115.** The truss shown in Fig. 5-97 supports two horizontal loads 20 kN each. Using the *shear method of sections,* calculate the force in members *AC* and *CD*.

*Ans. AC* = 25 kN C; *CD* = 50 kN

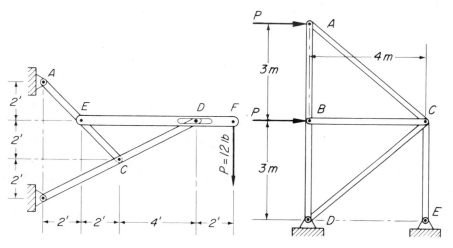

**Fig. 5-96    Prob. 5-114.**        **Fig. 5-97    Probs. 5-115 and 5-120.**

**5-116.** Using the *shear method of sections,* compute the *stresses* in members *DE, EG,* and *GF* of the truss shown in Fig. 5-70.

**5-117.** Using the *shear method of sections,* determine the *stresses* in members *AC, BC,* and *BD* of the Warren truss shown in Fig. 5-69.

*Ans. AC* = 18 kips T; *BC* = 3.6 kips C; *BD* = 16 kips C

**5-118.** Using the *shear method of sections,* find the *stresses* in members *CF, DF,* and *DG* of the loading-platform truss shown in Fig. 4-36.

**5-119.** Using the *shear method of sections,* determine the stress in the *active tension diagonal* at section 3-3 in Fig. 5-80 and the stresses in the chords *DF* and *EG.*

*Ans. DG* = 15 kips T; *DF* = 27 kips C; *EG* = 18 kips T

**5-120.** Calculate the stresses in members *CE* and *BD* of Fig. 5-97. $P$ = 20 kN.

**5-121.** Using the *component method of sections,* solve for the *stresses* in members *BD, CD,* and *CE* of the truss shown in Fig. 5-98. ($E_V$ = 2 kips↑; $E_H$ = 4 kips→.)

*Ans. BD* = 13.4 klps C; *CD* = 4 kips C; *CE* = 5.67 kips T

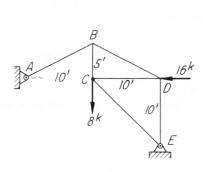

**Fig. 5-98   Prob. 5-121.**          **Fig. 5-99   Prob. 5-122.**

**5-122.** Using the *component method of sections,* determine the *stresses* in members *AB, BC,* and *CD* of the truss shown in Fig. 5-99. (SUGGESTION: Use part to the right of cut section and apply the *V* and *H* components of *AB* at *A,* and those of *BC* at *C.*)

*Ans. AB* = 10.12 kips C; *BC* = 5.81 kips T; *CD* = 7 kips T

**5-123.** Using the *component method of sections,* solve for the *stresses* in members *AC, BC,* and *BD* of the truss shown in Fig. 5-87. (See Prob. 5-100. HINT: Isolate right half.)

*Ans. AC* = 22.6 kN C; *BC* = 31.1 kN T; *BD* = 30.0 kN C

**5-124.** Using the *component method of sections,* find the *stresses* in members *BC, BE,* and *AE* of the truss shown in Fig. 5-40. (See Prob. 5-42. HINT: Isolate right half.)

**5-125.** Using the *component method of sections,* compute the *stresses* in members *CD, BD,* and *BE* of the truss shown in Fig. 5-30 (see Prob. 5-32).

*Ans. CD* = 11.2 kips T; *BD* = 20 kips C; *BE* = 21.2 kips T

**5-126.** In Fig. 5-80, which is the *active tension diagonal* in the third panel from the left, and what is its stress?

*Ans. DG* = 15 kips T

**5-127.** If, in Fig. 5-80, the two 24-kip loads are applied at panel points *C* and *E,* which will be the *active tension diagonal* in the third panel from the left, and what will be its stress?

*Ans. EF* = 15 kips T

## REVIEW QUESTIONS

**5-1.** Define a coplanar, nonconcurrent force system.

**5-2.** What items must be determined before the resultant of a coplanar, nonconcurrent force system is fully known?

**5-3.** In the string-polygon method, what do the rays represent and what do the strings represent?

**5-4.** When support reactions are determined by the string-polygon method, at which point must the string polygon be started? What is the only exception to this rule?

**5-5.** What is the direction of the reaction at the end of a two-force member?

**5-6.** What is known about the direction of the reaction at the end of a three-force member?

**5-7.** What is the direction of the reaction at a roller support and at a smooth-surface support?

**5-8.** What is meant by a pin reaction?

**5-9.** In problems to determine pin reactions, which forces must always be computed before the frame is dismembered?

**5-10.** What are the purpose and function of the method of sections?

**5-11.** To what type of truss is the shear method of sections especially applicable? Explain briefly the usual procedure.

**5-12.** When is the component method of sections generally used? Explain briefly the usual procedure.

**5-13.** Using the method of sections, is it always possible to determine any single stress in a truss independently of any other stress, provided not more than three members are cut at a section?

**5-14.** What is a counterdiagonal, and what purpose does it serve?

**5-15.** In any panel of a truss containing two diagonals, which of the two is considered to be active and to carry stress?

# Chapter 6
# Noncoplanar, Parallel
# Force Systems

## 6-1 INTRODUCTION

The most common system of noncoplanar, parallel forces is that of the gravity forces (weights) of the several parts of a structure or machine such as a building, an airplane, or an automobile. A common type of problem is that of determining the *resultant force* of the weights of all parts of an airplane and the location of its line of action, which, for proper balance of the plane, must pass through a given point in the plane. Other problems concern the *equilibrium of parallel forces in space;* such problems arise in the determination of loads on the columns of a building produced by the dead weight of the floors they support and the live loads thereon. These problems are discussed in the next two articles.

## 6-2 RESULTANT OF A NONCOPLANAR, PARALLEL FORCE SYSTEM

The magnitude of the resultant of a parallel force system in space is equal to the algebraic sum of the component forces. The **center of a system of parallel forces** is that point through which the resultant passes, or about which the algebraic sum of the moments of the forces is zero. The magnitude of a resultant force is readily determined by simple algebraic summation of the forces comprising the system, and the location of the center is readily found by moments, because *about any axis* **the moment of the resultant force equals the algebraic sum of the moments of the separate forces.** Summations of moments can be made in two planes, usually at right angles to each other.

## ILLUSTRATIVE PROBLEM

**6-1.** The magnitude and the location of the center $C$ of the system of parallel forces shown in Fig. 6-1$a$ are to be determined. The side of each square is 1 ft in length. All forces are in pounds.

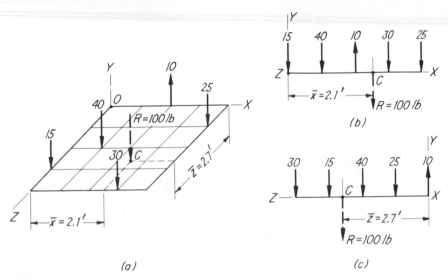

**Fig. 6-1** Resultant of a system of parallel forces in space. *(a)* XZ plane is horizontal and is 4 ft square. *(b)* Forces projected into XY plane. *(c)* Forces projected into YZ plane.

*Solution:* The center of these forces is fully located if we determine its position in each of two planes, usually perpendicular. First let all forces be projected back into the vertical $XY$ plane as shown in Fig. 6-1$b$. In this view the $Z$ axis appears as a point. Let $Z$ be the moment center, and let $\bar{x}$ (bar $x$) be the distance from $Z$ to the center of forces. The resultant $R$ is the algebraic sum of forces. Hence

$[R = \Sigma F]$   $R = 15 + 40 - 10 + 30 + 25$   or   $R = 100$ lb            *Ans.*
$[R\bar{x} = \Sigma M_Z]$   $100\bar{x} = (40)(1) - (10)(2) + (30)(3) + (25)(4)$
and                         $\bar{x} = 2.1$ ft            *Ans.*

Next let all forces now be projected into the vertical $YZ$ plane, as in Fig. 6-1$c$. Here, axis $X$ appears as a point. Let $X$ now be the moment center and let $\bar{z}$ be the distance from $X$ to the center of forces. Then

$[R\bar{z} = \Sigma M_X]$   $100\bar{z} = (25)(1) + (40)(2) + (15)(3) + (30)(4)$
and                         $\bar{z} = 2.7$ ft            *Ans.*

*Alternate Solution:* We can make our calculations for the location and the magnitude of the resultant of the parallel force system by reference to Fig. 6-1$a$ **only.** First let us take moments about the line $OZ$, recognizing that the moment arms must be measured perpendicular to $OZ$. We have, then,

$$100\bar{x} = 30(3) + 40(1) + 25(4) - 10(2) \quad \text{and} \quad \bar{x} = 2.1 \text{ ft}$$

In a similar manner, taking moments about $OX$,

$$100\bar{z} = 15(3) + 40(2) + 30(4) + 25(1) \quad \text{and} \quad \bar{z} = 2.7 \text{ ft}$$

## PROBLEMS

**6-2.** Solve Prob. 6-1 if the 30-lb force is upward-acting.

**6-3.** Solve Prob. 6-1 if a downward-acting force of 30 lb is added at $x = 3$ and $z = 3$.

*Ans.* $R = 130$ lb; $\bar{x} = 2.31$ ft; $\bar{z} = 2.77$ ft

**6-4.** Find the resultant $R$ of the force system shown in Fig. 6-2, and determine the location of its center $C$. The $XZ$ plane is horizontal and is 4 m square.

*Ans.* $R = 180$ N↓; $\bar{x} = 3.06$ m; $\bar{z} = 2.44$ m

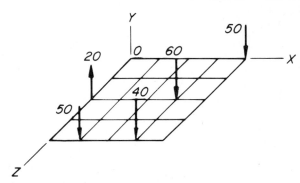

**Fig. 6-2** Probs. 6-4 and 6-6.

**6-5.** In Fig. 6-3, determine the resultant force $R$ and the location of its center $C$. The $XZ$ plane is horizontal and is 4 m square.

*Ans.* $R = 90$ N↓; $\bar{x} = 1.06$ m; $\bar{z} = 2.11$ m

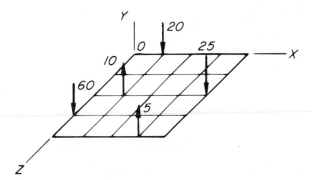

**Fig. 6-3** Probs. 6-5 and 6-7.

**6-6.** In Fig. 6-2, let the 40–N force be upward-acting. Compute $R$, $\bar{x}$, and $\bar{z}$.

**6-7.** In Fig. 6-3, let the 20–N force be upward-acting. Compute $R$, $\bar{x}$, and $\bar{z}$.

*Ans.* $R = 50$ N↓; $\bar{x} = 1.1$ m; $\bar{z} = 3.8$ m

## 6-3 EQUILIBRIUM OF NONCOPLANAR, PARALLEL FORCE SYSTEMS

Since their directions are known, only the magnitudes of the unknown forces of a parallel system in equilibrium need be determined. They are usually determined by summations of moments about two axes at right angles to each other. In statics, not more than three unknown forces may be so determined.

**ILLUSTRATIVE PROBLEM** _____

**6-8.** The square, horizontal plate shown in Fig. 6-4 supports a vertical load of 300 lb at $D$ and is, in turn, supported by vertical reactions at $A$, $B$, and $C$. Compute these vertical reacting forces at supports $A$, $B$, and $C$. The side of each square is 1 ft in length. Neglect weight of plate.

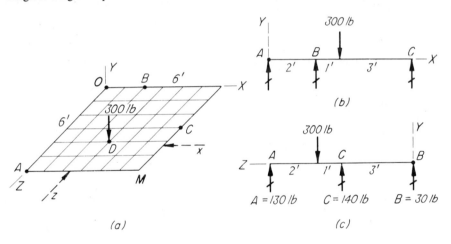

**Fig. 6-4   Determination of reactions. Parallel forces in space.** *(a)* XZ plane is horizontal and is 6 ft square. *(b)* Forces projected into XY plane. *(c)* Forces projected into YZ plane.

*Solution:* In Figs. 6-4b and 6-4c, the forces have been projected into the $XY$ and $YZ$ planes, respectively. In each view three forces $A$, $B$, and $C$ are unknown. With point $A$ as the moment center in both views, reaction $A$ is eliminated from consideration. Two independent equations may now be written which, then, are solved simultaneously, yielding reactions $B$ and $C$. Reaction $A$ may then be found by moments about $C$. Finally, a summation of vertical forces provides a sufficient check. Hence, from Fig. 6-4b, we have

$$[\Sigma M_A = 0] \qquad\qquad 2B + 6C = (300)(3) = 900 \qquad\qquad (a)$$

and from Fig. 6-4c we obtain

$$[\Sigma M_A = 0] \qquad\qquad 6B + 3C = (300)(2) = 600 \qquad\qquad (b)$$

Multiplying $(a)$ by 3 and subtracting $(b)$,

$$6B + 18C = 2700 \qquad\qquad (c)$$
$$\underline{6B + \ 3C = \ \ 600} \qquad\qquad (b)$$
$$15C = 2100 \quad \text{and} \quad C = 140 \text{ lb} \qquad Ans.$$

Substituting this value of $C$ in Eq. $(a)$ to find $B$, we have

$$2B + (6)(140) = 900 \quad \text{and} \quad B = 30 \text{ lb} \qquad Ans.$$

Taking moments about $C$, in Fig. 6-4c, to find $A$, we obtain

$[\Sigma M_C = 0]$ $\ 3A = (300)(1) + (30)(3) = 390 \quad$ and $\quad A = 130 \text{ lb} \qquad Ans.$
$[\Sigma V = 0] \qquad 300 = 140 + 30 + 130 \quad$ or $\quad 300 = 300 \qquad Check$

(NOTE: When two of three unknown forces overlie each other in one view, as do $A$ and $B$ in Fig. 6-5, the third force may be solved for directly.)

*Alternate Solution:* We may find the required reactions by the use of Fig. 6-4a **only.** Assume that $A$, $B$, and $C$ are upward-acting. A negative answer will indicate that the reaction is really opposite to that assumed. Taking moments about line $AO$ gives

$[\Sigma M_{AO} = 0] \qquad\qquad 2B + 6C = 300(3) \qquad\qquad (d)$

A moment summation about line $AM$ gives

$[\Sigma M_{AM} = 0] \qquad\qquad 6B + 3C = 300(2) \qquad\qquad (e)$

Multiplying Eq. $(e)$ by $-2$ gives

$$-12B - 6C = -1200 \qquad\qquad (f)$$

Adding Eqs. $(d)$ and $(f)$ gives

$$-10B = -300 \quad \text{or} \quad B = 30 \text{ lb} \qquad Ans.$$

Now from Eq. $(d)$

$$C = \frac{900 - 2(30)}{6} = 140 \text{ lb} \qquad Ans.$$

The reaction at $A$ can now be found by a moment summation about line $OX$

$[\Sigma M_{OX} = 0] \qquad 6A + 3(140) = 300(4) \quad$ or $\quad A = 130 \text{ lb} \qquad Ans.$

## PROBLEMS

**6-9.** The square, level plate shown in Fig. 6-5 weighs 120 lb, which may be concentrated at the center of the plate. Compute the vertical reactions at supports $A$, $B$, and $C$, due to this weight and to the two vertical loads at $D$ and $E$.

*Ans.* $A = 115$ lb; $B = 165$ lb; $C = 140$ lb

**6-10.** The horizontal plate shown in Fig. 6-6 is subjected to two vertical loads at $D$ and $E$. Neglecting the weight of the plate, compute the vertical reactions at supports $A$, $B$, and $C$. The side of each square is 1 m in length.

**6-11.** Solve Prob. 6-10 if the support $B$ is moved 1 m to the left.

*Ans.* $A = 250$ N; $B = 70$ N; $C = 130$ N

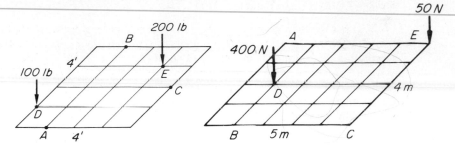

**Fig. 6-5   Probs. 6-9 and 6-17.**      **Fig. 6-6   Probs. 6-10, 6-11, and 6-18.**

**6-12.** A circular, level table, 4 ft in diameter (d) and weighing 150 lb, is supported on three equally spaced vertical legs at $A$, $B$, and $C$, as shown in Fig. 6-7. Compute the reacting force in each leg, if a vertical load of 180 lb is concentrated at $D$, which is 12 in. to the right of $A$.

*Ans.* $A = 170$ lb; $B = C = 80$ lb

**6-13.** If the 2-m-diameter circular table shown in Fig. 6-7 weighs 600 N, what maximum downward-acting vertical force $P$ may be applied at $E$ without tipping the table?

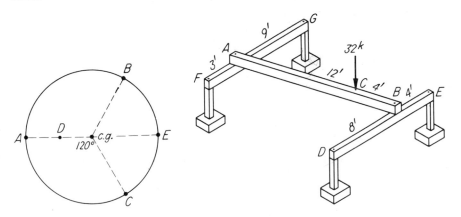

**Fig. 6-7   Probs. 6-12 and 6-13.**      **Fig. 6-8   Probs. 6-14, 6-15, and 6-19.**

**6-14.** Three horizontal beams, arranged as shown in Fig. 6-8 and resting on vertical posts, support a 32-kip vertical load at $C$. Neglecting the weights of the beams, compute the load on each of posts $D$, $E$, $F$, and $G$. Draw a separate free-body diagram of each of the three beams.

*Ans.* $D = 8$ kips; $E = 16$ kips; $F = 6$ kips; $G = 2$ kips

**6-15.** In Fig. 6-8, the beams are of uniform cross section and density. Beam $AB$ weighs 1200 lb, and each of beams $DE$ and $FG$ weighs 600 lb. Compute the loads on each of posts $D$, $E$, $F$, and $G$ due to these weights and the 32-kip load.

**6-16.** Solve Prob. 6-8 if the support $C$ is moved 2 ft further away from the $X$ axis (parallel to the $z$ axis) and if the load is moved to $x = 2$, $z = 3$.

**SUMMARY** (By Article Number) _____

**6-2.** The magnitude of the resultant of a system of parallel forces in space is the algebraic sum of those forces. Its location may be determined by summations of moments about two axes, usually perpendicular, since the moment of the resultant about any axis equals the algebraic sum of the moments of the separate forces.

**6-3.** The unknown forces (usually three reactions) of a system of noncoplanar, parallel forces in equilibrium may be determined by summations of moments about two axes, usually perpendicular.

**REVIEW PROBLEMS** _____

**6-17.** In Prob. 6-9, disregard the weight of the plate and compute the *reactions A*, *B*, and *C* required for equilibrium of the remaining system of noncoplanar, parallel forces.

$$Ans.\ A = 75\ lb;\ B = 125\ lb;\ C = 100\ lb$$

**6-18.** In Prob. 6-10, let *E* be an upward-acting force. Compute the vertical *reactions A*, *B*, and *C* required to maintain *equilibrium.*

**6-19.** In Prob. 6-14, let an additional downward-acting vertical force of 24 kips be acting on beam *AB* at a point 4 ft from *A*. Compute the *reactions* at beam ends *D*, *E*, *F*, and *G*.

$$Ans.\ D = 10\ kips;\ E = 20\ kips;\ F = 19.5\ kips;\ G = 6.5\ kips$$

**6-20.** Solve Prob. 6-8 if a 200-lb downward-acting force is added at the origin *0* (along the *Y* axis).

$$Ans.\ A = 170\ lb;\ B = 270\ lb;\ C = 60\ lb$$

**REVIEW QUESTIONS** _____

**6-1.** What is meant by the center of a system of parallel forces?

**6-2.** By what basic principle of statics may we determine the location of this center?

**6-3.** Describe briefly the method used to determine the location of the center of a noncoplanar, parallel system of forces.

**6-4.** Explain briefly the method used to determine the unknown forces of a noncoplanar, parallel force system in equilibrium.

# Chapter 7
# Noncoplanar, Concurrent Force Systems

## 7-1 INTRODUCTION

A simple example of a noncoplanar, concurrent force system is an ordinary tripod supporting a weight at the intersection of the three legs. Another example is the simple three-member frame shown in Fig. 7-4. The weight $W$ and the stresses in members $A$, $B$, and $C$ constitute such a system—concurrent because their action lines all meet at $O$ and noncoplanar because they do not all lie in one plane.

The resultant of such a system is seldom required in ordinary engineering practice. Problems in determining the unknown forces, reactions, or stresses required to maintain equilibrium are, however, frequently encountered.

## 7-2 COMPONENTS OF A FORCE IN SPACE

In Fig. 7-1, $F$ is a force in space represented as a vector of length $F$. For the purpose of reference, let $O$ arbitrarily be called its *near end* and $A$ its *far end*. Through $O$ are passed three axes, $X$, $Y$, and $Z$, all mutually perpendicular, of which $Y$ is vertical, and $X$ and $Z$ are horizontal.

The distances $x$, $y$, and $z$, which locate the far end $A$ of the force with respect to its near end $O$, are called the **space coordinates** of point $A$. These $x$, $y$, and $z$ coordinates are measured, respectively, parallel to the $X$, $Y$, and $Z$ axes from the origin $O$. They are positive when measured to the right, upward, and toward the observer.

We may now imagine Fig. 7-1 to represent a transparent box with visible edges whose dimensions are the coordinates $x$, $y$, and $z$. Force $F$ is the diagonal of this box, and $F_X$, $F_Y$, and $F_Z$ are, respectively, the $X$, $Y$, and $Z$ components of $F$, as a little study will show. Obviously, then, **the X, Y, and**

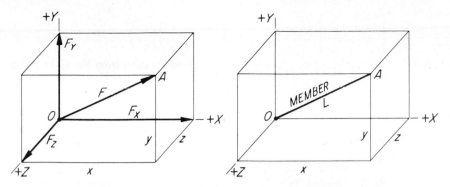

Fig. 7-1   A force in space.             Fig. 7-2   A two-force member in space.

$Z$ **components of a force in space are proportional to the** $x$, $y$, **and** $z$ **coordinates of its far end with respect to its near end.** The component always bears the sign, positive or negative, of its corresponding coordinate.

Since $F$ is the diagonal of a parallelepipedon,

$$F^2 = x^2 + F_Y^2 + F_Z^2 \qquad \text{or} \qquad F = \sqrt{F_X^2 + F_Y^2 + F_Z^2} \quad (7\text{-}1)$$

Figures 7-1 and 7-2 are identical. In both, the coordinates of the far end of the diagonal with respect to its near end are $x$, $y$, and $z$. In Fig. 7-2, however, let this diagonal represent a two-force structural member of length $L$. Let $F$, whose action line passes through points $O$ and $A$, be the *axial stress* in this member. Clearly, then, **when the near end of a two-force member in space is the origin** $O$ **of the rectangular** $X$, $Y$, **and** $Z$ **axes, the components** $F_X$, $F_Y$, **and** $F_Z$ **of the axial stress within it are proportional to the** $x$, $y$, **and** $z$ **space coordinates of its far end.** This proportionality is an important aid in the determination of axial stresses in space frames, discussed in Art. 7-3. In equation form, the proportions can be written

$$\frac{F_X}{x} = \frac{F}{L} \qquad \frac{F_Y}{y} = \frac{F}{L} \qquad \frac{F_Z}{z} = \frac{F}{L} \qquad \text{or} \qquad \frac{F_X}{x} = \frac{F_Y}{y} = \frac{F_Z}{z} \quad (7\text{-}2)$$

## ILLUSTRATIVE PROBLEM

**7-1.** In Fig. 7-1, let $F = 100$ lb, and let $x = 6$, $y = 4$, and $z = 3$. Then determine the components $F_X$, $F_Y$, and $F_Z$.

*Solution:* The components can be determined by Eq. (7-2). The length $L$ is

$$[L = \sqrt{x^2 + y^2 + z^2}] \qquad L = \sqrt{(6)^2 + (4)^2 + (3)^2} = \sqrt{61} = 7.81$$

Then

$$\left[ \frac{F_X}{x} = \frac{F}{L} \right] \qquad F_X = x\frac{F}{L} = 6\left( \frac{100}{7.81} \right) = 76.8 \text{ lb} \qquad\qquad Ans.$$

$$\left[ \frac{F_Y}{y} = \frac{F}{L} \right] \qquad F_Y = y\frac{F}{L} = 4\left( \frac{100}{7.81} \right) = 51.2 \text{ lb} \qquad\qquad Ans.$$

$$\left[ \frac{F_Z}{z} = \frac{F}{L} \right] \qquad F_Z = z\frac{F}{L} = 3\left( \frac{100}{7.81} \right) = 38.4 \text{ lb} \qquad\qquad Ans.$$

A check on these computations may be obtained from Eq. (7-1) as follows:

$$[F^2 - F_X{}^2 + F_Y{}^2 + F_Z{}^2] \qquad 10,000 = 5900 + 2623 + 1477 = 10,000 \qquad Check$$

## PROBLEMS

**7-2.** In Fig. 7-1, let $F = 680$ N, and let $x = 12$, $y = 9$, and $z = 8$. Determine the force components $F_X$, $F_Y$, and $F_Z$.

**7-3.** The member $OA$ shown in Fig. 7-2 has $x$, $y$, and $z$ components of length of 6, 8, and 24 ft respectively. Calculate the force in the member if the $X$ component of the force is 1200 lb.

$$Ans. \quad F = 5200 \text{ lb}$$

**7-4.** Solve Prob. 7-3 if the $Z$ component of the force is 600 lb.

## 7-3 EQUILIBRIUM OF NONCOPLANAR, CONCURRENT FORCE SYSTEMS

The usual problem is that of determining the unknown stresses in trusses similar to those shown in Probs. 7-5 and 7-6. A number of methods of solution are available, both analytical and graphical. Only analytical methods are presented here.

**The force method** is based on three algebraic summations of forces parallel to three rectangular $X$, $Y$, and $Z$ axes, thus requiring the simultaneous solution of three equations.

**The moment method** requires only two algebraic summations of moments to be made in two perpendicular planes, thus requiring the simultaneous solution of only two equations. This method, therefore, is generally preferred.

We may project the space frame and all forces acting on it onto one of the three planes associated with the coordinate axes. The planes used are the vertical $XY$ and $YZ$ planes and the horizontal $XZ$ plane, containing the three intersecting and mutually perpendicular $X$, $Y$, and $Z$ axes as illustrated in Fig. 7-3. An important point to note is that a force will not appear on a plane perpendicular to its line of action. In other words, on the $XY$ plane only forces parallel to the $X$ and $Y$ axes will be considered, only $Y$ and $Z$ forces are considered on the $YZ$ plane, and only $X$ and $Z$ forces are considered on the $XZ$ plane.

Instead of projecting all members and forces onto the mutually perpendicu-

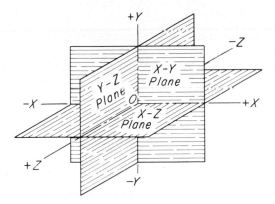

**Fig. 7-3  Planes formed by three intersecting and mutually perpendicular axes.**

lar planes, we may adopt an alternate viewpoint. The space frame and its loads and the reaction components may be visualized as a three-dimensional concept in which forces may be summed in any direction or moments can be taken about any axis. When taking moments, all forces whose lines of action *pass through* the moment axis and all forces *parallel* to the moment axis will not contribute to the summation.

Both the "projected on coordinate planes" and the "three-dimensional" concepts will be used in the following illustrative problems.

## ILLUSTRATIVE PROBLEMS

**7-5.** The three-member frame shown in Fig. 7-4 is supported at $A$, $B$, and $C$ on a vertical wall. Determine the stresses in the three members caused by the vertical load $W$.

*Solution by Moment Method:* Let each member, and the stress within it, be designated by the letter which also designates the support at its far end. Figures 7-5 and 7-6 show, respectively, a top view and a side view of the frame. Since $A$, $B$, and $C$ are *axial stresses*, they equal, respectively, the reactions at supports $A$, $B$, and $C$. The latter may, therefore, be solved for.

Let the direction of each member in these two views be indicated by a small triangle, the sides of which are proportional to the respective coordinates of the far end of the member and, therefore, proportional also to the components of the stress in the member. We then express the $Y$ and $Z$ components in terms of the $X$ component. In Fig. 7-6 the $Z$ components are perpendicular to the vertical $XY$ plane and hence produce no reactions in that plane. The components $A_X$ and $B_X$ overlie each other, and $C_Y$ is seen to equal $\frac{3}{4}C_X$. In Fig. 7-5 the $Y$ components act perpendicularly to the horizontal $XZ$ plane and hence produce no reactions in that plane. From the direction triangles, we see that $A_Z$ is $\frac{1}{3}A_X$ and $B_Z$ is $\frac{1}{2}B_X$. Clearly, $A$ and $B$ are tensile stresses, and $C$ is compressive. In this problem, one component of each of the three stresses $A$, $B$, and $C$ may now be solved for directly by moments. As usual, no sign convention is

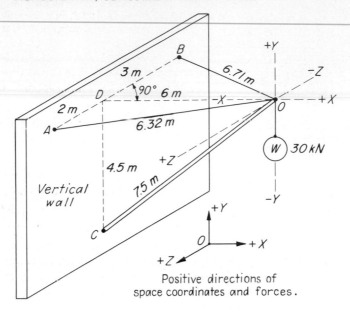

**Fig. 7-4   Determination of stresses in a space frame.**

necessary in moment solutions. Solving for $C_X$ by moments about point $AB$, in Fig. 7-6, we have

$$[\Sigma M_{AB} = 0] \qquad 4.5C_X = (30)(6) = 180 \qquad \text{and} \qquad C_X = 40 \text{ kN}$$

Then

$$C_Y = \frac{3}{4}\, C_X = \left(\frac{3}{4}\right) (40) = 30 \text{ kN and } C = \sqrt{(40)^2 + (30)^2} = 50 \text{ kN C } Ans.$$

Solving for $B_X$ by moments about point $A$, in Fig. 7-5, we obtain

$$[\Sigma M_A = 0] \qquad 5B_X = (40)(2) = 80 \qquad \text{and} \qquad B_X = 16 \text{ kN}$$

Then

$$B_Z = \frac{1}{2}\, B_X = \left(\frac{1}{2}\right) (16) = 8 \text{ kN and } B = \sqrt{(16)^2 + (8)^2} = 17.9 \text{ kN T } Ans.$$

Solving for $A_X$ by moments about point $B$, in Fig. 7-5, we get

$$[\Sigma M_B = 0] \qquad 5A_X = (40)(3) = 120 \qquad \text{and} \qquad A_X = 24 \text{ kN}$$

Then

$$A_Z = \frac{1}{3}\, A_X = \left(\frac{1}{3}\right) (24) = 8 \text{ kN and } A = \sqrt{(24)^2 + (8)^2} = 25.3 \text{ kN } Ans.$$

The positive signs of the computed stresses indicate that the type of each stress was correctly assumed. We must now check the results just obtained. To do so, we record all $V$ and $H$ components on the free-body diagrams, as was done in Figs. 7-5 and 7-6.

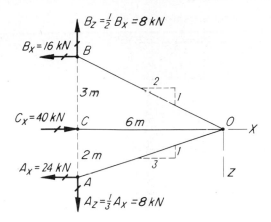

**Fig. 7-5   Top-view free-body diagram, horizontal XZ plane.**

If $\Sigma V = 0$ and $\Sigma H = 0$, a sufficient check is obtained, since the components were found by moments.

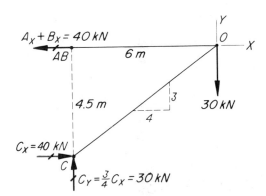

**Fig. 7-6   Side-view free-body diagram, vertical XY plane.**

*Alternate Solution by Moment Method:* The frame, the load, and the reactions are shown in Fig. 7-7. Members $AO$ and $BO$ have been assumed to be in tension, and member $CO$ is assumed to be in compression. We may take moments about the line $AB$. The reactions or the reaction components at $A$ and $B$ will not appear in the resulting equation because their lines of action pass through $AB$. The reaction component $C_Y$ will likewise not appear because its line of action also passes through $AB$. We have, then,

$$[\Sigma M_{AB} = 0] \qquad 6(30) = 4.5C_X \qquad \text{or} \qquad C_X = 40 \text{ kN}$$

The force $C$ can now be found from Eq. (7-2).

$$\left[\frac{F}{L} = \frac{F_X}{x}\right] \qquad \frac{C}{7.5} = \frac{40}{6} \qquad \text{or} \qquad C = \frac{40(7.5)}{6} = 50 \text{ kN C} \qquad\qquad Ans.$$

Taking moments about a vertical line through $A$ gives

$$[\Sigma M_{A_Y} = 0] \qquad 2C_X = 5B_X \qquad \text{or} \qquad B_X = \frac{2(40)}{5} = 16 \text{ kN}$$

The force $B$ is then

$$\left[\frac{F}{L} = \frac{F_X}{x}\right] \qquad \frac{B}{6.71} = \frac{16}{6} \qquad \text{or} \qquad B = \frac{16(6.71)}{6} = 17.9 \text{ kN T} \qquad \textit{Ans.}$$

Finally, taking moments about a vertical line through $B$ gives

$$[\Sigma M_{B_Y} = 0] \qquad 3C_X = 5A_X \qquad \text{or} \qquad A_X = \frac{3(40)}{5} = 24 \text{ kN T} \qquad \textit{Ans.}$$

The force $A$ is then

$$\left[\frac{F}{L} = \frac{F_X}{x}\right] \qquad \frac{A}{6.32} = \frac{24}{6} \qquad \text{or} \qquad A = \frac{24(6.32)}{6} = 25.3 \text{ kN T} \qquad \textit{Ans.}$$

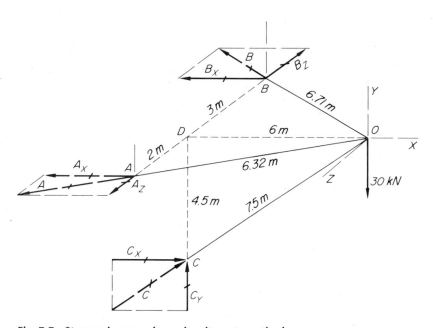

**Fig. 7-7   Stresses in space frame by alternate method.**

**7-6.** Figure 7-8 shows a three-member frame supported at $A$, $B$, and $C$. The locations of the three supports are given by $x$, $y$, and $z$ space coordinates measured from $O$, parallel to the rectangular $X$, $Y$, and $Z$ axes which pass through $O$. The frame is subjected to a 1200-lb load applied at $D$ and parallel to the $X$ axis. Determine the stresses in all members.

*Solution by the Moment Method:* Let the stress in each member be designated by $A$, $B$, and $C$, which letter also designates the support end of each member. Let the

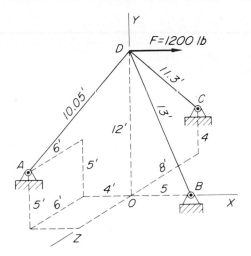

**Fig. 7-8   Determination of stresses in space frame.**

$X$, $Y$, and $Z$ components of all stresses be applied at the respective supports. Assume that members $AD$ and $CD$ are tensile and $BD$ compressive. This seems reasonable, and if the assumption is wrong, the member or members having stresses differing from those assumed will have negative computed components.

Figure 7-9 shows two free-body diagrams of the frame projected onto the vertical $XY$ and $YZ$ planes. The $x$, $y$, and $z$ components of the lengths of each member are shown in the appropriate sketch. The $X$, $Y$, and $Z$ force components of each reaction are shown in terms of the $Y$ component, using Eq. (7-2). We may now solve for the stresses by either the force method or the moment method. The solution by the moment method follows.

Using the free-body diagrams in Fig. 7-9, two independent moment equations can be written, one in the $XY$ plane and one in the $ZY$ plane, each with point $A$ as the moment center. All components of stress $A$ are thus automatically eliminated from both moment equations. Then, in the $XY$ plane,

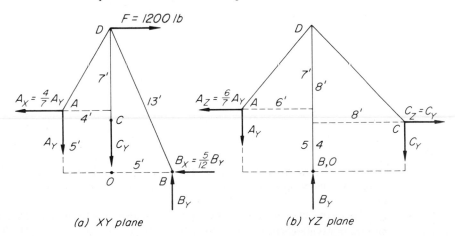

*(a) XY plane*    *(b) YZ plane*

**Fig. 7-9   Stresses in space frame. Projected members and forces. *(a)* XY plane. *(b)* YZ plane.**

$$[\Sigma M_A = 0] \qquad 9B_Y - 5\left(\frac{5}{12}B_Y\right) - 4C_Y - 7(1200) = 0$$

$$9B_Y - 2.08B_Y - 4C_Y = 8400$$
$$6.92B_Y - 4C_Y = 8400 \qquad\qquad (a)$$

Similarly, in the $YZ$ plane,

$$[\Sigma M_A = 0] \qquad\qquad 6B_Y + C_Y - 14(C_Y) = 0$$

$$6B_Y - 13C_Y = 0 \quad\text{or}\quad B_Y = \frac{13}{6}C_Y \qquad\qquad (b)$$

Substituting the value of $B_Y$ from Eq. $(b)$ into Eq. $(a)$ gives

$$6.92\left(\frac{13}{6}C_Y\right) - 4C_Y = 8400$$

$$15C_Y - 4C_Y = 8400 \quad\text{or}\quad C_Y = 765 \text{ lb}$$

By proportion,

$$\left[\frac{F}{L} = \frac{F_Y}{y}\right] \qquad \frac{C}{11.3} = \frac{765}{8} \quad\text{or}\quad C = 1080 \text{ lb T} \qquad\qquad Ans.$$

Also, from Eq. $(b)$, $B_Y = \dfrac{13}{6}C_Y = 1657 \text{ lb}$

$$\left[\frac{F}{L} = \frac{F_Y}{y}\right] \qquad \frac{B}{13} = \frac{1657}{12} \quad\text{or}\quad B = 1795 \text{ lb C} \qquad\qquad Ans.$$

We may find $A$ by a moment summation about $C$ in the $YZ$ plane.

$$[\Sigma M_C = 0] \qquad 8(1657) = \frac{6}{7}A_Y + 14A_Y \quad\text{or}\quad A_Y = 892 \text{ lb}$$

By proportion

$$\left[\frac{F}{L} = \frac{F_Y}{y}\right] \qquad \frac{A}{10.05} = \frac{892}{7} \quad\text{or}\quad A = 1280 \text{ lb T} \qquad\qquad Ans.$$

A partial check on the computations can be made by a force summation.

$$[\Sigma F_Y = 0] \qquad\qquad 1657 = 765 + 892 \qquad\qquad\qquad Check$$

Note that the computations showed all force components to be positive. This indicated that the tensile-compressive assumptions were correct.

*Alternate Solution by the Moment Method Using Fig. 7-10:* Note that in three-dimensional structures we take moments about an axis rather than a point. We may, then, take moments about both a vertical and a horizontal axis through $A$. The resulting equations can be solved for $C_Y$ and $B_Y$. A moment summation about a *vertical* line through $A$ in Fig. 7-10 gives

$[\Sigma M_{A_Y} = 0]$ $\qquad\qquad 6(1200) = 4C_Y + 6\left(\dfrac{5}{12}B_Y\right)$

$$4C_Y + 2.5B_Y = 7200 \qquad\qquad\qquad (c)$$

Next, a moment summation about a line *parallel* to the Z axis through $A$ gives

$[\Sigma M_{A_Z} = 0]$

$$7(1200) + 4C_Y + 5\left(\dfrac{5}{12}B_Y\right) - 9B_Y = 0 \quad \text{or} \quad 4C_Y = 6.92B_Y - 8400 \quad (d)$$

Substituting Eq. $(d)$ into $(c)$ gives

$$6.92B_Y - 8400 + 2.5B_Y = 7200$$

This gives

$$B_Y = \frac{15,600}{9.42} = 1657 \text{ lb}$$

Substituting this value of $B_Y$ in $(d)$ gives

$$4C_Y = 6.92(1657) - 8400 \quad \text{or} \quad C_Y = \frac{3060}{4} = 765 \text{ lb}$$

Summing moments about a vertical line through $B$ gives

$[\Sigma M_{B_Y} = 0]$ $\qquad\qquad 5(765) + 6\left(\dfrac{4}{7}A_Y\right) - 9\left(\dfrac{6}{7}A_Y\right) = 0$

$$3825 + 3.43A_Y - 7.73A_Y = 0 \quad \text{or} \quad A_Y = \frac{3825}{4.3} = 892 \text{ lb}$$

These are the same force components found in the first solution and so would yield $A = 1280$ lb $T$; $B = 1792$ lb $C$; and $C = 1080$ lb $T$ by proportion.

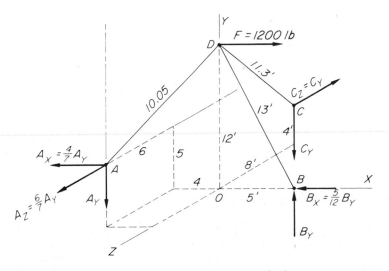

**Fig. 7-10   Stresses in space frame by alternate method.**

*Solution by the Force Method:* Summing forces in the $X$, $Y$, and $Z$ directions in Fig. 7-10 yields

$$[\Sigma F_X = 0] \qquad \frac{4}{7}A_Y + \frac{5}{12}B_Y = 1200 \qquad (e)$$

$$[\Sigma F_Y = 0] \qquad A_Y + C_Y - B_Y = 0 \qquad (f)$$

$$[\Sigma F_Z = 0] \qquad C_Y = \frac{6}{7}A_Y \qquad (g)$$

Equations $(e)$ to $(g)$ may be solved simultaneously For $A_Y$, $B_Y$, and $C_Y$. The stresses $A$, $B$, and $C$ may then be found by proportion.

## PROBLEMS

**7-7.** Determine the stresses in members $A$, $B$, and $C$ of the frame shown in Fig. 7-11 produced by the vertical load of 32 kips. The $X$ and $Z$ axes are horizontal.

*Ans.* $A = 40$ kips T; $B = C = 13.42$ kips C

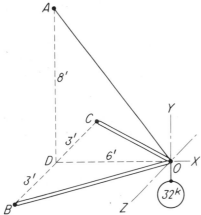

**Fig. 7-11   Prob. 7-7.**          **Fig. 7-12   Prob. 7-8.**

**7-8.** In members $A$, $B$, and $C$ of the frame shown in Fig. 7-12, find the stresses produced by the horizontal force of 3200 N. The $X$ and $Z$ axes are horizontal.

**7-9.** Solve for the stresses in members $A$, $B$, and $C$ of the frame shown in Fig. 7-13. (HINT: Solve by summations of moments in the $XY$ and $XZ$ planes.) The load is vertical.

*Ans.* $A = 26.7$ kips T; $B = 14.3$ kips T; $C = 45$ kips C

**7-10.** The frame shown in Fig. 7-14 supports a fixed load of 30 kips at $F$. Find the stresses in legs $A$ and $B$ when boom $DF$ and cable $OF$ are in the vertical plane $COFDC$. Determine also the stress in the mast $OD$. (SUGGESTIONS: Consider the entire frame in the $XY$ plane. Solve for the $X$ and $Y$ components at $AB$ and $D$. Then consider the frame in the $XZ$ plane.) Finally, isolate joint $D$ to find $OD$.

*Ans.* $A = B = 24.2$ kips T; $OD = 24$ kips C

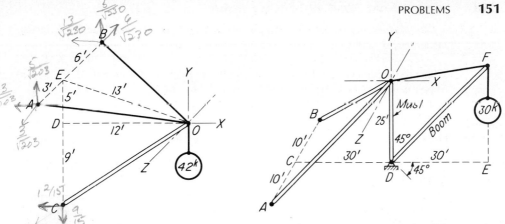

Fig. 7-13   Prob 7-9.

Fig. 7-14   Probs. 7-10 through 7-12.

**7-11.** The maximum stress in leg $A$ of the frame shown in Fig. 7-14 occurs when the vertical plane containing the boom is perpendicular to the vertical plane containing leg $B$. Solve for this maximum stress in $A$.

*Ans. A = 76.5 kips*

**7-12.** In Fig. 7-14 the boom of the frame shown can be swung horizontally 90° to either side of the vertical plane *COFED*. (*a*) Determine the stresses in legs $A$ and $B$ when the boom has been swung into a position which is 45° toward the observer. Will motion of the boom in a vertical plane affect the stresses (*b*) in the mast and (*c*) in the two legs? Will motion in a horizontal plane affect the stresses (*d*) in the mast and (*e*) in the two legs?

**7-13.** The frame shown in Fig. 7-15 is subjected to a force applied at $O$ having a vertical component of 12 kN and a horizontal component of 6 kN as shown. Find the stress in each of members $A$, $B$, and $C$.

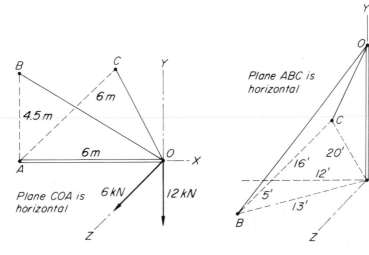

Fig. 7-15   Probs. 7-13 and 7-14.

Fig. 7-16   Probs. 7-15 and 7-16.

**7-14.** Interchange the 6- and 12-kN loads shown in Fig. 7-15 and solve for the stresses $A$, $B$, and $C$.

**7-15.** Member $OA$ in Fig. 7-16 is vertical. Find the stresses in members $A$, $B$, and $C$ due to the horizontal load of 3200 lb applied at $O$ as shown, parallel to the $X$ axis. Supports $A$, $B$, and $C$ all lie in the horizontal $XZ$ plane.

**7-16.** Determine the stresses in the frame shown in Fig. 7-16 if the 3200-lb load is tilted down at 30° to the $X$ direction.

Ans. $A = 6220$ lb C; $B = 4200$ lb T; $C = 1560$ lb T

**7-17.** Raise point $B$ in Fig. 7-4 vertically 4 ft and then solve for the stresses in the members.

Ans. $A = 21.5$ kN T; $B = 15.9$ kN T; $C = 42.5$ kN C

**SUMMARY**(By Article Number) ⎯⎯⎯⎯⎯⎯⎯⎯⎯⎯⎯⎯⎯⎯⎯⎯⎯⎯⎯⎯⎯

**7-2. The axial stress $F$ in a member of a three-dimensional space frame** equals the square root of the sum of the squares of its three rectangular $X$, $Y$, and $Z$ components. That is,

$$F = \sqrt{F_X^2 + F_Y^2 + F_Z^2} \tag{7-1}$$

If one of the three rectangular components of a stress $F$ is known, the other two components may readily be computed from the following relationship:

$$\frac{F_X}{x} = \frac{F_Y}{y} = \frac{F_Z}{z} \tag{7-2}$$

in which $x$, $y$, and $z$ are the space coordinates of the far end of the stressed member with respect to its near end.

**7-3.** Three unknown **axial stresses in members of space frames** may be determined by the **force method,** based on *algebraic summations of forces parallel to three rectangular $X$, $Y$, and $Z$ axes,* in which the three equations of equilibrium are

$$\Sigma F_X = 0 \qquad \Sigma F_Y = 0 \qquad and \qquad \Sigma F_Z = 0 \tag{7-5}$$

or by the **moment method,** in which *algebraic summations of moments are made in two planes at right angles to each other.* Since only two equations need be solved simultaneously, the moment method is preferred. An important fact is that **a force produces no reaction in a plane that is perpendicular to its line of action.**

**REVIEW PROBLEMS** ⎯⎯⎯⎯⎯⎯⎯⎯⎯⎯⎯⎯⎯⎯⎯⎯⎯⎯⎯⎯⎯⎯⎯⎯⎯

**7-18.** In Fig. 7-1, let $x = 12$, $y = 4$, and $z = 3$. Compute the *components* $F_X$, $F_Y$, and $F_Z$ when $F = 260$ lb.

Ans. $F_X = 240$ lb; $F_Y = 80$ lb; $F_Z = 60$ lb

**7-19.** In Fig. 7-1, let $x = 8$, $y = 9$, and $z = 12$. Compute the $x$ and $z$ components of the force if the $y$ component is 360 N. Also compute the magnitude of the force.

Ans. $F_X = 320$ N; $F_Z = 480$ N; $F = 680$ N

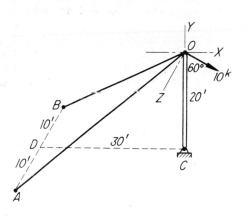

**Fig. 7-17  Prob. 7-21.**

**Fig. 7-18  Prob. 7-22.**

**7-20.** The line $OA$ in Fig. 7-2 represents a structural member of a space frame. The $x$, $y$, and $z$ coordinates of its far end $A$, measured from $O$, are, respectively, 12, 6, and 4. The $Y$ component of the stress in the member has been calculated and is 300 lb. Compute the *stress S* in the member.

*Ans. S = 700 lb*

**7-21.** Using the *moment method*, determine the stresses in the guy wires $OA$ and $OB$, and in the mast $OC$, of the frame shown in Fig. 7-17. The 10-kip force lies in the $XY$ plane and is tilted downward at an angle of 30° measured from the $X$ axis.

*Ans. OA = OB = 5.4 kips T; OC = 10.77 kips C*

**7-22.** Find the *stresses* in legs $A$ and $B$ and in the backstay $C$, of the frame shown in Fig. 7-18.

*Ans. A = B = 21 kips C; C = 21.63 kips T*

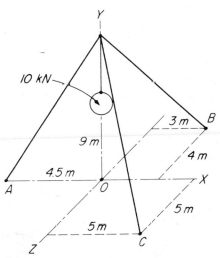

**Fig. 7-19  Prob. 7-23.**

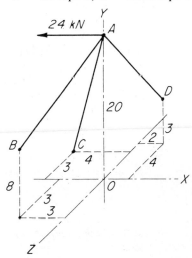

**Fig. 7-20  Prob. 7-24.**

**7-23.** The tripod shown in Fig. 7-19 supports a vertical load of 10 kN. Find the stresses in the three legs.

*Ans. A* = 5.2 kN C; *B* = 3.4 kN C; *C* = 3.0 kN C

**7-24.** The tripod shown in Fig. 7-20 is subjected to a horizontal load applied at *A* and parallel to the *X* axis. Find the stresses in the three legs.

*Ans. AB* = 17.6 kN T; *AC* = 97.7 kN C; *AD* = 80.8 kN

**7-25.** Let the 1200-lb load in Fig. 7-8 be in the *Z* direction. Solve for the stresses.

*Ans. A* = 534 lb C; *B* = 553 lb C; *C* = 1248 lb T

## REVIEW QUESTIONS

**7-1.** What is the function of space coordinates? From which point are they always measured?

**7-2.** How can the magnitude of a force be found if its $x$, $y$, and $z$ components are known?

**7-3.** What important relationship exists between the $x$, $y$, and $z$ space coordinates of the far end of a two-force member (with respect to its near end) and the $X$, $Y$, and $Z$ stress components within that member?

**7-4.** Outline briefly the essential steps in determining stresses in the three members of a space frame (*a*) by the force method and (*b*) by the moment method.

**7-5.** In the moment method, what forces and components of forces are considered (*a*) in the *XY* plane, (*b*) in the *YZ* plane, and (*c*) in the *XZ* plane?

# Chapter 8
# Noncoplanar,
# Nonconcurrent
# Force Systems

## 8-1 INTRODUCTION

The noncoplanar, nonconcurrent force system is most commonly encountered in the more advanced phases of design of structures and machines and is therefore treated only briefly here. The resultant of such a system may be a force, or a moment, or a force and a moment. The necessity of determining such a resultant is seldom encountered in ordinary engineering practice. The usual problem is that of determining the unknown reacting forces of a system in equilibrium.

## 8-2 EQUILIBRIUM OF NONCOPLANAR, NONCONCURRENT FORCE SYSTEMS

Unknown forces are determined by solving for their $X$, $Y$, and $Z$ components in three mutually perpendicular planes, usually the vertical $XY$ and $YZ$ planes and the horizontal $XZ$ plane. When signs of space coordinates and forces are used, they are positive in the directions indicated in Fig. 8-1. Much of the difficulty usually encountered in treating these force systems is overcome if it is clearly understood that *a force produces no reaction in a plane that is*

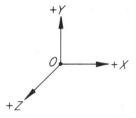

Fig. 8-1   Positive direction of space coordinates and forces.

155

*perpendicular to its line of action.* This point is clearly illustrated in the following example.

## ILLUSTRATIVE PROBLEM

**8-1.** A rectangular shelf, 12 by 20 in., of uniform thickness and weighing 80 lb, is supported in a horizontal position by *loose-fitting hinges* at $A$ and $B$ and by the cord $CD$, as is shown in Fig. 8-2. If the weight $W$ of the shelf is considered to be concentrated at its center, determine the magnitudes of the reacting forces at supports $A$, $B$, and $C$. Let it be assumed that all thrust parallel to the $Z$ axis is resisted by hinge $A$.

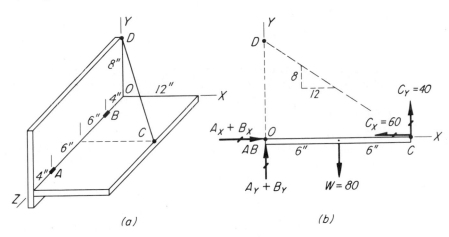

(a)                                (b)

**Fig. 8-2   Determination of reactions. Nonconcurrent forces in space.** *(a)* **Space diagram.** *(b)* **Free-body diagram, *XY* plane.**

*Solution:* The reaction at $C$ is clearly the stress in the cord $CD$. It has three components, $C_X$, $C_Y$, and $C_Z$. The magnitudes of these components are proportional to the $x$, $y$, and $z$ space coordinates of point $D$ which, with respect to point $C$, are $-12$, $+8$, and $-10$. Since only the magnitude of $C$ is desired, the signs of these space coordinates have no significance.

In Fig. 8-2b, the shelf has been isolated as a free body, and the $X$ and $Y$ components acting at $A$, $B$, and $C$ are shown projected onto the vertical $XY$ plane. In this plane, the $Z$ components are zero. Since $B$ is directly behind $A$, components $A_X$ and $B_X$ overlie each other, as do $A_Y$ and $B_Y$. $W$ is applied at the center of the shelf, and $C_X$ and $C_Y$ at $C$. $CD$ is clearly a tensile force. With $AB$ as the moment center, we may now find $C_Y$. That is,

$$[\Sigma M_{AB} = 0] \qquad\qquad 12C_Y = (80)(6) = 480$$
and
$$C_Y = 40 \text{ lb T}$$

Then, by proportion, when $x = 12$, $z = 10$, and $y = 8$, we have

$$\frac{C_X}{12} = \frac{C_Z}{10} = \frac{C_Y}{8} = \frac{40}{8} = 5$$

or$$C_X = (5)(12) = 60 \text{ lb}$$
and$$C_Z = (5)(10) = 50 \text{ lb}$$

Next, we project the forces into the vertical $YZ$ plane, as in Fig. 8-3$a$. In this plane, the $X$ components are zero. Both $C_Y$ and $C_Z$ are known forces. We may now find $A_Y$ by moments about $B$, and $B_Y$ by moments about $A$. Both are clearly upward-acting forces. Since we assumed originally that hinge $A$ takes all reaction in the $Z$ direction, $B_Z$ is zero and $A_Z$ is clearly opposite to $C_Z$. Then, by moments, we obtain

$$[\Sigma M_B = 0] \quad 12A_Y + (40)(6) = (80)(6) \quad \text{and} \quad A_Y = 20 \text{ lb} \uparrow$$
$$[\Sigma M_A = 0] \quad 12B_Y + (40)(6) = (80)(6) \quad \text{and} \quad B_Y = 20 \text{ lb} \uparrow$$

Then, to find $A_Z$,

$$[\Sigma F_Z = 0] \qquad\qquad A_Z = 50 \text{ lb} \leftarrow$$

Finally, to determine the $X$ components, we project the forces into the horizontal $XZ$ plane, as is shown in Fig. 8-3$b$. Both $C_X$ and $C_Z$ are known. By moments about $A$, we find $B_X$, whose sense is clearly as shown. By moments about $B$, we find $A_X$, whose sense apparently is as shown. That is,

$$[\Sigma M_A = 0] \quad 12B_X = (50)(12) + (60)(6) = 960 \quad \text{and} \quad B_X = 80 \text{ lb} \downarrow$$
$$[\Sigma M_B = 0] \quad 12A_X + (60)(6) = (50)(12) \quad \text{and} \quad A_X = 20 \text{ lb} \uparrow$$

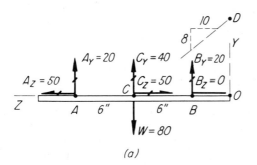

(a)

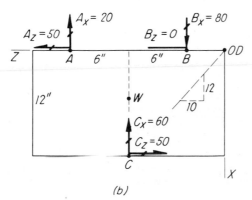

(b)

**Fig. 8-3   Free-body diagrams.** *(a) YZ plane. (b) XZ plane.*

Since all components of forces $A$, $B$, and $C$ are now known, we find these forces as follows:

$$A = \sqrt{A_X^2 + A_Y^2 + A_Z^2} = \sqrt{(20)^2 + (20)^2 + (50)^2} = 57.4 \text{ lb} \quad Ans.$$
$$B = \sqrt{B_X^2 + B_Y^2 + B_Z^2} = \sqrt{(80)^2 + (20)^2 + 0} = 82.5 \text{ lb} \quad Ans.$$
$$C = \sqrt{C_X^2 + C_Y^2 + C_Z^2} = \sqrt{(60)^2 + (40)^2 + (50)^2} = 87.8 \text{ lb} \quad Ans.$$

To emphasize again the importance to students of *visual equilibrium* in their work, all numerical values of components have been recorded in the several free-body diagrams. A careful study of the solution will, of course, indicate which forces in any free-body diagram are originally known and which forces must be solved for.

## PROBLEMS

**8-2.** Calculate the tension in the cord $AB$ and the $X$ and $Y$ components of the hinge forces at $C$ and $D$, shown in Fig. 8-4. The plate weighs 500 N and is subjected to a 150-N force at $E$.

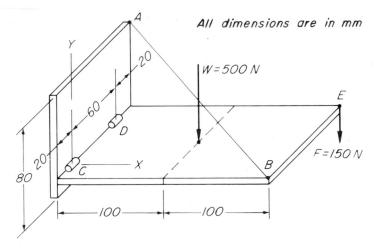

**Fig. 8-4    Prob. 8-2.**

**8-3.** A truck tailgate is rectangular, weighs 90 lb, is supported by loose-fitting hinges at $A$ and $B$ as shown in Fig. 8-5, and is held in the position shown by the chain $CD$. Compute the tensile force in the chain. Assume that the weight acts through $F$.

*Ans.* $T = 38.4$ lb

**8-4.** Figure 8-6 shows a circular water tank supported by a structure which is symmetrical in all planes about a vertical centerline. The tank has an inside diameter of 10 ft and a net inside depth of 12 ft; it weighs 5100 lb. Compute the total weight $W$ of the tank when filled with water (62.5 pcf), and the vertical forces *acting on the supporting truss* at $B$, $C$, $F$, and $G$. Disregard horizontal wind load $H$. (NOTE: Although no actual roller would exist at supports $H$ and $E$, freedom of lateral motion at these supports must be assumed, for the purpose of stress analysis.)

*Ans.* $W = 64,000$ lb; $R_B = R_C = R_F = R_G = 16,000$ lb↓

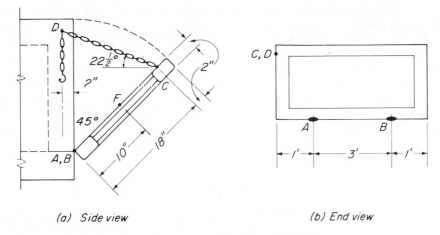

(a) Side view                    (b) End view

**Fig. 8-5   Probs. 8-3, 8-11, and 8-12. (a) Side view. (b) End view.**

**8-5.** Determine analytically the stresses in the sloping corner post $GH$, of the space frame shown in Fig. 8-6 and in the horizontal members $AH$, $HE$, $BG$, and $GF$ produced by the vertical load at $G$ only. (NOTE: See answer to Prob. 8.4.) All diagonal braces are inactive when only vertical loads are considered.

*Ans.* $GH = 16{,}970$ lb C; $AH = HE = 4000$ lb T; $BG = GF = 4000$ lb C

**8-6.** Calculate the vertical and horizontal forces produced *on the truss* at $B$, $C$, $F$, and $G$ by a maximum horizontal wind load $H$ of 2400 lb against the side of the tank, as shown in Fig. 8-6. In this problem, disregard the weight $W$. The usual manner of fastening the floor-system timbers with steel driftpins makes it reasonable to assume that one-fourth of $H$, or 600 lb, is resisted horizontally at each of supports $B$, $C$, $F$, and $G$. (SUGGESTIONS: Draw free-body diagrams of both tank and truss $ABGH$, separating tank from truss vertically by 2 in. Then solve for the reaction components *under the tank* and show the magnitude and sense of each force. One-half of the value of $H$ may be used to obtain the reaction components at $B$ and $G$ only.

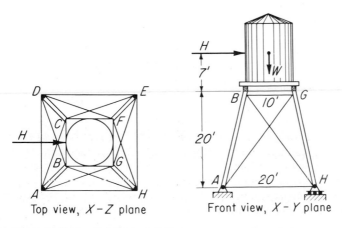

Top view, $X-Z$ plane          Front view, $X-Y$ plane

**Fig. 8-6   Probs. 8-4 through 8-8.**

These components may then be transferred to joints $B$ and $G$ of the truss *with opposite senses*.)

$$Ans. \ B_Y = C_Y = 840 \text{ lb↑}; \ F_Y = G_Y = 840 \text{ lb↓}; \ B_X = C_X = F_X = G_X$$
$$= 600 \text{ lb→}$$

**8-7.** Using the wind load forces on truss $ABGH$ (Fig. 8-6), as found in Prob. 8-6, compute the $X$ and $Y$ reaction components at supports $A$ and $H$. (Draw a complete front-view free-body diagram of the truss.) Then compute the stresses in members $GH$, $AH$, $HE$, $BG$, $GF$, and $AG$. Diagonal braces are slender steel rods and will resist tension only. Hence, brace $BH$ is now inactive. (Draw front-view and top-view free-body diagrams of each of joints $H$ and $G$.)

$$Ans. \ A_Y = 1620 \text{ lb↓}; \ H_Y = 1620 \text{ lb↑}; \ A_X = 1200 \text{ lb←}; \ GH = 1718 \text{ lb C}; \ AH$$
$$= HE = 405 \text{ lb T}; \ BG = 390 \text{ lb C}; \ GF = 210 \text{ lb C}; \ AG = 994 \text{ lb T}$$

**8-8.** Consider the stresses found in Probs. 8-5 and 8-7; what total stresses (dead load plus wind load) should be used in designing one corner post ($GH$), one upper horizontal member ($BG$), one lower horizontal member ($AH$), and one diagonal brace ($AG$)? (NOTE: The *design stress* for any member is *the sum of the stresses of similar type*, tensile *or* compressive, produced by all the load systems.)

$$Ans. \ GH = 18,690 \text{ lb C}; \ BG = 4390 \text{ lb C}; \ AH = 4405 \text{ lb T}; \ AG = 994 \text{ lb T}$$

## SUMMARY

**8-2.** The $X$, $Y$, and $Z$ *components* of a force vector in space are proportional to the $x$, $y$, and $z$ *coordinates* of its far end with respect to its near end (its point of application).

An important fact is that *a force produces no reaction in a plane that is perpendicular to its line of action.*

The unknown forces of a noncoplanar, nonconcurrent system of forces in equilibrium are determined by moment and force summations using the components of the forces.

## REVIEW PROBLEMS

**8-9.** In Fig. 8-7 is shown a solid rectangular block 300 mm long, 150 mm wide, and 100 mm high. A force $F$ of 1000 N is applied at $D$ and acts in the direction $DE$. Compute the *components* $F_X$, $F_Y$, and $F_Z$ of this force. Supports $A$, $B$, and $C$ lie in a horizontal plane.

$$Ans. \ F_X = 743 \text{ N}; \ F_Y = 1000 \text{ N}; \ F_Z = 557 \text{ N}$$

**8-10.** The solid rectangular block shown in Fig. 8-7 weighs 2000N and is supported at three points, $A$, $B$, and $C$, lying in one horizontal plane. The ball hinge at $A$ allows rotation but prevents displacement. The ball-point supports at $B$ and $C$ prevent vertical displacement only. Compute the *reaction components* at $A$, $B$, and $C$ caused by the weight of the block and the 1000–N force $F$.

$$Ans. \ A_X = 743 \text{ N}; \ A_Y = 1000 \text{ N}; \ A_Z = 557 \text{ N}; \ B = 124 \text{ N}; \ C = 1248 \text{ N}$$

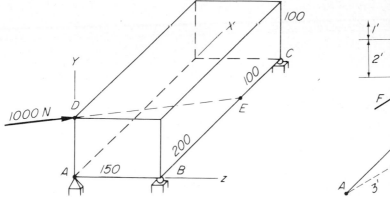

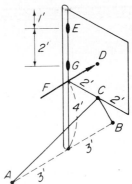

**Fig. 8-7   Probs. 8-9 and 8-10.**                 **Fig. 8-8   Prob. 8-13.**

**8-11.** Refer to Prob. 8-3. Move point $D$ in Fig. 8-5$a$ 6 in. toward the observer, and solve.

*Ans.* $T = 40.8$ lb

**8-12.** Solve Prob. 8-3 if the tailgate is lowered into a horizontal position.

**8-13.** The sign shown in Fig. 8-8 is 4 ft square and is supported by hinges $E$ and $G$ and the guy wires $AC$ and $BC$. Calculate the tension in wire $AC$ due to the lateral wind force $F$ of 400 lb. Note that wire $BC$ will be slack when the wind blows from the left.

**8-14.** Solve Prob. 8-1 if the lower end of the cord is attached to the shelf opposite to hinge $A$.

*Ans.* $A = 82.5$ lb; $B = 89.5$ lb; $CD = 107.75$ lb

## REVIEW QUESTIONS

**8-1.** Give the equation for obtaining the magnitude of a force $F$ in terms of its three components $F_X$, $F_Y$, and $F_Z$.

**8-2.** Explain briefly the moment method of determining the unknown forces of a noncoplanar, nonconcurrent force system in equilibrium.

# Chapter 9
# Friction

## 9-1 INTRODUCTION

When two objects are in contact under pressure, motion or attempted motion of either object with respect to the other in a direction *parallel to the contacting surfaces* will be resisted. This resistance is called *frictional resistance* or simply **friction.** The total frictional resistance between two contacting surfaces is called the **friction force,** since force is required to overcome it. It may be determined experimentally by measuring the force required to overcome it. The friction force always acts parallel to the two surfaces in contact, and its sense is such as to oppose any motion, or attempted motion, in that direction of either body with respect to the other.

Friction is generally thought of as being an adversary of mankind. This impression, of course, is not entirely correct; friction variously acts as friend and foe. But for friction between its wheels and the roadbed, an automobile could not start. On the other hand, friction consumes all the energy put out by the motor, in driving at uniform speed along a level road. Again, the stopping of the automobile depends on the frictional resistance developed by its brakes and between the wheels and the roadbed. When the surfaces in contact are at rest with respect to each other, this resistance is called **static friction;** when they are in motion with respect to each other, it is called **kinetic friction.**

Let the body shown in Fig. 9-1$a$ be resting on a level surface. The total pressure on the supporting surface equals the force $N$, the reaction of the supporting surface, which is called the **normal pressure.** Let $P$ be the force required to place the body on the verge of sliding (motion impending relative to its supporting surface), and let $F$ be the friction force resisting the tendency of $P$ to slide the body. Since frictional resistance is essentially a reaction, it cannot exist except when induced by a force producing or tending to produce

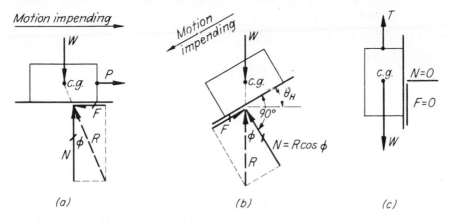

**Fig. 9-1** Friction forces and impending motion. *(a)* $\phi$ is angle of friction. *(b)* $\theta_H$ is angle of inclination. *(c)* $\theta_H = 90°$, $N = 0$.

motion. Evidently, therefore, $F$ equals $P$ for all values of $P$ from zero to the highest $F$ can reach without motion occurring. This highest value of $F$ for any given condition is called the **limiting friction.** The *reaction R* is the resultant of $N$ and $F$.

In Fig. 9-1$b$, a body of weight $W$ is resting on an inclined surface. Evidently the normal pressure $N$ varies with the *angle of inclination* $\theta_H$ of the supporting surface, being maximum when $\theta_H$ is zero and minimum when $\theta_H$ is 90 degrees, as is shown in Fig. 9-1$c$.

## 9-2 COEFFICIENT OF FRICTION, ANGLE OF FRICTION, AND ANGLE OF REPOSE

Experiments have shown that, if a body of known weight, as in Fig. 9-1$a$, is acted on by a force $P$ as shown, the magnitude of the friction force $F$ when motion impends, called the *limiting friction,* depends on (1) the normal pressure $N$ between the contacting surfaces, (2) the kinds of materials, and (3) the roughness of these surfaces. Since $N$ can be computed and $F$ may be determined experimentally, the ratio of the limiting friction to the normal pressure, $F/N$, for any two surfaces of like or unlike materials is readily found. This ratio is called the **coefficient of static friction** and is denoted by the symbol $f$ (Table 9-1). In Fig. 9-1$a$ and $b$ the angle $\phi$ between the reaction $R$ and the normal pressure $N$, *when motion is impending,* is called the **angle of friction,** denoted by the Greek letter $\phi$ (phi). Its tangent is seen to be $F/N$, which ratio was just defined as the coefficient of static friction. Therefore,

$$\text{Coefficient of static friction: } f = \frac{F}{N} = \tan \phi \qquad \textbf{(9-1)}$$

Table 9-1 AVERAGE
COEFFICIENTS OF STATIC
FRICTION FOR DRY
SURFACES

| | |
|---|---|
| Wood on wood | 0.30 – 0.60 |
| Wood on metal | 0.20 – 0.60 |
| Metal on metal | 0.15 – 0.30 |
| Leather on wood | 0.30 – 0.50 |
| Leather on metal | 0.30 – 0.60 |
| Stone on concrete | 0.50 – 0.70 |
| Rubber on concrete | 0.60 – 0.80 |

The coefficient of static friction $f$ at the contacting surfaces of bodies of different materials may also be determined experimentally as follows. A body is placed on an incline, as in Fig. 9-1$b$. The incline is then tipped until the body is on the verge of sliding. *The tangent of the angle* $\theta_H$, *when* *sliding impends,* *is the desired coefficient of static friction.* The angle of inclination $\theta_H$, when motion impends, is called the **angle of repose.**

These coefficients are for *dry surfaces* and are, of course, rather approximate because of the varying degrees of roughness of the contacting surfaces. Coefficients of kinetic (sliding) friction at low velocities for dry surfaces are about equal to those for static friction but tend to decrease at higher velocities. Frictional resistance between lubricated surfaces differs materially from that between dry surfaces, as is pointed out in Art. 9-3.

## 9-3  LAWS OF FRICTION

Experiments carried on by scientists and engineers for more than 150 years have led to the following general conclusions, known as the **laws of friction.**

### Dry Surfaces

1. The friction force *is dependent* on the kinds of the materials and the degree of roughness of the two surfaces in contact.
2. The maximum friction force (limiting friction) which can be developed *is dependent* on the normal pressure between the contacting surfaces and is proportional to it.
3. The friction force *is not dependent* on the areas of the surfaces in contact.
4. The friction force *is dependent* on the velocity (to a limited extent) and *decreases* as the velocity increases.

### Lubricated Surfaces

1. The friction force *is not dependent* on the kinds of materials nor (within limits) on the degree of roughness of the contacting surfaces.
2. The friction force *is not dependent* on the normal pressure between the contacting surfaces.

3. The friction force *is dependent* on the areas of the surfaces in contact.
4. The friction force *is dependent* on the velocity and *increases* as the velocity increases.

5. The friction force *is not dependent* on temperature to any important degree.

5. The friction force *is dependent* on temperature to a considerable degree.

These laws indicate clearly that the frictional effects produced on lubricated surfaces are almost the opposite of those produced on dry surfaces. This reversal of effect is due to the presence of a film of lubricant *between* the two surfaces, which therefore are no longer in contact, so that the frictional resistance is then largely dependent on the properties of the lubricant itself.

## 9-4 FRICTION PROBLEMS

Numerous problems involving friction, and relating to both static and kinetic friction, arise in engineering practice. This subject, therefore, deserves careful study.

Few solid bodies remain in continuous motion. The problem of setting into motion a body that is at rest is therefore a common one. In such problems, we are naturally interested in *the total frictional resistance at the moment the body is on the verge of motion.* These, then, are problems in *static friction when motion is impending.* Graphical solutions are especially adapted to problems involving static friction, as is illustrated in the following examples. These examples, and the problems that follow, indicate the importance of a good understanding of static friction.

### ILLUSTRATIVE PROBLEMS _____

**9-1.** In Fig. 9-2a is shown a block $B$ weighing 1000 N which is being moved slowly to the right by the action of the vertical force $P$ on the wedge $A$. Determine the force $P$ required barely to move the block, if the coefficient of friction $f$ is 0.3 for all contacting surfaces. (The friction, and hence $P$, may be considered to be the same for impending or very slow motion.) Weight of wedge may be neglected.

*Graphical Solution:* When the free-body diagrams of the wedge and the block are drawn, as in Figs. 9-2b and 9-2c, we recognize that the problem must be solved in two parts, beginning with the object on which a *known* force acts, which here is block $B$. Let the reacting forces at the contacting surfaces be $R$, $Q$, and $S$. Since only three forces act on each free body, they must be concurrent. By careful reasoning we may now determine the *proper directions* of these forces. Apparently, from Fig. 9-2a, block $B$ tends to move to the right. Hence reaction $S$ will be inclined to the left, so as to *oppose* the motion. Since the tangent of the angle of friction $\phi$ equals the coefficient of friction $f$, which is 0.3, the true direction of $S$ may be shown by laying off to scale 10 units *normal* to the surface of the body and 3 units *parallel* to it. Clearly, wedge $A$ in Fig. 9-2b tends to move downward; $R$ and $Q$ are therefore inclined upward. Of course, force $Q$ on block $B$ is equal, opposite, and collinear with $Q$ on wedge $A$. We are ready now to solve for $P$. Since only $W$ is known, the force diagram (Fig. 9-2c) begins with forces $W$, $S$, and $Q$, acting on block $B$. With respect to block $A$, the force

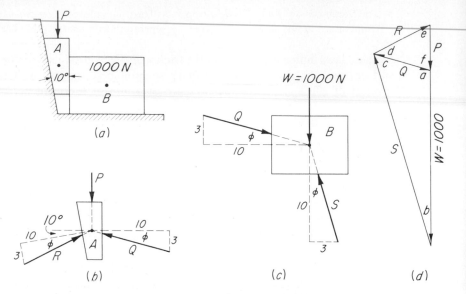

**Fig. 9-2   Friction forces. Wedge and block problem. (a) Wedge and block. (b) Wedge A as free body. (c) Block B as free body. (d) Force diagram.**

$Q$ is oppositely directed, and $P$ is solved for by completing the polygon of forces $Q$, $R$, and $P$ ($P$ scales 265 N).

*Trigonometric Solution:* For more precise results, we may obtain a trigonometric solution by determining the angles at $a$, $b$, $c$, $d$, $e$, and $f$ in the force diagram in Fig. 9-2$d$; then $P$ is found through two applications of the law of sines. Since $\tan \phi = 0.3$, $\phi = 16°42'$. Then $a = 90° + \phi = 106°42'$; $b = \phi = 16°42'$; $c = 90° - 2\phi = 56°36'$; $d = 10° + 2\phi = 43°24'$; $e = 90° - 10° - \phi = 63°18'$; $f = 90° - \phi = 73°18'$. As a check on these angles, $a + b + c = 180°$ and $d + e + f = 180°$. Solving first for $Q$ and then for $P$ by the law of sines, we have

$$\frac{Q}{\sin b} = \frac{W}{\sin c} \qquad \text{and} \qquad Q = \frac{W \sin b}{\sin c} = \frac{(1000)(0.287)}{0.835} = 344 \text{ N}$$

Also,

$$\frac{P}{\sin d} = \frac{Q}{\sin e} \qquad \text{and} \qquad P = \frac{Q \sin d}{\sin e} = \frac{(344)(0.687)}{0.893} = 265 \text{ N} \qquad Ans.$$

**9-2.** The ladder shown in Fig. 9-3$a$ is 12 ft long and is supported by a horizontal floor and a vertical wall as shown. The coefficient of friction $f$ at the wall is 0.2 and at the floor, 0.4. The weight $W$ of the ladder is 36 lb, considered as concentrated at its center $B$. The ladder supports a vertical load $P$ of 180 lb, at $C$. Determine the $V$ and $H$ reaction components at $A$ and $D$, and compute the lowest value of the angle $\alpha$ at which the ladder may be placed without slipping to the left.

*Analytical Solution:* The $V$ and $H$ reaction components at supports $A$ and $D$ are shown in the free-body diagram, Fig. 9-3$b$. Let $N_A$ be the *normal pressure,* or $V$ component, at $A$. The $H$ component at $A$ is the friction force, which, since the ladder is on the verge of slipping, is $f_A \cdot N_A$ or $0.4N_A$, and is directed to the right, since the

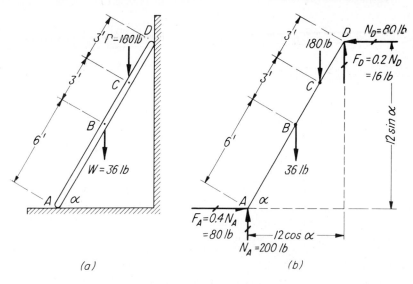

**Fig. 9-3** Friction forces. Ladder problem, analytical solution. (a) Space diagram. (b) Free-body diagram.

bottom of the ladder tends to slide to the left. At $D$, the horizontal normal pressure is $N_D$, and the friction force is $f_D \cdot N_D$ or $0.2N_D$, directed upward, since at $D$ the ladder tends to move downward. We now have two unknown forces, $N_A$ and $N_D$, which must be determined through two force equations, since $\alpha$ is also unknown. Therefore, $\alpha$ is found by a summation of moments. The force equations are

$$[\Sigma H = 0] \qquad\qquad N_D = 0.4N_A \qquad\qquad\qquad (a)$$
$$[\Sigma V = 0] \qquad\qquad 0.2N_D + N_A = 180 + 36 = 216 \qquad\qquad (b)$$

Substituting Eq. ($a$) in Eq. ($b$), we have

$$(0.2)(0.4N_A) + N_A = 216$$

or $\qquad\qquad\qquad 1.08N_A = 216 \qquad$ and $\qquad N_A = 200 \text{ lb} \qquad\qquad$ *Ans.*

Using this value in Eq. ($a$), we obtain

$$N_D = 0.4N_A = (0.4)(200) \qquad \text{and} \qquad N_D = 80 \text{ lb} \qquad\qquad Ans.$$

The friction forces then are

$$F_A = (0.4)(200) = 80 \text{ lb} \qquad \text{and} \qquad F_D = (0.2)(80) = 16 \text{ lb} \qquad Ans.$$

When these values are recorded on the free-body diagram, equilibrium is seen to exist. Finally, we determine $\alpha$ by moments about $A$. That is,

$$[\Sigma M_A = 0] \quad (36)(6 \cos \alpha) + (180)(9 \cos \alpha) = (80)(12 \sin \alpha) + (16)(12 \cos \alpha)$$

Hence,

$$216 \cos \alpha + 1620 \cos \alpha = 960 \sin \alpha + 192 \cos \alpha$$

or $\qquad\qquad\qquad\qquad 1644 \cos \alpha = 960 \sin \alpha$

Then,

$$\frac{\sin \alpha}{\cos \alpha} = \tan \alpha = \frac{1644}{960} = 1.712 \quad \text{and} \quad \alpha = 59°43' \qquad Ans.$$

*Graphical Solution:* In Fig. 9-4a, the ladder is drawn in its true length, and all forces acting on it are shown. The tangents of the angles of friction at $A$ and $D$ are 0.4 and 0.2, equaling the respective coefficients of friction. These tangents establish the directions of the reacting forces at $A$ and $D$. Consequently, we may now draw the complete force diagram, thus solving for $R_A$ and $R_D$. The $V$ and $H$ components are then scaled as indicated in Fig. 9-4b.

To determine the angle $\alpha$, we may resort to a simple trial-and-error type of solution. The reacting forces at $A$ and $D$, and the resultant of $W$ and $P$, form a system of three forces in equilibrium which, then, must be concurrent. The point of concurrence $O$ is selected at random. From $O$ the lines of action of the three forces are drawn in their true directions, as shown in Fig. 9-4c.

By the inverse-proportion method (Fig. 9-4a), we determine the point of application $E$ of the resultant of $W$ and $P$. Then, from the free-body diagram, we transfer the three points of application, $A$, $E$, and $D$, to the straight edge of a strip of paper. These three points must naturally lie on their respective lines of action. The strip may then be adjusted on the action line diagram until they do, as shown in Fig. 9-4c. If a line is then drawn through the points along the edge of the strip, $\alpha$ is the angle desired. This graphical solution is, of course, a special one applicable only to this and similar problems. However, the reasoning by which the solution is arrived at is so fundamental that its presentation here seems justified.

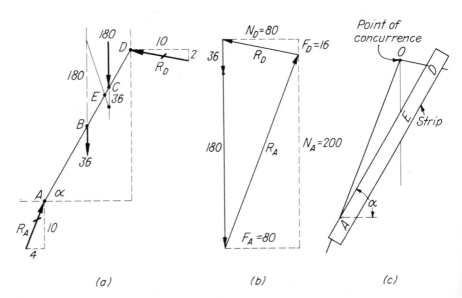

**Fig. 9-4   Friction forces. Ladder problem, graphical solution. (a) Free-body diagram. (b) Force diagram. (c) Action-line diagram.**

**PROBLEMS**

**9-3.** The right end of the inclined plane surface shown in Fig. 9-5 is slowly raised until the block slips when $\alpha$ is 19°. Determine the coefficient of friction $f$.

*Ans.* $f = 0.344$

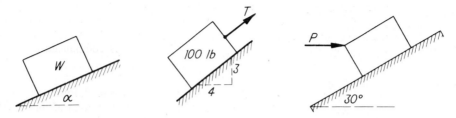

**Fig. 9-5   Probs. 9-3 and 9-4.   Fig. 9-6   Prob. 9-5.   Fig. 9-7   Prob. 9-6.**

**9-4.** If the coefficient of friction of cardboard on smooth sheet metal is 0.3, at what angle $\alpha$, as shown in Fig. 9-5, should a metal chute be placed so that cardboard boxes will slide on it at a slow uniform speed?

**9-5.** In Fig. 9-6, the coefficient of friction $f$ between the block and incline is 0.4. Determine the force $T$ which will cause impending motion (*a*) up the incline and (*b*) down the incline. Compute also the friction force $F$ (*c*) when $T = 40$ lb, (*d*) when $T = 60$ lb, and (*e*) when $T = 70$ lb. (NOTE: Separate free-body diagrams should be drawn for each of the five parts.)

*Ans.* (*a*) $T = 92$ lb; (*b*) $T = 28$ lb; (*c*) $F = 20$ lb ↗ ; (*d*) $F = 0$; (*e*) $F = 10$ lb ↙

**9-6.** Compute the horizontal force $P$ required to cause motion of the block shown in Fig. 9-7 to impend up the incline, if $W = 100$ N and $f = 0.25$.

**9-7.** A jackscrew has a threaded screw which turns in a threaded base (see Fig. 9-8 *a*). The load $P$ is applied to an arm at a distance $a$ from the centerline. For analysis,

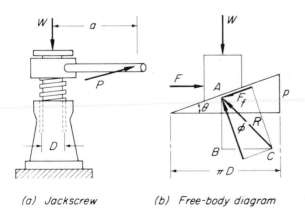

*(a) Jackscrew          (b) Free-body diagram*

**Fig. 9-8   Probs. 9-7, 9-8, and 9-38.** *(a)* **Jackscrew.** *(b)* **Free-body diagram.**

a single thread may be shown as an incline up which the load $W$ is pushed (see Fig. 9-8$b$). If the pitch $p$ is 1 in. and the pitch diameter $D$ is 3 in., calculate the required force $P$ to raise $W$ = 4 tons. Assume $a$ = 20 in. and $f$ = 0.12. HINT: Angle $CAB$ = $\phi + \theta$, $BC$ = $F$, and $AB$ = $W$.

*Ans. P* = 137.5 lb

**9-8.** Solve Prob. 9-7 if $W$ = 6000 N, $a$ = 250 mm, $f$ = 0.15, $D$ = 25 mm, and $p$ = 7.5 mm.

**9-9.** Block $A$ in Fig. 9-9 is acted on by a force $F$ of 100 lb. Assume $f$ to be 0.2 between blocks $A$ and $B$, and 0.3 between the other contacting surfaces. Determine the *minimum* horizontal force $P$ required to prevent block $B$ from slipping to the left. Disregard weights of blocks.

*Ans. P* = 128.4 lb

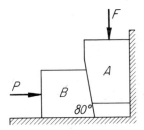

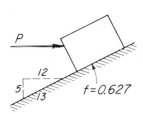

**Fig. 9-9   Probs. 9-9 and 9-39.**          **Fig. 9-10   Probs. 9-10 and 9-11.**

**9-10.** Compute the horizontal force $P$ required to cause motion of the block shown in Fig. 9-10 to impend down the incline if $W$ = 156 lb and $f$ is assumed to be 0.627.

*Ans. P* = 26 lb←

**9-11.** Solve Prob. 9-10 if motion is to impend up the incline.

*Ans. P* = 221 lb→

**9-12.** Block $A$ in Fig. 9-11 weighs 200 lb. Find the force $P$ required to lift block $A$ if $f$ = 0.3 for all surfaces of contact. Disregard the weight of the wedge.

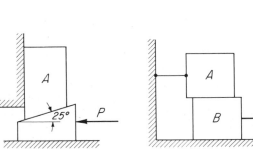

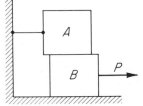

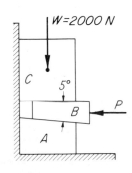

**Fig. 9-11   Prob. 9-12.**          **Fig. 9-12   Prob. 9-13.**          **Fig. 9-13   Prob. 9-14.**

**9-13.** Block $A$ in Fig. 9-12 weighs 200 N and block $B$ weighs 300 N. Find the force $P$ required to move block $B$. Assume the coefficient of friction $f$ for all surfaces is 0.3.

*Ans.* $P = 210$ N

**9-14.** In Fig. 9-13, the heavy rectangular stone block $C$ weighs 2000 N. It is being raised slightly by means of two wooden wedges $A$ and $B$ and by sledgehammer blows $P$ on wedge $B$. The angle between the contacting surfaces of the wedges is 5°. If $f = 0.3$ for all surfaces, compute the value of $P$ required to cause upward motion of the block to impend. Neglect weight of wedges.

**9-15.** Two wooden blocks $A$ and $C$, as shown in Fig. 9-14, are held in position through the pressures exerted on their sides by two steel plates held together by the tensile force in the bolt $B$. Assume that the centers of pressure between plates and blocks are as indicated by the dots. (*a*) Compute the maximum value which the load $P$ may reach without having either block slip vertically between the plates when the tensile force in the bolt is 500 lb and $f = 0.4$. (*b*) If $P$ is increased, on which block will slipping occur first?

*Ans.* (*a*) $P = 160$ lb; (*b*) on block $C$

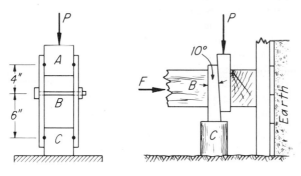

**Fig. 9-14   Probs. 9-15**     **Fig. 9-15   Prob. 9-16.**
**and 9-40.**

**9-16.** In Fig. 9-15 two opposing timber walls used as shoring in an excavated trench are held tightly against the earth by the horizontal strut $B$, which is wedged tightly by sledgehammer blows on one of two like wedges. Block $C$ is used merely to give vertical support to the left wedge while the right wedge receives the blow. The angle between the two faces of each wedge is 10°, and $f$ between all contacting surfaces is 0.4. Determine the horizontal compressive force $F$ in the strut when the force $P$ of a 500-lb blow drives the wedge slightly. Disregard any possible friction between wedge and block $C$. Consider reaction at block $C$ to be vertical. (Force scale: 1 in. = 200 lb.)

**9-17.** The steel bracket shown in Fig. 9-16 may slide freely on the vertical shaft. Let $A$ and $B$ be the points of application of the resultant pressures on the shaft, caused by the force $P$. Assume that $f = 0.2$ and that the friction forces at $A$ and $B$ are equal. (*a*) Compute the *minimum* distance $x$ from the edge of the shaft at which a force $P$ can be applied without having the bracket slide on the shaft, if its weight is neglected. (*b*) Find $x$ if $P$ is 40 lb and if the weight $W$ of the bracket is 20 lb, concentrated at its center of gravity (*cg*).

*Ans.* (*a*) $x = 14$ in.; (*b*) $x = 19$ in.

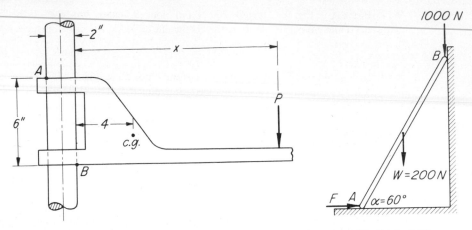

Fig. 9-16    Probs. 9-17, 9-18, and 9-41.          Fig. 9-17    Prob. 9-19.

**9-18.** Solve Prob. 9-17 graphically. (HINT: In the space diagram, consider only the points of application of the forces. Scales on 8½ by 11: 1 in. = 4 in.; 1 in. = 20 lb.)

**9-19.** A ladder 5 m long, weighing 200 N, is placed against a wall as shown in Fig. 9-17 and is prevented from slipping at the bottom by friction and by an additional horizontal force $F$. The coefficient of friction at the wall is 0.25 and at the floor, 0.35. (*a*) Find the minimum force $F$ that will prevent slipping, if the weight of the ladder is concentrated at its midpoint and if it supports at the top a person weighing 1 kN. (*b*) Find the minimum angle $\alpha$ at which the ladder may be placed to prevent slipping, if the force $F$ is removed.

*Ans.*  (*a*) $F = 183.6$ N; (*b*) $\alpha = 68.95°$

**9-20.** A ladder resting on a horizontal floor leans against a vertical wall at an angle of 60° with the floor. If $f$ is 0.25 at all surfaces and the weight of the ladder is neglected, what percentage of its length can a person ascend without causing the ladder to slip? Solve graphically.

*Ans.*  46.6 percent

## 9-5  BELT FRICTION

The transmission of power by belt drives and the braking effect obtained by band-type brakes depend on the frictional resistance developed between a flexible band and a cylindrical surface. Now consider power transmission by means of a flat belt (see Fig. 9-18*a*). A driving force is transmitted from the surface of the driving pulley to the belt because of the frictional resistance between the two surfaces. The tension in the belt will vary throughout its length of contact with the pulley (see Fig. 9-18*b*).

The normal force exerted by the pulley on a small increment of belt clearly depends on the tension in the belt (see Fig. 9-18*c*). If slipping is impending, the friction force between the belt and pulley will depend on the normal force

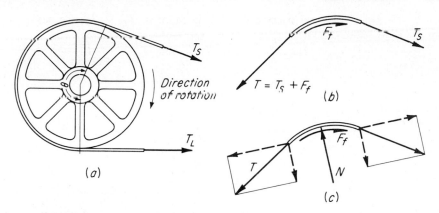

**Fig. 9-18  Belt friction. Flat belt.** *(a)* **Driving pulley.** *(b)* **Tension varies along the belt.**
*(c)* **Normal force varies with belt tension.**

and the coefficient of friction $f$ for the two surfaces. The friction force per unit
of length of belt will vary along the surface of contact because of the variation
in normal force, which increases in magnitude in the direction from $T_S$ to $T_L$.

It can be shown that, when slipping is impending,

$$\log_{10} \frac{\mathbf{T}_L}{\mathbf{T}_S} = \frac{f\theta}{132} \tag{9-2}$$

This equation may be written in the alternate forms

$$\log_{10} \mathbf{T}_L - \log_{10} \mathbf{T}_S = \frac{f\theta}{132} \quad \text{or} \quad \frac{\mathbf{T}_L}{\mathbf{T}_S} = e^{(f\theta/57.3)} \tag{9-3}$$

In the above equations

$T_L$ = larger belt tension
$T_S$ = smaller belt tension
$f$ = coefficient of friction
$\theta$ = angle of contact between belt and cylinder, degrees
$e$ — base of natural logarithms

Small belt drives are often of the V-belt type (see Fig. 9–19). The wedging
action of the belt tension increases the normal force on the belt and hence also
increases the maximum allowable friction force.

From Fig. 9-19$c$, it can be seen that the normal force $N$ is given by

$$N = \frac{F/2}{\sin \beta}$$

Since the friction force, when motion is impending, is given by $f \cdot N$,

$$\text{side } F_f = \frac{fF}{2 \sin \beta} \quad \text{or} \quad F_f = \frac{fF}{\sin \beta}$$

Since the total friction force $F_f$ is equal to $T_L - T_S$, we have

$$\log_{10}\frac{T_L}{T_S} = \log_{10}T_L - \log_{10}T_S = \frac{f\theta}{132\ \sin\ \beta}$$

or $\qquad\qquad\qquad \dfrac{T_L}{T_S} = e^{(f\theta/57.3\ \sin\beta)}$ $\qquad\qquad\qquad$ (9-4)

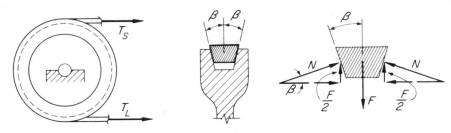

(a) Pulley and belt        (b) V-belt in groove      (c) Free-body diagram of V-belt

Fig. 9-19   V-belt pulley and belt. Wedging action. (a) Pulley and belt. (b) V belt in groove. (c) Free-body diagram of V belt.

## ILLUSTRATIVE PROBLEMS

**9-21.** The coefficient of friction between a certain kind of belting and a wooden pulley is to be determined by the apparatus shown in Fig. 9-20a. The spring indicates a belt tension of 180 lb in the horizontal portion of the belt when slipping is impending. Determine the coefficient of friction $f$.

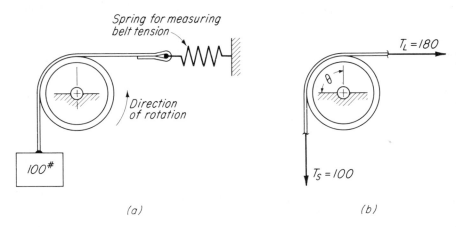

(a)                                                   (b)

Fig. 9-20   Determination of coefficient of friction between belt and pulley. (a) Apparatus. (b) Belt tensions.

*Solution:*  The ratio of the two belt tensions is known (see Fig. 9-20b). Equation (9-2) may be used to find the unknown coefficient of friction $f$. The angle of contact is 90°.

$$\left[ \log_{10} \frac{T_L}{T_S} = \frac{f(\theta)}{132} \right] \qquad\qquad \log_{10} \frac{180}{100} = \frac{f(90)}{132}$$

The log of 1.8 is 0.255; therefore

$$0.255(132) = 90(f) \qquad \text{or} \qquad f = 0.374 \qquad\qquad Ans.$$

**9-22.** What is the maximum weight that can be slowly lowered by a person who can exert a 500-N pull on a rope if the rope is wrapped 1¼ turns around a horizontal spar? Assume that the coefficient of friction between the surfaces is 0.3 (see Fig. 9-21).

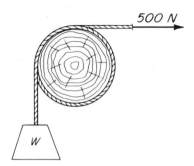

500 N

W

**Fig. 9-21   Heavy weight lowered by 1¼ turns on spar.**

*Solution:* The smaller tension in the rope is 500-N, the angle of contact is 1¼ turns or $(1.25)(360) = 450°$, and the friction coefficient $f$ is 0.3. The larger tension $T_L$ may be found from Eq. (9-3).

$$\left[ \frac{T_L}{T_S} = e^{(f\theta/57.3)} \right] \qquad \frac{T_L}{500} = e^{(0.3)(450)/57.3}$$

or
$$T_L = 500\ e^{2.36} = 5274\ N = W \qquad\qquad Ans.$$

**PROBLEMS** _____

**9-23.** Find the tension in the horizontal portion of the belt of Prob. 9-21 if the direction of rotation of the pulley is reversed. Assume that the coefficient of friction $f$ is 0.374 as found in Prob. 9-21.

$$Ans.\ T = 55.6\ lb$$

**9-24.** Solve Prob. 9-22 if the rope is given 2¼ turns on the spar.

**9-25.** A rope makes three turns around a capstan on a pier. What is the maximum permissible force on the loaded end of the rope if the pull on the other end of the rope is not to exceed 60 lb? Assume that the coefficient of friction $f$ is 0.35.

**9-26.** The coefficient of friction between the brake drum and the flexible brake band shown in Fig. 9-22 is 0.44. What is the maximum load $W$ that can be lowered slowly if the force $P$ is 25 lb?

$$Ans.\ W = 112.3\ lb$$

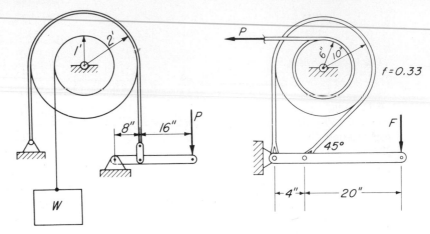

Fig. 9-22   Probs. 9-26 and 9-27.          Fig. 9-23   Probs. 9-28 and 9-44.

**9-27.** Solve Prob. 9-26 if the line supporting the weight comes off the opposite side of the drum. (The direction of the tendency for rotation will be reversed from that of Prob. 9-26.)

*Ans.* $W = 447$ lb

**9-28.** What force $F$ must be exerted on the brake arm shown in Fig. 9-23 to hold the drum against rotation if $P = 800$ lb?

*Ans.* $F = 77.8$ lb

**9-29.** A belt makes contact with a driving pulley through one half its circumference. If $T_S = 10$ lb and $f = 0.3$, calculate the value of $T_L$ for impending slippage if the belt is (*a*) flat, and (*b*) V type with $\beta = 20°$.

*Ans.* (*a*) $T_L = 25.7$ lb; (*b*) $T_L = 157.1$ lb

**9-30.** A V-belt pulley, $\beta = 15°$ and $f = 0.4$, has belt tensions $T_S = 10$ N and $T_L = 100$ N. What is the minimum value of $\theta$ that will prevent slipping?

## 9-6 ROLLING RESISTANCE

Our everyday experiences tell us that rolling resistance of objects is less than sliding resistance. This at least partially explains the common use of wheels on moving vehicles, and of ball and roller bearings in machinery. If sliding surfaces are separated by a lubricant, the molecules of the lubricant may act as small "balls" between the surfaces and thus reduce the frictional resistance.

Consider a wheel on a "rigid" surface as shown in Fig. 9-24. We wish to determine the force $P$ required to maintain a small constant velocity to the right. A moment summation about point $O$ will show that force $F$ must equal zero. Therefore, since $\Sigma H = 0$, the force $P$ must equal zero. Evidently, once

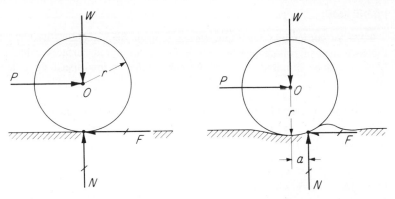

Fig. 9-24   "Rigid" surface.          Fig. 9-25   Deformation under wheel.

started, the wheel would roll forever on such a surface. We know, however, that no material is perfectly rigid. Any force, no matter how small, applied to a material will cause some deformation. The weight of the wheel and any load that it might be supporting will cause the material under the wheel to yield. The deformation shown in Fig. 9-25 seems realistic, though admittedly greatly exaggerated, for motion of the wheel to the right. A moment summation about point $O$ gives

$$[\Sigma M_O = 0] \qquad Fr = Na \qquad \text{or} \qquad a = \frac{Fr}{N} \tag{9-5}$$

The dimension $a$, in inches or millimeters, is called the **coefficient of rolling resistance.**

## ILLUSTRATIVE PROBLEM

**9-31.** A railway freight car is supported by eight wheels. The coefficient of rolling resistance $a$ for steel wheels on steel rails is about 0.02 in. What force $P$ is required to keep an 80-ton car rolling on a straight, level track? The wheels are 32 in. in diameter, and $P$ is applied parallel to the rails.

*Solution:* We may assume that the 160,000-lb load is distributed equally to all eight wheels. The load per wheel then is 20,000 lb. The radius of each wheel is 16 in. and the coefficient of rolling resistance is 0.02. From Eq. (9-4) we then obtain

$$\left[ F = \frac{Na}{r} \right] \qquad F = \frac{20,000(0.02)}{16} = 25 \text{ lb}$$

For all eight wheels, the total required force $P = (25)(8) = 200$ lb.   *Ans.*

Note that the entire weight could have been assumed to be concentrated on one wheel without changing the final answer. Hence the distribution of weight among several wheels is immaterial.

## PROBLEMS

**9-32.** An automobile has 30-in. wheels (outside diameter) and weighs 3200 lb. If the coefficient of rolling resistance $a$ is $\frac{3}{8}$ in. for paved surfaces, what horizontal force would be required to keep the car rolling at low speed on a level paved street?

**9-33.** A horizontal pull of 1000 N is required to tow a 30-kN car on a level gravel road. What is the coefficient of rolling resistance $a$ if the tires have a diameter of 600 mm?

*Ans. a = 10 mm*

## SUMMARY (By Article Number)

**9-1. Friction** is the resistance offered by two surfaces, in contact under pressure, to motion or attempted motion of one surface on the other. Frictional resistance is measured in terms of force which always acts parallel to the contacting surfaces and is so directed as to oppose the motion.

**9-2.** The **coefficient of friction** $f$ is the numerical ratio of the *limiting friction force* $F$ to the *normal pressure* $N$ between the contacting surfaces. That is, $f = F/N$, where the limiting friction $F$ is the maximum that can be and is developed when *motion is impending*. The *reaction* $R$ is the resultant of $F$ and $N$. The angle $\phi$ between the forces $R$ and $N$ is called the *angle of friction*, and its tangent is $F/N$, or $f$. The highest angle with the horizontal to which an inclined plane may be tipped without causing a body resting thereon to slip is called the *angle of repose;* hence the angle of repose equals the angle of friction.

**9-3.** According to the **laws of static friction** for dry surfaces, the limiting friction force (1) is dependent on the kinds of materials and on the degree of roughness of the contacting surfaces, (2) is independent of the areas in contact, and (3) is proportional to the normal pressure.

**9-5.** The maximum friction force that may be developed between a flexible belt and a cylindrical surface depends on the angle of contact $\theta$ and the coefficient of friction $f$ between the contacting surfaces. The difference between the belt tensions at the initial and final points of contact is equal to this friction force. Flat-belt tensions are related by

$$\log_{10} \frac{T_L}{T_S} = \log_{10} T_L - \log_{10} T_S = \frac{f\theta}{132}$$

or

$$\frac{T_L}{T_S} = e^{f\theta/57.3} \qquad \textbf{(9-2 and 9-3)}$$

and V-belt tensions by

$$\log_{10} \frac{T_L}{T_S} - \log_{10} T_L - \log_{10} T_S = \frac{f\theta}{132 \sin \beta}$$

or

$$\frac{T_L}{T_S} = e^{(f\theta)/(57.3 \sin \beta)} \qquad \textbf{(9-4)}$$

**9-6.** The **coefficient of rolling resistance** $a$ measures the resistance of a rolling object to continued motion. The coefficient is large for materials that yield considerably under load. Theoretically, rigid surfaces offer no resistance to rolling. Practically, they do offer relatively low resistance.

## REVIEW PROBLEMS

**9-34.** A person is standing on a roof which rises 5 ft vertically in every 10 ft measured horizontally. (*a*) What must be *the minimum coefficient of friction f* between the roof and the person's shoes to prevent slipping? (*b*) If $f$ is 0.36, what maximum slope could the roof have without slipping?

*Ans.* (*a*) $f = 0.5$; (*b*) $\theta_H = 19°48'$

**9-35.** The blocks $A$ and $B$ shown in Fig. 9-26 are connected with a flexible cable passing over a sheave $C$, the friction of which may be neglected. Let $f = 0.2$, between the blocks and their supporting surfaces. If block $A$ weighs 100 lb, compute *the minimum weight $W_B$ of block B which will barely prevent sliding.*

*Ans.* $W_B = 220$ lb

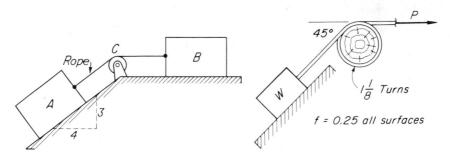

**Fig. 9-26   Probs. 9-35 and 9-36.**          **Fig. 9-27   Prob. 9-37,**

**9-36.** If block $B$ in Fig. 9-26 weighs 480 N and if $f = 0.25$, find the weight $W_A$ of block $A$ in order that *motion will be impending.* Neglect rope and sheave friction.

*Ans.* $W_A = 300$ N

**9-37.** Calculate the pull $P$ in Fig. 9-27 that must be exerted to slowly lower the weight $W = 1000$ lb down the incline.

*Ans.* $P = 90.6$ lb

**9-38.** Calculate the load that can be raised by a 50-lb force applied to the arm of the jackscrew shown in Fig. 9-8 if $a = 15$ in., $D = 2$ in., $p = ½$ in., and $f = 0.1$.

**9-39.** In Fig. 9-9, let the force $F = 100$ lb, and let the weight of each of blocks $A$ and $B$ be 100 lb. Determine graphically *the least value of the force P* that will prevent block $B$ from sliding to the left, if $f = 0.3$.

*Ans.* $P = 182$ lb

**9-40.** Assume in Prob. 9-15 (Fig. 9 14) that another bolt is inserted through the side plates 3 in. below bolt $B$, and that the tension in each bolt is now 300 lb. (*a*) Compute $P$. (*b*) On which block will slipping occur?

*Ans.* (*a*) $P = 216$ lb; (*b*) on block $A$

**9-41.** Solve part (*b*) of Prob. 9-17 if *P* is 40 lb *upward-acting* instead of downward-acting, as illustrated in Fig. 9-16. (HINT: Points *A* and *B* now move to opposite sides of the shaft.)

*Ans.* $x = 9$ in.

**9-42.** A rope is thrown over a horizontal overhead pipe. (*a*) How heavy a weight can a person lift with a 100-lb pull? (*b*) What weight can be lowered slowly if the person can maintain a 100-lb pull on his or her end of the rope while the weight is being lowered? Assume the friction coefficient *f* is 0.4.

*Ans.* (*a*) $W = 28.5$ lb; (*b*) $W = 351$ lb.

**9-43.** A power-driven capstan can maintain a 10,000-lb pull on a hawser. What is the minimum number of turns that the hawser must make on the capstan drum if the tension in the slack end of the hawser is to be 10 lb and the coefficient of friction is 0.25?

*Ans.* $\theta = 4.4$ turns

**9-44.** The cable drum of Fig. 9-23 is to be held against rotation by the brake shown. What is the maximum permissible cable tension *P* if the brake force *F* is limited to 50 lb?

*Ans.* $P = 515$ lb

**9-45.** The cylinders shown in Fig. 9-28 have a total weight of 100 N. Assume that $f = 0.2$ for all surfaces of contact. Calculate the pull *P* to slowly rotate the cylinders.

*Ans.* $P = 31.6$ N

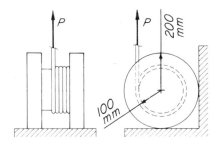 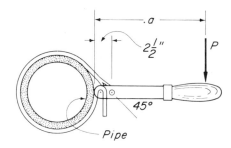

**Fig. 9-28   Prob. 9-45.**          **Fig. 9-29   Prob. 9-46. Strap wrench.**

**9-46.** A strap wrench is shown in Fig. 9-29. Calculate the minimum arm *a* that will prevent slipping if $f = 0.1$. Include the additional friction at the left end of the wrench due to the radial force at that point.

## REVIEW QUESTIONS

**9-1.** What is meant by (*a*) frictional resistance and (*b*) friction force?
**9-2.** Explain the meaning of normal pressure.
**9-3.** What is meant by impending motion?
**9-4.** Define limiting friction.

**9-5.** What is generally meant by angle of inclination?

**9-6.** What is the coefficient of friction?

**9-7.** Define the angle of friction.

**9-8.** What relation exists between the coefficient of friction and the tangent of the angle of friction?

**9-9.** Explain the meaning of angle of repose

**9-10.** Does the friction force at a dry surface depend on (*a*) the degree of roughness of the contacting surfaces? (*b*) the normal pressure? (*c*) the areas of the contacting surfaces? (*d*) velocity (explain)? (*e*) temperature?

**9-11.** Does Eq. (9-2) relating the belt tensions on opposite sides of a pulley apply when slipping is not impending?

**9-12.** What is the advantage of a V-belt system?

**9-13.** What is the common unit of the coefficient of rolling resistance?

**9-14.** What is the approximate coefficient of rolling resistance for rigid surfaces of contact?

# Chapter 12
# Miscellaneous Problems

## 12-1 INTRODUCTION

Some common engineering applications of the principles of statics will be presented in this chapter of miscellaneous problems. Hydrostatic pressure and some forms of hydrostatic loads will be discussed. Granular materials cause lateral pressures similar to those of true fluids. Therefore, the principles of hydrostatics can and will be used to evaluate the loads against and the stability of some simple retaining walls. Finally, flexible cables and arches will be considered.

## 12-2 INTENSITY AND DIRECTION OF HYDROSTATIC PRESSURES

When a liquid is contained and restrained by the surface of a solid object, force is exerted on that surface by the liquid. The intensity of this force per unit area is called *pressure*. A characteristic of liquid pressure is that it is exerted equally in all directions. In engineering practice, pressure is usually expressed in pounds per square foot, abbreviated psf, or in pounds per square inch, abbreviated psi. For fluids at rest, the pressure is always exerted perpendicular to the resisting surface (see Fig. 12-1a).

The pressure $p$ at any point in a liquid is

$$p = wh \qquad (12\text{-}1)$$

in which $w$ is the weight per unit volume of the liquid and $h$ is the vertical distance from the free surface. That is, liquid pressure is zero at the free surface and increases at a uniform rate from there downward (see Fig. 12-1b).

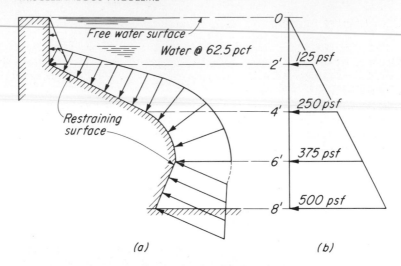

**Fig. 12-1    Intensity and direction of hydrostatic pressure.** *(a)* **Pressure is always perpendicular to restraining surface.** *(b)* **Intensity of pressure** $p = wh$.

## 12-3  BUOYANCY

The weight $w$ per unit volume of a fluid is called its *weight density* and, in most engineering practice, is generally expressed in pounds per cubic foot, abbreviated pcf, or newtons per cubic meter, abbreviated N/m³. An object of lower weight density than the fluid in which it is immersed or floats will rise or float because the upward pressure of the fluid on the object exceeds the downward pressure of the object on the fluid. The total upward force exerted on such a body by the fluid is called its **buoyancy** and is equal to the weight of the fluid displaced. The load required to barely submerge a floating body will herein be referred to as its **load capacity;** it is equal to the difference between the weight of object and the weight of the fluid it can displace by complete submergence.

## 12-4  HYDROSTATIC LOADS

The total hydrostatic force $P$ acting against a submerged plane area is equal to the *average pressure* $\bar{p}$ against the area multiplied by the area $A$ over which the pressure acts. That is,

$$P = \bar{p}(A) \tag{12-2}$$

The average pressure $p$ is always found at the centroid of the submerged plane area. The method of computing the hydrostatic loads and the resultant reacting forces on some simple structures will be illustrated in the following three problems. In SI units, one pascal (1 Pa) is the equivalent of one newton per square meter (1 N/m²).

## ILLUSTRATIVE PROBLEMS

**12-1.** The sides of the water flume whose cross section is shown in Fig. 12-2a are supported by steel rods passing through opposite vertical timber posts spaced 4 ft on centers along the flume. Calculate the tension $T$ in one steel rod and the reacting force $R$ at the bottom of one post.

*Solution:* The *pressure diagram* shown in Fig. 12-2b illustrates the uniform variation of hydrostatic pressure from the free water surface to the bottom. The hydrostatic pressure varies uniformly from zero at the top of the wall to 375 psf at the bottom. The average pressure is

$$\bar{p} = \frac{0 + 375}{2} = 187.5 \text{ psf}$$

The hydrostatic force acting against a section of wall 6 ft high and extending 2 ft each side of a post and rod is

$$[P = \bar{p}A] \qquad\qquad P = 187.5(6)(4) = 4500 \text{ lb}$$

This resultant hydrostatic force acts through the centroid of the pressure prism as shown in Fig. 12-2c. Taking moments about the bottom of the wall (point $B$ in the load diagram), we obtain

$$[\Sigma M_B = 0] \qquad T(7) = 4500(2) \quad \text{or} \quad T = 1286 \text{ lb} \qquad\qquad Ans.$$

The reacting force $R$ at the bottom of the post should now be found by a summation of moments about $A$. That is,

$$[\Sigma M_A = 0] \qquad R(7) = 4500(5) \quad \text{or} \quad R = 3214 \text{ lb} \qquad\qquad Ans.$$

A check on some of the numerical work may be obtained by a force summation.

$$[\Sigma H = 0] \qquad\qquad 1286 + 3214 = 4500 \qquad\qquad Check$$

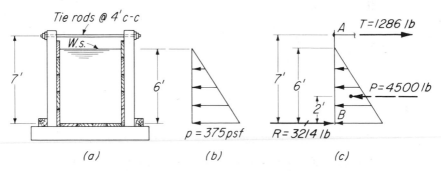

**Fig. 12-2  Hydrostatic pressure and load against a vertical surface.** *(a)* **Water flume.** *(b)* **Pressure diagram.** *(c)* **Load diagram.**

**12-2.** In the cross section of the water flume shown in Fig. 12-3a, the sloping timber posts with their supporting steel rods are spaced 1.5 m on centers along the flume. Calculate the tension $T$ in one rod and the $V$ and $H$ components of the reaction forces at the bottom of one post.

*Solution:* The hydrostatic pressure $p$ at the bottom of the flume is $2(9789) = 19\,578$ Pa, and acts perpendicular to the sloping restraining surface, as shown in Fig. 12-3b.

The total hydrostatic force $P$ acting against the contributing area $A$ for one post is the average pressure $\bar{p}$ multiplied by the area $A$ on which it acts. Hence

$$[P = \bar{p}A] \qquad\qquad P = \left(\frac{19\ 578}{2}\right)(1.5)(2.5) = 36.71 \text{ kN}$$

The line of action of this force passes through the centroid of the pressure prism as shown in Fig. 12-3c and intersects the wall at a distance of 3.6/3 or 1.2 ft from the bottom. The tensile force $T$ in the rod may be found by a moment summation about $B$.

$$[\Sigma M_B = 0] \qquad T(2.5) = 36.71(0.8333) \qquad \text{or} \qquad T = 12.24 \text{ kN} \qquad\qquad Ans.$$

The force $P$ may be resolved into its vertical and horizontal components. By similar triangles

$$\frac{P}{5} = \frac{P_V}{3} = \frac{P_H}{4} \qquad \text{or} \qquad P_V = 22.03 \text{ kN} \qquad \text{and} \qquad P_H = 29.37 \text{ kN}$$

Summing forces in the vertical and horizontal directions gives

$$[\Sigma V = 0] \qquad\qquad B_V - P_V = 0 \qquad \text{or} \qquad B_V = 22.03 \text{ kN} \qquad\qquad Ans.$$
$$[\Sigma H = 0] \qquad B_H + 12.24 - 29.37 = 0 \qquad \text{or} \qquad B_H = 17.13 \text{ kN} \qquad\qquad Ans.$$

A check may be obtained by a moment summation about $C$.

$$[\Sigma M_C = 0] \qquad 12.24(2.5 - \tfrac{2}{3}) = 17.13(\tfrac{2}{3}) + 22.03(0.5)$$
$$\text{or} \qquad\qquad\qquad\qquad 22.44 = 22.44$$

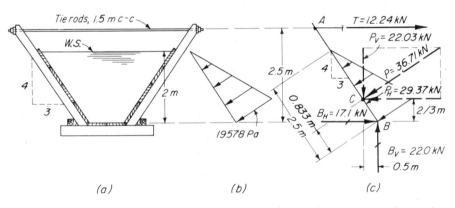

(a)     (b)     (c)

**Fig. 12-3   Hydrostatic pressure and load against a sloping surface. (a) Water flume. (b) Pressure diagram. (c) Load diagram.**

**12-3.** An opening in the vertical side of a wood-stave water tank is closed by means of a door hinged at the bottom and held in place by a bolted crossbar $A$, as illustrated in Fig. 12-4a. Calculate the force on each hinge and on each bolt due to the hydrostatic pressure caused by a depth of 10 ft of water in the tank, if the hinges are at the bottom of the tank.

*Solution:* With 10 ft of water in the tank, the top of the 4-ft-high door is 6 ft below the free water surface, and the pressure at that point is 6(62.5) or 375 psf. At the bottom of the door, the pressure is (62.5)(10) or 625 psf, as shown in the pressure diagram, in Fig. 12-4b. The two bolts will have equal tensions, and the forces on the two hinges

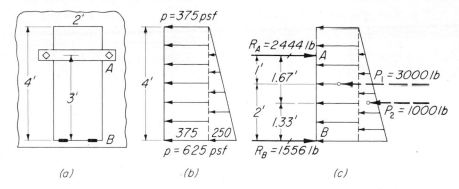

**Fig. 12-4  Hydrostatic pressure and load against submerged surface. *(a)* Full view of door. *(b)* Pressure diagram. *(c)* Load diagram.**

will be alike. An edge view of the door is shown in Fig. 11-4c. The total hydrostatic force acting against the door is

$$[P = \bar{p}A] \qquad\qquad P = \frac{375 + 625}{2}(2)(4) = 4000 \text{ lb}$$

Since the line of action of this force is not self-evident, it will be expedient to break the trapezoidal pressure prism into two prisms, one a rectangular prism representing a uniform pressure of 375 psf against the door and the other a triangular prism under which the pressure varies uniformly from zero at the top of the door to 250 psf at the bottom.

Considering each pressure prism separately,

$$[P = \bar{p}A] \qquad\qquad P_1 = 375(2)(4) = 3000 \text{ lb}$$
$$\text{and} \qquad\qquad\qquad P_2 = \tfrac{1}{2}(250)(2)(4) = 1000 \text{ lb}$$

To find the total reaction force at $A$, take moments about $B$

$$[\Sigma M_B = 0] \qquad 3R_A - 3000(2) - 1000(1.333) = 0 \qquad \text{or} \qquad R_A = 2444 \text{ lb}$$

Similarly,

$$[\Sigma M_A = 0] \qquad 3R_B - 3000(1) - 1000(1.67) = 0 \qquad \text{or} \qquad R_B = 1556 \text{ lb}$$

Then the tension in each bolt is

$$T = \frac{2444}{2} = 1222 \text{ lb} \qquad\qquad\qquad Ans.$$

and the force in each hinge is

$$F = \frac{1556}{2} = 778 \text{ lb} \qquad\qquad\qquad Ans.$$

## PROBLEMS

**12-4.**  The sides of the water flume shown in Fig. 12-5 are supported by steel tie-rods through vertical timber posts spaced 2 m on centers along the flume. Calculate the tension $T$ in one rod and the reaction $R$ at the bottom of one post, due to the water pressure.

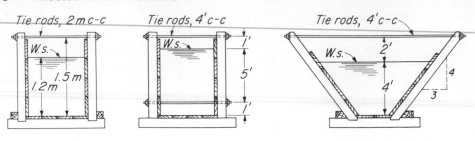

Fig. 12-5  Probs.          Fig. 12-6   Prob. 12-6.          Fig. 12-7  Probs. 12-7 and 12-8.
12-4 and 12-5.

**12-5.** Solve Prob. 12-4 when the depth of water is 2 m, the distance from the bottom to the rod is 2.5m, and the tie-rod spacing is 1.5 m

*Ans.* $T = 7.83$ kN; $R = 21.54$ kN

**12-6.** Calculate the tensions $T_1$ and $T_2$ in the upper and lower tie rods, due to water pressure against one side of the flume shown in Fig. 12-6. Neglect any attachment of the bottom of the post to the horizontal beam.

**12-7.** Determine the tension $T$ in the tie rod of the flume shown in Fig. 12-7, caused by the water pressure against the sloping side, and the $V$ and $H$ components of the reaction at the bottom $B$ of the post.

*Ans.* $T = 695$ lb; $B_V = 1500$ lb; $B_H = 1305$ lb

**12-8.** Solve Prob. 12-7 when the depth of water is 4.5 ft, the distance from the bottom to the rod is 6 ft, the slope of the post is 3/2 (3 units vertical to 2 horizontal), and the tie-rod spacing is 5 ft.

**12-9.** In Fig. 12-8 is shown a temporary timber flashboard used on a concrete diversion dam to raise the water level. Planks are laid sloping from the bottom to the horizontal beam at the top, which, in turn, is supported by square timber posts spaced 6 ft on centers along the dam. Calculate the compressive force $C$ in one post when depth of water $h$ is 4 ft and $\theta$ is 45°.

*Ans.* $C = 1417$ lb

**12-10.** Solve Prob. 12-9 when the slope of the planks is 4/3 (4 units vertical to 3 horizontal), the spacing of the posts is 3 m on centers, and $h$ is 2 m.

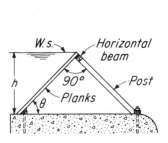

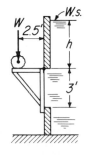

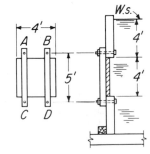

Fig. 12-8   Probs. 12-9 and          Fig. 12-9   Probs.          Fig. 12-10   Prob. 12-13.
12-10.                               12-11 and 12-12.

**12-11.** A simple form of self-opening gate, used for automatic control of the water level behind a small timber dam, is shown in Fig. 12-9. The gate is hinged at its upper edge and is kept closed by the weight $W$ on the bracket. The opening in the dam covered by the gate is 2 ft wide and 3 ft high. If $W$ is 1800 lb, calculate the level $h$ above the top of the gate to which the water may rise before the pressure against it will open it.

*Ans.* $h = 6$ ft

**12-12.** In Prob. 12-11, let the gate be 3 ft wide and 3 ft high, and let W be 2025 lb. Then solve for $h$.

**12-13.** An opening in the side of a tank is temporarily closed as illustrated in Fig. 12-10. Calculate the tensions $T$ produced in each of the four bolts by the water pressure.

*Ans.* $T_A = T_B = 1367$ lb; $T_C = T_D = 1633$ lb

**12-14.** The concrete dam whose cross section is shown in Fig. 12-11 weighs 16 000 kN per linear meter along the dam. This weight may be considered as concentrated at the center of gravity $C$ of the section. The factor of safety against overturning is the resisting moment $M_R$ of the weight of the dam about point $B$ at the base divided by the overturning moment $M_O$ due to the water pressure on the upstream face of the dam. (*a*) Calculate these moments for a 1-m section. (*b*) Calculate the factor of safety against overturning.

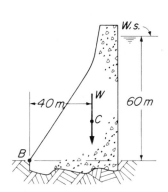

Fig. 12-11    Prob. 12-14.

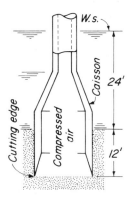

Fig. 12-12    Probs. 12-15 and 12-16.

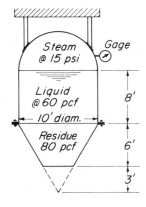

Fig. 12-13    Probs. 12-17 and 12-18.

**12-15.** Underwater excavation is accomplished by use of a heavily constructed concrete caisson, a cross section of which is shown in Fig. 12-12. By maintaining an air pressure inside the caisson equal to the upward or lateral pressure of the mud at its bottom, the mud is prevented from creeping under the cutting edge. Assuming that the mud acts as a fluid weighing 90 pcf, calculate the minimum air pressure $p$ (in psi) required to maintain this equilibrium when the cutting edge is 36 ft below the free water surface.

*Ans.* $p = 17.95$ psi

**12-16.** Solve Prob. 12-15 when the cutting edge has been lowered a total of 24 ft into the mud.

**12-17.** The pressure vessel shown in Fig. 12-13 is cylindrical in cross section. The bottom portion collects residue and is in the form of a truncated cone. Calculate the pressure intensity on the bottom. Assume that the residue acts as a true fluid weighing 80 pcf.

*Ans.* $p = 21.65$ psi

**12-18.** Refer to Prob. 12-17. The bottom section is bolted to the cylindrical portion with 96 equally spaced bolts. The empty bottom weighs 2400 lb. To effect a tight seal, the bolts are to be tensioned an additional 500 lb beyond that required to overcome the loads. Calculate the required total force per bolt.

**12-19.** A log weighing 6000 N/m³ floats in still water. What percentage of its volume is above the water surface?

**12-20.** A lightly constructed houseboat is to rest on several 3-ft-di logs, 20 ft long, floating in water. The houseboat is estimated to weigh 11,300 lb, and the logs weigh 42.5 pcf. Let the load required to barely submerge a free-floating log be known as its load capacity. If only two-thirds of the load capacity of the logs may be used, how many are required?

*Ans.* 6 logs

**12-21.** At a large construction job, a light steel pipeline carrying air runs horizontally through a water reservoir. The outside diameter of the pipe is 400 mm and it weighs 200 N per linear meter. Every 5 m along its length the pipe is fastened to a concrete-block anchor by means of a short, light steel cable. Calculate the tension $T$ in each cable.

*Ans.* $T = 5151$ N

## 12-5 STABILITY OF RETAINING WALLS

Materials such as loose sand or gravel, granular soil, or mud, for example, are considered to cause pressures against retaining walls or other restraining surfaces in a manner *similar* to those of true fluids. Since the retained materials may be (1) dry and well compacted, producing low lateral pressures, (2) semidry like ordinary soil, or (3) wet and fluid like mud or freshly poured concrete, producing high lateral pressures, the pressures resulting are not always proportional to the actual weights of the materials and are therefore expressed in terms of the weight of an "equivalent fluid" which will produce the given hydrostatic pressure. This equivalent weight $w'$ is called the **fluid weight** of the material, and the pressure $p'$ that it will produce is called the **equivalent fluid pressure.** These equivalencies are used to determine the forces acting on retaining walls, sheet piling, caissons, and other similar structures subjected to various types of pressures.

A retaining wall is usually analyzed to determine (1) the factor of safety against sliding, (2) the factor of safety against overturning, and (3) the maximum soil pressure under the foundation. The problems of this chapter will be limited to a study of the resistance of the wall to overturning. The factor of safety against overturning may be defined as the moment of the resisting forces divided by the moment of the overturning forces, both with respect to the axis about which the wall would overturn.

## ILLUSTRATIVE PROBLEM

**12-22.** The retaining wall shown in Fig. 12-14 is of reinforced concrete weighing 150 pcf. Lateral pressure is exerted against the retaining wall by earth having an actual weight of 100 pcf and a *fluid weight* of 25 pcf. (*a*) Determine the factor of safety against overturning of the wall about an axis through the toe $A$. (*b*) Does soil pressure exist over the entire area of the base?

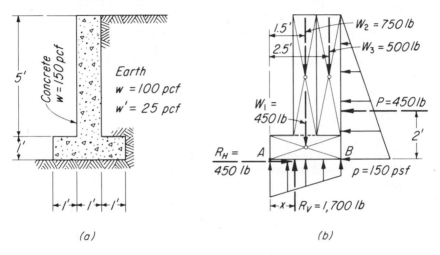

(a)                    (b)

**Fig. 12-14   Analysis of retaining wall against overturning. (a) Retaining wall. (b) Free-body diagram.**

*Solution:* For convenience, the retaining wall will be broken into two sections, the base and the wall. Consider a 1-ft length of wall. The weight of the footing is $W_1 = (150)(1)(3) = 450$ lb. The weight of the wall is $W_2 = (150)(1)(5) = 750$ lb. The weight of soil over the footing is $W_3 = (100)(1)(5) = 500$ lb. The lateral earth pressure varies from zero at the top of the wall to $(6)(25) = 150$ psf at the base of the footing. The lateral force $P$ is

$$[P = \bar{p}A] \qquad P = \tfrac{1}{2}(150)(1)(6) = 450 \text{ lb}$$

If the wall overturns, it will do so about point $A$ of Fig. 12-14*b*. The moment of the forces above the base that resist overturning ($RM$) about point $A$ is

$$RM = (1.5)W_1 + (1.5)(W_2) + (2.5)(W_3)$$
or
$$RM = (1.5)(450) + (1.5)(750) + (2.5)(500) = 3050 \text{ lb·ft}$$

The moment of the overturning forces ($OM$) above the base, with respect to $A$, is

$$OM = (2)(450) = 900 \text{ lb·ft}$$

The factor of safety ($FS$) against overturning is

$$FS = \frac{RM}{OM} = \frac{3050}{900} = 3.39 \qquad\qquad Ans.$$

The soil pressure under the base will not be uniform. The pressure will be greatest at point $A$ of Fig. 12-14$b$ and will be least at point $B$. If the resultant vertical force caused by this soil pressure falls within the middle one-third of the base, soil pressure will exist over the entire base as indicated in Fig. 12-14$b$. By a force summation in the vertical direction

$$[\Sigma V = 0] \qquad\qquad R_V = 450 + 750 + 500 = 1700 \text{ lb}$$

The line of action of this force can be located by a moment summation about $A$

$$[\Sigma M_A = 0] \quad (1.5)(450) + (1.5)(750) + (2.5)(500) - 2(450) - (x)(1700) = 0$$

or $\qquad\qquad x = \dfrac{675 + 1125 + 1250 - 900}{1700} = \dfrac{2150}{1700} = 1.265 \text{ ft}$

*Ans.* Since the resultant vertical force $R_V$ falls in the middle one-third of the base, compressive soil pressure will exist over the entire base

### PROBLEMS

**12-23.** The retaining wall shown in Fig. 12-15 is of concrete weighing 150 pcf. The wall is subjected to a lateral pressure due to soil having an equivalent *fluid weight* of 30 pcf. (*a*) Find the factor of safety against overturning and (*b*) determine if the foundation pressure will extend over the entire base.

> *Ans.* (*a*) FS $= 1.53$; (*b*) The foundation soil pressure will *not* exist over the entire base. (Compressive pressure will exist on 63.5 percent of the base.)

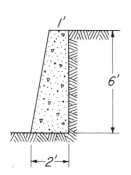

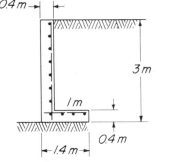

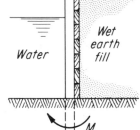

**Fig. 12-15   Prob. 12-23.**   **Fig. 12-16   Prob. 12-24.**   **Fig. 12-17   Prob. 12-25.**

**12-24.** The retaining wall shown in Fig. 12-16 is constructed of reinforced concrete weighing 23.7 kN/m$^3$. It is subjected to a lateral pressure due to soil with an actual weight $w$ of 20 kN/m$^3$ and a *fluid weight* $w'$ of 4.8 kN/m$^3$. (*a*) Determine the factor of safety against overturning and (*b*) comment on the pressure distribution on the base.

> *Ans.* (*a*) FS $= 2.83$; (*b*) Compressive soil pressure will *not* exist over entire base.

**12-25.** A timber retaining wall supporting water on one side and loose wet soil with an equivalent *fluid weight* of 90 pcf on the other is shown in Fig. 12-17, it has planking nailed solidly to deeply driven round timber piles spaced 6 ft on centers. If the depth of water to the solid ground line is 6 ft and the depth of the supported soil is 9 ft, compute the reacting moment $M$ in the pile at the ground line due to the lateral pressures of these materials

*Ans. M* = 52,100 lb·ft

## 12-6 FLEXIBLE CABLES; RIGID ARCHES; CONCENTRATED VERTICAL LOADS

A cable is considered to be perfectly flexible when its resistance to bending is relatively so small that it may be disregarded in practical problems. In such a cable the internal force at any point, due either to its own weight or to externally applied loads, always parallels a tangent to the cable axis at that point. Telephone wires, power wires, and suspension-bridge cables are, for practical purposes, considered to be of this type. A cable supporting traffic lights is a common example of a cable subjected to concentrated forces.

A flexible cable is a nonrigid "structure" whose configuration is determined largely by its length and the loads it supports; when the loads are changed, the configuration of the cable will immediately change until equilibrium is established. An arch, on the other hand, is a rigid structure whose configuration, or shape, is predetermined, depending on the loads it is to support. However, because the arch is rigid, a change in loads will not change its shape except to a very minute extent. The graphical solution of both types of problems is somewhat similar, since a flexible cable may be regarded as an "upside-down" arch.

When supports are not at the same level or when the loading is "unsymmetrical," as illustrated in Prob. 12-26, both types of problems become somewhat more complex. When supports are at the same level and the loading is "symmetrical," about a vertical center line, as illustrated in Prob. 12-31, the problems are rather simple.

Figure 12-18a shows a flexible cable supporting concentrated loads. The supports are at different levels, and the loads have different magnitudes, as do the horizontal dimensions between their lines of action. This case, then, would represent the most general type of problem and is the most difficult to solve.

Because a flexible cable is not a rigid body, its final shape under given loads cannot be predetermined, as in the case of the rigid arch. However, we do know beforehand the locations of the end supports $A$ and $E$ (dimensions $L$ and $y$), and usually we also know the horizontal dimensions $a$, $b$, $c$, and $d$. This leaves four unknown quantities to be determined: the vertical and horizontal reaction components at $A$ and $E$. The end reactions at either end may be found if we know the cable slope at that end. For example, if we know $v$ (i.e., the location of point $b$), the reaction components at $A$ may be found. The unknown

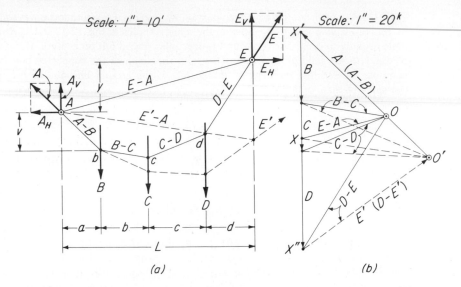

**Fig. 12-18  Flexible cable, concentrated loads, graphical solution. (a) Space diagram. (b) Force diagram or polygon.**

reaction components $A_V$ and $A_H$ may be determined analytically by two moment summations, one about $E$ and the other about $b$, both involving $A_V$ and $A_H$, which may then be solved simultaneously. A force summation of the entire cable system in both the vertical and horizontal directions will now yield $E_V$ and $E_H$. A step-by-step graphical solution is outlined below.

### PROCEDURE FOR GRAPHICAL SOLUTION ─────────────────

A study of Prob. 5-17 is recommended.

1. In a space diagram, from given dimensions, locate supports $A$ and $E$, point $b$, and the vertical action lines of loads $B$, $C$, and $D$. Then draw known string $AB$.
2. In a force diagram, lay off known loads $B$, $C$, and $D$. Draw ray $AB$ (whose direction is known), and select trial pole $O'$ at random but lying on ray $AB$. Next draw trial rays (dashed) $BC$, $CD$, and $DE'$.
3. In the space diagram, draw trial strings (dashed) $BC$, $CD$, and $DE'$, parallel respectively to the corresponding trial rays. Then draw trial closing string $E'A$ (dashed). Also draw final closing string $EA$, which must pass through supports $E$ and $A$.
4. In the force diagram, draw ray $E'A$ (dashed) parallel to trial closing string $E'A$, thus locating point $X$. From $X$ draw final ray $EA$ parallel to final string $EA$, thus locating final pole $O$. Then draw final rays $BC$, $CD$, and $DE$ (solid lines).

5. The stresses in cable segments $AB$, $BC$, $CD$, and $DE$ are now found by scaling the corresponding rays. Note that the scaled value of ray $AB$ gives the stress in cable segment $AB$ as well as the reaction $A$ and that ray $DE$ gives the stress in $DE$ as well as reaction $E$.

The following facts should be observed: ($a$) the configuration of the cable can be controlled by the arbitrary location of point $b$; ($b$) when dimension $v$ is least, the cable stresses are greatest, whereas when $v$ is greatest, the stresses are least; and ($c$) the scaled values $XX'$ and $XX''$ would be, respectively, the vertical reactions at supports $A$ and $E$, if $AE$ were a compression strut and if the resulting "frame" (strut plus cable) were then hinged at $A$ and supported by rollers on a horizontal surface at $E$ (in which case the reactions at those supports would be vertical).

### ILLUSTRATIVE PROBLEM

**12-26.** In Fig. 12-18$a$ let dimensions $L = 20$ ft, $y = 5$ ft, $v = 4$ ft, $a = 4$ ft, $b = 5$ ft, $c = 6$ ft, and $d = 5$ ft; and let loads $B = 15$ kips, $C = 10$ kips, and $D = 20$ kips. Solve ($a$) graphically for the four stresses in cable segments $AB$, $BC$, $CD$, and $DE$ (scales on 8½ by 11: 1 in. = 5 ft, 1 in. = 10 kips) and ($b$) analytically for the support reactions $A$ and $E$ which equal, respectively, cable stresses $AB$ and $DE$. (NOTE: A complete graphical solution by students is recommended, as well as the analytical solution to determine cable stresses $BC$ and $CD$ which is called for in Prob. 12-29.)

*a. Graphical solution:* When the above procedure is followed, the following values may be scaled from the force diagram, Fig. 12-18$b$:

Reaction $A$ = stress $AB$ = 25.4 kips    Reaction $E$ = stress $DE$ = 32.5 kips
Stress $BC$ = 18.2 kips                        Stress $CD$ = 19.3 kips

*b. Analytical solution:* Clearly, reaction $A$ is parallel to cable segment $AB$. Therefore, a relation between $A_V$ and $A_H$ may be established either by a moment equation about point $b$ or by proportion. The latter relation is by proportion

$$\frac{A_V}{4} = \frac{A_H}{4} \qquad \text{or} \qquad A_V = A_H$$

Then, by moments about $E$, we obtain

$[\Sigma M_E = 0]$    $20A_V + 5A_V - 20(5) - 10(11) - 15(16) = 0$
or        $25A_V = 100 + 110 + 240 = 450$    and    $A_V = A_H = 18$ kips

from which $A = \sqrt{(18)^2 + (18)^2} = \sqrt{648}$    or    $A = 25.4$ kips    *Ans.*

Then, to obtain $E_V$, $E_H$, and $E$, we have

$[\Sigma V = 0]$    $E_V + 18 - 15 - 10 - 20 = 0$    or    $E_V = 27$ kips
$[\Sigma H = 0]$    $E_H = 18$ kips

and

$$E = \sqrt{(27)^2 + (18)^2} = \sqrt{729 + 324} = \sqrt{1053} \qquad \text{or} \qquad E = 32.5 \text{ kips} \quad \textit{Ans.}$$

## PROBLEMS

**12-27.** Solve Prob. 12-26 when $L = 33$ ft, $y = 14$ ft, $v = 6$ ft, $a = 8$ ft, $b = 12$ ft, $c = 7$ ft, $d = 6$ ft, and loads $B = 42$ kips, $C = 21$ kips, and $D = 12$ kips. (Scales on 8½ by 11: 1 in. $= 10$ ft, 1 in. $= 20$ kips.)

**12-28.** Solve Prob. 12-26 when $L = 20$ ft, $y = 5$ ft, $v = 2$ ft, $a = 4$ ft, $b = 5$ ft, $c = 6$ ft, $d = 5$ ft, and loads $B = 15$ kips, $C = 20$ kips, and $D = 10$ kips. (Scales on 8½ by 11: 1 in. $= 4$ ft, 1 in. $= 10$ kips.)

*Partial Ans.* $A = 38$ kips, $E = 44$ kips

**12-29** Complete the analytical solution of Prob. 12-26 by solving for the stresses in cable segments $BC$ and $CD$. Calculate also the vertical distances $y_c$ and $y_d$ from support $A$ to point $c$ and $d$. Check all results against the graphical solution. (Use free-body diagrams as shown in Fig. 12-19.)

*Ans.* $BC = 18.2$ kips, $y_c = 4.83$ ft; $CD = 19.3$ kips; $y_d = 2.5$ ft

**12-30.** Solve analytically for cable stresses $BC$ and $CD$ in Prob. 12-28 and for the vertical distances $y_c$ and $y_d$ from support $E$ to points $c$ and $d$.

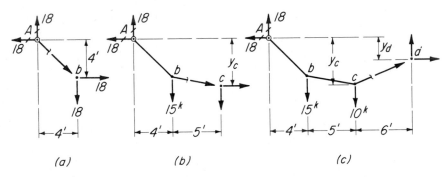

**Fig. 12-19  Flexible cable, concentrated loads, analytical solution. (a)** Free-body diagram, segment *AB*. **(b)** Free-body diagram, segment *ABC*. **(c)** Free-body diagram, segment *ABCD*.

## ILLUSTRATIVE PROBLEM

**12-31.** Figure 12-20 shows a typical tied, sectional roof-arch rib supporting loads $B$, $C$, and $D$. (*a*) Determine graphically the rise $r$ and the stresses in sections $AB$, $BC$, $CD$, and $DE$, when $B = C = D = 5$ kN, $a = 10$ m, and $\theta = 30°$. (*b*) Determine the shape of the arch and the stresses in all sections when $r = 8$ m. (NOTE: It is recommended that students complete the solution as explained below.)

*Solution:* (*a*) The string and force polygons for this solution are shown in solid lines in Fig. 12-20. Note that for an upward-curving arch the pole must be located at the left of the load line.

(*b*) When the rise $r$ is given a fixed value, an indirect graphical solution can be made by several trial locations of the pole $O'$. A simpler and direct analytical-graphical solution is to make a vertical cut through section $BC$ of the rib and horizontal tie, as shown in Fig. 12-20a, and to calculate the stress in the tie by moments about point $C$. When this stress is then laid off in the force polygon (line $O'X$), both string and force

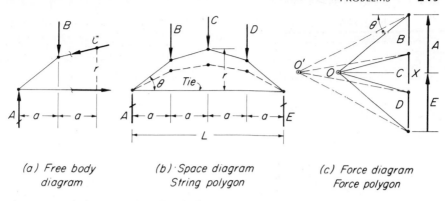

(a) Free body
diagram

(b) Space diagram
String polygon

(c) Force diagram
Force polygon

**Fig. 12-20   Tied, sectional roof-arch rib.**

polygons (dashed lines) are readily completed. The force $OB$ is 15 kN C, and the force $OC$ is 11.5 kN C.

## PROBLEMS

**12-32.** The cable shown in Fig. 12-21 supports four equal loads as shown. The cable is to be inclined at an angle of 45° to the horizontal at the supports. Determine graphically the maximum cable sag $s$ and the maximum cable tension $T$ if $W$ = 2 kips. Let $a$ = 5 ft.

*Ans. s* = 7.5 ft; *T* = 5.66 kips

**12-33.** The weights $W$ in Fig. 12-21 are 50 lb each and $a$ = 12 ft. What is the sag $s$ if the horizontal pull at the supports is limited to 120 lb?

*Ans. s* = 15 ft

**12-34.** The cable shown in Fig. 12-21 supports four equal loads. Determine graphically the allowable cable sag $s$ if $W$ = 800 N and the maximum cable tension $T$ is limited to 4000 N. Check solution analytically. Assume $a$ = 2 m.

**12-35.** An arch is to support the five 2-kip loads, as shown in Fig. 12-22. Determine graphically the rise $r$ of the arch and the reactions at $A$ and $B$ if the arch is constructed

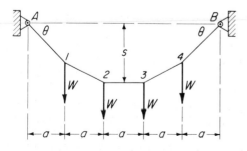

**Fig. 12-21   Probs. 12-32 through 12-34.**

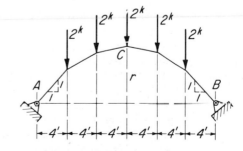

**Fig. 12-22   Probs. 12-35 and 12-36.**

along the load line $ACB$. The load line is to have a 1 to 1 slope at supports. Check analytically.

*Ans.* $r = 7.2$ ft; $R_A = R_B = 7.07$ kips

**12-36.** An arch is constructed along the load line $ACB$, shown in Fig. 12-22. Determine the reactions $A$ and $B$ and the compressive force $C$ in the segment adjacent to the crown point $C$, if the load line makes an angle of 30° with the horizontal at $A$ and $B$. Check analytically.

## 12-7 FLEXIBLE CABLES, HORIZONTALLY UNIFORM LOADS

A common problem is that of determining the maximum internal force $T$ in a cable, due either to its own weight or to an externally applied load uniformly and horizontally distributed between the ends. A uniform horizontal distribution of load is generally assumed in telephone wires, and in power and trolley wires in which the ends are lying at, or near, the same level and in which the maximum sag does not exceed about 10 percent of the distance between supports. The weight of the roadway of a suspension bridge is another example of uniform horizontal distribution of load on the supporting cables. If the suspension cable weighs little in comparison with the roadway it supports, its weight may be disregarded, and the distribution of loads is then horizontal, regardless of the amount of cable sag. The shape assumed by such cables is parabolic. A cable under only its own weight or a load that is uniformly distributed along the axis of the cable will assume the shape of a catenary. Only the case of uniform horizontal loading will be considered here.

Consider the flexible cable shown in Fig. 12-23$a$. Let $L$ be the horizontal distance (span) between supports $A$ and $B$. When these supports are at the same elevation, as in this case, the maximum vertical sag $s$ of the cable will occur at the center $C$. At this point $C$ of maximum sag, a tangent to the axis of the cable will always be horizontal, and so will the stress $H$ within the cable.

If the uniform horizontal load is $w$ lb/ft, the total resultant load on section $CD$ is $wx$, which is considered to be concentrated at the centroid of the load

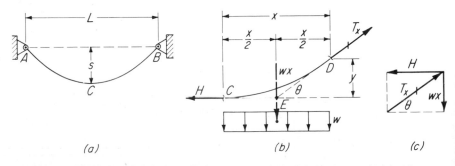

**Fig. 12-23  Flexible cable, horizontally uniform load. Supports at same level.** *(a)* **Flexible cable.** *(b)* **Free-body diagram of portion of loaded cable.** *(c)* **Force parallelogram.**

area. The tension $T_X$ in the cable at any point $D$ is always directed parallel to the cable axis at $D$, as shown in Fig. 12-23b.

Now, since only three forces, $H$, $wx$, and $T_X$ act on section $CD$ of the cable, they must be in equilibrium and also concurrent (see Art. 2-15). Forces $H$ and $wx$, whose action lines are known, intersect at point $E$, through which point the action line of $T_X$ then also must pass.

The magnitude of stress $H$ for any given section of cable $CD$ is established by a summation of moments about $D$. That is,

$$[\Sigma M_D = 0] \quad Hy - wx\left(\frac{x}{2}\right) = 0 \quad \text{or} \quad H = \frac{wx^2}{2y}$$

From this equation, we note that the maximum value of $H$ occurs when $x = L/2$ and $y = s$. Then

$$H = \frac{wL^2}{8s} \tag{12-3}$$

From Fig. 12-23c we note that $T_X$ is equal and opposite to the resultant of $H$ and $wx$. Hence the cable tension at any horizontal distance $x$ from the center $C$ is

$$T_X = \sqrt{H^2 + (wx)^2}$$

From the above equation, we note than when $x = 0$ (at the center of the span) $T_X = H$, and that as the value of $x$ increases, so does the value of stress $T_X$, which becomes greatest at the supports $A$ and $B$.

The maximum value of $T_X$ then also occurs when $x = L/2$ and $y = s$. That is,

$$T_{\text{max}} = \sqrt{H^2 + \left(\frac{wL}{2}\right)^2} = \sqrt{\frac{w^2L^4}{64s^2} + \frac{w^2L^2}{4}}$$

or

$$T_{\text{max}} = \frac{wL}{2}\sqrt{\frac{L^2}{16s^2} + 1} \tag{12-4}$$

From Fig. 12-23c, the angle $\theta$ of the cable at any point is given by

$$\tan\theta = \frac{wx}{H}$$

At the supports $x = L/2$. Therefore

$$\tan\theta_{\text{max}} = \frac{wL}{2H} = \frac{wL}{2(wL^2/8s)} = \frac{4s}{L} \quad \text{or} \quad \theta = \tan^{-1}\frac{4s}{L} \tag{12-5}$$

**ILLUSTRATIVE PROBLEM** _____

**12-37.** A power wire weighing 2 lb/ft is stretched between two supports at equal elevations, 100 ft apart. If the sag at the center is 10 ft, compute the tension $H$ at the center and the maximum tension $T$ at one support.

*Solution:* From Eq. (12-3)

$$\left[ H = \frac{wL^2}{8s} \right] \qquad H = \frac{2(100)^2}{8(10)} = 250 \text{ lb} \qquad\qquad Ans.$$

From Eq. (12-4) the maximum tension is

$$\left[ T_{max} = \frac{wL}{2} \sqrt{\frac{L^2}{16s^2} + 1} \right] \qquad T_{max} = \frac{2(100)}{2} \sqrt{\frac{(100)^2}{16(10)^2} + 1} = 269 \text{ lb } Ans.$$

## PROBLEMS

**12-38.** Compute the maximum tension in a flexible wire weighing 0.3 lb/ft stretched between two supports 160 ft apart and at equal levels if the sag is 16 ft.

**12-39.** Determine the sag in the wire of Prob. 12-38 if it is stretched until the tension reaches the maximum allowable value of 480 lb.

*Ans. s* = 2.01 ft

**12-40.** In order to control the sag in a radio aerial wire, one end $A$ is fastened to the top of a tower, as shown in Fig. 12-24, while the other passes over a small sheave $B$ and supports a weight $W$. If the wire weighs 0.75 N/m, and if $A$ and $B$ are 30 m apart and at the same level, compute the weight $W$ required to keep the sag at 0.5 m. Neglect the diameter of the sheave $B$ and the weight of the wire between $B$ and $W$.

**12-41.** If the maximum allowable tension in the aerial wire of Prob. 12-40 is 200 N, compute the minimum permissible sag.

**12-42.** A footbridge is supported by two like suspension cables, shown in Fig. 12-25. The tops of the towers are at the same level and are 300 ft apart. The weight of the bridge on each cable is 200 lb per horizontal foot, and the sag at the center is 50 ft. (The weight of the cable would probably be less than 4 lb/ft and may, therefore, be neglected.) Determine graphically the tensions in the middle, $C$, and at the ends, $B$ and $D$, of the cable, and the angle $\theta$ the cable makes with the horizontal at the tower. Check the results analytically using the equations of Art. 12-7.

When the end supports of flexible wires and cables are *not at the same level,* which generally is the case with communication and power wires in rolling or rough terrain, the point of greatest sag will *not* be midway between supports, but can be located as shown in Prob. 12-43. The maximum stress in the cable will then occur at the higher of the two supports. When the difference in elevation of the end supports does not exceed about 20 percent of the horizontal span and the maximum sag does not exceed about 10 percent of this span,

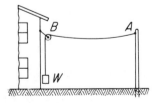

**Fig. 12-24   Probs. 12-40 and 12-41.**

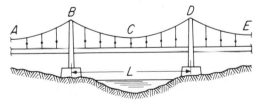

**Fig. 12-25   Probs. 12-42 and 12-56.**

measured vertically from a straight line between the end supports, the weight of the cable may, without appreciable error, be assumed to be a horizontally uniformly distributed load. The problem may then be solved as follows.

## ILLUSTRATIVE PROBLEM

**12-43.** A power cable spans a river and is supported at towers $A$ and $B$, as shown in Fig. 12-26. Support $A$ is 120 ft below $B$, and the lowest point of the cable is $s' = 30$ ft below $A$. The horizontal span $L$ is 600 ft. During winters, when coated with ice, the cable weighs 10 lb/ft. Solve for the distance $x$ from $A$ to the point of lowest sag $C$, and for cable stresses $H$, $T_A$, and $T_B$ under these conditions.

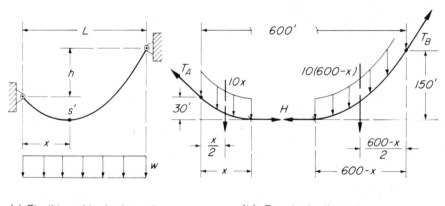

(a) Flexible cable, horizontally          (b) Free-body diagrams
        uniform load

**Fig. 12-26   Flexible cable, horizontally uniform load. Supports at different levels.**

*Solution:* The cable is cut at the point of maximum sag $C$, and free-body diagrams of both portions are drawn, as in Fig. 12-26b. The load on the left portion is clearly $10x$, and on the right portion is $10(600 - x)$, both concentrated at the midpoints (horizontally) of the respective sections. To solve for $x$, two equations of moment equilibrium may now be written, one about $A$ and the other about $B$, thus

$$[\Sigma M_A = 0] \qquad 30H = 10x\left(\frac{x}{2}\right) = 5x^2 \quad \text{or} \quad H = \frac{x^2}{6} \tag{a}$$

$$[\Sigma M_B = 0] \qquad 150H = (6000 - 10x)\left(300 - \frac{x}{2}\right)$$

$$\text{or} \qquad 150H = 5x^2 - 6000x + 1{,}800{,}000 \tag{b}$$

Substituting Eq. (*a*) for $H$ in Eq. (*b*), we obtain

$$150\left(\frac{x^2}{6}\right) = 25x^2 = 5x^2 - 6000x + 1{,}800{,}000$$

from which    $20x^2 + 6{,}000x = 1{,}800{,}000$    or    $x^2 + 300x = 90{,}000$

Completing the square gives

$$x^2 + 300x + (150)^2 = 90,000 + (150)^2 = 112,500$$

or       $x + 150 = \pm \sqrt{112,500} = 335$   and   $x = 185$ ft       *Ans.*

Substituting this value of $x$ in Eq. ($a$), we obtain

$$H = \frac{x^2}{6} = \frac{(185)^2}{6} = \frac{34,225}{6} = 5,704 \text{ lb} \qquad Ans.$$

The vertical component of $T_A$ is clearly $10x = 10(185) = 1850$ lb, and the vertical component of $T_B$ is $10(600 - 185) = 4150$ lb. The horizontal component of both $T_A$ and $T_B$ is $H = 5704$ lb. Hence

$$T_A = \sqrt{(1850)^2 + (5704)^2} = 6000 \text{ lb} \qquad Ans.$$

and       $$T_B = \sqrt{(4150)^2 + (5704)^2} = 7054 \text{ lb} \qquad Ans.$$

If desired, the slopes of the cable at $A$ and $B$ are easily determined, since $\cos \theta_A = H/T_A$ and $\cos \theta_B = H/T_B$.

## PROBLEMS

**12-44.** Solve Prob. 12-43 when $L = 400$ ft, $h = 80$ ft, and $s' = 20$ ft (see Fig. 12-26).

**12-45.** Solve Prob. 12-43 (see Fig. 12-26) when $L = 100$ m, $h = 20$ m, $s' = 5$ m, and $w = 200$ N/m.

Still another type of flexible cable problem is that of the cableway, often used to transport people and/or materials short distances over irregular terrain, or to transport material such as freshly mixed concrete from the mixing plant to the point of placement in a large concrete dam. For large spans, and where supports are not at the same level, such problems become rather complex, and will not be dealt with here. However, for spans not exceeding a few hundred feet where supports are at, or near, the same level, and the carriage load does not exceed a few thousand pounds, the approximate method shown in Prob. 12-46 will give results sufficiently accurate for practical purposes. The maximum cable tension occurs at the supports when the carriage is at the midpoint. In this type of cableway, however, it is customary to use a cable that is continuous between the points of anchorage ($D$ and $E$ in Fig. 12-27) over the supporting towers to which the cable is completely fastened to prevent slipping. Under these conditions, the greatest stress occurs in the backstay sections of the cable, being greatest there when dimension $d$ is comparatively small and decreasing as $d$ increases.

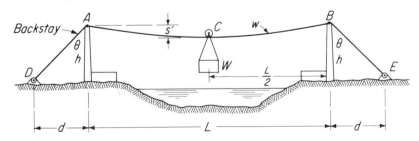

**Fig. 12-27   Simple cableway supporting movable concentrated load.**

## ILLUSTRATIVE PROBLEM

**12-46.** At a forest camp a cableway spans a river, as shown in Fig. 12-27. The tower supports $A$ and $B$ are at the same elevation. The horizontal span $L$ is 200 ft; the weight $w$ of the cable is 4 lb/ft, the height $h$ of the towers is 20 ft, and the weight $W$ of the fully loaded carriage is 800 lb. For convenient operation, the maximum sag $s'$ of the cable must equal 6 ft when the loaded carriage is at midspan. (a) Determine the maximum tension $T$ in section $AB$ of the cable under these conditions. (b) If the cable is continuous from $D$ to $E$, calculate the tension in the backstay portion $BE$, and the vertical compression in tower $B$, when $\theta$ is 45° (disregard weight of cable in backstay). (c) If the maximum allowable safe tension in the cable is 12,000 lb, calculate the required horizontal dimension $d$ from base of tower $B$ to anchorage $E$.

*Solution:* (a) The cable is cut at the midpoint $C$ and a free-body diagram of either half is drawn, as shown in Fig. 12-28a. The cable weight of 400 lb is applied at the midpoint of the right half as shown, and one-half of the carriage weight, equaling 400 lb, is applied at the midpoint $C$ of the entire cable, the other half of the carriage weight being supported by the left half of the cable not shown. Clearly, then,

$$[\Sigma V = 0] \qquad\qquad T_V = 400 + 400 = 800 \text{ lb}$$

A moment equation about $C$ gives $T_H$ as follows:

$$[\Sigma M_C = 0] \qquad\qquad 6T_H + 400(50) = 800(100)$$

from which $\quad 6T_H = 80,000 - 20,000 = 60,000 \quad$ and $\quad T_H = 10,000$ lb

Then, $\qquad\qquad T_B = \sqrt{(800)^2 + (10,000)^2} = 10,032 \text{ lb} \qquad\qquad Ans.$

(b) The free-body diagram of the backstay portion $BE$ of the cable is shown in Fig. 12-28b. Clearly,

$$[\Sigma H = 0] \qquad\qquad T_H = E_H = 10,000 \text{ lb}$$

and, when $\theta = 45°$, $\qquad\qquad E_H = E_V = 10,000$ lb

Hence, $\quad BE = \sqrt{(10,000)^2 + (10,000)^2} = \sqrt{200,000,000} = 14,140 \text{ lb} \qquad Ans.$

From Fig. 12-28 it is clear that the vertical compressive stress in the tower is

$$T_V + B_V = 800 + 10,000 = 10,800 \text{ lb} \qquad\qquad Ans.$$

(c) The stress in $BE$ exceeds the allowable 12,000 lb, but can be reduced to 12,000 lb by moving anchorage $E$ to the right, thereby reducing the vertical component of

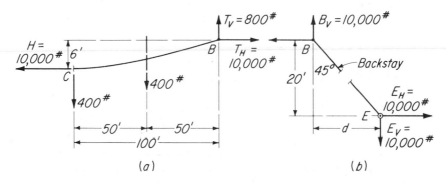

**Fig. 12-28    Free-body diagrams. (a) Right half of cable AB. (b) Right backstay BE.**

~~BE~~ while the horizontal component remains constant. The calculations may be as follows:

$$\text{Square of allowable stress} = (12,000)^2 = 144,000,000$$
$$\text{Square of component } E_H = (10,000)^2 = 100,000,000$$
$$\text{Square of component } E_V = (E_V)^2 = 44,000,000$$

from which
$$E_V = \sqrt{44,000,000} = 6630 \text{ lb}$$

Then, by proportion
$$\frac{E_V}{E_H} = \frac{6630}{10,000} = \frac{20}{d}$$

or
$$6630d = 200,000 \quad \text{and} \quad d = 30.2 \text{ ft} \qquad Ans.$$

## PROBLEMS

**12-47.** Solve parts (*a*), (*b*), and (*c*) of Prob. 12-46 when $L = 400$ ft, $w = 8$ lb/ft, $h = 30$ ft, $\theta = 60$, $W = 2000$ lb, $s' = 10$ ft, and the maximum allowable tension in the cable is 38,000 lb.

$$Ans. \quad T_B - 36,000 \text{ lb}; \quad BE = 41,600 \text{ lb}; \quad d = 88.8 \text{ ft}$$

**12-48.** Solve parts (*a*), (*b*), and (*c*) of Prob. 12-46 when $L = 100$ m, $w = 30$ N/m ft, $h = 20$ m, $\theta = 45°$, $W = 3$ kN, $s' = 10$ m, and the maximum allowable tension in the cable is 14 kN.

## SUMMARY (By Article Number)

**12-2.** The intensity of **hydrostatic pressure** at any point in a fluid is $wh$. This pressure is equal in all directions. The symbol $h$ represents the vertical distance below the *free surface*.

**12-3.** The **buoyancy** of a body is equal to the weight of the fluid displaced.

**12-4.** The *hydrostatic force* acting on a submerged plane area is equal to the product of the average pressure against the plane and the area of the submerged plane.

**12-5.** Lateral pressures caused by loose earth, sand, or mud against retaining walls and surfaces are similar to those caused by fluids, and are expressed in terms of the weight of an "equivalent fluid" which would produce these pressures, called the **equivalent fluid weight.**

**12-6.** The stress in a flexible cable or in an arch designed to be free of bending stresses under load is always parallel to the axis of the cable or arch. The stresses can be found graphically by drawing a force polygon.

**12-7.** A flexible cable under uniform horizontal load will hang as a parabola. The maximum cable tension is given by

$$T_{max} = \frac{wL}{2} \sqrt{\frac{L^2}{16s^2} + 1} \qquad (12\text{-}4)$$

and the angle of the cable at the support is given by

$$\theta = \tan^{-1}\left(\frac{4s}{L}\right) \qquad (12\text{-}5)$$

## REVIEW PROBLEMS

**12-49.** Figure 12-29 shows a partial section of a timber flume subjected to water pressure. The planking is supported by sloping posts $A$ spaced 4 ft on centers which are held in place partly by the struts $B$. Compute the total *hydrostatic load P* against one post, and the compressive force $C$ in one strut.

*Ans.* $P = 10,000$ lb; $C = 6670$ lb

**12-50.** The vertical planking of the timber flashboard shown in Fig. 12-30 is supported partly by a horizontal beam at midheight bolted to inclined steel rods. If the rods are spaced 1.5 m on centers, compute the tension $T$ in one rod.

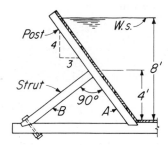
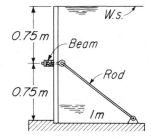
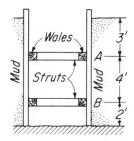

**Fig. 12-29   Prob. 12-49.**      **Fig. 12-30   Prob. 12-50.**      **Fig. 12-31   Prob. 12-51.**

**12-51.** The trench shoring shown in Fig. 12-31 supports mud with an *equivalent fluid weight* of 80 pcf. Horizontal struts are wedged between the wales and are spaced 3 ft on centers along the trench. Compute the compressive stresses, in each of struts $A$ and $B$, due to the pressure of the mud.

*Ans.* $A = 2430$ lb; $B = 7290$ lb

**12-52.** A temporary timber cofferdam will be subjected to the mud and water loads shown in Fig. 12-32. Assuming that the water weighs 62.5 pcf, the mud 120 pcf, and that the fluid weight of the mud is 120 pcf, determine the compressive force in the timber struts behind the dam. The struts are spaced at 6-ft centers. Neglect the weight of the timber structure in these calculations.

**12-53.** A pontoon bridge uses welded steel pontoons, 4 ft deep by 6 ft wide and 20 ft long, supporting a timber deck. Each pontoon weighs 5000 lb, and the timber deck

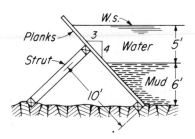
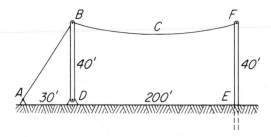

**Fig. 12-32   Prob. 12-52.**      **Fig. 12-33   Prob. 12-54.**

weighs 300 lb per foot of bridge. The pontoons are spaced 16 ft on centers. (a) How deep in the water will the pontoons float when they support only the dead weight of the bridge? (b) If the top of the pontoons must remain 1 ft out of the water at maximum load, how much *additional* total deck load $W$ can be supported by each pontoon?

*Ans.* (a) $d = 1.31$ ft; (b) $W = 12,700$ lb

**12-54.** A *flexible cable* is stretched between the tops of two poles which are 200 ft apart, as shown in Fig. 12-33. Pole $BD$ is pin-connected at the bottom, and its top is supported by the guy wire $AB$, lying in the plane $ABDE$. The lower end of the pole $FE$ is buried solidly in the ground. If the weight of the cable is 0.3 lb/ft and the sag at the center $C$ is 5 ft, compute the tension in the guy wire $AB$, and the reacting moment $M$ at $E$ in the pole $FE$. Neglect the weight of the guy wire.

*Ans.* $AB = 500$ lb; $M_E = 12,000$ lb·ft

**12-55.** A pipeline is supported over a river by means of wire hangers suspended from a *flexible cable,* similar to the main span shown in Fig. 12-25. The backstays are unloaded and make an angle of 45° with the ground. The tops of the towers are at the same height and are 60 m apart. The cable sag is 6 m. The pipeline weighs 10 kN/m, including an allowance for the weight of the cable and hangers. Determine graphically the tension at the center of the span and in the backstays. Check all results analytically.

*Ans.* $H = 750$ kN; $T_{max} = 807.8$ kN; $T_{BSTY} = 1061$ kN

**12-56.** In Fig. 12-25 the span $L$ of the footbridge is 50 m, the maximum sag $s$ of the cable at $C$ is 7.5 m, and the weight of the cable and deck is 3 kN per horizontal meter. Determine graphically the tension in the cable at $C$ and $D$. Check analytically.

## REVIEW QUESTIONS

**12-1.** Define *pressure,* and state the usual units of pressure.

**12-2.** What is the direction of the pressure of a fluid at rest on a submerged surface?

**12-3.** How is the intensity of liquid pressure at any point below the free surface obtained?

**12-4.** What is meant by (a) buoyancy and (b) load capacity?

**12-5.** How may the hydrostatic force against a submerged plane area be obtained?

**12-6.** What is meant by (a) fluid weight and (b) equivalent fluid pressure?

**12-7.** The outside rays in the force polygon for a flexible cable or arch give us what information?

**12-8.** What is the effect of moving the pole of the force polygon of a flexible cable farther from the loads?

**12-9.** What is the shape of a cable under uniformly distributed load if the load is (a) uniform per horizontal foot of span and (b) uniform per foot along the cable axis?

**12-10.** What is true of the horizontal component $H$ of the stress $T$ at any point in a flexible cable supported at its ends?

# Part 2
# Strength of Materials

# Chapter 1
# Stress and
# Deformation

## 1-1 SCOPE OF TEXT

The title of this text, *Applied Strength of Materials,* unfortunately does not disclose very well the actual contents between its covers, especially so to the student and reader not already fairly well acquainted with the subject matter. Such a reader would, quite naturally, perhaps assume that at least the major part of the text was devoted to discussions of the actual testing of various materials to determine their strengths, together possibly with extensive experimental test data and the final conclusions drawn from these data.

As a matter of fact, less than 5 percent of this text deals with the actual testing of materials to determine strength and other properties. The balance is devoted (1) to discussions and explanations of the relation existing between the externally applied forces and the internally induced stresses in various types of structural members and parts, such as bolts, rivets, shafts, pressure tanks, beams, columns, and so on, and (2) to explanations of the relation existing between these same externally applied forces and the resulting deformations, such as the elongation of a steel rod in tension, the shortening of a wood post in compression, the amount of twist in a shaft due to a torque, the deflection of a beam due to transverse loads, and so forth. The above text material, together with such closely allied subjects as riveted and welded connections, centroids of areas, moments of inertia of areas, analysis of statically indeterminate beams, combined axial and bending stress, and others, normally constitutes the content of a textbook on strength of materials.

Other titles, such as *Mechanics of Materials* and *Resistance of Materials,* that have been used by a few authors for many years do not seem to explain the scope of the text material any better. Most authors and professional engineers have therefore accepted the titles *Strength of Materials* or *Applied Strength of Materials.*

1

## 1-2 DEFINITION OF STRESS

**Stress** is defined as *resistance to external forces*. It is measured in terms of the *force exerted per unit of area*. In engineering practice, the force is generally given in pounds or newtons, and the resisting area in square inches or millimeters. Consequently stress is usually expressed in pounds per square inch, often abbreviated to psi, or newtons per square millimeter (MPa). Stress, then, is produced in all bodies upon which forces act.

In practice the word "stress" is often given two meanings: (1) force per unit area, or *intensity* of stress, generally referred to as **unit stress;** and (2) total internal force within a single member, generally called **total stress.** When the word stress is used alone, the context should make it clear which of the two meanings is intended.

## 1-3 THE BASIC STRESSES

Only two basic stresses exist: (1) **normal stresses,** which always act normal (perpendicular) to the stressed surface under consideration; and (2) **shearing stresses,** which act parallel to the stressed surface. (Normal stresses may be either tensile or compressive.) Other stresses either are similar to these basic stresses or are a combination of them. For example, the stresses in a bent beam, in a general way referred to as "bending stresses," actually are a combination of tensile, compressive, and shearing stresses; torsional stress, as encountered in the twisting of a shaft, is a shearing stress.

When the external forces acting on a member are parallel to its major axis and the member is of constant cross section, or substantially so, the resulting internal stresses are likewise parallel to that axis. Such forces are called **axial forces,** and the stresses are referred to as **axial stresses.**

**Tensile Stress.** When a pair of axial forces *pull* on a member, and thus tend to stretch or elongate it, they are said to be **tensile forces,** and they produce axial **tensile stresses** internally in the member on a plane lying perpendicular, or normal, to its axis. An example of such a condition is illustrated in Fig. 1-1a. An eyebar is subjected to a pair of tensile forces *P*. If a section is cut out of this bar, the internal tensile stresses are disclosed and, being *tensile* stresses, they are, of course, *pulling* on each of the two remaining ends of the bar as well as on the section removed.

The stressed area considered is a plane surface lying normal to the direction of stress. In general, it is the cross-sectional area of the stressed member, and the stress is generally assumed to be uniformly distributed over that area.

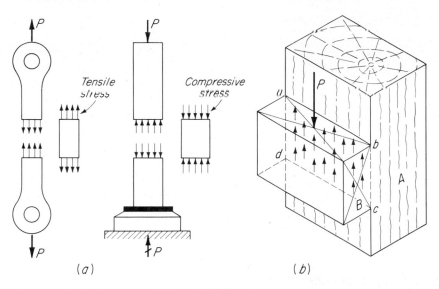

**Fig. 1-1**   The basic stresses. (*a*) Normal. (*b*) Shearing.

**Compressive Stress.** When a pair of axial forces *push* on a member and shorten or compress it, they are called **compressive forces** and they produce axial **compressive stresses** internally in the member on a plane perpendicular, or normal, to its axis. This condition is also illustrated in Fig. 1-1*a*. A short post is supporting at its top an axial load *P*, which, of course, is counteracted by the reacting force *P* at its base. When a section is cut out of this post, the internal compressive stresses are disclosed, pushing on the two remaining ends of the post as well as on the section removed.

As in the case of tension, the stressed area considered is a plane surface lying normal to the direction of stress. In general, it is the cross-sectional area of the stressed member, and the stress is assumed to be uniformly distributed over that area.

**Shearing Stress.** This type of stress differs from tensile and compressive stresses in that the stressed plane (the shear plane) lies parallel with the direction of stress rather than perpendicular to it, as in the cases of tensile and compressive stresses.

An illustration of shear is seen in Fig. 1-1*b*. The vertical-grained block of wood *A* is clamped tightly into a testing machine. A vertical force *P*, applied to the upper surface of the projecting portion *B* of the block, will tend to shear it off the main body of the block along the shear plane *abcd*. Such action is resisted by the shearing strength of the wood fibers. The unit shearing stresses produced by force *P* on surface *abcd* are indicated by the small arrows.

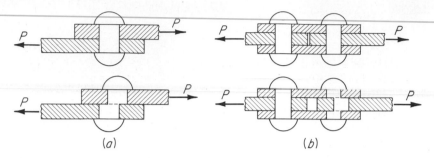

Fig. 1-2   Shear failures of rivets. (a) Single shear. (b) Double shear.

Two examples of shearing failure of simple riveted connections are shown in Fig. 1-2. In the lap joint on the left, failure will occur on the single plane through the rivet coinciding with the plane of the contacting surfaces. This case is referred to as **single shear,** and the area in shear is the cross-sectional area of the rivet. In the butt joint on the right, the right rivet is the smallest, and it will therefore fail first in shear, while the other rivet will hold. But shear failure must of necessity occur simultaneously on the two planes through the rivet which coincide with the two surfaces of contact between the plates. This case is called **double shear,** and the total shear area is naturally twice the cross-sectional area of the rivet.

In the case of bolts and rivets and in beams, as will be shown later, the forces producing shearing stresses generally act *transversely* to the longitudinal axis of the member. This is in contrast to tensile and compressive forces, which generally act *parallel* to that axis.

In shear, the stressed area considered is a plane surface lying parallel to the direction of stress. In the case of bolts and rivets and in beams, it is generally the cross-sectional area of the member. Depending on the conditions involved, shear stresses may or may not be considered to be uniformly distributed over the stressed area.

**Bearing Stress.** Compressive stress exerted on an *external* surface of a body, as, for example, when one object presses against another, is referred to as **bearing stress.** Examples of bearing stresses can be found in the middle sketch of Fig. 1-1. Let us assume that we have here a timber post resting on a steel baseplate supported on a concrete footing, which in turn is ultimately supported by the soil. The post *bears* on the plate, the plate *bears* on the footing, and the footing *bears* on the soil. The resulting compressive stresses at the external surfaces of contact between these bodies are *bearing stresses*. In the examples just mentioned, the stressed areas are plane surfaces, all lying normal to the direction of stress. Bearing on curved surfaces, such as occur on bolts and rivets, is discussed in Chap. 3.

## 1-4 ULTIMATE STRESS, ALLOWABLE STRESS, FACTOR OF SAFETY, UNITS

Before a material can safely be used for making structural parts or members, a number of its properties must be known. In the case of metals such as steel, cast iron, bronze, aluminum, and others which are widely used as structural members and machine parts we must have reliable information concerning such properties as strength, elasticity, stiffness, ductility, malleability, toughness, and so forth.

The production of these metals from the raw materials, and the subsequent processes required for their formation into structural parts, have been so well standardized that the properties found through tests to be possessed by a random sample may safely be assumed to be possessed in like degrees by similar materials produced and processed under similar conditions. Methods used to determine some of these properties are discussed in Chap. 2.

**Ultimate stress,** or strength, is defined as *the greatest unit stress a material can withstand without rupture.* In the practical design of structures and machines, we would not, of course, for obvious reasons of safety, attempt to use the full strength of any material.

**Allowable stress** is that portion of the ultimate strength which may safely be used in design. The terms *working stress* and *safe working stress* have been given the same meaning and both are widely used. However, the term *allowable stress* is most generally accepted and is used in this text.

**Factor of safety** is *the ratio of "failure" stress to allowable stress.* Failure may mean exactly that, or quite often, it may simply mean excessive deformation. The appropriate factor of safety for any given design depends upon a number of considerations. Among these are the following:

1. The degree of safety required. That is, what dangers to human life and property are involved?
2. The degree of economy desired. A tendency would exist to use a high percentage of the strength of costly materials.
3. Dependability of the material. Are flaws easily detected? Do apparently similar pieces behave similarly under stress? Does the material give ample warning of impending failure? Hard, brittle materials fail without warning; "soft," tough materials deform and thus give warning.
4. Permanency of design. Lower factors of safety are often permitted in temporary designs.
5. Load conditions, whether steady, varying, or shock loads.
6. The exactness with which probable loads and stresses can be predicted or determined.

7. Accessibility of parts for inspection and maintenance.
8. Relative difficulty in preventing deterioration of the material, such as decay of timber, rusting of steel, disintegration of concrete, and so forth.

Designers of structures and machines generally work with a predetermined allowable stress. But in aircraft, building, and bridge design, the ultimate strength and a predetermined factor of safety are often used.

Allowable stresses which may safely be used in various designs are as a rule based upon extensive laboratory tests made to disclose the various properties of the materials used and upon a vast amount of accumulated experience. Most large cities have established building codes which govern all structural design within their respective municipalities. The federal government, the Army, the Navy, cities, and many counties all have their separate codes.

The factors of safety suitable in aircraft design are generally determined by the severity of the conditions under which the aircraft is expected to be flown. Machines are usually designed to withstand all reasonable wear and tear. A factor of safety of 2 might be sufficient for a temporary structure subjected only to steady loads and endangering neither life nor property, while a factor of safety as high as 10 might be required in a machine subjected to unpredictable shock loads whose failure would endanger life and cause loss of other property. For steady loads, a factor of safety between 3 and 4 is commonly used.

Tables giving properties of various materials and suitable allowable stresses are found throughout the text. A common test to determine the properties of strength and elasticity of steel is described in Art. 1-12.

**Units.** When dealing with quantities such as forces, areas, and volumes, we must establish a system of *units* of these quantities and their accepted abbreviations. The English abbreviations shown in Table 1-1 agree with those recommended by the American National Standards Institute (ANSI). A **kip** (shortened from *kilopound*) represents 1000 lb.

In the SI system abbreviations for units are not capitalized unless the unit is derived from a proper name. Examples are m (meter), rad (radian), mm (millimeter), but N (Newton) and Pa (Pascal). The prefixes M (mega), k (kilo), and m (milli) have the significance indicated below.

$$1 \text{ MPa} = 1000 \text{ kPa} = 1\,000\,000 \text{ Pa} \qquad \text{and} \qquad 1000 \text{ mm} = 1 \text{ m}$$

Note that commas are *not* used to separate groups of zeros. Additional examples are:

$$2\,400\,000 \text{ Pa} \qquad \text{and} \qquad 0.000\,078\,5 \text{ m}$$

We could write these quantities as $2.4 \times 10^6$ Pa or 2.4 MPa and $78.5 \times 10^{-6}$ m or 0.0785 mm.

Table 1-1 ENGINEERING UNITS AND ABBREVIATIONS

| Item | Common Units | Abbreviations |
|---|---|---|
| Force | Pounds, kips, tons, newtons | lb, k, N, kN |
| Length | Feet, inches, miles, meters, milli meters, kilometers | ft, in., mi, m, mm, km |
| Area | Square inches, square feet, square meters, square millimeters | in.$^2$, ft$^2$, m$^2$, mm$^2$ |
| Force (or load) per unit length | Pounds per foot, newtons per meter | lb/ft, N/m, kN/m |
| Force (or load) per unit area | Pounds per square inch, kips per square inch, newtons per square meter or pascals | psi, ksi, N/m$^2$ or Pa, N/mm$^2$ or MPa, kN/m$^2$ |
| Volume | Cubic inch, cubic meter, cubic millimeter | in.$^3$, m$^3$, mm$^3$ |
| Weight per unit volume | Pounds per cubic inch, pounds per cubic foot, newtons per cubic meter | lb/in.$^3$, lb/ft$^3$, N/m$^3$ |

The gravitational pull (that is, the weight) on a 1-kg mass is 9.806 N. A beam whose mass is 100 kg/m weighs 980.6 N/m. If the dimensions in a sketch represent millimeters, no units are required on the sketch.

## 1-5 THE DIRECT-STRESS FORMULA

The types of stresses described in the preceding articles are often referred to as **direct stresses,** in order to distinguish them from torsional, bending, and other stresses.

In cases similar to the three illustrated in Fig. 1-1, the reasonable assumption is usually made that the intensity of stress is about the same on all units of stressed area. That is, the maximum and minimum unit stresses do not differ greatly from the average stress. While this assumption in some instances may be incorrect in varying degrees, such instances are usually recognized and appropriate modifications can readily be made.

Using the assumptions stated above, we may now establish a relation between the average intensity of stress $s$ produced by an externally applied force $P$ on each unit of stressed area $A$. The total internal force is clearly the product of the average unit stress and the stressed area, which is $sA$. And this total internal force $sA$ must, according to the law of equilibrium, be equal to the external force $P$ producing it, since the two forces are collinear. Consequently,

$$P = sA \qquad \text{or} \qquad s = \frac{P}{A} \qquad\qquad (1\text{-}1)$$

where $s$ = average unit stress, psi or MPa

$P$ = total external force or load, lb or N

$A$ = stressed area, in.$^2$ or mm$^2$

Study of Eq. (1-1) discloses that (1) given $P$ and $A$, we may obtain $s$; (2) given $s$ and $A$, we may obtain $P$; and (3) given $s$ and $P$, we may obtain the required area $A$. Of these cases, (1) and (2) are problems in *analysis;* that is, a given member is analyzed to determine either its unit stress or its load capacity. Case (3) is one of *design;* that is, a member of cross-sectional area $A$ must be selected or designed to withstand the given load $P$.

The unit of force in SI is the newton (N). The English system of units does not have a special name for the *force per unit area.* The force per unit area is usually expressed as psi (pounds per square inch) or ksi (kips per square inch). The corresponding quantities in SI are newton per square meter (N/m$^2$) or newton per square millimeter (N/mm$^2$). However, in SI, the force per unit area has a special name, the pascal (Pa). One newton per square meter is equivalent to one pascal. We should note the following relations:

$$1 \text{ m}^2 = 10^6 \text{ mm}^2 \quad \text{and} \quad 1 \text{ N/mm}^2 = 10^6 \text{ Pa} = 1 \text{ MPa}$$

In Eq. (1-1) the load $P$ may be expressed in newtons (N) and the area in square millimeters (mm$^2$). The force per unit area is then in N/mm$^2$ or its equivalent, MPa.

## 1-6 SIMPLE STRUCTURAL MEMBERS

Many of the structural members used in common engineering practice, such as timber beams and columns, steel bolts, rods, pipes, beams, and columns, and many others, are commercially produced in a wide variety of sizes whose cross-sectional areas and other properties can readily be determined from tabulated information furnished by the manufacturers. A number of such tables are given in Appendix E.

**Threaded Steel Rods.** In many problems involving analysis or design, several details must be considered. For example, if a steel rod is used as a tension member in a truss, as illustrated in Fig. 1-3, its ends are usually threaded to facilitate its being fastened to other members. In such a case, the threading of a 1-in.-diameter rod with 8 threads per inch would reduce its effective diameter from 1 in. to 0.8376 in. at the root of the threads and its area from 0.7854 in.$^2$ to a net **root area** of 0.551 in.$^2$, a reduction of about 30 percent.

The method of calculating the allowable tensile load on threaded rods and bolts varies. Among engineers, some use the *nominal* bolt diameter in calculating the stressed area, some use the *root* area as the stressed area, and some use a tensile stress area between these two extremes (see Appendix E-3). We will use the *root area* as the stressed areas, unless otherwise stated in the problem.

Two series of threads are in general use: (1) American Standard Coarse Series for structural work and (2) American Standard Fine Series, mostly used in machine work. If the root area is used to determine the allowable tensile load, the material between the threaded ends of a rod is stressed to only part of its usable strength. To overcome this inefficiency in long rods, **upset ends** are often used; that is, the part to be threaded is enlarged so that its root diameter after threading will not be smaller than the diameter of the unthreaded portion of the rod.

**Standard timber** is cut in rectangular shapes to *rough* cross-sectional dimensions which generally are multiples of 2 in. (or 50 mm), such as 2 by 4 in., 6 by 10 in., or 12 by 18 in. The actual sizes of surfaced timber are less than these nominal dimensions. Most structural timber is dressed, or surfaced, on all four sides, a treatment denoted S4S (see Table E-2).

**Structural steel shapes,** such as I beams, wide-flange sections, channels, angles, and pipe sections, all widely used in structural work, have been standardized as to size and weight per unit length. The cross-sectional area, weight, and several other useful properties of these sections are given in Tables E-4 to E-9 inclusive.

## 1-7  PROBLEMS IN ANALYSIS AND DESIGN

In engineering practice the term **analysis** includes within its meaning (1) all work relating to the investigation required to estimate correctly all external loads and forces to which a structure or machine which is to be designed may possibly be subjected and (2) the determination of the resulting internal forces in the various members and other component parts. **Applied mechanics,** then, is essentially analysis.

When applied mechanics is combined in such investigations with a knowledge of details and aspects of structural elements, as those of buildings, bridges, and aircraft frames, for example, the process is called **structural analysis.** For economic and other reasons, several differing analyses may be made before one is finally selected from which the actual design may proceed.

**Design** is the final determination of the required sizes of all the various elements, members, and joint details which constitute the structure as a whole and from which determination the actual construction may be accomplished. Design is, then, essentially a problem of *materials,* involving their strength and other properties. The unit stresses which may safely be used in design are called **allowable stresses,** and their values depend upon several properties of the materials in question. The bases for their determination are discussed in Art. 1-4.

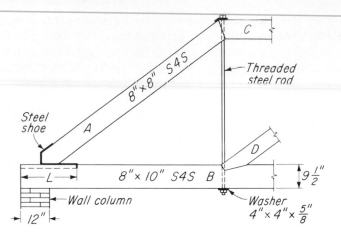

**Fig. 1-3   Unit stresses. Analysis.**

### ILLUSTRATIVE PROBLEMS

**1-1.** Figure 1-3 shows a part of a common type of roof truss, constructed mainly of timbers and steel rods. Let us assume that the forces to be resisted by the various members, and the sizes of the members, have already been determined. The timbers are surfaced (S4S) and the steel rod is threaded. Determine (*a*) the average unit compressive stress in the 8- by 8-in. diagonal timber *A* if the total stress in it is 20 kips (1 kip = 1,000 lb); (*b*) the maximum average unit stress in the ¾-in. threaded steel rod if the total stress in it is 4 kips; (*c*) the average unit bearing stress between the timber and the 4- by 4-in. square steel washer if the hole in it is ⅞ in. in diameter; (*d*) the average unit bearing stress between the brick wall column and the 8- by 10-in. timber if the total load on the column is 15 kips; (*e*) the unit shearing stress along the dashed line in the 8- by 10-in. timber if the horizontal thrust against the steel shoe is 16 kips and *L* = 22 in.

*Solution:* We note that both timbers are surfaced (S4S) and that the steel rod is threaded. Hence from Table E-2 we find the net area of the 8- by 8-in. timber to be 56.25 in.² (7½ by 7½ in.). Similarly, from Table E-3, we find the root area of the ¾-in.-diameter rod to be 0.302 in.² Consequently

(*a*)   $P = 20,000$ lb   $A = 56.25$ in.²   $s_c = \dfrac{P}{A} = \dfrac{20,000}{56.25} = 356$ psi   *Ans.*

(*b*)   $P = 4000$ lb   $A = 0.302$ in.²   $s_t = \dfrac{P}{A}\dfrac{4000}{0.302} = 13,260$ psi   *Ans.*

(*c*) The net area of the 4- by 4-in. washer is $16 - 0.601 = 15.399$ in.². Hence

$P = 4000$ lb   $A = 15.399$ in.²   $s_b = \dfrac{P}{A} = \dfrac{4000}{15.399} = 260$ psi   *Ans.*

(*d*) The 8- by 10-in. timber is 7½ in. wide and bears 12 in. on the brick wall column, giving a bearing area of 90 in.². Therefore

$$P = 15{,}000 \text{ lb} \qquad A = 90 \text{ in.}^2 \qquad s_b = \frac{P}{A} = \frac{15{,}000}{90} = 167 \text{ psi} \qquad Ans.$$

(*e*) The horizontal component of the total stress in the 8- by 8-in. timber is 16 kips. This force tends to shear off the part of the timber above the dashed line. The width of the shear area is 7½ in., and its length is 22 in., a total area of 165 in.². Then, since $P = 16{,}000$ lb and $A = 165$ in.²,

$$s_s = \frac{P}{A} = \frac{16{,}000}{165} = 97 \text{ psi} \qquad Ans.$$

**1-2.** A typical warehouse-floor support is shown in Fig. 1-4. The short timber post $C$ is capped by a section of steel channel in order to provide greater bearing area for the ends of beams $A$ and $B$. These beams are 200 by 400 mm, with the greatest dimension in a vertical plane. The steel baseplate $P$ likewise provides greater bearing area on the concrete footing $F$. The load transmitted to post $C$ from each of beams $A$ and $B$ is 70 kN, making a total load on the post of 140 kN. The weights of the cap, post, plate, and footing may here be neglected.

Find (*a*) the minimum length $L$ of channel required to support the beams if the maximum bearing stress allowed in the timber *perpendicular to the grain* is 2.2 MPa (*b*) the minimum available nominal size of square timber post required if the unit stress parallel to the grain is not to exceed 7.0 MPa; (*c*) the required size of the square steel baseplate $P$ if the permissible unit bearing stress on the concrete is 3.5 MPa; (*d*) the minimum dimension $d$ of the square footing $F$ if the allowed pressure on the soil is 0.4 MPa.

*Solution:* When the weights of all parts are disregarded, the load $P$ will be 140 000 N. Then

(*a*)     $P = 140\ 000$ N     $s = 2.2$ MPa     $A = \dfrac{P}{s} = \dfrac{140\ 000}{2.2} = 63\ 600 \text{ mm}^2$

$$A = 200L \qquad \text{and} \qquad L = \frac{63\ 600}{200} = 318 \text{ mm} \qquad Ans.$$

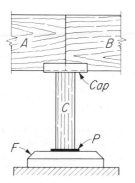

**Fig. 1-4   Unit stresses. Design.**

(b)   $P = 140\ 000$ N      $s = 7.0$ MPa      $A = \dfrac{P}{s} = \dfrac{140\ 000}{7.0} = 20\ 000$ mm²

Use 150 by 150 mm post ($A = 22\ 500$ mm²).          *Ans.*

(c)   $P = 140\ 000$ N      $s = 3.5$ MPa      $A = \dfrac{P}{s} = \dfrac{140\ 000}{3.5\ \text{MPa}} = 40\ 000$ mm²

Use 200 by 200 mm baseplate.          *Ans.*

(d)   $P = 140\ 000$ N      $s = 0.4$ MPa      $A = \dfrac{P}{s} = \dfrac{140\ 000}{0.4\ \text{MPa}} = 350\ 000$ mm²

Use 600 by 600 mm footing.          *Ans.*

PROBLEMS

**1-3.** In Prob. 1-1 assume that (a) the diagonal member is 6 by 6 in. S4S and that the total stress in it is 18 kips; (b) the steel rod is of 1-in. diameter and the total stress in it is 7.5 kips; (c) the washer is 5 by 5 in. in size with a 1⅛-in. hole; (d) timber B is 6 by 10 in. S4S (9½ in. vertical) and the load on the column is 15 kips; (e) the horizontal thrust against the steel shoe is 12 kips and $L = 18$ in. Then solve the five parts.

*Ans.* (a) $s_c = 594$ psi;   (b) $s_t = 13,600$ psi;   (c) $s_b = 312$ psi;   (d) $s_b = 227$ psi;   (e) $s_s = 121$ psi

**1-4.** Assume in Prob. 1-2 that beams A and B are 8 by 20 in. S4S (19½ in. vertical), that each beam transmits a load of 16 kips to the post, and that the allowable stresses are as follows: in timber, 320 psi perpendicular to the grain and 1100 psi parallel to the grain; on concrete, 500 psi; on soil, 8000 psf. Then solve the four parts.

**1-5.** A round, threaded steel rod of 30-mm diameter supports an axial load $P$ in tension. Calculate (a) the maximum unit tensile stress $s_t$ in the rod if the axial load $P$ is 50 kN, and (b) the maximum allowed load $P$ on this rod if the unit tensile stress must not exceed 138 MPa. (c) What are the corresponding answers for a rod with upset ends?

*Ans.* (a) $s_t = 96.3$ MPa; (b) $P = 71.6$ kN; (c) $s_t = 70.8$ MPa, and $P = 97.4$ kN

**1-6.** Determine the required nominal diameter $d$ of a threaded steel rod to carry an axial load of 20,000 lb in tension if a unit stress of 20,000 psi is permitted. What is the corresponding answer for a rod with upset ends?

*Ans.* $d = 1⅜$-in. threaded rod and $d = 1⅛$-in. upset rod (less than 1 percent overstress)

**1-7.** A short 8- by 8-in. S4S timber post supports an axial compressive load. Calculate (a) the maximum unit compressive stress $s_c$ in the post if the load $P$ is 40,000 lb and (b) the maximum allowed load $P$ on this post if the permissible stress in it is 1000 psi. (c) What are the corresponding answers for a post that is rough instead of S4S?

**1-8.** Determine the required nominal size of a short, square, rough timber post to withstand an axial compressive load of 245 kN if the maximum allowable unit stress is 7 MPa.

*Ans.* 200 by 200 mm

**1-9.** The threaded steel rod shown in Fig. 1-5 supports a vertical post of a timber earth-retaining wall. (*a*) Determine the smallest suitable diameter *d* which may safely be used if the allowable unit stress in the rod is 20,000 psi. (*b*) Find also the required side dimensions of a square steel-plate washer if stress in bearing perpendicular to the grain of the timber is 257 psi and the hole in the washer is ¾ in. in diameter.

*Ans.* (*a*) ⅝ in.; (*b*) 4 by 4 in.

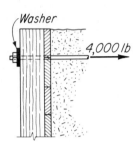

Fig. 1-5    Prob. 1-9.

**1-10.** The short steel column *C* shown in Fig. 1-6 supports an axial compressive load of 100,000 lb and is welded onto a steel baseplate *B* resting on a concrete footing *F*. (*a*) Select the lightest (most economical) wide-flange section to use if the unit stress is not to exceed 13,500 psi. (*b*) Determine the required side dimension *b* of the square baseplate *B* if the allowable bearing stress on the concrete is 525 psi. (*c*) Calculate the required side dimension *f* of the base of the square concrete footing if the pressure on the soil is not to exceed 5000 psf. Disregard weight of column, baseplate, and footing.

*Ans.* (*a*) W 10 × 26; (*b*) 14- by 14-in. plate; (*c*) $f$ = 4 ft 6 in.

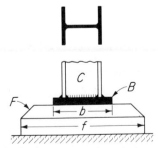

Fig. 1-6    Prob. 1-10.

**1-11.** A square, hollow tube of wrought aluminum alloy of 40-mm outside dimension with a wall thickness of 4 mm is used as a structural member in an airplane fuselage. Calculate the unit stress in this member caused by an axial tensile load of 50 kN.

*Ans.* $s_t = 86.8$ MPa

**1-12.** A piece of schedule 80 steel pipe used as a structural column is to support an axial load of 45,000 lb. If the allowable unit stress in the column is 12,000 psi, what size pipe should be used?

*Ans.* 4 in. nominal diameter

**1-13.** Figure 1-7 shows the cross section of a heavy structural steel column built up of one 12- by ½-in. web plate, two 13- by ⅜-in. cover plates, and four 6- by 4- by ⅝-in. angles. If the allowable unit stress in the column is 14,000 psi, what total load *P* can it support?

*Ans.* $P = 549,000$ lb

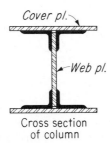

Cover pl.
Web pl.
Cross section
of column   **Fig. 1-7   Prob. 1-13.**

**1-14.** Two steel straps, each 50 mm wide by 10 mm thick, are held together with an M20 bolt, as shown in Fig. 1-8. (*a*) If the allowable shearing stress in the bolt is 70 MPa, what maximum axial load *P* can the bolt resist in shear? (*b*) If the bolt hole is of 22-mm diameter and the allowable unit tensile stress in the straps is 100 MPa, what maximum load *P* is permitted in tension? Possible friction between the straps may be disregarded.

*Ans.* (*a*) $P = 22$ kN; (*b*) $P = 28$ kN

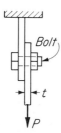

Bolt
*t*
*P*   **Fig. 1-8   Probs. 1-14 and 1-15.**

**1-15.** The tensile load *P* in Fig. 1-8 is 5600 lb. Calculate (*a*) the required diameter *d* of the bolt and (*b*) the required width *w* of the straps. (Select bolt diameter of the nearest higher multiple of ⅛ in.) Assume that bolt-hole diameter is ⅛ in. larger than bolt and that thickness of strap is ⅜ in. Allowable stresses: shear, 10,000 psi; tension, 15,000 psi. Disregard possible friction between the straps.

**1-16.** An eyebar is connected to a fitting by means of a 15-mm-diameter finished pin, as shown in Fig. 1-9. (*a*) What maximum pull *P* in the bar can this pin resist in shear if its allowable shearing stress $s_s$ is 100 MPa? (*b*) If a tensile stress $s_t$ of 112 MPa is permitted in the round bar, what diameter *d* will give the bar the same strength in tension as the pin has in shear?

*Ans.* (*a*) *P* = 35.3 kN; (*b*) *d* = 20 mm

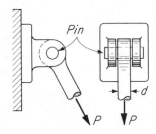

Fig. 1-9   Probs. 1-16 and 1-17.

**1-17.** The pull *P* on the eyebar in Fig. 1-9 is 3000 lb. Determine the required diameters *d* of (*a*) the pin and (*b*) the eyebar. Allowable stresses: shear, 15,000 psi; tension, 20,000 psi. (Select diameters of nearest higher multiple of $^1/_{16}$ in.)

*Ans.* (*a*) *d* = $^3/_8$ in.; (*b*) $^7/_{16}$ in.

**1-18.** In Fig. 1-10 a 1-in.-diameter square-headed bolt passes through a heavy supporting steel plate. Under the tensile pull *P* the bolt might fail in three ways: (1) in tension across the 1-in.-diameter shank, (2) in shear on the cylindrical surface formed if the shank of the bolt should pull straight out of its head, and (3) in bearing on the contacting surfaces between the bolt head and the supporting plate. Calculate the average unit tensile, shearing, and bearing stresses resulting from a pull *P* of 15,000 lb.

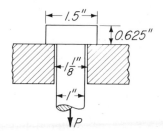

Fig. 1-10   Prob. 1-18.

## 1-8 AXIAL DEFORMATION OR STRAIN

The meaning of the term **strain,** as used in engineering practice, is often misunderstood. *Strain is deformation,* although the word is often used by laypersons to designate the *force* which produces deformation in some object. For example, in an expression such as "the fast-moving train subjected the bridge to a *terrific strain*," the word *strain* is undoubtedly used to mean *force* rather than deformation, since, in the case envisioned, the force may be "terrific," while the strain or deformation may actually be comparatively small.

All material bodies subjected to external forces, with internal stresses produced as a consequence, are necessarily deformed (strained). For example, a long rod subjected to an axial tensile load would be stretched or elongated, while a column supporting an axial load would be compressed or shortened. The *total deformation* produced in a member is designated by the Greek letter δ (delta). If the length of the member is $L$, the deformation per unit of length, expressed by the Greek letter $\epsilon$ (epsilon), is

$$\textbf{Unit deformation} = \frac{\textbf{total deformation}}{\textbf{length}} \quad \text{or} \quad \epsilon = \frac{\delta}{L} \quad \textbf{(1-2)}$$

The quantities total deformation δ and length $L$ are generally given in inches. Consequently, unit deformation $\epsilon$ will be in inches per inch (or millimeters per millimeter). Obviously, then, the quantity *unit deformation* carries no mathematical unit, since inches/inches = 1. In a perfectly straight bar of homogeneous material and of constant cross section, $\epsilon$ would be the *actual* unit deformation. Since such perfection does not exist, we must realize that in practice $\epsilon$ is the average unit deformation.

The amount of deformation produced in a given member by a given force will vary with the stiffness of the material of which the member is made. In some instances, a considerable deformation may not be objectionable; in others, even small deformations may produce serious results. For example, a deflection of as much as 150 mm of a 15-m airplane wing under certain flight conditions might well correspond with expectations, while so high a deflection of a 15-m beam in a building would probably cause its condemnation. Also, in many precision machines even exceedingly small deformations of certain parts may render a machine unserviceable.

## 1-9 ELASTICITY. ELASTIC LIMIT

Another important property of a structural material is that having been deformed by a force, it must be capable of complete recovery of its original form upon removal of the deforming force. A material having this property is said to be **elastic.**

Considerable misconception exists regarding the true technical meaning of elasticity. A material is popularly thought to be "elastic" if it can withstand a high percentage of deformation without injury. Thus, rubber is considered to be highly elastic. Technically speaking, however, a material is elastic only if it has the ability to return to its original form after removal of a deforming force. Within this meaning, both steel and glass are elastic.

Good examples of the value of the elastic property of steel are found in the wide use of steel springs: the balance and main springs of a watch, the coil springs used in furniture and beds and as wheel springs on many auto-mobiles, and the spring in a door check which automatically shuts the door after it has been opened.

In all these instances, the satisfactory performance of the intended duty of each spring depends upon the complete recovery of its original form after removal of a deforming force. Failure to recover completely would render such a spring useless.

However, in order to preserve the elastic property of a material having limited physical strength, deformations and the stresses which accom-pany the deformations must not exceed a certain limit, appropriately re-ferred to as its **elastic limit.** A material stressed beyond its elastic limit will return only partially to its original form upon complete removal of the de-forming force. The deformation remaining is called **permanent set.**

## 1-10 HOOKE'S LAW. MODULUS OF ELASTICITY

According to Hooke's law, *stress is proportional to strain.* To this should be added, *within the elastic limit.*

The various properties of materials are now quite readily being deter-mined in testing laboratories under well-standardized conditions, and Hooke's law can easily be verified.

For example, to verify the proportionality of stress to strain, and to de-termine the elastic limit of so-called soft steel, a standard bar is clamped tightly into a testing machine capable of exerting a gradual pull on the bar. At various intervals, the pull on the bar and the resulting elongation are measured simultaneously. Equal increments in stress are found to pro-duce equal increments in deformation up to a stress of about 36,000 psi (250 MPa), thereby substantiating the theory of proportionality of stress to strain. For stresses beyond this point the elongations will increase at a much faster rate than the stresses, indicating thereby that the elastic limit has been passed.

The proportionality of stress to strain is expressed as *the ratio of unit stress to unit strain,* or deformation. In a stiff but elastic material, such as steel, we find that a given unit stress produces a relatively small unit de-formation. In a softer but still elastic material, such as bronze, the defor-mation caused by the same intensity of stress is about twice that of steel, and in aluminum it is three times that of steel.

Consequently the ratio of unit stress $s$ to unit strain $\epsilon$ of any given mate-rial, which may be determined experimentally, then gives us a measure of its stiffness or elasticity, which we call the **modulus of elasticity** of the material, denoted by the symbol $E$. That is,

$$\text{Modulus of elasticity} = \frac{\text{unit stress}}{\text{unit deformation}} \quad \text{or} \quad E = \frac{s}{\epsilon} \quad \text{(1-3)}$$

We now have a fixed relation between modulus of elasticity, unit stress, and unit deformation. The values of the moduli of elasticity of most engineering materials are given in various handbooks (see also Tables 1-2, 8-1, and E-1). For example, for steel, $E$ is about 29,000,000 psi; for bronze, $E$ is about 15,000,000 psi; for brass, $E$ is about 13,500,000 psi; for timber, which is somewhat less elastic, $E$ varies from 1,100,000 to 1,760,000 psi; and for concrete, still less elastic, $E$ varies roughly from 1,500,000 to 3,000,000 psi.

In Eq. (1-2) the relation $\epsilon = \delta/L$ (or $\delta = \epsilon L$) was established, and from Eq. (1-3), $\epsilon = s/E$. Also, $s = P/A$. We may now formulate an equation to determine the total deformation $\delta$ in a prismatic member subjected to an axial load $P$. That is,

$$\delta = \epsilon L = \frac{s}{E} \cdot L = \frac{P}{A} \cdot \frac{1}{E} L$$

or
$$\delta = \frac{PL}{AE} \tag{1-4}$$

where $\delta$ = total axial deformation, in. or mm
  $P$ = total axial load, lb or N
  $L$ = length of member, in. or mm
  $A$ = cross-sectional area of member, in.$^2$ or mm$^2$
  $E$ = modulus of elasticity of material, psi or MPa (N/mm$^2$)

Equation (1-4) holds only for stresses *below* the elastic limit of the material. A discussion of the determination of elastic limits of various materials is given in Chap. 3.

ILLUSTRATIVE PROBLEM

**1-19.** Determine the total elongation $\delta$ of a steel rod 1 in. in diameter and 20 ft long produced by a tensile load $P$ of 15,000 lb if the modulus of elasticity $E$ of the steel is 29,000,000 psi.

*Solution:* According to the list of values following Eq. (1-4), the length of 20 ft must be converted to inches. A bar of 1-in. diameter has a cross-sectional area of 0.7854 in.$^2$. Then, since $P$ = 15,000 lb, $L$ = (20)(12) = 240 in., $A$ = 0.7854 in.$^2$, and $E$ = 29,000,000 psi,

$$\left[ \delta = \frac{PL}{AE} \right] \qquad \delta = \frac{(15,000)(240)}{(0.7854)(29,000,000)} = 0.158 \text{ in.} \qquad Ans.$$

PROBLEMS

**1-20.** A surveyor's steel tape with a cross-sectional area of 0.006 in.$^2$ must be stretched with a pull of 16 lb when in use. If the modulus of elasticity of this steel is 30,000,000 psi, (*a*) what is the total elongation $\delta$ in the 100-ft tape; (*b*) what unit tensile stress is produced by the pull?

*Ans.* (*a*) $\delta$ = 0.1069 in.; (*b*) $s_t$ = 2670 psi

Table 1-2 SOME AVERAGE PROPERTIES OF STRUCTURAL MATERIALS UNDER STEADY LOADS, ksi (MPa)

| Material | Allowable Stress | | | Modulus of Elasticity | Shearing Modulus of Elasticity |
| | Tension | Compression | Shear | | |
| --- | --- | --- | --- | --- | --- |
| Steel, carbon* | 22 (152) | 22 (152) | 14.5 (100) | 29,000 (200 000) | 12,000 (83 000) |
| Low alloy | 25–30 | 25–30 | 17–20 | 29,000 (200 000) | 12,000 (83 000) |
| | (172–207) | (172–207) | (117–138) | | |
| Cast iron, gray | 5–8 | 20–30 | 7.5–12.5 | 14,000–18,300 | 5,600–7,400 |
| | (35–55) | (117–207) | (52–86) | (96 500–126 000) | (39 000–51 000) |
| Aluminum alloy | 15–22 | 15–22 | 10–14 | 10,000 (69 000) | 3,800 (26 200) |
| | (103–152) | (103–152) | (69–97) | | |
| Brass | 8 (55) | 15 (103) | | 13,500 (93 000) | 5,100 (35 200) |
| Copper rods, bolts | 8 (55) | 8 (55) | | 16,000 (110 000) | 6,000 (41 400) |

* ASTM A36 steel.

**1-21.** A round, hollow cast-iron column of 200-mm outside diameter and 150-mm inside diameter, 5 m long, supports an axial compressive load of 1000 kN. Calculate the total shortening produced in this column if the modulus of elasticity of cast iron is 100 000 MPa.

**1-22.** A ¾ by 4-in. steel eyebar 20 ft long between centers of the pins at its ends is used as a diagonal member in a light bridge, as shown in Fig. 1-11. Assume that the bar is of constant cross-sectional area throughout its length of 20 ft. If the total tensile stress in the bar is 60,000 lb and if $E$ is 30,000,000 psi, calculate (a) the unit tensile stress $s_t$; (b) the unit elongation $\epsilon$; (c) the total elongation $\delta$.

Ans. (a) $s_t$ = 20,000 psi; (b) $\epsilon$ = 0.000 67 in./in.; (c) $\delta$ = 0.16 in.

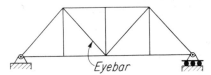

Fig. 1-11   Prob. 1-22.

**1-23.** A vertical timber column 6 by 8 in. of Douglas fir, dense No. 1, S4S, and 10 ft 8 in. long is subjected to a vertical axial load of 32,000 lb. Calculate (a) the unit stress in the timber; (b) the unit axial deformation (see Table 8-1 for values of $E$); (c) the total axial deformation.

Ans. (a) $s_c$ = 774 psi; (b) $\epsilon$ = 0.000 44 in./in.; (c) $\delta$ = 0.0563 in.

**1-24.** A section of a heavy concrete foundation wall 3 m high and of uniform thickness is subjected to a vertical load which produces a unit stress $s$ of 3.5 MPa. If $E$ for this concrete is 14 000 MPa, calculate (a) the unit vertical deformation, and (b) the total vertical deformation.

Ans. (a) $\epsilon$ = 0.000 25 mm/mm; (b) $\delta$ = 0.75 mm

**1-25.** The ends of the laminated-wood roof arch shown in Fig. 1-12 are tied together with a horizontal steel rod 90 ft 10 in. long which must withstand a total stress of 60,000 lb. Two turnbuckles are used. All threaded rod ends are upset. (a) Determine the required diameter $d$ of rod if the maximum allowable stress is 20,000 psi. (b) If the unstressed length of the rod is 90 ft 10 in. and there are 4 threads per inch on the upset ends, how many turns of one turnbuckle will bring it back to its unstressed length after it has elongated under full allowable tensile stress? Use $E$ = 29,000,000 psi.

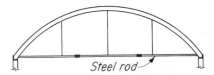

Fig. 1-12   Prob. 1-25.

**1-26.** A copper aerial wire 160 ft long, weighing 0.0396 lb/ft, and having a cross-sectional area of 0.010 28 in.², is suspended vertically from its upper end; $E$ = 15,000,000 psi. (a) Calculate the *average* unit strain in the wire due to its own weight. (The average unit strain will be one-half the maximum unit strain.) (b) Calculate the total amount $\delta$ this wire will stretch because of its own weight. (c) Find the total elongation $\delta$ if a weight of 50 lb is attached to its lower end. (*d*) Determine what maximum weight *W* this wire can safely support at its lower end if the unit stress in it must not exceed its elastic limit of 10,000 psi.

*Ans.* (*a*) $\epsilon = 20.5 \times 10^{-6}$ in./in.  (*b*) $\delta = 0.0394$ in.;  (*c*) $\delta = 0.662$ in.;
(*d*) $W = 96.46$ lb

**1-27.** A piston rod of a steam engine is of 4-in. diameter and is alternately subjected to equal tensile and compressive stresses of 10,000 psi. If the distance between two points *A* and *B* on the unstressed rod is 4 ft 10 in., calculate the minimum and maximum distances between *A* and *B* when the rod is fully stressed to 10,000 psi. Assume $E = 29,000,000$ psi.

*Ans.* Min $L = 57.98$ in.; max $L = 58.02$ in.

## 1-11 SHEARING DEFORMATION AND POISSON'S RATIO

A shearing force causes a shearing deformation, just as an axial force causes a shortening or a lengthening of a member. A shearing deformation is angular (see Fig. 1-13). The shearing force *F* causes the material to deform into the dashed position. The total deformation is $\delta_s$, and this deformation occurs over the length *L*. The deformation per unit of length is $\delta_s/L = \tan \phi$. For very small angles the tangent of the angle is practically equal to the angle expressed in radians. Therefore the shearing unit strain is the angle $\phi$ (Greek letter phi) in radians. Now, just as normal unit stress is related to unit strain by Hooke's law ($s = E\epsilon$), so is shearing unit stress related to shearing unit deformation. The relation is

$$s_s = G\phi \qquad (1\text{-}5)$$

where *G* is the shearing modulus of elasticity. Values of *G* for common metals are given in Table 1-2. The relative motion between the ends of a shaft under torsion is due to shearing deformation. Such deformations will be studied in Chap. 4.

The ratio of unrestrained unit lateral contraction (or expansion) to the unit longitudinal elongation (or contraction) is called **Poisson's ratio** $\mu$. We are aware of the Poisson effect although we may not have associated a name with the phenomenon. Pulling on a rubber band not only lengthens

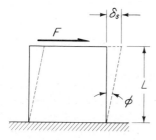

**Fig. 1-13  Shearing deformation.**

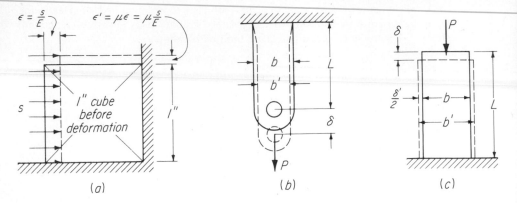

**Fig. 1-14**   Poisson's ratio. (a) Deformation of unit cube. (b) Tension. (c) Compression.

it in the direction of the pull but also visibly contracts the lateral dimensions of the cross section. In the unit cube shown in Fig. 1-14a the applied load causes the stress $s$ and the horizontal unit deformation $\epsilon$. The vertical unit deformation is Poisson's ratio times the unit strain $\epsilon$, or $\mu\epsilon$ as shown. Refer to Fig. 1-14b or c. Poisson's ratio is

$$\mu = \frac{\epsilon}{\epsilon'} = \frac{\delta'/b}{\delta/L} \qquad \text{(1-6a)}$$

where $\delta' = |b - b'|$ and $b$ is the original lateral dimension. Since $\delta/L$ is the unit deformation $\epsilon$, Eq. (1-6a) yields

$$\delta' = \mu\epsilon b \qquad \text{(1-6b)}$$

Numerical values of Poisson's ratio $\mu$ generally vary between 0.25 and 0.35. Values can be found in handbooks or can be determined by Eq. (1-7) if the appropriate moduli of elasticity are known. The shearing modulus $G$ and the modulus of elasticity $E$ are related by

$$2G = \frac{E}{1 + \mu} \qquad \text{or} \qquad \mu = \frac{E}{2G} - 1 \qquad \text{(1-7)}$$

**ILLUSTRATIVE PROBLEM**

**1-28.** Compute the numerical value of Poisson's ratio for copper. If a 20-in.-long copper rod of ½-in. diameter is subjected to an axial pull $P$ of 800 lb, determine the change in diameter.

*Solution:* From Table 1-2, $E = 16{,}000{,}000$ psi and $G = 6{,}000{,}000$ psi. Substitution of these values into Eq. (1-7) gives

$$\left[\mu = \frac{E}{2G} - 1\right] \qquad \mu = \frac{16{,}000{,}000}{(2)(6{,}000{,}000)} - 1 = 0.333 \qquad \textit{Ans.}$$

The unit tensile stress $s_t$ is

$$\left[ s_t = \frac{P}{A} \right] \qquad\qquad s_t = \frac{800}{0.196} = 4080 \text{ psi}$$

The unit elongation can now be found:

$$\left[ E = \frac{s}{\epsilon} \right] \qquad\qquad \epsilon = \frac{s}{E} = \frac{4080}{16,000,000} = 0.000\ 255$$

Now from Eq. (1-6b)

$$[\delta' = \mu\epsilon b] \qquad\qquad \delta' = (0.333)(0.000\ 255)(0.5 = 0.000\ 0425 \text{ in.} \qquad\qquad Ans.$$

## PROBLEMS

**1-29.** Electric strain gages mounted on the surface of a cylindrical-shaped block of metal subjected to axial compressive load show a unit longitudinal strain of 0.0012 and a unit lateral expansion of 0.0004. If the block is 50 mm in diameter and the load $P$ is 178 kN, determine (a) Poisson's ratio; (b) the modulus of elasticity $E$; and (c) the shearing modulus of elasticity $G$.

**1-30.** What pressure in psi must be exerted along the edges of a 10-in.-wide aluminum plate if the 10-in. dimension is to remain constant when the longitudinal compressive stress is raised by 10,000 psi?

*Ans.* $s = 3160$ psi

## 1-12 DETERMINATION OF ALLOWABLE STRESS AND MODULUS OF ELASTICITY. THE STRESS-STRAIN DIAGRAM

The allowable stresses that are considered as safe to use in design are usually specified by the governmental agency having jurisdiction over the particular design, such as the federal, state, county, or city government, and the Army, Navy, Department of Commerce, etc. They are published in the form of codes and are readily available to designers.

As previously mentioned, these allowable stresses are based upon extensive laboratory tests together with a vast amount of accumulated experience. Such tests furnish information concerning the strength, elasticity, ductility, and so forth, of the material tested. One of the most common of these, the tension test of a soft-steel bar, is described below.

**The Stress-Strain Diagram.** In the tension test most commonly used for metals, a testing machine applies a controlled and gradually increasing tensile force to a round bar until rupture finally occurs. The total pull on the bar at any time during the test is measured by means of scales which are a part of the machine. Extensometers capable of measuring as little as 0.0001 in. are used to measure deformations. At stated intervals, the total tensile force on the bar and the total deformation in an 8-in. section of it

are measured simultaneously. From these measurements, apparent unit
stresses and strains are calculated and are then plotted to give the
stress-strain diagram shown in Fig. 1-15. For soft steel, the stress is found
to be proportional to the strain up to a stress of about 30,000 psi, shown
as point *A* on the curve, thus proving Hooke's law. For stresses beyond
point *A*, the strain increases at a faster rate than the stress. Consequently
*A* is the **proportional limit.** At a stress of about 33,000 psi, the **elastic limit**
of the material is reached. That is, if subjected to stresses beyond this
point, the bar will no longer return to its original length after removal of
the load. In other words, it has acquired a permanent set.

From a practical standpoint most designers make little or no distinction
between the proportional limit and the elastic limit, both generally being
known by the latter term. Since permanent set due to stress beyond the
elastic limit is generally objectionable, the allowable stress of a material
never exceeds its elastic limit but is usually fixed at some reasonable
lower point.

Shortly after the elastic limit has been reached (at *B* on the curve), a
sudden elongation takes place, while the load on the bar actually drops.
That is, the material suddenly *yields*. The stress at which this occurs is
called the **yield point.** However, the bar again immediately picks up ability
to resist increasing stress, but the elongations now increase at a much
faster rate than the stresses.

As the test is continued, a point of maximum stress—the **ultimate
strength** of the material—that the bar is capable of resisting is finally
reached at *E*. That is, the ordinate to *E* on the curve represents the *ulti-*

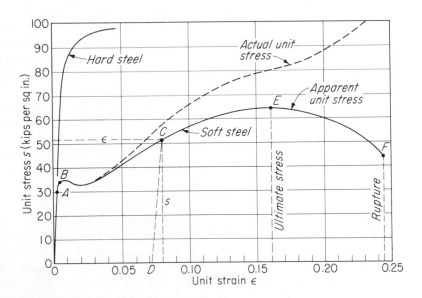

Fig. 1-15   **Stress-strain diagram. Soft steel.**

*mate strength* of the material. Beyond point $E$, elongations will continue but at gradually decreasing stresses until finally rupture of the bar occurs. After passing the point of ultimate strength at $E$, a more rapid local failure starts and the bar in Fig. 1-16a begins to "neck," as illustrated in Fig. 1-16b, with a consequent reduction in cross sectional area.

All the unit stresses plotted in the stress-strain diagram are based on the original cross-sectional area of the unstressed bar. This is common practice, but it results in a curve that does not give the true stresses. Obviously, as the bar begins to elongate, there is a consequent reduction in its diameter, which becomes measurable soon after the yield point has been passed and, of course, becomes obvious when necking starts. If the actual rather than the original area is used, the actual calculated stresses give a curve following approximately the dashed line in Fig. 1-15.

If the load on the bar should gradually be decreased where point $C$ on the curve is reached, and if simultaneous readings of load and elongations are taken as before and are plotted, the points will fall approximately on line $CD$. That is, an elongation equal to $OD$ will remain after complete removal of all load, thus indicating that amount of *permanent set*.

Line $CD$ is found to be parallel to $OA$, thereby indicating no change in the modulus of elasticity of the steel. If loads are then again applied as previously, the new points, if plotted, will also fall approximately along line $DC$, thus indicating that the drawing out of the material of the bar has had the effect of increasing both its proportional and its elastic limits. In fact, by such a process of cold drawing, the usable elastic limit of soft steel can be increased almost to its ultimate strength.

In a soft-steel bar, the **percentage of elongation** in the 8-in. gage length might average about 30 percent. The reduction in area at the ruptured section would be about 50 percent. These percentages show that soft steel is a highly *ductile* material. This property of ductility is exceedingly valuable in steel used for buildings, bridges, derricks, and so forth, since noticeable deformations often would give ample warning of impending failure.

A similar test on a specimen of *hard steel* (0.6 percent carbon) would show it to have greater strength, higher elastic limit, and no yield point, as indicated by the hard-steel curve in Fig. 1-15. Obviously, hard steel is brittle and would give little warning of impending failure.

A number of other materials, such as cast iron, concrete, and wood, commonly used in the design of machines and structures, have elastic properties somewhat less well defined than those of steel. Study of their

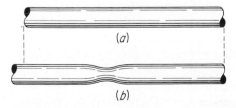

(a)

(b)                              Fig. 1-16   Necking of soft-steel bar.

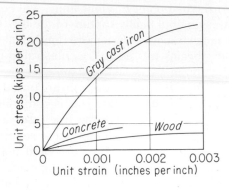

Fig. 1-17   Stress-strain diagram. Gray cast iron, concrete, and wood.

stress-strain diagrams, as shown in Fig. 1-17, reveals chiefly their lack of proportionality of stress to strain. This lack has not been found to be especially objectionable. It does, however, complicate the matter of determining the modulus of elasticity of these materials. In such cases the slope of a tangent to the curve at the average allowable stress is often used to determine the modulus of elasticity.

While most steels have approximately equal strengths in tension and compression, wood is slightly stronger in tension than in compression. But copper, brass, cast iron, and concrete are much stronger in compression than in tension. Concrete has practically no usable strength in tension, and is therefore generally reinforced with steel rods, which then resist the tensile stresses.

## 1-13 STRESS CONCENTRATION

The relation $s = P/A$ gives the average stress at any cross section of a slender *prismatic* tension member subjected to axial load. However, an abrupt change in cross section (such as often occurs near the end of a member because of rivet holes, for example) changes radically the distribution of stress at that section. As indicated in Fig. 1-18a, a concentration of stress, which often exceeds twice the average stress, will occur at the edge of a hole in a bar, the stress depending on the ratio of the radius $r$ of the hole and the net width $b$ of the bar, as shown in Fig. 1-18.

The reasons for such stress concentrations are readily understandable. Let us say that Fig. 1-18a shows only a small section of a much longer bar under stress. We may reasonably assume that a short distance below the hole, such as at section *HI*, the stress is fairly uniformly distributed over the cross-sectional area of the bar. If now we think of the stress as *flowing* upward through the resisting material of this member, it becomes clear that the stress in the material directly below the hole, as it approaches $E$, is diverted around the hole through the material at and near edges $D$ and $F$ and that it then again converges somewhere above $E'$, probably reaching fairly uniform distribution again at section $AB$. This diversion naturally

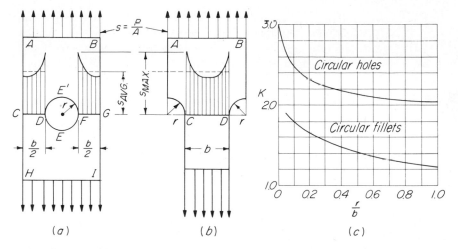

Fig. 1-18   **Stress concentrations. (a) Circular hole. (b) Circular fillets. (c) Stress-concentration factors for flat bars.**

causes the material at and near $D$ and $F$ to be more highly stressed. The ordinates shown in the diagram represent probable stresses on a plane through $CDFG$.

Figure 1-18$b$ shows similar stress concentrations at another type of change in cross section. The ordinates represent stress on a plane through $CD$ if flow of stress is downward.

These stress concentrations should be carefully considered by designers, especially in connection with hard and brittle materials, since such materials are apt to begin to crack without much warning when stresses exceed the elastic limit.

In softer metals, such as ordinary structural steel, the material at points of high stress will yield after the elastic-limit stress has been exceeded, thereby shifting to adjacent material the burden of carrying some of the excess stress. Undoubtedly, excessive stresses, yielding, and consequent redistribution of stresses are frequent occurrences in designs using structural materials capable of yielding.

The maximum stress adjacent to a hole or fillet is given by the equation

$$s_{\max} = K \frac{P}{A_{\min}} \tag{1-8}$$

where $K$ is a stress-concentration factor which can be obtained from the graph in Fig. 1-18$c$.

**ILLUSTRATIVE PROBLEM**

**1-31.** A 15-mm-diameter hole is drilled on the centerline of a steel strap 45 mm wide by 5 mm thick, subjected to an axial pull $P = 10$ kN. Calculate ($a$) the unit

stress $s_t$ in the strap several inches away from the hole, and (b) the maximum unit
stress adjacent to the hole.

*Solution:* The stress some distance away from the hole is fairly uniformly distributed and is given by the equation

$$\left[ s_t = \frac{P}{A} \right] \qquad s_t = \frac{10\ 000}{(5)(45)} = 44.4 \text{ MPa} \qquad\qquad \textit{Ans.}$$

The net width $b$ at the hole is $45 - 15 = 30$ mm, and the ratio of the hole radius
$r$ to the net width $b$ is $r/b = 7.5/30 = 0.25$. Now, from Fig. 1-18c, for $r/b = 0.25$
we find for a circular hole that the stress-concentration factor $K$ is 2.27. Then from
Eq. (1-8)

$$\left[ s_{max} = K \frac{P}{A_{min}} \right] \qquad s_{max} = 2.27 \frac{10\ 000}{(5)(30)} = 151 \text{ MPa} \qquad\qquad \textit{Ans.}$$

PROBLEMS

**1-32.** Solve Prob. 1-31 when $P = 2,000$ lb, $t = \frac{1}{4}$ in., $w = 1\frac{1}{2}$ in., and the
hole diameter is $\frac{3}{4}$ in.

*Ans.* (a) $s_t = 5333$ psi; (b) $s_{max} = 22,600$ psi

**1-33.** A flat steel bar $3\frac{1}{2}$ in. wide and $\frac{3}{8}$ in. thick is reduced in width to $2\frac{1}{2}$ in.
by circular fillets of $\frac{1}{2}$-in. radius on each side. If the bar is subjected to an axial
pull of 10,000 lb, calculate (a) the stress $s_t$ in the wider part of the bar some distance from the change in section, (b) the average tensile stress in the narrow part
of the bar, and (c) the maximum tensile stress adjacent to the circular fillet.

*Ans.* (a) $s_t = 7620$ psi; (b) $s_t = 10,670$ psi; (c) $s_{max} = 17,600$ psi

**1-34.** A 2-in.-diameter hole is drilled on the centerline of a steel bar 4 in. wide
by $\frac{1}{2}$ in. thick. What axial pull $P$ can be applied to the bar without exceeding a
maximum stress at any point in the bar of 20,000 psi?

*Ans.* $P = 9390$ lb

**1-35.** A long rectangular steel bar has a 25- by 150-mm cross section and is subjected to an axial tensile force of 200 kN. What is the maximum local stress near a
50-mm-diameter hole drilled on the centerline of the flat side of the bar?

## 1-14 AXIAL STRESSES IN MEMBERS OF TWO MATERIALS

When a member subjected to axial tensile or compressive stress is made
of two materials so joined that both must deform equally under stress,
their unit stresses will be proportional to their moduli of elasticity.

Let the subscripts 1 and 2 designate the materials having the lowest and
the highest modulus of elasticity, respectively. Since $E = s/\epsilon$ and
$\epsilon = s/E$, and since the two materials are subjected to equal deformations,

$$\epsilon = \frac{s_1}{E_1} = \frac{s_2}{E_2} \qquad\qquad (a)$$

thus proving that for equal unit deformations, the stresses are proportional to the moduli of elasticity of the materials. By cross multiplication we have

$$s_2 = \frac{E_2}{E_1} s_1 \qquad \text{or} \qquad s_2 = n s_1 \qquad\qquad (b)$$

where $n = E_2/E_1$. Clearly, the total load $P$ equals the load $s_1 A_1$ carried by the one material plus the load $s_2 A_2$ carried by the other material. That is

$$P = s_1 A_1 + s_2 A_2 \qquad\qquad (c)$$

or, substituting $n s_1$ for $s_2$,

$$P = s_1 A_1 + n s_1 A_2 = s_1 (A_1 + n A_2) \qquad\qquad \textbf{(1-9)}$$

The quantity $n A_2$ may be considered as an "equivalent area" of the lower-stressed material, replacing the higher-stressed material but carrying the same portion of the total load, since $s_2 A_2 = n s_1 A_2$.

## ILLUSTRATIVE PROBLEMS

**1-36.** A heavy rigid iron casting weighing 1964 lb is suspended from three ¼-in.-diameter wires which are symmetrically spaced and are 24 ft 2 in. long. The two outer wires are of steel having a modulus of elasticity $E$ of 29,000,000 psi and the middle wire is of bronze, for which $E$ is 14,500,000 psi. Calculate (a) the unit stresses in the bronze and the steel wires; (b) the total elongation $\delta$ of each wire. The weight of the wires may be neglected.

*Solution:* Study of Fig. 1-19 discloses that because the bar is rigid and the wires symmetrically support the bar, all three wires are subjected to the same total and unit elongations. Let the subscripts $B$ and $S$ denote bronze and steel, respectively.

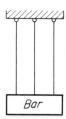

**Fig. 1-19  Prob. 1-36.**

Since the bronze carries the lowest stress, let $s_B$, $A_B$, and $E_B$ designate the bronze and $s_S$, $A_S$, and $E_S$ the steel; $n = E_S/E_B = 29,000,000/14,500,000 = 2$. Then, since $P = 1,964$ lb, $A_B = 0.0491$ in.$^2$, $n = 2$, and $A_S = 2(0.0491) = 0.0982$ in.$^2$,

$$[P = s_B(A_B + nA_S)] \qquad 1964 = s_B[0.0491 + 2(0.0982)]$$

or, the unit stress in the bronze is

$$s_B = \frac{1964}{0.0491 + 0.1964} = \frac{1964}{0.2455} = 8000 \text{ psi} \qquad\qquad Ans.$$

and the unit stress in the steel is

$$s_S = ns_B = (2)(8000) = 16,000 \text{ psi} \qquad \text{Ans.}$$

To determine the total elongation $\delta$, we may use the relations $\epsilon = s/E$ and $\delta = Ls/E$. Either the bronze or the steel may be used. Since $L = 24$ ft 2 in. $= 290$ in., using the bronze,

$$\left[\delta = \frac{Ls}{E}\right] \qquad \delta = \frac{(290)(8000)}{14,500,000} = 0.16 \text{ in.} \qquad \text{Ans.}$$

or, using the steel,

$$\delta = \frac{(290)(16,000)}{29,000,000} = 0.16 \text{ in.} \qquad \text{Ans.}$$

**1-37.** A short column has an inner core of brass ($E = 95\,200$ MPa) and an outer shell of aluminum ($E = 70\,000$ MPa), as shown in Fig. 1-20. Calculate the unit stress in each material due to an axial load $P$ of 100 kN.

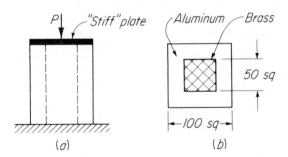

Fig. 1-20   Prob. 1-37. (a)
Column of two materials.
(b) Section.

*Solution:* Let the ratio of the modulus of elasticity of brass to the modulus of elasticity of aluminum be denoted by $n$. Then $n = E_B/E_A = 95\,200/70\,000 = 1.36$. Also, $A_B = 2500$ mm² and $A_A = 7500$ mm². Then from Eq. (1-9), and noting that subscript 1 applies to the aluminum,

$$[P = s_1(A_1 + nA_2)] \qquad 100\,000 = s_A\,[7500 + (1.36)(2500)]$$

or

$$s_A = 9.17 \text{ N/mm}^2 = 9.17 \text{ MPa} \qquad \text{Ans.}$$

The stress in the brass is $n$ times that in the aluminum, and therefore,

$$[s_2 = ns_1] \qquad s_B = 1.36(s_A) = 1.36(9.17) = 12.5 \text{ MPa} \qquad \text{Ans.}$$

A partial check on the answers may be obtained as follows:

$$[P = s_1A_1 + s_2A_2] \qquad P = 9.17(7500) + 12.5(2500) = 100\,000 \text{ N} \qquad \textit{Check}$$

## PROBLEMS

**1-38.** A concrete column measuring 10 in. square and reinforced with four 1-in. round steel bars, as shown in Fig. 1-21, supports an axial compressive load $P$ of

**Fig. 1-21    Prob. 1-38.**

91,200 lb. Steel bearing plates at top and bottom of column assure equal deformation of steel and concrete. Calculate the unit stress in each material, assuming that
for concrete $E = 2,000,000$ psi and for steel $E = 30,000,000$ psi

**1-39.** Solve Prob. 1-36 when the casting weighs 1500 lb, the wires are 10 ft long,
the outer wires are aluminum, and the middle wire is of steel.

*Ans. (a)* $s_A = 6235$ psi; $s_S = 18,080$ psi; *(b)* $\delta = 0.0748$ in.

**1-40.** Figure 1-22 shows a timber post measuring 150 by 200 mm and reinforced
with two 10- by 150-mm steel plates thoroughly fastened to the post. The compressive load $P$ on the post is 500 kN. Calculate the stress in each material. Use
$E_W = 12\ 000$ MPa and $E_S = 204\ 000$ MPa.

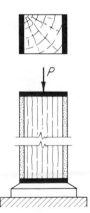

**Fig. 1-22    Probs. 1-40 and 1-41.**

**1-41.** The short wood post in Fig. 1-22 is of lodgepole pine and measures 2 in.
square. It is reinforced on two sides with copper plates, measuring ⅛ by 2 in.,
which are thoroughly fastened to the wood. The post supports an axial compressive load of 9500 lb. Calculate the unit stress in the wood and the copper. See
Table 8-1 for the modulus of elasticity of pine and use 14,500,00 psi for copper.

*Ans.* $s_w = 897$ psi (wood) and $s_c = 11,800$ psi (copper)

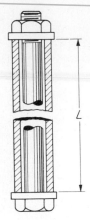

**Fig. 1-23  Probs. 1-42 and 1-43.**

**1-42.** Figure 1-23 shows a structural-steel rod extending through a bronze tube ($E = 15,000,000$ psi). The cross-sectional area of the steel rod is 0.6 in.², and its upper end has 20 threads per inch. The bronze tube is 25 in. long, and its area is 1.8 in.². Calculate the unit stresses in the unthreaded length of the steel rod and in the bronze tube which would result from a quarter turn of the nut on the bolt. (HINTS: The elongation of the rod plus the shortening of the tube equals the change in position of the nut along the rod. Also, the load $P_S$ on the bolt equals the load $P_B$ on the tube.)

*Ans.* $s_S = 8820$ psi; $s_B = 2940$ psi

**1-43.** In the preceding problem, what will be the results if the rod is of steel and the tube is of aluminum?

## 1-15 TEMPERATURE STRESSES

When a material body is subjected to changes in temperature, it will expand when the temperature increases and contract when it decreases.

Through experiments, it is possible to measure the change in length per unit of length of a bar of material for each degree of change in temperature. This change, called the **linear coefficient of thermal expansion** of the material, is denoted by the Greek letter $\alpha$ (alpha). Some average values of $\alpha$ for structural materials are given in Table 1-3.

Table 1-3 AVERAGE COEFFICIENTS OF
THERMAL EXPANSION (PER °F)*

| Wood | 0.000 0030 | Wrought iron | 0.000 0067 |
|------|------------|--------------|------------|
| Glass | 0.000 0044 | Copper | 0.000 0093 |
| Concrete | 0.000 0060 | Bronze | 0.000 0100 |
| Cast iron | 0.000 0061 | Brass | 0.000 0104 |
| Steel | 0.000 0065 | Aluminum | 0.000 0128 |

* $\alpha$ per degree Celsius = $\frac{9}{5}\alpha$ per degree Fahrenheit.

If a straight bar of material is free to change its length with variations in temperature, no change will occur in its internal stress. If, however, the bar is restrained, partially or fully, so as to restrain or prevent change in its length due to temperature variations, internal stresses will be produced in it equal to those required to produce the prevented change in length.

If $\alpha$ is the unit deformation per degree change in temperature, $t$ the number of degrees of change, $L$ the length of a bar, and $\delta$ the total change in length, then the change per unit length is $\alpha t$, and the total change in length $L$ is

$$\delta = \alpha t L \qquad \textbf{(1-10)}$$

But change in unit length $\alpha t$ is also denoted by $\epsilon$, in Art. 1-8, and $E = s/\epsilon$, or $s = E\epsilon$. Hence the unit stress produced in a fully restrained bar by a temperature change $t$ is

$$s = E\,\alpha t \qquad \textbf{(1-11)}$$

## ILLUSTRATIVE PROBLEM

**1-44.** The straight bottom chord of a riveted steel bridge is 200 ft long. (a) Calculate the change in its length caused by a change in temperature of 70°F (A possible difference between day and night temperatures). (b) What unit stress will be produced in this chord if both of its ends are fully restrained against linear expansion of the chord? (c) If the lower chord is fully restrained between two virtually "immovable" concrete piers and its average cross-sectional area contains 20 in.² of steel, what total force $P$ would it exert against its fastenings to the piers?

*Solution:* $\alpha = 0.000\ 0065$, $t = 70°$; $L = 200$ ft $= 2400$ in. Also $E = 29,000,000$ psi. Hence

| | | |
|---|---|---|
| $[\delta = \alpha t L]$ | $\delta = (0.000\ 0065)(70)(2400) = 1.092$ in. | *Ans.* |
| $[s = E\alpha t]$ | $s = (29,000,000)(0.000\ 0065)(70) = 13,195$ psi | *Ans.* |
| $[P = sA]$ | $P = (13,195)(20) = 263,900$ lb | *Ans.* |

## PROBLEMS

**1-45.** A surveyor's steel tape measures exactly 100 ft between end markings at 70°F. What error is made in measuring a distance of 1000 ft when the temperature of the tape is 32°F?

**1-46.** A continuous straight concrete pavement was laid in 40-ft rectangular sections. The expansion joints between ends of sections measured ½ in. at 70°F. What width $w$ does a joint measure (a) at 30°F and (b) at 110°F?

*Ans.* (a) $w = 0.6152$ in.; (b) $w = 0.3848$ in.

**1-47.** The steel rails of a continuous, straight railroad track are each 15 m long and are laid with spaces between their ends of 5 mm at 20°C. (a) At what temperature will the rails touch end to end? (b) What unit compressive stress will be produced in the rails if the temperature rises to 65°C ($E = 200\ 000$ MPa)?

**1-48.** Solve Prob. 1-44 when length of chord is 160 ft, change in temperature is 80°F, and the average cross-sectional area of the chord is 6 in.²

**1-49.** A locomotive tire of hard steel with an inside diameter of 59.972 in. at 70°F is to be expanded by heat to a diameter of 60.050 in. (*a*) What temperature will produce this expansion? (*b*) What unit stress will exist in the tire after it has been placed upon a wheel with a diameter of 60 in. and is allowed to cool to 70°F? Let $E = 30,000,000$ psi.

*Ans.* (*a*) 270°F; (*b*) $s = 14,000$ psi

**SUMMARY** (By Article Number) ———————————————

**1-2. Stress** is resistance to external forces. **Unit stress** is force per unit of area. Total stress is the total force within a member.

**1-3. The two basic stresses** are normal (tensile or compressive) and shearing. **Tensile stresses** pull on the fibers and tend to elongate them. **Compressive stresses** push on the fibers and tend to compress them. A stress directed parallel to the longitudinal axis of a member is called an **axial stress. Shearing stresses** are generally caused by forces acting transversely on a member. **Bearing stresses** are compressive stresses on external surfaces in contact under pressure.

**1-4. Ultimate stress** is the greatest stress a material is capable of resisting. **Allowable stress** is a predetermined stress considered safe for use in design. **Factor of safety** is the ratio of ultimate stress to allowable stress; it is a measure of the probable safety of a design.

**1-5. The direct-stress formula** establishes a relation by which we may find the unit stress $s$ produced by a total force $P$ on a stressed area $A$. That is,

$$s = \frac{P}{A} \tag{1-1}$$

**1-7. Analysis** relates chiefly to the preliminary investigations which precede design. **Design** is the final determination of size, shape, and strength of a structure and of its component parts.

**1-8. Deformation,** or **strain,** is a measure of the change in form of a body caused by externally applied forces. In a prismatic bar subjected to axial stress,

$$\textbf{Unit deformation} = \frac{\textbf{total deformation}}{\textbf{length}} \quad \text{or} \quad \epsilon = \frac{\delta}{L} \tag{1-2}$$

**1-9. Elasticity** is that property of a material which enables it, within certain limits of stress, to recover its original form after removal of a deforming force. The **elastic limit** is that unit stress beyond which a material will no longer recover its original form. **Permanent set** is that deformation permanently sustained by an elastic material by being stressed beyond the elastic limit.

**1-10.** According to **Hooke's law: below the elastic limit, stress is proportional to strain (deformation).**

The measure of the elasticity of a given material, called the modulus of elasticity and denoted by $E$, is the ratio of its unit stress $s$ to unit strain $\epsilon$. In practical terms, $E$ is actually a measure of the stiffness of an elastic material, or its resistance to deformation under load. That is,

$$\text{Modulus of elasticity} = \frac{\text{unit stress}}{\text{unit deformation}} \quad \text{or} \quad E = \frac{s}{\epsilon} \quad \text{(1-3)}$$

Since $\delta = \epsilon L$, $\epsilon = s/E$, and $s - P/A$, the total elongation $\delta$ of a prismatic bar subjected to axial stress is

$$\delta = \epsilon L = \frac{s}{E} \cdot L = \frac{P}{A} \cdot \frac{L}{E} \quad \text{or} \quad \delta = \frac{PL}{AE} \quad \text{(1-4)}$$

**1-11.** **Shearing stress** is related to unit shearing deformation by

$$s_s = G\phi \quad \text{(1-5)}$$

The ratio of unrestrained lateral unit deformation to longitudinal unit deformation is called **Poisson's ratio.**

$$\mu = \frac{\delta'/b}{\delta/L} \quad \text{or} \quad \delta' = \mu\epsilon b \quad \text{(1-6)}$$

**1-13.** **Stress concentrations** occur at sudden changes in cross section in members under axial load. The maximum stress at this point is given by

$$s_{\max} = K\frac{P}{A_{\min}} \quad \text{(1-8)}$$

**1-14.** In a prismatic member made of two materials so joined that they deform equally under axial stress, the unit stresses in the two materials are proportional to their respective moduli of elasticity. The total load $P$ on such a member is

$$P = s_1 A_1 + n s_1 A_2 = s_1(A_1 + nA_2) \quad \text{(1-9)}$$

where $n = E_2/E_1$, $s_1$ is lowest of the two unit stresses, and $E_1$ is the lowest of the two moduli of elasticity.

**1-15.** Material bodies expand or contract with changes in temperature. For any material the change in length per unit of length per degree Fahrenheit change in temperature is called its **coefficient of thermal expansion** and is denoted by the Greek letter $\alpha$. The total linear deformation $\delta$ in an unrestrained bar of length $L$ caused by a change in temperature $t$ is

$$\delta = \alpha t L \quad \text{(1-10)}$$

The unit axial stress $s$ produced in a fully restrained bar by a temperature change $t$ is

$$s = E\alpha t \quad \text{(1-11)}$$

## REVIEW PROBLEMS

**1-50.** A steel rod 20 ft in length must resist a total axial stress of 60,000 lb in tension. What diameter $d$ is suitable $(a)$ if ends are threaded and $(b)$ if ends are upset? Allowable stress is 20,000 psi.

*Ans.* $(a)\, d = 2\frac{1}{4}$ in.; $(b)\, d = 2$ in.

**1-51.** A steel wire 450 ft long is of ⅛-in. diameter and weighs 0.042 lb/ft. If the wire is suspended vertically from its upper end, calculate (*a*) the maximum tensile stress due to its own weight, and (*b*) the maximum weight *W* which it can safely support at its lower end if the unit tensile stress is not to exceed 24,000 psi.

**1-52.** If the maximum force *F* in Fig. 1-24 which must be exerted by the punch *P* to shear a hole 1 in. in diameter in a steel plate is 30,000 lb, calculate the average unit compressive stress $s_c$ in the punch.

*Ans.* $s_c = 38{,}200$ psi

**1-53.** In punching a 1-in.-diameter hole in steel plate ½ in. thick, the punch *P* in Fig. 1-24 causes a cylindrical plug of steel to be sheared out of the plate. Calculate the required force *F* if the ultimate shearing strength of the steel is 40,000 psi.

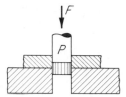

Fig. 1-24   Probs. 1-52 and 1-53.

**1-54.** The bolt *C* shown in Fig. 1-25 has a 1 in. diameter. (*a*) Calculate the pull *P* to cause a unit shearing stress of 13,000 psi in bolt *C*. (*b*) Using this calculated force *P*, find the required diameter of the two threaded bolts *B* if the allowable tensile stress at the root of the threads is 16,000 psi. (*c*) If *t* = ¾ in., calculate the bearing stress on bolt *C*.

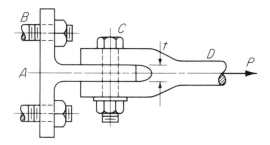

Fig. 1-25   Probs. 1-54 and 1-55.

**1-55.** The load *P* shown in Fig. 1-25 is 150 kN. (*a*) Calculate the shearing stress in the 40-mm-diameter bolt *C*. (*b*) Calculate the required diameter of rod *D* if the allowable tensile stress in the rod is 200 MPa. (*c*) Calculate the required dimension *t* if the allowable bearing stress is 250 MPa.

**1-56.** A structural-steel rod 1.5 in. in diameter and 25 ft long in supporting a balcony carries a tensile load of 29,000 lb. (*a*) Find the total elongation δ of this rod and (*b*) the diameter *d* required if the total elongation must not exceed 0.1 in.

**1-57.** A steel bar 3 in. wide and ⁵/₁₆ in. thick must be reduced abruptly to a 1½ in. width. What is the minimum fillet radius *r* that can be used if the stress next to the fillet must not be greater than 1½ times the average stress in the narrow portion of the bar when under axial load?

*Ans.* $r = 0.54$ in.

**1-58.** An initially straight aluminum alloy strap ¼ in. thick is bent into the form of a circular hoop with a mean diameter of 200 in. The inner fibers are in compression, the outer fibers are in tension, and those midway between the two faces are unchanged in length and therefore unstressed. Compute (a) the unit elongation of the outer fibers and (b) the unit tensile stress in these fibers.

*Ans.* (a) $\epsilon = 0.001\ 25$; (b) $s_t = 12,500$ psi

**1-59.** A spanner wrench (see Fig. 1-26) is used to tighten the brass couplings on a fire hose. Compute the minimum diameter of the projecting brass pins if the allowable shearing stress in the pins is 5000 psi and if the pull on the wrench will not exceed the value shown. Assume a smooth frictionless surface at $A$.

*Ans.* $d = 0.366$ in.

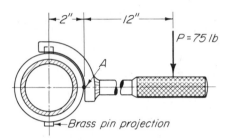

Fig. 1-26   Spanner wrench of Prob. 1-59.

**1-60.** The water-pump pliers shown in Fig. 1-27 are gripped with the hand, thus causing the 30-lb forces shown. Find the shearing stress in pin $A$ if its diameter is $5/16$ in. Assume that the crushing force exerted on the pipe is in the direction of line $BC$.

*Ans.* $s_s = 2565$ psi

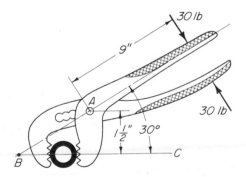

Fig. 1-27   Water-pump pliers of Prob. 1-60.

**1-61.** A structural-steel eyebar 10 mm thick and 8 m long between centers of pins carries a total stress of 125 kN in tension. If this bar must not elongate more than 5 mm and if $E$ for this steel is 200 000 MPa, calculate (a) the required width $w$ of the bar and (b) the average unit stress at this maximum elongation.

*Ans.* (a) $w = 100$ mm; (b) $s_t = 125$ MPa

**1-62.** Two steel rods of ¾ in. diameter are joined by a turnbuckle (American Standard coarse threads) and are fastened at their outer ends to virtually immovable concrete piers spaced 40 ft apart. If the rods are already stressed, calculate the increase $P$ in the total stress caused by one complete tightening turn of the turnbuckle. Assume that $E$ is 29,000,000 psi and that the cross-sectional area is constant throughout the entire 40 ft.

*Ans.* $P = 5340$ lb

**1-63.** To construct a building column of high strength, the inside of a steel cylinder is filled with concrete. The inside diameter of the cylinder is 12 in., and its wall thickness is ⅜ in. The axial load is applied so as to bear correctly on the entire cross-sectional area of the top of the column. Calculate the stress in each material due to an axial load $P$ of 240,000 lb. Use $E_S = 30,000,000$ psi and $E_C = 2,000,000$ psi.

**1-64.** The cast-brass collar shown in Fig. 1-28 just slides on the 1-in.-diameter steel shaft ($T = 60°F$) when the collar temperature is 100°F and the shaft temperature is 60°F. If both the shaft and the collar are at 60°F, what pull $P$ on the steel shaft will allow the collar to slide?

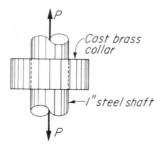

Fig. 1-28   Prob. 1-64.

**1-65.** A rigid bar is supported horizontally by two rods of equal diameter, as shown in Fig. 1-29. Disregard the weight of the bar and the rods. (*a*) If rod $A$ is of aluminum and $B$ of hard steel and both are of equal original length, what distance $d$ from $A$ should load $P$ be placed for equal total elongations of rods $A$ and $B$? (*b*) Assume now that rod $A$ is three times as long as rod $B$ but that the bar remains horizontal. If rod $A$ is of structural steel and $B$ of bronze ($E_B = 14,500,000$ psi), what distance $d$ from rod $A$ should $P$ be placed for equal total elongations of the rods?

*Ans.* (*a*) $d = 8.92$ in.; (*b*) $d = 7.30$ in.

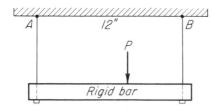

Fig. 1-29   Prob. 1-65.

**1-66.** An aluminum wire is stretched tightly between two "virtually immovable" supports. If the unit tensile stress in the wire is 40 MPa at 20°C, what is it (*a*) at 5°C and (*b*) at 30°C?

**1-67.** A cylindrical water tank is constructed of vertical wood staves tightly held in place by horizontal encircling ¾-in.-diameter steel hoops. The hoops are tightened to a unit stress of 8000 psi at 70°F, when their diameter is exactly 12 ft. Assuming that the tank does not contract (illogical, but useful for investigative purposes), what will be the unit stress in the hoops at 10°F?

*Ans.* $s_t = 19{,}310$ psi

**1-68.** A surveyor's tape of hard steel having a cross-sectional area of 0.004 in.² is exactly 100 ft long when a pull of 16 lb is exerted upon it at 70°F. (*a*) With the same pull on it, how long will the tape be at 40 and at 100°F? (*b*) What pull *P* on it will maintain its length of 100.0000 ft at 40 and at 100°F? Use $E = 30{,}000{,}000$ psi.

**1-69.** A tension member in a steel bridge truss carries a load of 90 kips. The member (see Fig. 1-30) was fabricated from three eyebars. Each bar was ½ in. thick and 3 in. wide. The length from center to center of eyes was intended to be 30 ft 0 in., but the two outside bars were fabricated ⅛ in. long and the center bar was fabricated ⅛ in. short. All eyebars were forcibly assembled, end pins inserted, and the member placed in the truss. Calculate the unit stress in each bar under the design load of 90 kips tension.

Fig. 1-30  Prob. 1-69.

**1-70.** The outer cylinder shown in Fig. 1-31 is of steel and has an inside diameter of 4 in., a thickness of ¼ in., and a length of 10 in. The inner solid cylinder is of cast brass, has a diameter of 2 in., and is 10.005 in. long when unloaded. A load of 80 kips is applied as shown. Calculate the stress in each member.

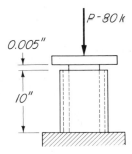

Fig. 1-31  Prob. 1-70.

**1-71.** Calculate the change in length δ of a copper aerial wire 50 m long due to a change in temperature of 50°C.

*Ans.* $\delta = 41.9$ mm

## REVIEW QUESTIONS

**1-1.** What are stress, unit stress, total stress?

**1-2.** Explain the meaning of tensile stress and compressive stress.

**1-3.** What is meant by shearing stress, bearing stress, and axial stress?

**1-4.** In connection with pins, bolts, and rivets, what is meant by single shear and double shear?

**1-5.** Define ultimate stress, allowable stress, and factor of safety.

**1-6.** What are some of the factors that govern the determination of a reasonable factor of safety?

**1-7.** Who determines the allowable stresses which may safely be used in design? Upon what information must such determination necessarily rely?

**1-8.** Give the direct-stress formula and explain each term in it.

**1-9.** What is meant by (*a*) root area and (*b*) S4S, and when must they be considered?

**1-10.** Explain the general meaning of the terms analysis and design.

**1-11.** Define strain, elasticity.

**1-12.** What is meant by shearing deformation and how is it related to shearing stress?

**1-13.** Define Poisson's ratio. Give approximate numerical limits.

**1-14.** Sketch two examples of axially loaded members with local stress concentrations.

**1-15.** What is meant by elastic limit, permanent set?

**1-16.** State Hooke's law and define modulus of elasticity.

**1-17.** What is a stress-strain diagram?

**1-18.** What condition in a member will generally produce a concentration of stress? What may be the result of a high concentration of stress in a soft steel and in a hard steel?

**1-19.** When two materials are so combined in a member that they deform equally under load, what is true of their unit stresses?

**1-20.** How do changes in temperature affect material bodies?

**1-21.** Define coefficient of thermal expansion.

**1-22.** How can the coefficient of thermal expansion for use with the Celsius temperature scale be obtained from a known coefficient for use with the Fahrenheit scale?

# Chapter 2
# Engineering
# Materials and
# Their Properties

## 2-1 INTRODUCTION

A considerable number of so-called engineering materials—materials used in the building of structures and machines—are available: the various woods and metals and inert materials such as brick, stone, and concrete. Some of the most commonly used of these materials, the processes by which they are made or obtained, and some of their properties are briefly described in this chapter.

## 2-2 DEFINITIONS OF SOME PROPERTIES OF MATERIALS

The proper and efficient use of engineering materials requires considerable knowledge of their mechanical properties. Among the most important of these are strength, elasticity, and stiffness. Other properties are ductility, malleability, hardness, resilience, toughness, creep, and machinability.

The methods used for determining these and other properties of materials have been evolved over a long period of years and they have, by now, been carefully standardized. The American Society for Testing and Materials (ASTM) publishes complete directions for making such tests together with the limits, agreed upon by engineers, scientists, and manufacturers, within which the test results must fall for any given classification of a material. Numerical values representing the relative properties of strength and elasticity of many engineering materials are given in the several tables included in this text.

**Strength** is the ability of a material to resist stress without failure. Several materials, such as structural steel, wrought iron, aluminum, and copper, have approximately equal strength in tension and compression, while their strength in shear is about two-thirds of that in tension. In other materials, such as gray cast iron, brick, and concrete, the strength in tension and in shear is only a fraction of the strength in compression. The measure of the strength of a material is its ultimate stress, or the greatest force per unit area it can withstand without rupture.

**Elasticity** is that property of a material enabling it to return to its original size and shape after removal of a deforming force. This property is important in all structures subjected to varying loads and is exceedingly important in precision tools and machines.

**Stiffness** is the property by virtue of which a material can resist deformation. This property is desirable in materials used in beams, columns, machines, and machine tools. The measure of the stiffness of a material is its modulus of elasticity, obtained by dividing the unit stress by the unit deformation caused by that stress.

**Ductility** is that property of a material that enables it to be drawn permanently through great change of shape without rupture, as, for example, copper being drawn into wire. Copper, aluminum, and wrought iron are among the ductile metals. The measure of ductility is the percent elongation before rupture of a test specimen. Ductility is desirable in a member or part which may be subjected to sudden and severe loads, since evident excessive deformation would give warning of impending failure. Brittleness is the opposite of ductility. Brittle materials fail suddenly without warning when stressed beyond their strength.

**Malleability** is that property of a material that enables it to undergo great change in shape without rupture under compressive stress, as, for example, copper, aluminum, or wrought iron being hammered into various shapes or steel being rolled into structural shapes or sheets.

**Hardness** is the ability of a material to resist indentation or abrasion. It is most commonly measured by the Brinell test, in which a hardened-steel ball 10 mm in diameter is forced into a flat surface of a test specimen by a force of 29 420 N. The **Brinell hardness number** can be calculated from the measured size of the indentation.

**Resilience** is that property of a material which enables it to absorb, without being permanently deformed, the energy produced by the impact of a suddenly applied load or blow. This property is essential in steel used for all types of springs, as in an automobile, on railroad cars, in watches, and so forth, where energy must be absorbed quickly without causing permanent deformation. Resilience is sometimes referred to as "elastic toughness," since the energy must be absorbed without stressing the material beyond its elastic limit. The measure of resilience is the amount of energy that a unit volume of the material has absorbed after being stressed to its elastic limit.

**Toughness** is that property of a material which enables it to absorb energy at high stress without fracture, usually above the elastic limit. Being above the elastic limit, the stress will cause permanent deformation. Wrought iron, for example, is tough because it can be bent into most any shape without fracture. The measure of toughness is the amount of energy that a unit volume of the material has absorbed after being stressed up to the point of fracture.

**Creep** is that property which causes some materials under constant stress to deform slowly but progressively over a period of time. Creep occurs to a slight extent at room temperatures in timber and concrete subjected to constant stress over a long period of years. In the softer metals, such as lead and zinc, subjected only to moderate, steady stress, creep will occur in a relatively short period of time. Creep occurs also in steel under constant stress, but not in any measurable degree until the temperature of the steel exceeds about 600°F. It assumes importance, however, in the design of steam boilers and turbines, of internal-combustion engines, and of other machines which operate at temperatures in excess of 600°F. In such cases, lower allowable stresses must be used. The maximum stress that may be steadily applied to a material at a given temperature and during a given period of time without causing a specified deformation to be exceeded is called the **creep limit** for that material. The creep limit generally used is the maximum unit stress at which the creep will not exceed 1 percent in 100,000 h of constant stress, equal to a little over 11 years.

**Machinability** is the readiness with which a material yields to shaping with cutting tools.

## 2-3 METHODS OF MANUFACTURE

Of the engineering materials mentioned earlier in this chapter, only wood and stone are used in the state as found in nature, except for shaping. All the metals are refined and manufactured from raw materials called **ore** found in various forms at or near the earth's surface. Brick and concrete are manufactured from the clays, limestones, and gravels found in great quantities on the earth's surface. The following descriptions of some of the manufacturing processes used serve only as brief introductions to the subject.

# The Woods

## 2-4 WOOD

Wood is a product of nature and in most cases is used in its natural state. It requires only to be shaped into suitable form for practical use. This process of the sawmill is well known. Among the woods most commonly used for structural purposes are Douglas fir, southern pine, southern cypress, and mountain hemlock. A vast number of experiments have shown, however, that the mechanical properties of these woods, especially strength and stiffness, vary considerably; for example, fir and pine are both stronger and stiffer than cypress and hemlock. Hence for the guidance of engineers and architects who include wood as a structural material in their designs, the lumber industry has adopted standard grading rules for the purpose of classifying structural woods according to their mechanical properties and their recommended uses. The most commonly used of these woods, their classifications, and some properties are listed in Table 8-1.

## 2-5 PRESERVATION OF WOOD

One of the disadvantages of wood as a structural material is that it is subject to decay. However, if used under proper conditions, the wood in a structure will generally outlive the useful life of the structure itself. Among the causes of decay are the following: (1) alternating wetness and dryness, resulting in wet rot, (2) lack of ventilation, resulting in dry rot, and (3) the destructive action of fungi, worms, and insects. In most structures, the effects of these causes are readily controlled.

When it is known to the designer that the causes of decay cannot be eliminated, two solutions to the problem are available: (1) to allow the decaying process to proceed and to replace decayed timbers when necessary, in which case an average life of about 6 to 10 years can be expected, or (2) to use timber treated with creosote or some other suitable preservative, in which case the average life is 10 to 14 years more, depending on the degree of exposure to decay. In the creosote treatment, the timber, after having been properly seasoned, is placed in a closed cylindrical chamber, and steam is introduced to soften the wood fibers. Air and moisture in the timber are then removed by a vacuum pump. Finally, creosote is injected into the cyclinder under pressure, resulting in almost complete penetration of the wood fibers by the creosote preservative. Other simpler and less expensive preservative treatments are also being used.

## The Metals

### 2-6 IRON AND STEEL

Iron ore is found in huge quantities in the earth's crust, at or near the surface. It occurs mostly in the form of iron oxide containing small amounts of silica and other impurities. The iron is separated from the oxide in a **blast furnace,** which consists of a cylindrical steel shell from 20 to 30 ft in diameter and from 150 to 200 ft high, lined inside with firebrick capable of withstanding temperatures higher than any produced in the furnace. Coke is used as a fuel to produce heat. The "charge," consisting of alternate layers of coke and ore into which a "flux," usually limestone, is mixed, is fed into the furnace stack through the top, beginning with a layer of coke from 20 to 25 ft deep at the bottom of the stack. Air under considerable pressure and preheated to a high temperature is blown into the furnace at the bottom through a series of nozzles called tuyeres. There it unites with the carbon of the coke to form carbon monoxide at a temperature exceeding 3000°F. As the hot carbon monoxide rises through the ore in the furnace, melting it, it combines with the oxygen of the iron oxide ore, escaping at the top as carbon dioxide, leaving the iron to trickle down to the bottom through the bed of coke. It is in this bottom high-temperature zone that smelting takes place. As the molten iron trickles down through the bed of coke, it combines with about 3 to 5 percent of carbon from the coke. The limestone flux combines with the silica to form molten slag, which floats above the molten iron near the bottom and is drawn off as a waste. The molten iron which collects and is drawn off at the bottom of the furnace is cast into small bars called "pigs." This pig iron cannot be used directly but requires additional processing and refining before it becomes cast iron, wrought iron, or steel.

## 2-7  CAST IRON

The term **cast iron** covers a series of materials possessing a broad range of mechanical and physical properties which are the result chiefly of the differing compositions of the materials. Pig iron, mixed with scrap cast iron and a limestone flux and, for special purposes, with scrap steel, chromium, copper, nickel, molybdenum, and so forth, is melted in a cupola furnace fired with coke and air blast. The molten metal is then poured into sand or metal molds, producing castings of desired forms and dimensions. The widely used gray cast iron, recognizable by the color and appearance of its fracture, contains carbon ranging from 2 to 4 percent and silicon from 1 to 3.5 percent. The structure of gray cast iron contains free graphite plates that act to impart properties peculiar to this material such as resistance to wear and corrosion, dampening ability, and machinability. Other types are chilled, or white, cast iron, which is highly resistant to abrasion; malleable cast iron, which has strength and ductility; and nodular cast iron, the structure of which contains spheroidal graphite that tends to develop attributes approaching those of carbon-steel castings.

## 2-8  WROUGHT IRON

Wrought iron is an iron product so made and refined that it possesses the important properties of ductility, malleability, and toughness. It is, therefore, especially suitable for machine parts to be shaped by forging. Also, its range of usefulness is considerably increased because of the ease with which it can be welded. Its ultimate strength is about three-quarters of that of structural steel.

Several processes are used in the production of wrought iron. Of these, the **puddling process** is perhaps most commonly used. In this process pig iron, together with a quantity of flux, is melted in a reverberatory furnace lined with iron oxide, the oxygen of which reduces the carbon content of the iron to less than 0.15 percent and also removes some sulfur. The mixture is then "puddled" by being stirred with steel rods. Toward the end of the process of purifying the iron, the heat of the furnace is insufficient to keep the purified iron in a molten state, and it therefore forms into balls of pasty iron, which contain a small amount of slag. These balls are removed from the furnace and are squeezed into "blooms" in a press, thus removing some excess slag. The blooms are finally rolled into smaller bars suitable for commercial use. The presence of slag in wrought iron gives it a fibrous texture. Other methods of producing wrought iron use the "charcoal-hearth" type of furnace, by which a finer grade can be produced.

## 2-9 STEEL

Steel is an iron product containing varying amounts of carbon, ranging up to about 1.7 percent. This product is technically referred to as **carbon steel.** An **alloy steel** is a carbon steel to which has been added one or more additional elements in amounts sufficient to give it recognizable properties not originally possessed by the carbon steel. Because of the marked differences in the properties of the carbon and the alloy steels, each group will be discussed separately.

**The Carbon Steels.** Pure iron (ferrite) contains no carbon, of course. It is relatively soft and is both ductile and malleable, but it does not possess great strength. Almost pure iron has an ultimate tensile strength of about 40,000 psi. The addition of carbon to pure iron, in amounts ranging from 0.05 to 1.7 percent, produces what is known as **steel.** The carbon has the effect of increasing strength and hardness, but it also decreases ductility and toughness. A carbon content of about 0.1 percent produces a so-called **soft steel** with an ultimate tensile strength of about 50,000 psi, suitable for rolling into plates. **Structural steel,** commonly used for rolling into structural shapes such as angles, beams, and columns, contains about 0.25 percent carbon, giving a fairly ductile and tough steel with an ultimate strength of about 64,000 to 72,000 psi. An increase to 0.40 percent of carbon produces a **machine steel** with an ultimate tensile strength of about 80,000 psi, while 0.75 percent carbon produces a hard **spring steel** with an ultimate tensile strength of 100,000 psi and 0.90 to 1 percent carbon produces a very hard **tool steel** with an ultimate tensile strength of 120,000 to 130,000 psi.

**The Alloy Steels.** When small amounts of one or more metals are added to a carbon steel in sufficient quantities so as to produce definite recognizable new properties in the steel, the product is called an **alloy steel.** The most commonly used alloying metals are nickel, manganese, chromium, vanadium, and molybdenum. Silicon, an inert material, is also used as an alloy to produce notch toughness. The effect of the addition of these alloys to carbon steel, singly or in combinations such as chromium and nickel, chromium and vanadium, and chromium and molybdenum, is, in general, to increase its strength and hardness, sometimes without sacrificing other desirable properties already possessed by the carbon steel.

## 2-10 THE MANUFACTURE OF STEEL

The making of steel from pig iron is essentially a refining process. It consists in first removing from the molten pig iron the excess carbon together with several impurities, such as silicon, manganese, and sulfur, and then adding carbon and/or any of the alloying elements required to produce a steel having the desired qualities and properties. The three most commonly used processes are (1) the *Bessemer process,* (2) the *open-hearth process,* and (3) the *electric-furnace process.* Of these, the open-hearth process produces the greatest quantity of steel in the United States, and the electric-arc furnace produces the most highly refined steel. The particular process used depends upon the chemical analysis of the pig iron to be refined and also upon the desired quality of the steel to be produced. The finished molten steel is poured into ingots in sizes suitable for use by the rolling mills.

**The Bessemer Process.** A Bessemer furnace, or converter, consists of a steel cylinder lined with firebrick. It is open at the top and is so constructed and supported that it can be revolved horizontally and tipped vertically to receive or discharge its contents. It is charged with 10 to 20 tons of molten pig iron, often directly from a blast furnace. Several thousand cubic feet per minute of unheated air is then blown through the molten iron for from 10 to 15 min. This is called the **blow,** and its action is to remove, in the form of burning gases, the silicon, manganese, carbon, and sulfur. After this, the elements required to give the steel the desired chemical properties are added to the molten mass. The Bessemer process is the most rapid of the three, but this rapidity makes control of the chemical properties more difficult.

**The Open-Hearth Process.** In this process, preheated air and gas are used to melt the charge consisting generally of scrap iron, pig iron, iron oxide, and a limestone flux. The excess carbon, manganese, and phosphorus are removed by oxidation with the oxygen from the iron oxide, after which the elements required to give the desired chemical properties are added. The capacities of open-hearth furnaces range up to nearly 200 tons. This process is much slower than the Bessemer process, requiring about 10 h for each charge, but the slowness of the process makes possible a high degree of control of the quality of the steel.

**The Electric-Furnace Process.** Two types of electric furnaces are in general use. They are distinguished by the method used to produce heat. In the *electric-arc furnace*, the charge is heated by an electric arc passing between electrodes. In the *electric-induction* furnace, the charge is placed in a crucible encircled by coils and wire through which a high-frequency current is passed. This "induces" secondary eddy currents in the iron and heats it by virtue of its resistance to the passage of these currents. Both methods permit a high degree of control of the steel product, and the *induction* method is especially suitable for the making of high-grade alloy steels.

## 2-11 ALUMINUM

This metal is very widely used in the manufacture of items in which lightness is an advantage or is required, most notably in aircraft. The ultimate tensile strength of nearly pure aluminum is less than 10,000 psi, but when alloyed with 4 to 5 percent copper and smaller percentages of manganese, magnesium, and/or other metals, various grades of wrought aluminum can be produced whose ultimate tensile strengths vary from about 20,000 to 60,000 psi or better. Aluminum is also extensively used for castings after being alloyed with varying percentages of silicon, copper, iron, zinc, manganese, and magnesium, either singly or in combinations.

Aluminum is obtained principally from an ore called **bauxite,** which, chemically, is hydrated oxide of aluminum, found in great abundance on the earth's surface. When this ore is treated with caustic soda, aluminum oxide is produced, which is then mixed with molten cryolite and is reduced electrolytically to metallic aluminum.

Although aluminum is one of the most chemically active of the commonly used metals, it is comparatively resistant to corrosive influences in its purer forms but is much less resistant in its alloyed forms, especially when alloyed with copper. The destructive effect of corrosion is most serious in thin sheets of the metal, so widely used in aircraft construction. In order to combine the high strength of aluminum alloyed with copper with the corrosion resistance of pure aluminum, sheets and other forms of the high-strength metal are coated with thin surface layers of the pure metal. The trade name Alclad is one which designates products so treated.

**Wrought aluminum** is easily drawn into wire, rolled into sheets, or forged into machine parts. Tubing and small structural shapes such as angles and beams are readily formed by extrusion or rolling. Large **aluminum castings** are usually formed in sand molds. Smaller castings are made either in metal molds or by die-casting, a process in which liquid metal is forced into a die under high pressure.

## The Inert Materials

### 2-12 STONE, BRICK, AND CONCRETE

These materials are generally used in large masses in structures where only compressive stresses are to be resisted, such as in foundations, piers, and walls.

A property which is common to all of the materials in this group is their relatively low strength in tension and in shear. Consequently, if any part of such a structure is subjected to tensile stress, as is the case in concrete beams and floor slabs of buildings, the part must be reinforced by steel bars placed in the areas where tensile or shear stresses must be resisted. In geographical areas subjected to earthquake shocks, walls of stone, brick, or concrete must be reinforced with networks of steel bars.

**Stone.** This natural material was still used extensively in heavy foundations and structures until half a century ago. Since then it has gradually been replaced with steel and concrete. Its use at present is confined almost entirely to architectural ornamentation.

**Brick.** Two types of bricks are extensively used: (1) clay bricks and (2) sand-lime bricks. Clay bricks consist of clay mixed to a given consistency and molded to shape by machine. They are then "burned" in a kiln at temperatures below the melting point of the clay until they become semi-vitrified. Sand-lime bricks are made of a mixture of common sand mixed with lime, which acts as a binder. They are molded by machine under high pressure. Both types of brick are made under controlled conditions, and their strength is purposely made about equal. Depending on grade, the compressive strength of brick varies from about 1000 psi to 5000 psi or more. Brick is usually very brittle and has relatively low strength in tension and shear.

**Concrete.** Concrete is a mixture of cement, sand, and gravel with enough water added to bring about complete chemical action of the cement and to make the mixture workable, that is, capable of being poured into forms leaving reasonably smooth outer surfaces after having dried. Because of its strength and fire-resisting qualities and the ease with which it can be mixed and molded into any desired form, concrete has, in the past half-century, become one of the most important of the structural materials. In addition, the necessary ingredients are readily available in huge quantities and at low cost almost everywhere on the face of the earth.

**Cement.** Several types of cement are being manufactured. Of these, *Portland cement* has greatest strength and hardens most quickly and is, therefore, most widely used. *Natural cement* is less strong and hardens less quickly.

The main ingredients of cement are calcium carbonate, silica, and alumina. Limestone is the chief source of calcium carbonate. Silica and alumina are obtained from clay, shale, or argillaceous limestone. These raw materials are first ground and mixed in suitable proportions. Next the mixture is burned to a clinker, just short of fusion, after which it is ground to a fine powder.

The aggregates most commonly used are sand and gravel or crushed rock, clean and free of alkali and organic matter. To produce concrete ready for pouring into forms, cement, sand, and gravel are mixed with water in such proportions by volume that the desired strength is obtained. If the proportions are $1:3:5$, a concrete with an ultimate strength of 1500 psi can be expected; a $1:2:4$ mix should give a 2500-psi concrete; and a $1:1^1/_2:2^1/_2$ should give a 3500-psi concrete. Low-strength concrete is used in heavy masses, such as foundations, piers, dams, while higher-strength concrete is generally used in bridges and in the reinforced-concrete frames of buildings.

The amount of water used per sack of cement varies from 5 to 8 gal. A low water-cement ratio gives high strength but poor workability, while a high ratio gives low strength and good workability. The concrete will *set* in a few hours after pouring and the expected strength is usually reached in 28 days, during which period it "cures." During hot, dry weather, the concrete must be kept damp to permit proper curing. Too quick drying interferes with the chemical action of the cement and results in loss of strength. Freezing of concrete before it has set will materially destroy its strength.

## REVIEW QUESTIONS

2-1. What is meant by the term "properties of materials"?

2-2. Define in general the properties of strength and elasticity.

2-3. What is meant by ductility and malleability?

2-4. Explain the property of hardness and how it is measured.

**2-5.** What is meant by resilience? How is it measured?

**2-6.** What is meant by toughness and how is it measured?

**2-7.** Explain the terms "creep" and "machinability."

**2-8.** Name at least four kinds of wood commonly used as structural materials. Discuss some of their properties.

**2-9.** How may wood be preserved against decay?

**2-10.** From what ore is iron most generally obtained?

**2-11.** Describe the common process of smelting iron ore. What is the final product of this process called?

**2-12.** Explain the process of producing iron castings from pig iron. What are some of the usual properties of cast iron?

**2-13.** What are the three most important properties of wrought iron? Name some uses for wrought iron.

**2-14.** Describe what is known as the "puddling process" of making wrought iron.

**2-15.** Describe in general terms the product steel, and explain the distinction between carbon steel and alloy steel.

**2-16.** Name at least four elements with which carbon steel is commonly alloyed.

**2-17.** Name three processes by which steel is commonly produced.

**2-18.** From what ore is metallic aluminum usually obtained and how is it generally produced?

**2-19.** What are some of the metals with which aluminum is commonly alloyed?

**2-20.** Discuss some differences in properties of pure aluminum and aluminum alloyed with copper.

**2-21.** Describe one process by which a copper-aluminum alloy is commonly made more resistant to corrosion.

**2-22.** Name some commercial shapes into which wrought aluminum is generally formed.

**2-23.** Discuss processes by which bricks are made.

**2-24.** What are some of the strength characteristics of brick?

**2-25.** Describe concrete and how it is produced.

**2-26.** Discuss some strength properties of concrete.

**2-27.** Describe in general the manufacture of cement.

# Chapter 3
# Riveted and
# Welded Joints.
# Thin-Walled
# Pressure Vessels

## 3-1 RIVETED JOINTS

Steel structures, such as building frames composed of beams and columns, roof and bridge trusses, cranes, railroad cars, boilers, water tanks, and many others, are built up of a number of separate pieces, which then must be effectively joined together. A common method of joining is that of **riveting.**

A **rivet** is a short bar of metal with a half-spherical *head* already formed on one end. Matching holes are punched or drilled in the two pieces to be joined. After the pieces have been placed in their relative final positions, rivets are inserted into the holes and are then *driven,* usually with a pneumatic hammer. That is, the head end of a rivet is held tightly against the joined pieces while the opposite end is hammered until another similar head is formed there, thus gripping the pieces tightly together. The force to be resisted is then transmitted from the one piece to the rivets and by the rivets to the other piece. Rivets may be driven hot or cold.

Steel rivets, as used in most structural and boiler work, when heated to a bright red heat, "soften" greatly and are then more easily driven. During cooling a hot-driven rivet will shrink, thus producing tension in the rivet and consequently gripping the joined pieces even more tightly together. Copper and aluminum rivets, which are frequently used in aircraft work, are generally driven cold. Such rivets are usually small and the metal is somewhat softer than steel.

When rivets thus grip plates tightly together, a considerable amount of frictional resistance is developed between the plates, which, of course, increases the safety of a joint. In practical design, however, this frictional resistance is often disregarded.

52

## 3-2 TYPES OF RIVETED JOINTS

Two types of riveted joints in general use are the **structural joint** and the **boiler joint.** The structural joint, as the term implies, is generally used for connecting members of structures such as building frames, trusses, cranes, and so forth. The boiler joint is used in all cases requiring an absolutely tight joint, such as in boilers, water tanks, compressed air tanks, and vessels of all types subjected to pressure. When several rivets are used at a joint, each rivet is presumed to resist its proportional part of the total load on the joint. Certain differences which exist in the methods of design of structural and boiler joints are explained in later paragraphs.

**Structural Joints.** Figure 3-1 shows two typical riveted structural joints. Such joints are generally referred to as **connections.** In the beam-to-girder connections, two beams frame into a large supporting girder from opposite sides. The connection is made possible by riveting two short opposite angles to the web of each beam, using four rivets $B$ in each. The two sets of angles are then fastened to the web of the girder by eight common rivets $G$. Note here that all these rivets are in double shear.

The beam-to-column connection in Fig. 3-1$b$ is very similar. Here, a single beam frames into the flange of a column. Two short opposite angles are riveted to the web of the beam with three rivets $B$. These angles are then fastened to the column by means of six rivets $C$ through its flange. Note here that while the three rivets $B$ through the web of the beam are in double shear, the six rivets $C$ through the column flange are in single shear. If all the rivets have the same diameter, we see that the three in double shear have the same shear strength as the six in single shear. A "seat" angle is often used, thus giving the joint greater strength and rigidity and greatly simplifying the task of erection.

## 3-3 FAILURES OF RIVETED JOINTS

When a riveted joint or connection is subjected to a load which is beyond its capacity to resist, it may fail in one of several ways, as indicated in Fig. 3-2. While failure may occur entirely in shear, in bearing, or in tension, it may also occur simultaneously or progressively in several combinations of these.

**Shearing Failure.** As shown in Fig. 3-2$a$, a rivet may fail in shear, either in *single* shear on *one* cross-sectional area of the rivet or in *double* shear on *two* cross-sectional areas, depending on the type of joint.

**Compression or Bearing Failure.** The material of the plate in front of the rivet where it bears against the plate may crush, as indicated in Fig. 3-2$b$.

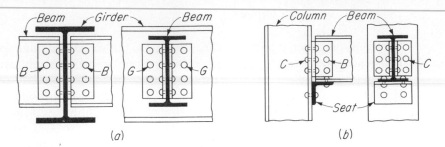

Fig. 3-1    Riveted structural joints. (a) Beam-to-girder connection. (b) Beam-to-column connection.

The rivet itself, of course, is subjected to the same bearing stresses, and the rivet metal may fail, or failure may be partly in plate and partly in rivet.

**Tension Failure.**  Failure in tension occurs when a plate tears apart along some line of least resistance as, for example, along line *a-a* in Fig. 3-2*c*.

**Other Types of Failure.**  A riveted joint subjected to a load beyond its capacity to withstand will very likely fail either in shear, in bearing, or in tension, as described in preceding paragraphs. Certain other types of failures are possible, however, but are usually preventable if correct design procedures are followed.

Figure 3-3 illustrates two other possible types of failure, **end shearing** and **end tearing.** Such failures result only when the *end distance e in line of stress* is insufficient. Failure of this type is most often a combination of crushing and shearing or of crushing and tearing. Because of this the computed shear stress along the failure planes in Fig. 3-3*a* is an unreliable indicator of adequate design. Hence, in considering the safety of a design, it is best to follow the empirical rule that the minimum value of *e* should not

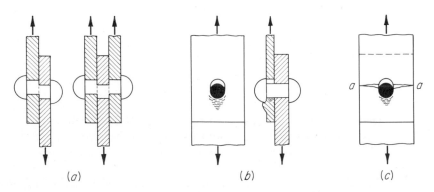

Fig. 3-2    Failures of riveted joints. (a) Shearing failure. (b) Bearing failure. (c) Tearing.

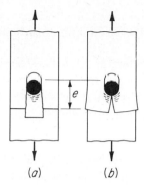

Fig. 3-3   **End failures. (a) End shearing. (b) End tearing.**

be less than the shearing area of the rivet, single or double, divided by the thickness of the plate.

## 3-4 STRESSES IN RIVETED JOINTS

The discussion of possible failures in the preceding article has indicated the stresses—shearing, bearing, and tensile—existing in riveted joints, However, rivets and plates are also subjected to other stresses which are not so apparent. For example, in the simple lap joint shown in Fig. 3-4$a$, it is clear that the forces $P$ in the plates are transmitted to the rivet at approximately the midpoints of the plates, the centerlines of which are a distance $e$ apart.

These forces tend to bend both the rivet and the plates. If a substantial rivet is used to fasten together two rather light plates and if a sufficiently great force is then applied, the plates will bend somewhat as illustrated in Fig. 3-4$b$, until the plate forces $P$, at short distances from the rivet, are nearly in line. The forces presumably acting on the rivet are shown in Fig. 3-4$c$. The two forces $P$ form a counterclockwise couple $Pb$, which must be and is counteracted by an opposed clockwise couple $P'a$. Clearly, the forces $P$ produce shearing stresses in the rivet and also bend it; the forces

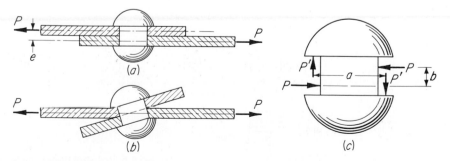

Fig. 3-4   **Bending in lap joints.**

$P'$ likewise bend the rivet and, in addition, produce tensile stresses in it. Naturally, the actual forces acting on the rivet are nonuniformly distributed, and $P$ and $P'$ are merely resultants.

Despite their known existence, these bending and tensile stresses are not directly considered in the design of riveted joints. Their intensities would be very difficult to determine. However, allowable stresses permitted in design are sufficiently conservative to compensate for disregarding them.

**Shearing Stresses.** The shearing stresses on a cross-sectional area of a rivet are not uniformly distributed. Consequently, the use of the direct-stress formula $s = P/A$ gives only the *average unit shearing stress*. The stressed area in single shear is the cross-sectional area and in double shear is twice the cross-sectional area.

**Bearing Stresses.** The area in bearing between a plate and a rivet is semicylindrical. The actual distribution of stress over this area is rather complex. Therefore, as in the case of shear, the use of the direct-stress formula gives only an *average unit bearing stress*. The area in bearing commonly used for the purpose of determining this average stress is the **projected area,** which is the diameter of the rivet times the thickness of the plate against which it bears.

In some joints, either structural or boiler, two bearing stresses of different intensities exist and must be considered. For example, in all pinned, bolted, or riveted connections where a main or inside piece or plate is *enclosed* and is held in place by two outside pieces or plates, the pressure between the enclosed or inside piece and the pin, bolt, or rivet is called **inside bearing.** The pressure between the two outside pieces and the pin, bolt, or rivet is called **outside bearing.** These conditions are clearly illustrated in Fig. 3-5. When only two pieces are used, as, for example, to form a lap joint, as in Fig. 3-2*a*, *b*, and *c*, only outside bearing exists. In some literature on the subject, inside bearing is called **enclosed bearing** and outside bearing is called **plain bearing.** These terms, however, especially the latter, seem inadequately explanatory.

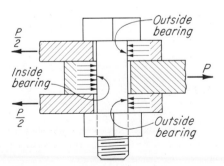

Fig. 3-5    Bolted structural connection. Inside and outside bearing.

**Tensile Stresses.** The tensile stresses here dealt with are those in the plates or in the connected structural members that are subjected to direct tensile stress. Obviously, to determine the unit tensile stress, the *net area in tension* must be used. That is, only the area of material actually resisting stress may be included.

In **structural joints,** rivet holes are generally laid out and punched separately in the parts to be joined. To avoid difficulty in assembly of the parts, the holes are punched with a diameter $1/16$ in. greater than the rivet diameter. The punching operation damages the material of the wall of the cylindrical hole to some slight depth, perhaps $1/32$ in., thus making that much of the material of the plate or part incapable of resisting tensile stress. Therefore, *for the purpose of calculating the net area in tension of punched plates or parts, the diameter of the hole is considered to be $1/8$ in. (3 mm) greater than the diameter of the rivet to be used.* This specification applies also when *unfinished bolts* are used instead of rivets. *In structural work, areas in shear and bearing are based on the original nominal diameter of the rivet.* Steel rivets are available in diameters which are multiples of $1/8$ in. from $3/8$ to $1\frac{1}{4}$ in.

**Allowable Stresses.** Safe allowable stresses for use in design are generally based on the results of a large number of tests in which bolts, rivets, plates, and made-up joints are tested to failure to determine their ultimate strength. The allowable stress is then arrived at by dividing the ultimate

Table 3-1 ALLOWABLE STRESSES FOR STRUCTURAL RIVETS, BOLTS, AND PINS. AISC SPECIFICATIONS. RECOMMENDED FOR STRUCTURAL JOINTS.

| | Allowable Tensile Stress,* ksi (MPa) | Allowable Shearing Stress, ksi (MPa) | | Bearing on ASTM A36 Steel‡ ksi (MPa) |
|---|---|---|---|---|
| | | Friction-Type Connections, Standard-Size Holes,† ksi (MPa) | Bearing-Type Connections, ksi (MPa) | |
| A502, Grade 1, hot-driven rivets | 23.0 (160) | ................. | 17.5 (120) | 43.5 (300) |
| A307 bolts | 20.0 (140) | ................. | 10.0 (70) | 43.5 (300) |
| A325 bolts, when threads *not* excluded from shear plane | 44.0 (300) | 17.5 (120) | 21.0 (145) | 43.5 (300) |
| A325 bolts, threads excluded from shear plane | 44.0 (300) | 17.5 (120) | 30.0 (205) | 43.5 (300) |

* Static loads only.
† Hole diameter = Nominal size + $1/16$ in. (1.6 mm). Clean mill scale between surfaces.
‡ With preferred end distance. May be very conservative where edge and end distances are much greater than usual values.

strength by a reasonable and predetermined factor of safety, generally varying between 3 and 5. The allowable stresses recommended for use in design of riveted joints are specified by various codes and vary somewhat according to the type of joint. The most generally accepted allowable stresses for structural joints are listed in Table 3-1.

**Allowable Stresses for Structural Joints.** According to the American Institute of Steel Construction, *the safe allowable stress in tension in* ASTM A36 structural steel is 22 ksi (152 MPa) on the gross section or 29 ksi (200 MPa) on the net section, except for pin-connected or threaded members. The allowable tensile stress for pin-connected members of ASTM A36 steel is 16.2 ksi (112 MPa) on the net section. The AISC also recommends the allowable stresses given in Table 3-1 for riveted and bolted joints.

When slip between the connected parts cannot be tolerated and must therefore be prevented by high clamping force, the allowable shear values are those given for *friction-type connections*. When a small amount of initial slip is permissible (to allow the rivets or bolts to seat on the bearing surfaces), the allowable stresses are those for *bearing-type connections*.

## 3-5 ANALYSIS OF RIVETED STRUCTURAL JOINTS

An analysis of a given joint is made to determine one of two things: (1) the average unit stresses, $s_s$, $s_c$, and $s_t$, in shear, bearing, and tension produced in a connection of given dimensions by a given load $P$ or (2) the maximum load $P$ which can safely be resisted by a connection of given dimensions. Both are illustrated in Probs. 3-1 and 3-2.

**Allowable Loads on Bolts and Rivets.** In design practice, it is customary to work with the *allowable load* on a rivet in shear and bearing, inside and outside, rather than the allowable unit stress. This allowable load is the product of the allowable stress and the area under stress. Thus the allowable load on a $3/4$-in.-diameter A325 bolt in a friction-type connection is $(17,500)(0.442) = 7,735$ lb and in double shear is $(17,500)(0.442)(2) = 15,470$ lb. In the structural joint of Fig. 3-5, let the inside plate be $3/8$ in. thick and each side plate be $1/4$ in. thick. All plates are of ASTM A36 steel. The allowable load on an A325 bolt in a friction-type connection is the least of the following calculated values: $(17,500)(0.442)(2) = 15,470$ lb or $(43,500)(3/8)(3/4) = 12,234$ lb or $(43,500)(2)(1/4)(3/4) = 16,312$ lb. Since the least, or controlling, value is that calculated for bearing, the allowable load for this bolt is 12,234 lb. In any connection, the *least allowable load on a bolt or rivet* is called its value.

## ILLUSTRATIVE PROBLEMS____

**3-1.** Figure 3-6 shows a simple structural connection composed of two steel straps held together by two bolts 1 in. in diameter. The upper strap is $^3/_8$ in. thick, and the lower strap is $^1/_2$ in. thick. Both straps are 3 in wide. Calculate the maximum unit stresses in shear, bearing, and tension caused by a pull $P$ of 15,000 lb.

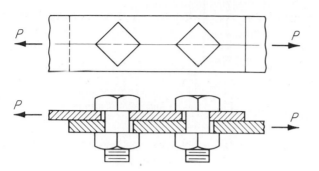

**Fig. 3-6   Bolted connection.**

*Solution:* In each case we must calculate the size of the area resisting the pull. By dividing this area into the pull $P$, we obtain the average unit stress.

*Shear.* The bolt is in single shear. Hence, $A_s = 2(0.7854) = 1.57$ in.$^2$ and

$$\left[ s = \frac{P}{A} \right] \qquad\qquad s = \frac{15,000}{1.57} = 9550 \text{ psi} \qquad\qquad Ans.$$

*Bearing.* The upper strap bears against the right side of each bolt, and the area in bearing is presumed to be the bolt diameter times the thickness of the strap. That area is $(2)(1)(0.375) = 0.75$ in.$^2$. The lower strap bears against the left side of each bolt, and that bearing area is $(2)(1)(0.5) = 1.0$ in.$^2$. Clearly, since each strap is subjected to a pull of 15,000 lb, the maximum stress will be exerted on the smallest of these two areas and is

$$\left[ s = \frac{P}{A} \right] \qquad\qquad s_c = \frac{15,000}{0.75} = 20,000 \text{ psi} \qquad\qquad Ans.$$

*Tension.* When either rivets or unfinished bolts are used in *structural* joints, the hole is considered to be of $^1/_8$ in. larger diameter than the rivet or bolt. The maximum tensile stress will be found in the upper strap, where the net resisting area is smaller than in the lower strap. This area is $(3.0 - 1.125)(0.375) = 0.703$ in.$^2$. Hence

$$\left[ s = \frac{P}{A} \right] \qquad\qquad s_t = \frac{15,000}{0.703} = 21,300 \text{ psi} \qquad\qquad Ans.$$

**3-2.** An ASTM A36 steel beam, W 21 × 62, is connected to a steel column, W 14 × 99, with $^3/_4$-in.-diameter rivets, as illustrated in Fig. 3-7. All connecting angles are $^3/_8$ in. thick. Calculate the allowable load $R$ that may safely be transmitted from the beam to the column through the bearing-type connection.

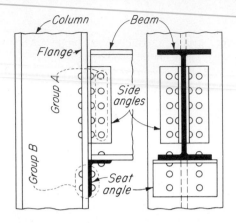

Fig. 3-7 Beam-to-column connection.

*Solution:* In order to determine the allowable load on this connection, we must first find the allowable load on one rivet in shear, single or double, and in bearing. In this connection, however, there are two groups of rivets which must be considered separately: group A of five rivets through the web of the beam and group B of four rivets which connect the seat angle to the column flange. The two rivets fastening the beam flange to the seat angle do not resist any vertical load and must be disregarded.

From Table E-4 we find that the web of the W 21 × 62 beam is 0.40 in. thick. The side angles are $3/8$ in. (0.375 in.) thick. The rivets through the beam web are in double shear, in bearing against 0.40-in. web and two $3/8$-in. angles.

Allowable load on one $3/4$-in. rivet in group A (through beam web):

$$\text{In double shear } (17.5)(2)(0.442) \qquad = 15.47 \text{ kips}$$

$$\text{In bearing on 0.40-in. plate } (43.5)(0.4)\left(\frac{3}{4}\right) = 13.05 \text{ kips} \leftarrow \text{Governs}$$

$$\text{In bearing on } (2)(0.375) = 0.75 \text{ in.} \qquad = 24.47 \text{ kips}$$

In group B, the four rivets are in single shear, in bearing against one $3/8$-in. angle and also in bearing against the $3/4$ in.-thick column flange.

Allowable load on one $3/4$-in. rivet in group B (through seat angle):

$$\text{In single shear } (17.5)(0.442) \qquad = 7.774 \text{ kips} \leftarrow \text{Governs}$$

$$\text{In bearing on } \tfrac{3}{8}\text{-in. angles, } 43.5\left(\frac{3}{8}\right)\left(\frac{3}{4}\right) = 12.23 \text{ kips}$$

$$\text{In bearing on } \tfrac{3}{4}\text{-in. flange} \qquad = 24.47 \text{ kips}$$

The total allowable load on the connection then is

Allowable load on 5 group A rivets = (5)(13.05) = 65.25 kips
Allowable load on 4 group B rivets = (4)(7.74)  = 30.96 kips
Total allowable load R on connection     = 96.21 kips    *Ans.*

## ANALYSIS PROBLEMS

NOTE: In the following problems assume ASTM A36 steel and A325 bolts in bearing-type connections unless otherwise stated. Furthermore, assume care will be taken to exclude threads from the shear planes.

**3-3.** In the bolted structural joint shown in Fig. 3-6, assume that each strap is 10 mm thick and 60 mm wide and that they are held together by two unfinished M16 bolts. Calculate the allowable pulls $P_s$, $P_c$, and $P_t$ in shear, bearing, and tension on this connection.

**3-4.** Assume in Fig. 3-7 that the beam is a W 21 × 101, the column is a W 12 × 65, all angles are $3/8$ in. thick, and 1-in.-diameter rivets are used. Then calculate the allowable load $R$ on the connection.

**3-5.** In Fig. 3-1*a* assume that the beam is a W 18 × 46, the girder is a W 21 × 111, the angles are $3/8$ in. thick, and $3/4$-in.-diameter rivets are used. Then calculate the allowable load $R$ that may safely be transmitted from the beam to the girder.

*Ans.* 47.0 kips

**3-6.** Assume in Fig. 3-1*b* that the beam is a W 14 × 61, the column is a W 10 × 33, all rivets are of $7/8$-in. diameter, and all angles are $1/4$ in. thick. Calculate the value of the maximum load $R$ (end reaction) that may safely be transmitted from the beam to the column.

**3-7.** A typical chain link is shown in Fig. 3-8. The side straps are 10 mm thick by 40 mm wide between the pins and 60 mm wide across the pins. The finished pins are 15 mm in diameter and are close fitting. Bushings are used to space the inner straps. Calculate the unit stresses in shear, bearing, and tension in this link when it is subjected to a pull of 15 kN.

*Ans.* $s_s = 42.44$ MPa; $s_c = 50.0$ MPa; $s_t = 18.75$ MPa

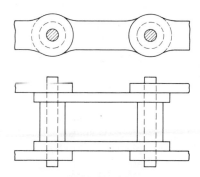

Fig. 3-8   Probs. 3-7 and 3-8.

**3-8.** Calculate the value of the maximum load $P$ to which the chain link in the preceding problem may be subjected. The side straps are of structural steel having an allowable bearing stress of 250 MPa and an allowable tensile stress of 125 MPa. The allowable shear stress in the pin is 100 MPa.

**3-9.** The bell crank shown in Fig. 3-9 is held in the position shown by the two forces $H$ and $T$. The pin $A$ is $1/2$ in. in diameter and fits snugly in a reamed hole.

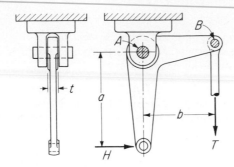

Fig. 3-9   Probs. 3-9 and 3-10.

(*a*) Calculate the unit shearing stress in pin *A* when *H* = 2400 lb, *a* = 4 in., and *b* = 3 in. (*b*) What force *T* will produce a shearing stress in pin *A* equal to 15,000 psi?

**3-10.** In Fig. 3-9, the pin *A* is ³/₄ in. in diameter and fits snugly into the reamed hole. Calculate the thickness *t* of the bell crank which will be required to give the pin equal strength in shear and bearing under the conditions shown. Use allowable stresses of 15,000 psi for shear and 42,000 psi for bearing.

## 3-6 DESIGN OF RIVETED STRUCTURAL JOINTS

The problem of designing a riveted joint to resist a given load *P* involves the use of predetermined allowable stresses *s* in order to evaluate the required area *A* in shear, bearing, and, in some connections, tension. In design, then, the basic equation $s = P/A$ is used in the form $A = P/s$.

A complete design would involve also consideration of the shearing and tensile stresses in the gusset plate. If, however, the usual recommended rivet spacings and minimum edge distances are used, the gusset-plate stresses will invariably be less than those allowed, thus making design of gusset plates unnecessary except in special cases.

### ILLUSTRATIVE PROBLEMS

**3-11.** A typical structural joint of a light roof truss is shown in Fig. 3-10. Each member is made up of two ⁵/₁₆-in.-thick angles placed back to back and riveted to a gusset plate. (*a*) Calculate the minimum thickness *t* of gusset plate for each rivet to be approximately as strong in bearing as it is in shear. (*b*) If force *A* = 26,100 lb, *B* = 43,500 lb, and *C* = 34,800 lb, how many ³/₄-in.-diameter rivets are required in each member?

*Solution:* We note that all rivets are in double shear and inside bearing. From Table 3-1 the allowable stresses in rivets are shear, 17,500 psi; bearing, 43,500 psi. We may now calculate the "value" of a ³/₄-in.-diameter rivet in double shear, which is

$$(2)(0.4418)(17,500) = 15,460 \text{ lb}$$

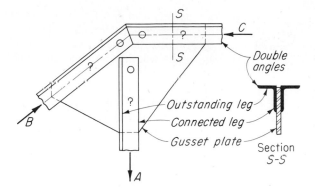

**Fig. 3-10   Riveted structural joint.**

The area in bearing on a $^3/_4$-in. rivet is $0.75t$ in.$^2$, and its value in bearing then is $(0.75t)(43,500) = 32,620t$. Then for equal strength in shear and bearing

$$32,620t = 15,460 \quad \text{or} \quad t = 0.474 \text{ in.}$$

$$\text{Use } t = \tfrac{1}{2} \text{ in.} \qquad\qquad Ans.$$

If we use a $^1/_2$-in. gusset plate, the strength of a $^3/_4$-in. rivet in bearing is $(0.75)(0.5)(43,500) = 16,300$ lb, which is approximately the same as its strength in double shear. Now, for equal or greater strength in bearing, each angle must be at least half as thick as the gusset plate. Since each angle is $^5/_{16}$ in. thick, this requirement is satisfied. Consequently, we may disregard bearing on the angles.

The value in shear, being less than the value in bearing, will then determine the number of rivets required. That is,

$$\text{Member } A: \quad N = \frac{26,100}{15,460} = 1.69 \quad \text{Use 2 rivets.} \qquad Ans.$$

$$\text{Member } B: \quad N = \frac{43,500}{15,460} = 2.81 \quad \text{Use 3 rivets.} \qquad Ans.$$

$$\text{Member } C: \quad N = \frac{34,800}{15,460} = 2.25 \quad \text{Use 3 rivets.} \qquad Ans.$$

**3-12.** Design a bolted structural connection of the type shown in Fig. 3-11 to withstand a pull $P = 25$ kips. Use high-strength ASTM A325 bolts. The steel straps are of ASTM A36 steel. No slippage between plates is to be permitted, and care will be taken to exclude threads from the shear planes.

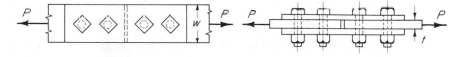

**Fig. 3-11   Bolted structural connection.**

*Solution:* The design requires the determination of (a) the bolt diameter $d$, (b) the thickness $t$ and the width $w$ of the main straps, and (c) the thickness and width of the splice plates. Item $c$ is easily determined after the thickness and width of the main strap are computed. Each splice plate will be made the same width and at least half as thick as the main strap.

The required size of the main strap cannot be determined until the size of the bolt hole is known. The bolt diameter is obtained from the required shear area. From Table 3-1, for this friction-type joint the allowable stresses are $s_s = 17.5$ ksi and $s_c = 43.5$ ksi. Since each bolt must resist half the total load, or 12.5 kips, the required shear area is

$$\left[A_s = \frac{P}{s_s}\right] \qquad A_s = \frac{12.5}{17.5} = 0.71 \text{ in.}^2$$

Since each bolt has two shear planes, the required area per shear plane is 0.36 in.² and the diameter is $3/4$ in. $(A = 0.442$ in.²$)$. *Ans.*

We may now determine the thickness $t$ from the required bearing area, which is $td$. That is,

$$\left[A_c = \frac{P}{s_c}\right] \qquad A_c = \frac{12.5}{43.5} = 0.287 \text{ in.}^2 \qquad \text{and} \qquad t = \frac{0.287}{0.75} = 0.383 \text{ in.}$$

$$\text{Use } t = \frac{7}{16} \text{ in. } (0.4375 \text{ in.}). \qquad \textit{Ans.}$$

The width $w$ may now be determined by the required net area in tension. The critical section is through the exterior bolt. The tensile force at this section is 25 kips. The required area is

$$\left[A_t = \frac{P}{s_t}\right] \qquad A_t = \frac{25}{22} = 1.14 \text{ in.}^2$$

and the net width then is

$$\frac{A_t}{t} = \frac{1.14}{0.4375} = 2.61 \text{ in.}$$

To obtain the gross width $w$, we must add the bolt diameter $d$ plus $1/8$ in. The required width $w$ is then

$$w = 2.61 + 0.75 + 0.125 = 3.48 \text{ in.}$$

$$\text{Use } w = 3\frac{1}{2} \text{ in.} \qquad \textit{Ans.}$$

As previously stated, the thickness of each side plate should be at least half that of the main strap.

$$\text{Use } \frac{1}{4} \times 3\frac{1}{2}\text{-in. splice plates.} \qquad \textit{Ans.}$$

**3-13.** In Fig. 3-12, the load to be transmitted from each of the opposite W 16 × 89 steel beams to the W 21 × 101 girder is 100 kips. All connecting angles are $3/8$ in. thick, and all rivets are of $7/8$-in. diameter. If only four rivets can be used

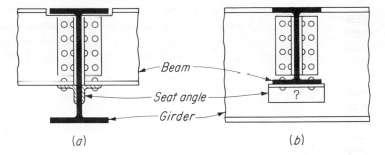

**Fig. 3-12   Structural beam-to-girder connection. (a) Section through girder. (b) Section through beam.**

through the web of each beam, how many rivets are required to fasten the two opposite seat angles to the web of the girder?

*Solution:* We must first determine the amount of load carried by the four rivets through the web of the beam. The remainder then must be carried by the seat-angle rivets. The rivets connecting the beam flange to the seat angle do not resist any of the vertical load. From Table E-4, we find that the web thickness of each beam is 0.525 in. and of the girder is 0.500 in.

Allowable load on one $7/8$-in. rivet through beam web:

In double shear $(2)(0.601)(17.5)$ $\qquad = 21.04$ kips

In bearing on web $(0.500)\left(\frac{7}{8}\right)(43.5)$ $\qquad = 19.03$ kips 9 ← Governs

In bearing on angle $(2)(0.375)\left(\frac{7}{8}\right)(43.5) = 28.55$ kips

Then load carried by the four beam-web rivets $= (4)(19.98) = 79.92$ kips, and the load to be carried by the seat-angle rivets $= 100 - 79.92 = 20.08$ kips from each side or $2(20.08) = 40.16$ kips from both sides.

Allowable load on one $7/8$-in. rivet through seat angle and girder web:

In double shear $(2)(0.601)(17.5)$ $\qquad = 21.04$ kips

In bearing on web $(0.500)\left(\frac{7}{8}\right)(43.5)$ $\qquad = 19.03$ kips ← Governs

In bearing on angle $(2)(0.375)\left(\frac{7}{8}\right)(43.5) = 28.55$ kips

The number of rivets through the seat angle and girder web is

$$N = \frac{40.16}{19.03} = 2.11 \qquad \text{Use 3 rivets.} \qquad\qquad\qquad Ans.$$

## DESIGN PROBLEMS. STRUCTURAL JOINTS

NOTE: The following problems are all of the structural type. Assume ASTM A36 steel and A325 bolts in bearing-type connections unless otherwise stated. Also assume that threads will be excluded from the shear planes.

**3-14.** In the structural joint in Fig. 3-10, three double-angle members are riveted to a gusset plate. All angles are $^3/_8$ in. thick. Rivets of $^7/_8$-in. diameter are to be used. (*a*) Calculate the required thickness *t* of gusset plate which will give the $^7/_8$-in. rivets approximately equal strength in shear and bearing. (Give *t* as a multiple of $^1/_{16}$ in.) (*b*) If force *A* = 70 kips, *B* = 118 kips, and *C* = 95 kips, how many rivets are needed in each member?

**3-15.** Design a bolted structural connection of the type shown in Fig. 3-11 capable of safely resisting a tensile load *T* of 50,000 lb.

**3-16.** The steel bar in Fig. 3-13 is held in place by two bolts. The bar and bolts must be designed to support a tensile load *T* of 60 kips. (*a*) Based on their strength in shear, what diameter of bolts should be used? (*b*) If a $^5/_8$-in.-thick ASTM A36 bar is used, what width *w* should it be?

*Ans.* (*a*) $^7/_8$ in.; (*b*) *w* = $5^1/_2$ in.

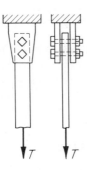

$T$   $T$    **Fig. 3-13   Probs. 3-16 and 3-17.**

**3-17.** The vertical steel bar in Fig. 3-13 is 15 mm thick and must be designed to withstand a tensile load *T* of 200 kN. Two high-strength A325 bolts will be used. Calculate (*a*) the required diameter *d* of the bolts and (*b*) the required width *w* of the bar. Assume hole 3 mm larger than bolt.

*Ans.* (*a*) *d* = 25 mm; (*b*) *w* = 125 mm

**3-18.** Assume, in Fig. 3-12, that the beams are W 14 × 61, the girder is a W 21 × 73, three rivets are used through each beam web, all angles are $^1/_2$ in. thick, and all rivets are 1 in. in diameter. How many rivets will be required to connect the seat angle to the girder web if the load to be transmitted from each beam to the girder is 85 kips?

**3-19.** A typical structural beam-to-column connection is shown in Fig. 3-14. The beam is a W 16 × 67 and the column is a W 10 × 49. All angles are $^3/_8$ in.

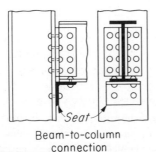

Beam-to-column
connection                **Fig. 3-14   Probs. 3-19 and 3-20.**

thick. If the connection must be capable of transmitting a load of 63 kips from the beam to the column, what diameter of rivets should be used?

*Ans.* $^3/_4$-in. rivets

**3-20.** A structural connection similar to the one shown in Fig. 3-14 must be capable of transmitting 90 kips from the beam to the column. The beam is a W 18 × 46, and the column is a W 12 × 53. All angles are $^3/_8$ in. thick, and all rivets are $^7/_8$ in. in diameter. If two rivets connect the seat angle to the column flange, how many rivets are required through the beam web?

**3-21.** Figure 3-15 shows a typical splice often used in the lower chord of double-angle trusses. A splice plate is used to connect the outstanding legs, thus enabling stress in them to be transmitted from one side of the splice to the other. Thicknesses: angles, $^5/_{16}$ in.; gusset plate, $^7/_{16}$ in.; splice plate, $^3/_8$ in. If $T =$ 85 kips and the splice plate is fastened as shown with $^3/_4$-in. rivets, how many $^3/_4$-in. rivets should be used through the connected legs on each side of the center?

*Ans.* Four rivets

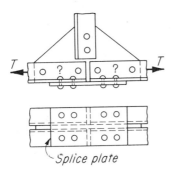

Splice plate          **Fig. 3-15  Probs. 3-21 and 3-22.**

**3-22.** The type of connection shown in Fig. 3-15 is used to join two sections of the lower chord of a double-angle truss. The angles are 10 mm thick, the gusset plate 15 mm, and the splice plates 10 mm. The bolts are M20 × 2.5. If four bolts are used through the connected legs on each side of where they join, how many bolts will be required on each side through the splice plate to resist a pull of 500 kN?

## 3-7 RIVETED BOILER JOINTS

Generally speaking, a boiler joint is one in which two plates are fastened together in order to make a joint that will remain airtight under pressure, such as normally found in steam boilers and in compressed-air tanks. The simplest way to accomplish this is to lap the two plates and to fasten them together with rivets fitted into matched holes. Such a joint is called a **lap joint,** three examples of which are shown in Fig. 3-16*a, b,* and *c.* When greater strength or efficiency is required, the main plates to be fastened together are butted, and cover plates, or splice plates, riveted to both

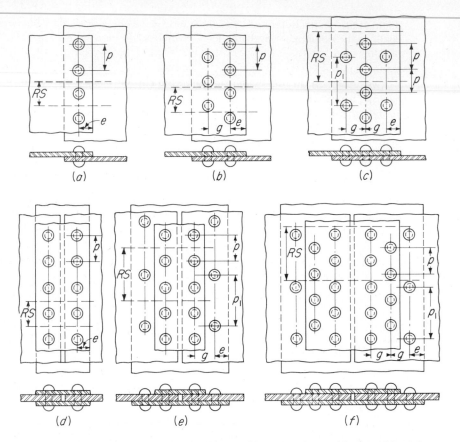

**Fig. 3-16   Riveted boiler joints.** (*a*) Single-riveted lap joint. (*b*) Double-riveted lap joint. (*c*) Triple-riveted lap joint. (*d*) Single-riveted butt joint. (*e*) Double-riveted butt joint. (*f*) Triple-riveted butt joint.

main plates then fasten them together. Such a joint is called a **butt joint,** of which Fig. 3-16*d, e,* and *f* are examples.

If absolute tightness of a joint is required, as perhaps in the case of joints in water tanks and steam boilers, the edge of the outer plate is beveled and the joint is then calked. That is, with the aid of a blunt-edged chisel-shaped calking tool, the lower edge of the outer plate is hammered into the seam, thus sealing the joint tightly. Occasionally, inner seams are calked also.

The single-riveted lap joint is commonly used in low-pressure tanks. It is most suitable with plates less than $1/2$ in. (12 mm) in thickness. In the double-riveted lap joint, greater tightness is secured by staggering the rivets in the two rows. In the triple-riveted lap joint, both the strength and the "efficiency" of the joint are increased by making the pitch $p$ of the two outer rows of rivets twice as great as the pitch of the middle row. This can be seen when we consider the strength in tension of these joints. Given the same plate thicknesses, rivet diameters, and pitch $p$ for three such lap joints and assuming that all are weakest in tension, failure in tension of the main plates would occur along a line between the rivet holes in the outer row. The joint with the smallest pitch in the outer row has the least net resisting area and therefore is the weakest.

A disadvantage of increased rivet spacing in the outer rows is that the effectiveness of calking decreases as the rivet pitch increases, since the most effective calking naturally is secured when rivets are closely spaced.

For boilers and tanks subjected to higher loads and pressures, and when plates $1/2$ in. (12 mm) or more in thickness are used, the butt joint illustrated in Fig. 3-16$d$, $e$, and $f$ is most suitable. The main plates are butted together and are held in place by thinner cover plates, with most of the rivets passing through all three plates. When only one cover plate is used, its thickness must be equal to or greater than the thickness of the main plates. When two cover plates are used, the thickness of each should be not less than one-half the thickness of the main plates in the case of boiler joints, and not less than five-eighths in the case of similar structural joints. When these specifications are adhered to, the factor of safety of the cover plates will be as high as or higher than that of the main plates and the cover plates need not be separately designed or investigated for stresses.

**Definitions.** We must define here certain terms that are commonly used in connection with riveted joints:

**Pitch** is the distance $p$ between centers of adjacent rivets in a row. In joints using two or more rows of rivets, each row may use a different pitch. The longer pitch $p_1$, as in Fig. 3-16$c$ and $f$, is generally twice as great as the shorter pitch $p$. The minimum recommended pitch is generally three times the rivet diameter.

**Gage** is the distance $g$ between centerlines of parallel rows or rivets.

**Repeating Section.** In continuous riveted joints the rivets are generally so spaced as to form a pattern which repeats itself along the rows of rivets. The length of joint required by one complete pattern is called a **repeating section.** In Fig. 3-16$a$, $b$, and $d$, the length of a repeating section equals the pitch $p$. In Fig. 3-16$c$, $e$, and $f$ the length of a repeating section equals the longer pitch $p_1$, which in both these instances equals $2p$.

**Cover plate** is the longitudinal plate, as in a butt joint, that is the element connecting the two main plates. A single cover plate may be sufficient at a butt joint, but two are generally used.

**Edge distance** is the distance from the center of a rivet to the edge of the plate. This is distance $e$ in Fig. 3-16, generally a minimum of $1^1/2$ times the rivet diameter.

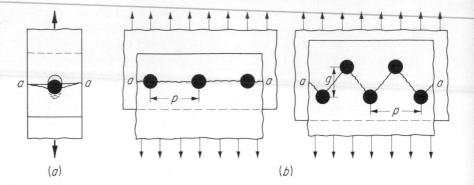

**Fig. 3-17** Tension failures of riveted boiler joints. (a) Tearing. (b) Tearing of plate between rivet holes.

**Tension Failures of Boiler Joints.** Failure in tension occurs when a plate tears apart along some line of least resistance as, for example, along line *a-a* in Fig. 3-17.

In a continuous joint with two or more rows of staggered rivets, a possibility exists of failure along the diagonal lines between the closest adjacent rivets. This type of failure is easily prevented by making the gage $g$ not less than $1^3/_4$ times the rivet diameter.

## 3-8 STRENGTH AND EFFICIENCY OF RIVETED BOILER JOINTS

The strength of a riveted joint is the maximum load it can resist without failure. Since a joint may fail in several ways, its strength in each must be calculated. The strength having the least value, then, is the strength of the joint.

The efficiency of a riveted boiler joint is the ratio of the strength of the joint to the strength in tension of the unpunched plate. That is,

$$\text{Efficiency} = \frac{\textbf{strength of joint}}{\textbf{strength of unpunched plate}} \tag{3-1}$$

To determine the strength of a continuous joint, a section equal in length to a repeating section is generally used.

## 3-9 STRESSES IN THIN-WALLED PRESSURE VESSELS

The most common form of pressure vessel in which riveted (or welded) joints are used is the cylinder, as used for steam boilers, air-compressor tanks, and water tanks. See Fig. 3-18. Spheres are also used to some extent for containing gas under pressure. Liquids and gases causing internal pressure in a closed vessel are both referred to as **fluids.** When the fluid is a gas,

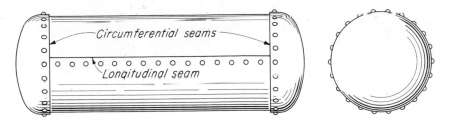

**Fig. 3-18**   Riveted cylindrical pressure vessel.

the pressure is constant in all parts of the vessel. When it is liquid, the pressure is lowest at the top and increases roughly $1/2$ psi per foot of depth of liquid. Because of its comparative insignificance, this increase is generally disregarded except when the depth exceeds, say, 10 ft. When pressure is caused by water, as in open water tanks, it is equal at any point to the weight of a column of water extending from the point vertically to the level of the free water surface. This vertical distance is called the **head** $h$. Assuming that the water weighs 62.5 pcf (pounds per cubic foot) the pressure $p$ at any point would be $62.5h$ psf, or $62.5h/144 = 0.434h$ psi. That is, a head of 10 ft of water would cause a pressure of $(0.434)(10) = 4.34$ psi. In SI units, $p = 9790h$ for water, where $p$ is in pascals $(N/m^2)$ and $h$ is in meters.

In order that joints or seams in these vessels, longitudinal or circumferential, may be designed, the force to be resisted per unit length of seam must first be determined. We consider here only cylinders and spheres in which the walls are thin relative to their diameters. For example, a wall thickness not exceeding 10 percent of the diameter of the vessel may safely be considered to be thin. In such vessels, the intensity of stress between inner and outer surfaces is approximately constant. In thick-walled vessels, the stress variation becomes more complex, being highest at the outer surface.

**Stress on Longitudinal Seam of Cylinder.**   Consider a thin-walled cylinder, shown in cross section in Fig. 3-19, to be subjected to internal fluid

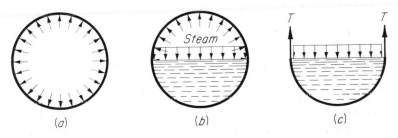

**Fig. 3-19**   Stress on longitudinal seam of cylindrical pressure tank. (*a*) Internal pressure due to gas or liquid. (*b*) Steam and water under pressure. (*c*) Stress in wall of tank shell.

pressure. A law of fluid mechanics is that *fluid pressure at any point is equal in all directions and is always directed perpendicular to the resisting surface,* as indicated in Fig. 3-19a.

In Fig. 3-19b, a steam boiler is half filled with water with steam under pressure in the upper half. The steam pressure is exerted against the semicylindrical surface of the upper half of the boiler and also against the level water surface in the lower half, perpendicular to those resisting surfaces. The water acts as a solid surface resisting the downward steam pressure.

If we now cut this vessel at the water line and isolate the lower half as a free body, it becomes apparent that the total downward pressure $P$ must be balanced by the two forces $T$.

A section of length $L$ of the lower half has been isolated in Fig. 3-20. Since the intensity of pressure $p$ is constant over the area $dL$, $P = pdL$. Similarly, $T = s_t tL$. Then

$$pdL = 2s_t tL$$

and

$$\textbf{Unit wall tensile stress } s_t = \frac{pd}{2t} \qquad \textbf{(3-2)}$$

where $p$ = internal pressure, psi or MPa
  $d$ = inside diameter of cylinder, in. or mm
  $t$ = thickness of wall material, in. or mm
  $s_t$ = tensile stress, psi or MPa

In Fig. 3-20, water and steam were used for explanatory purposes. Identical results will be obtained, however, with any medium properly calssified as a fluid.

We should note here that the stress on a **longitudinal** seam is a **circumferential** stress, since it is directed around the circumference. Similarly, the stress on a **circumferential** seam is a **longitudinal** stress, since it is directed parallel to the longitudinal axis of the cylinder. See Fig. 3-20.

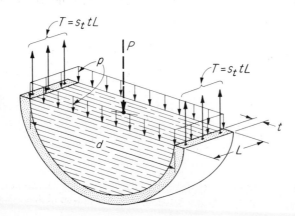

Fig. 3-20 Stress on longitudinal seam of cylinder.

When designing riveted or welded seams in pressure vessels, we must know the force $T_L$ to be resisted per linear inch of seam. If the force on a longitudinal section $L$ inches long is $s_t t L$, as previously found, the force $T_L$ per inch of length is $s_t t L / L$, or $T_L = s_t t$. But $s_t = pd/2t$. Hence

$$\text{Force per inch of longitudinal seam } T_L = \frac{pd}{2} \qquad (3\text{-}3)$$

To obtain the minimum required wall thickness of a cylinder to resist a given pressure $p$, Eq. (3-2) may be rewritten thus:

$$\text{Required wall thickness of cylinder } t = \frac{pd}{2s_t} \qquad (3\text{-}4)$$

Equation (3-4) may be used to determine the required wall thickness of tanks where welded seams with an efficiency of 100 percent can be obtained. Since the efficiencies of riveted boiler joints usually vary between 65 and 85 percent, somewhat thicker plate must be used in riveted cylinders.

**Stress on Circumferential Seam of Cylinder or Sphere.** We will show now that the force $T_C$ per linear inch of a circumferential seam of a cylinder is only half that in a longitudinal seam.

Consider the end portion of a cylinder as shown in Fig. 3-21a. Let this portion be filled with liquid under pressure $p$ per square inch. Since the cross-sectional area of the cylinder is $\pi d^2 / 4$, the total force against the end is $p\pi d^2 / 4$. This force is equaled by the total resistance offered by the material of the cylinder wall around the entire circumferential seam. If $T_C$ is the resisting force per linear inch of the circumference, which is $\pi d$, then the total resisting force is $T_C \pi d$. But $T_C \pi d = p\pi d^2 / 4$, or

$$\text{Force per inch of circumferential seam } T_C = \frac{pd}{4} \qquad (3\text{-}5)$$

which is seen to be just half the force exerted on the longitudinal seam.

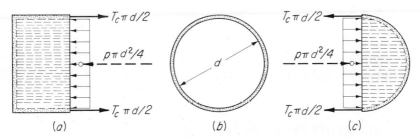

**Fig. 3-21** Force on circumferential seams of cylinders and spheres. (*a*) Pressure on end section of cylinder. (*b*) Cross section of cylinder or sphere. (*c*) Pressure on half of sphere.

Table 3-2 ULTIMATE AND ALLOWABLE STRESSES. ASME BOILER CODE. RECOMMENDED FOR BOILERS AND TANKS.

| Kind of Stress | Symbol | Ultimate Strength, psi (MPa) | Factor of Safety | Allowable Stress, psi (MPa) |
|---|---|---|---|---|
| Tension | $s_t$ | 55,000 (380) | 5 | 11,000 (76) |
| Shear | $s_s$ | 44,000 (300) | 5 | 8,800 (60) |
| Bearing | $s_c$ | 95,000 (650) | 5 | 19,000 (130) |

The stress on the circumferential seam is then

$$s_t = \frac{T_C}{t} = \frac{pd}{4t}$$   (3-6)

which is seen to be one-half that on the longitudinal seam as given by Eq. (3-2).

## 3-10 ALLOWABLE STRESSES FOR BOILER JOINTS

The Boiler Code of the American Society of Mechanical Engineers (see Table 3-2) gives the *ultimate* strengths of boiler steel to be used for boilers and tanks and also recommends a factor of safety of 5.

In the case of boiler joints, no distinction is made between inside and outside bearing. Areas in shear and bearing are based on the rivet-hole diameters rather than on the smaller rivet diameters, as in structural joints.

## 3-11 ANALYSIS AND DESIGN OF BOILER JOINTS

The design of riveted boiler joints so as to obtain the greatest efficiency consistent with the practical aspects of the design is a somewhat laborious problem and is beyond the scope of this text. Various handbooks list a great many standard designs of both lap and butt joints whose efficiencies have been calculated and range from about 50 percent for some single-riveted lap joints to as high as 94 percent for a butt joint of great strength. In most cases, these standard designs are used. The following examples, however, indicate how strength and efficiency of a given boiler joint are calculated.

### ILLUSTRATIVE PROBLEMS

**3-23.** A cylindrical air-compressor tank with an inside diameter of 600 mm is to be designed to withstand safely a working pressure of 1 MPa. Boiler plate with an allowable tensile stress of 75 MPa will be used (see Table 3-2). All seams will be butt welds with an expected efficiency of 100 percent. Determine (*a*) the force $T_L$ per linear mm of longitudinal seam; (*b*) the required thickness $t$ of wall; (*c*) the unit tensile stress in the circumferential seam (see Fig. 3-22).

Fig. 3-22   Design of welded air-compressor tank.

*Solution:* The force per linear mm is obtained from Eq. (3-3). That is,

$$\left[ T_L = \frac{pd}{2} \right] \qquad\qquad T_L = \frac{(1)(600)}{2} = 300 \text{ N/mm} \qquad\qquad\qquad Ans.$$

Equation (3-4) gives the required thickness of wall. That is,

$$\left[ t = \frac{pd}{2s_t} \right] \qquad\qquad t = \frac{(1)(600)}{(2)(75)} = 4 \text{ mm} \qquad\qquad\qquad Ans.$$

To obtain the unit wall stress caused by internal pressure of 1 MPa, we first find from Eq. (3-5) the force $T_C$ per mm of circumference. That is,

$$\left[ T_C = \frac{pd}{4} \right] \qquad\qquad T_C = \frac{(1)(600)}{4} = 150 \text{ N/mm}$$

and

$$\left[ s_t = \frac{T_C}{t} \right] \qquad\qquad s_t = \frac{150}{4} = 37.5 \text{ MPa} \qquad\qquad\qquad Ans.$$

This last result was to be expected, since the wall thickness was designed to resist 75 MPa and the longitudinal stress is half the circumferential stress.

**3-24.** The longitudinal seam of a boiler having an inside diameter of 5 ft is a double-riveted butt joint similar to that shown in Fig. 3-23. The main boiler plate is $3/4$ in. thick, the cover plates (butt straps) are $1/4$ in. thick, and the rivets are 1 in. in diameter in $1^1/_{16}$-in. holes. The pitch $p$ is $3^1/_4$ in., and $p_1$ is $6^1/_2$ in. Calculate (*a*) the strength and efficiency of a repeating section and (*b*) the maximum allowable internal steam pressure, if a factor of safety of 5 is required.

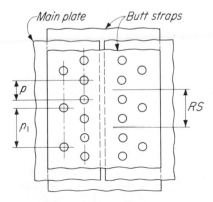

Fig. 3-23   Strength and efficiency of riveted boiler joint.

*Solution:* The strength of a repeating section $6\frac{1}{2}$ in. in length and containing three rivets, two in the inner row and one in the outer row, is its least ultimate strength in shear, bearing, or tension, or in whatever combination of these is least. In shear and bearing we find that

$$P_S = (3)(2)(0.8866)(44,000) = 234,000 \text{ lb}$$
$$P_B = (3)(0.75)(1.0625)(95,000) = 227,000 \text{ lb}$$

In tension, the main plate may fail by tearing the material between the rivets in the outer row. But it may also fail by tearing the lesser area of material between the rivets in the inner row. This latter can happen only if simultaneously therewith one rivet in the outer row fails either in shear or in bearing. The calculated values of $P_S$ and $P_B$ indicate that failure will occur in bearing. Then

$$\text{Outer row: } P_T = (6.5 - 1.0625)(0.75)(55,000) = 224,000 \text{ lb}$$
$$\text{Inner row: } P_T = (6.5 - 2.125)(0.75)(55,000) + \left(\frac{1}{3}\right)(227,000)$$
$$= 180,500 + 75,700 = 256,200 \text{ lb}$$

Clearly, now, tension failure would occur between rivets in the outer row. Consequently, this being least,

$$\text{Strength of repeating section: } P = 224,000 \text{ lb} \qquad \textit{Ans.}$$

The efficiency of the joint is the ratio of its strength to the strength of the solid unpunched plate. That is,

$$\text{Efficiency} = \frac{224,000}{(6.5)(0.75)(55,000)} = 0.835 \qquad \text{or} \qquad 83.5 \text{ percent } \textit{Ans.}$$

To determine the maximum allowable internal steam pressure $p$, we must first find the allowable force $T_L$ per linear inch of longitudinal seam. With a factor of safety of 5

$$\text{Allowable } T_L = \frac{224,000}{(6.5)(5)} = 6880 \text{ lb}$$

Then from Eq. (3-3)

$$\left[ p = \frac{2T_L}{d} \right] \qquad \qquad p = \frac{(2)(6880)}{60} = 230 \text{ psi} \qquad\qquad \textit{Ans.}$$

PROBLEMS _____

**3-25.** Assume that the tank in Prob. 3-23 must have an inside diameter of 29.5 in. and that it must withstand a pressure of 165 psi. Then solve. Use stresses from Table 3-2.

**3-26.** In Prob. 3-24 assume the following: boiler diameter, 4 ft; main boiler plate, 1 in. thick; cover plates, $5/8$ in. thick; rivets, $1\frac{1}{8}$-in. diameter in $1\frac{3}{16}$-in. holes; $p = 4$ in. and $p_1 = 8$ in. Then solve.

**3-27.** A seamless-tube steam pipe with an inside diameter $d$ of 8.071 in. and a wall thickness $t$ of 0.277 in. is subjected to internal steam pressure $p$ of 200 psi. (*a*) Calculate the unit tensile stress $s_t$ in the wall material. (*b*) What maximum internal pressure $p$ can this pipe safely stand if the allowable tensile stress is 11,000 psi?

*Ans.* (*a*) $s_t = 2910$ psi; (*b*) $p = 755$ psi

**3-28.** A 400-mm seamless-tube water pipe must be capable of withstanding a pressure head of 300 m. What minimum wall thickness is required if allowable tensile stress is 75 MPa?

**3-29.** A lubricating hoist is supported by a hollow cylindrical shaft whose inside diameter is 9 in. The pressure of compressed air against its closed upper end raises and lowers the shaft. What minimum air pressure $p$ must be supplied to raise a load of 8000 lb?

*Ans.* $p = 126$ psi

**3-30.** A 20-in.-diameter air-compressor tank must safely withstand a working pressure of 145 psi. Boiler plate $^1/_4$ in. thick will be used. The longitudinal seam will be a single-riveted lap joint. What diameter of rivet and what pitch should be used? Minimum rivet spacing is 3 diameters.

**3-31.** A spherical tank 20 m in diameter is used to contain gas under pressure. The plate is 15 mm thick and is butt-welded to give a joint efficiency of 100 percent. (*a*) What maximum interior pressure can the tank safely stand if allowable tensile plate stress is 75 MPa? (*b*) If the tank must stand a pressure of 0.5 MPa, what should the plate thickness be?

**3-32.** The longitudinal seam in a boiler whose inside diameter is 66 in. is a double-riveted butt joint with an efficiency of 80 percent. The boiler plate is $^3/_4$ in. thick. Calculate the maximum safe internal pressure $p$ if allowable tensile plate stress is 11,000 psi.

*Ans.* $p = 200$ psi

**3-33.** A seamless-tube water pipe with an inside diameter of 10.192 in. runs from a reservoir to a small nearby town. (*a*) Calculate the pressure $p$ and the force $T_L$ per linear inch of (imaginary) longitudinal seam in the pipe at a point where it is 300 ft vertically below the open (free) water surface of the reservoir. (*b*) If the wall thickness is 0.279 in. and allowable tensile stress is 11,000 psi, what maximum head $h$ may this pipe safely withstand?

**3-34.** A large water pipe has an inside diameter of 4 ft. It is made of $^3/_8$-in. steel plate using double-riveted lap joints and $^7/_8$-in.-diameter rivets in $^{15}/_{16}$-in. holes (see Fig. 3-16*b*), with a pitch of 4 in. in both rows. (*a*) Calculate the strength and efficiency of the joint. (*b*) Calculate the maximum head $h$ which this pipe may safely withstand if a factor of safety of 5 is used.

*Ans.* (*a*) 60,700 lb, 73.6 percent; (*b*) 291 ft

**3-35.** Assume in Prob. 3-34 that the pipe diameter is 2 m, the steel plate is 15 mm thick, 30-mm-diameter rivets are used in 32-mm holes, and the pitch is 125 mm. Then solve.

## 3-12 ECCENTRICALLY LOADED RIVETED JOINTS

As in the case of riveted boiler joints, most riveted structural connections have been standardized so as to provide the maximum possible strength and efficiency along with the simplicity of standardized fabrication.

Maximum strength and efficiency are obtained when the line of action of the load to be resisted by the connection passes through the centroid $C$ of the cross-sectional resisting shear areas of the rivets in the group, as illustrated in Figs. 3-24a and b and 3-26a and c. All such connections are said to be **concentrically loaded,** and, although perhaps not strictly true, because of imperfections of material and in fabrication, the very reasonable assumption is made that each rivet in the group carries its proportional part of the total load, as illustrated in Fig. 3-26a and c.

When the line of action of the load on the connection fails to pass through the centroid $C$, as it does in Figs. 3-25 and 3-26b and d, the connection is said to be **eccentrically loaded.** This condition results in additional forces $Q$ on the rivets due to the twisting action on the connection caused by the eccentric *moment Pe* of the load with respect to the centroid, as discussed in Prob. 3-36 and illustrated in Fig. 3-28. Consequently, designers carefully avoid eccentricity in connections when possible to do so. There are, however, many special-purpose types of connections in which eccentricity cannot be avoided. A typical case is illustrated in Prob. 3-36.

The **location of the centroid of the rivet areas** in riveted connections therefore becomes a matter of importance. Since in most structural connections all rivets of a group have the same diameter and are generally arranged in a symmetrical pattern, the location of the centroid $C$ of the group can usually be seen at a glance, as in Figs. 3-24 to 3-27. For rivet groups not symmetrically spaced, the centroid may be located by the method outlined in Chap. 5.

Consider the eccentrically loaded connection shown in Fig. 3-26b. Two equal but oppositely directed forces may be added at the centroid $C$ without altering the equilibrium condition. Let each of these forces have the magnitude of the applied load $P$ (see Fig. 3-27a). The three forces $P$

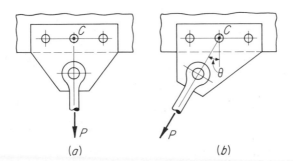

$(a)$        $(b)$      **Fig. 3-24   Concentrically loaded connections.**

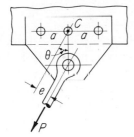

Fig. 3-25   Eccentrically loaded connections.

may be grouped as shown, that is, as a couple and a single concentric force. The rivet forces due to the concentric force $P$ are obviously $P/2$, as shown in Fig. 3-27$b$. The couple, or moment, $Pa$ must be balanced by another couple. This couple is formed by the rivet forces shown in Fig. 3-27$c$. Equating the couples gives $(F)(4a) = Pa$, or $F = P/4$. The values shown in Fig. 3-26$d$ could well be checked by the technique suggested here.

Figure 3-28$a$ shows a typical case of a plate supporting a crane rail and riveted to the flange of a column. Clearly, the eccentric moment $Pe$ will twist, or rotate, the plate about some center. We will now show that *the center of rotation is the centroid of the group of rivet areas.*

The plate is shown isolated in Fig. 3-28$b$. The centroid of the rivet areas is $C$, and $O$ is the center of rotation whose position is unknown. Let $c$ be the distance from $O$ to the rivet farthest away and $r$ the distance to any other rivet. If $s$ is the unit shearing stress in the rivet distant $c$ from $O$, then the unit shearing stress in any rivet distant $r$ from $O$ is $(r/c)s$, since *the intensity of the shearing stress in any rivet is proportional to its distance $r$ from $O$.* Proof of this statement may be found in Chap. 4, Torsion, Art. 4-2 and Fig. 4-3 (see also Observation and Conclusion 2 in Art. 7-1). The resisting force $Q$ of *any* rivet due to the moment $Pe$ then is $(r/c)sA$, where $A$ is the cross-sectional area of the rivet. The line of action of $Q$ is always perpendicular to the radial distance $r$. The horizontal component

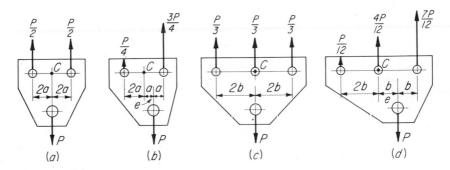

Fig. 3-26   Rivet loads in concentrically and eccentrically loaded connections.

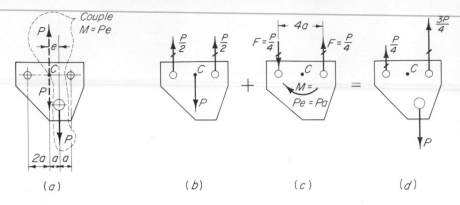

**Fig. 3-27** Rivet loads in eccentric connection by superposition. (*a*) Equal and opposite loads *P* added at *C*. (*b*) Rivet forces due to direct load. (*c*) Rivet forces due to couple *Pe*. (*d*) Rivet forces by superposition.

of $Q$ is $(r/c)sA \sin \theta$, or $Q_X = (s/c)Ay$, since $r \sin \theta = y$. But, since the force system acting on the plate is necessarily in equilibrium, the sum of all horizontal components is zero. That is, $\Sigma Q_X = 0$, or $(s/c)\Sigma Ay = 0$. Since $s/c$ is not zero, it follows that $\Sigma Ay$ must be zero, which it can be only when the $X$ axis passes through the centroid $C$. We may similarly show that the $Y$ axis must also pass through $C$, thus proving that *the center of rotation is the centroid C.*

The remaining problem now is to evaluate these moment-resisting forces $Q$. It has already been pointed out that (1) the rivet forces developed in resisting the applied twisting moment will each be perpendicular to a line joining the rivet and the center of rotation $C$, and (2) the rivet forces will be proportional to their distances from $C$. We can state this second fact mathematically by saying $Q = kr$, where $Q$ is the rivet force, $r$ is the distance from the centroid $C$, and $k$ is a proportionality constant.

If we take moments about the centroid $C$, we have

$$Pe = Q_1 r_1 + Q_2 r_2 + Q_3 r_3 + \cdots$$

$Q_1$, $Q_2$, and $Q_3$ are the forces acting on rivets 1, 2, and 3, respectively. The distance from each rivet to the centroid of the rivet group is denoted by $r_1$, $r_2$, $r_3$, and so forth. Since each rivet force is equal to $kr$,

$$Pe = kr_1^2 + kr_2^2 + kr_3^2 + \cdots = k(r_1^2 + r_2^2 + r_3^2 + \cdots)$$

This can be abbreviated $Pe = k\Sigma r^2$. The $\Sigma$ sign indicates "the sum of." Now, since $x$ and $y$ are the space components of $r$, and since $r^2 = x^2 + y^2$,

$$Pe = k\Sigma(x^2 + y^2) = k(\Sigma x^2 + \Sigma y^2)$$

or

$$k = \frac{Pe}{\Sigma x^2 + \Sigma y^2} \tag{3-7}$$

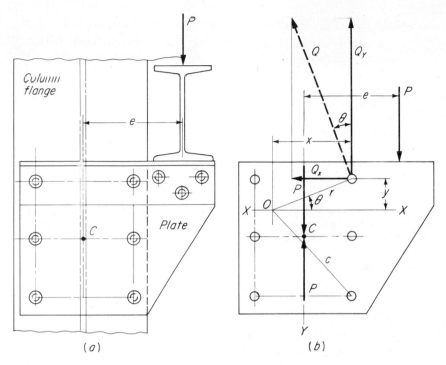

**Fig. 3-28**  Proof that center of rotation $O$ coincides with centroid $C$ of rivet group under rotational couple $Pe$. (a) Column bracket. (b) Bracket plate as free body.

When $k$ has thus been evaluated, the twisting force $Q$ on any rivet in the group is easily calculated. That is,

$$Q = kr \qquad \text{(3-8)}$$

in which $r$ is the distance to the rivet from the centroid of the group.

When the *twisting* force $Q$ on any given rivet has thus been evaluated, it must be combined vectorially with the direct force $P/N$ on that rivet. To do so algebraically, we find it convenient to use the $x$ and $y$ components of $Q$, which are easily obtained by simply multiplying $Q_1$ by the appropriate $y$ or $x$ distance from $O$. That is,

$$Q_X = Q \sin \sigma = kr\,\frac{y}{r} \qquad \text{or} \qquad Q_X = ky \qquad \text{(3-9)}$$

and

$$Q_Y = Q \cos \sigma = kr\,\frac{x}{r} \qquad \text{or} \qquad Q_Y = kx \qquad \text{(3-10)}$$

Note carefully here that the force components $Q$ are proportional to the *opposite* space components ($x$ or $y$), not to the corresponding space components.

ILLUSTRATIVE PROBLEM

**3-36.** The bracket plate shown in Fig. 3-29*a* supports a crane-rail load of 12 kips at a distance of 9 in. from a vertical axis passing through the centroid of the group of rivets which fasten the plate to a column flange. Calculate the total force $F$ on each of the six rivets. All rivets will have same diameter. (NOTE: In solving this type of problem, the plate, and not the rivets, is isolated as a free body. Consequently, instead of solving for the force exerted by the plate on each rivet, we solve for the equal but opposite force exerted by each rivet on the plate.)

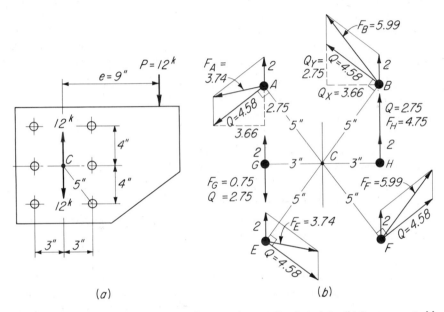

**Fig. 3-29  Eccentrically loaded riveted connection. (a) Bracket plate. (b) Forces exerted by rivets on plate.**

*Solution:* As a first step in solving this problem, we locate at a glance the centroid $C$ of the rivet group. Next, we add two equal and opposite 12-kip forces at $C$, acting parallel to $P$. This we may do without changing the final results. Now, we have a downward-acting 12-kip force at the centroid resisted equally by all six rivets ($P/N = 12/6 = 2$ kips), and a clockwise couple or twisting moment, which is equal to $Pe = (12)(9) = 108$ kip·in. and is resisted by the twisting forces $Q$. We then evaluate $\Sigma x^2$ and $\Sigma y^2$ and calculate $k$. That is, measuring from the centroid,

$$\Sigma x^2 = (-3)^2 + (-3)^2 + (-3)^2 + 3^2 + 3^2 + 3^2 = 54$$
$$\Sigma y^2 = 4^2 + 4^2 + 0 + 0 + (-4)^2 + (-4)^2 = 64$$

and

$$k = \frac{Pe}{\Sigma x^2 + \Sigma y^2} = \frac{(12)(9)}{54 + 64} = 0.915 \text{ kip/in.}$$

Each of the four corner rivets is 5 in. from the centroid $C$, and the two middle rivets are 3 in. from $C$. Hence

$$[Q = kr] \qquad\qquad Q_5 = (0.915)(5) = 4.58 \text{ kips}$$

and

$$Q_3 = (0.915)(3) = 2.75 \text{ kips}$$

Since $P$ is downward-acting, the six 2-kip resisting forces, referred to as *direct* forces, are upward-acting. Also, since the moment $Pe$ is clockwise with respect to the centroid $C$, the moments of the resisting twisting forces are counterclockwise. This must be clearly understood in order for these forces to be properly directed, as in Fig. 3-29$b$. If this diagram is laid out to scale, the total force $F$ on each rivet can readily be determined graphically. Analytically we find for the four corner rivets that the vertical and horizontal components of $Q$ are

$$[Q_X = ky] \qquad\qquad Q_X = (0.915)(4) = 3.66 \text{ kips}$$

and

$$[Q_Y = kx] \qquad\qquad Q_Y = (0.915)(3) = 2.75 \text{ kips}$$

When these components are now added vectorially to the direct forces, we obtain the force $F$ on each of the rivets. That is,

$$\begin{aligned}
F_A = F_E &= \sqrt{(3.66^2) + (2 - 2.75)^2} = 3.74 \text{ kips} \\
F_B = F_F &= \sqrt{(3.66^2) + (2 + 2.75)^2} = 5.99 \text{ kips} \\
F_G &= 2.75 - 2 \qquad\qquad\quad = 0.75 \text{ kip} \\
F_H &= 2.75 + 2 \qquad\qquad\quad = 4.75 \text{ kips} \qquad\qquad \textit{Ans.}
\end{aligned}$$

## PROBLEMS

**3-37.** Calculate the resultant forces exerted by each of the two rivets on the angle clip shown in Fig. 3-30 when $P = 6625$ lb. Check $F$ against each rivet by moments.

*Ans.* $F = 12,600$ lb; $F = 5960$ lb

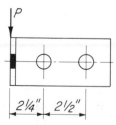

**Fig. 3-30    Prob. 3-37.**

**3-38.** In Fig. 3-31, calculate the maximum force $F$ exerted by any one of the four rivets on the angle clip when $P = 13,250$ lb. Check the $Q$ force against each rivet by moments.

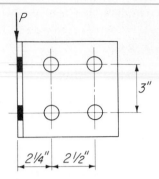

Fig. 3-31    Prob. 3-38.

**3-39.** The angle clip shown in Fig. 3-32 is acted upon by a force $P$ of 70 kN. Calculate the maximum force $F_{max}$ exerted on the angle by any one of the three rivets. Check the $Q$ force against each rivet by moments.

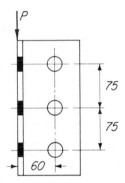

Fig. 3-32    Prob. 3-39.

**3-40.** Isolate the crane-rail bracket plate shown in Fig. 3-33 and calculate the force $F$ exerted on the plate by the most highly stressed rivet in the group: $a =$ 3 in., $b = 4$ in., $e = 10$ in., and $P = 22,000$ lb. What diameter rivets should be used, assuming bearing will not control?

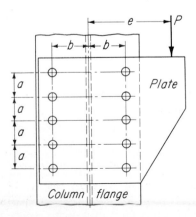

Fig. 3-33    Probs. 3-40 and 3-68.

**3-41.** In the riveted connection shown in Fig. 3-25, $P$ = 9000 lb, $e$ = 2 in., the rivet spacing $a$ = 3 in., and $\theta$ = 30°. Calculate the forces $F$ exerted on the gusset plate by each of the three rivets.

**3-42.** Each of the two plates in Fig. 3-34 is 10 mm thick. (*a*) If $a$ = 100 mm, $e$ = 125 mm, and $P$ = 125 kN, what diameter of rivets should be used if they are in single shear?

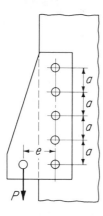

**Fig. 3-34   Probs. 3-42 and 3-43.**

**3-43.** If both plates in Fig. 3-34 are $^5/_{16}$ in. thick, $a$ = 4 in., $e$ = 5 in., and $^3/_4$-in.-diameter rivets are used (in single shear), what maximum load $P$ may be applied as shown?

*Ans. P* = 20,700 lb

**3-44.** In Fig. 3-35, let $a$ = 4 in., $e$ = 6 in., $\theta$ = 30°, and $P$ = 16,000 lb. (*a*) What diameter of rivets should be used? (*b*) Calculate the maximum shearing and bearing stresses on the most highly stressed rivet if each plate is $^5/_{16}$ in. thick.

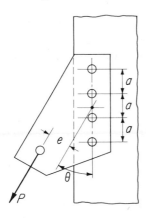

**Fig. 3-35   Prob. 3-44.**

## 3-13 WELDED JOINTS

Welding is a process of joining two pieces of metal by fusion. That is, the two parts to be welded are securely held in place and are heated by an

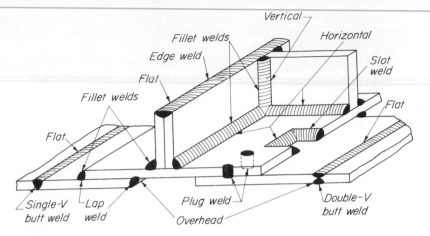

Fig. 3-36    Types of welds. Examples of locations.

electric arc or an oxyacetylene torch. Fused metal from a slender welding rod is then deposited between them and allowed to cool.

While the strength of the welding-rod metal is usually higher than that of the parts to be welded, the strength of the finished welded joint depends largely upon the skill of the welder. An improperly done weld may be dangerous, although shop welding done under carefully supervised conditions is very dependable. The quality of a weld is difficult to determine. Where small factory-made welded parts are produced in large quantities, it is usually feasible and practical to x-ray the welded joint. By this means, flaws in a weld can generally be detected. For most large parts and for welds made in the field, the x-ray method is seldom feasible.

Welding is especially suitable for repairing broken parts and for joining many of the steel parts of shop-fabricated roof trusses, derricks, and frames for tractors, railroad cars, and many other types of heavy equipment and machinery.

### 3-14 TYPES OF WELDS

Various types of welds are illustrated in Fig. 3-36 and also, in more detail, in Figs. 3-37 to 3-42.

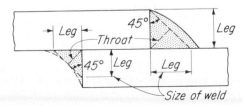

Fig. 3-37    Fillet welds.

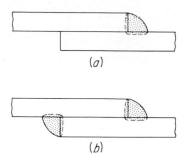

Fig. 3-38   Fillet-weld joints. (*a*) Single-fillet lap joint. (*b*) Double-fillet lap joint.

The two types of welds most frequently used are **fillet welds** and **butt welds.** Detailed illustrations and explanations of terminology are shown in Figs. 3-37 to 3-39.

The critical section in a fillet weld is the **throat.** The size of a fillet weld is the leg length of the largest inscribed right isosceles triangle, as is illustrated in Fig. 3-37, and the throat is 0.707 times that leg length. Fillet welds are usually made with equal legs and with a somewhat convex rather than a flat surface, since such a weld is most efficient and is also easiest to make. It is customary to add about $1/4$ in. to the length of each weld for starting and stopping.

Three types of butt welds are shown in Fig. 3-39. The square butt weld is generally used with plates not exceeding $1/4$ in. in thickness, the single V with plates from $1/4$ to $1/2$ in., and the double V with plates over $1/2$ in. thick. Butt welds are usually subjected mainly to tension and compression, not to shear. The **throat** of a butt weld joining two plates is the thickness of the thinnest of the plates.

Three other types of welds are illustrated in Figs. 3-40 to 3-42. In **plug welds,** punched holes in one of the two plates are filled with weld metal, which fuses partly into the other plate. **Slot welds** are sometimes used when other types of welds are not suitable and also at times to provide additional strength to a fillet-welded joint. **Spot welding** is used extensively in the fabrication of sheet-metal parts, merely as a simple and quick method of fastening light pieces together at intervals along a seam. Opposite metal rods clamp the pieces together tightly for an instant, during which a high electric current fuses the metal of both pieces at a small spot, as illustrated in Fig. 3-42.

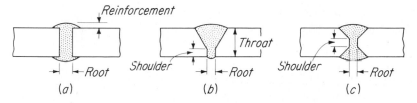

Fig. 3-39   Butt-welded joints. (*a*) Square butt weld. (*b*) Single-V butt weld. (*c*) Double-V butt weld.

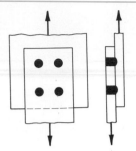

Fig. 3-40   Plug weld.

## 3-15 STRENGTH OF WELDED JOINTS

The Code of the American Welding Society (AWS) recommends the following allowable unit stresses in welds (E70XX electrodes and ASTM A36 steel)

Butt welds: 21,000 psi (145 MPa) in tension on section through throat

Fillet welds: 21,000 psi (145 MPa)

in shear on throat plane but not more than 14,500 psi (100 MPa) in shear on baseplate.

Since the throat length of a fillet weld is 0.707 times its leg length (size), the allowable load $F$ on a $^1/_2$-in. fillet weld 1 in. long would be $F = (^1/_2)(0.707)(1)(21,000) = 7,420$ lb. However, the allowable load for A36 steel is limited to $^1/_2(14,500) = 7,250$ lb. Table 3-3 lists the recommended sizes of fillet welds along edges, together with the allowable load per unit of length on each. However, when required by design conditions and when specifically designated on drawings, fillet welds equal in size to the thickness of a plate or rolled section may be used. Nevertheless, the effective sizes listed in Table 3-3 are generally used in good practice.

The net required length of a fillet weld should be specified on drawings. The welder will put down the length of weld specified and will then taper it off at the ends, usually for an additional length equal to the size of the weld.

In practice, it is advisable to maintain the same fillet-weld size throughout the connection, since a change of fillet size necessitates a

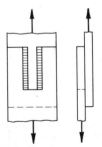

Fig. 3-41   Slot weld.

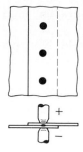

Fig. 3-42    Spot weld.

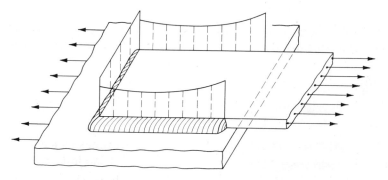

Fig. 3-43    Ordinates show variations in stress in fillet welds.

Table 3-3 RECOMMENDED FILLET WELD SIZES AND
ALLOWABLE LOAD $F$ PER UNIT LENGTH*

| Thickness of Thinnest Plate | | Minimum Weld Size | | Maximum Weld Size | | Allowable Shear Force $F$ per Unit Length — Maximum Size Weld | |
|---|---|---|---|---|---|---|---|
| in. | mm | in. | mm | in. | mm | lb/in. | N/mm |
| $^{11}/_{16}$ | 18 | $^1/_4$ | 6 | $^5/_8$ | 16 | 9060 | 1600 |
| $^5/_8$ | 16 | $^1/_4$ | 6 | $^9/_{16}$ | 14 | 8150 | 1400 |
| $^9/_{16}$ | 15 | $^1/_4$ | 6 | $^1/_2$ | 13 | 7250 | 1300 |
| $^1/_2$ | 14 | $^3/_{16}$ | 5 | $^7/_{16}$ | 12 | 6340 | 1200 |
| $^7/_{16}$ | 12 | $^3/_{16}$ | 5 | $^3/_8$ | 10 | 5440 | 1000 |
| $^3/_8$ | 10 | $^3/_{16}$ | 5 | $^5/_{16}$ | 8 | 4530 | 800 |
| $^5/_{16}$ | 8 | $^3/_{16}$ | 5 | $^1/_4$ | 6 | 3620 | 600 |
| $^1/_4$ | 6 | $^1/_8$ | 3 | $^1/_4$ | 6 | 3620 | 600 |
| $^3/_{16}$ | 5 | $^1/_8$ | 3 | $^5/_{16}$ | 5 | 2720 | 500 |
| $^1/_8$ | 3 | $^1/_8$ | 3 | $^1/_8$ | 3 | 1810 | 300 |

* E70XX electrodes and ASTM A36 steel. Values shown assume that the baseplate is at least as thick as the weld size.

change of welding rods and therefore slows the work and may cause errors.

Experiments have shown that the ends of a fillet weld, lying parallel to the line of action of the load, carry higher unit stresses than the midportion of the weld, as illustrated in Fig. 3-43. Also, when an end weld is combined with side welds, the unit stress in the end weld will be somewhat higher than those in the side welds. For this reason, designers prefer not to combine end and side welds unless lack of space makes it necessary.

## ILLUSTRATIVE PROBLEM

**3-45.** A 10- by 100-mm square-edged structural-steel strap is to be welded to a heavy steel plate in the manner illustrated in Fig. 3-44. Calculate the length $L$ of fillet weld required to withstand safely a tensile pull $P$ of 100 kN.

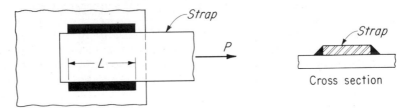

**Fig. 3-44    Required length of fillet weld.**

*Solution:* From Table 3-3, the recommended size of fillet weld along the square edge of the 10-mm-thick strap is 8 mm, on which the allowable load is 800 N/mm. Since the total length of fillet weld is $2L$,

$$2L = \frac{100\ 000}{800} = 125 \text{ mm} \quad \text{and} \quad L = 62.5 \text{ mm} \qquad Ans.$$

## 3-16 ECCENTRICITY IN WELDED JOINTS

One of the most common examples of eccentrically loaded welded joints is that of a structural angle welded to a gusset plate, as shown in Fig. 3-45. The total load $P$ in the angle is presumed to act along its gravity axis, the location of which may be obtained from a handbook. Consequently, a condition of eccentric loading exists, since $d_1$ is smaller than $d_2$.

In riveted joints, because of rather rigid adherence to standard patterns of placing rivets, the rivets are often overstressed because of eccentricity of loading, which is not always taken into account. In welded joints, however, we often find it possible to place the welds so that these excess stresses are eliminated. The following problem illustrates a common case.

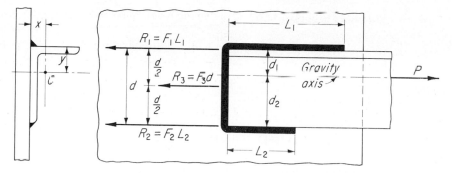

**Fig. 3-45**   Balancing effect of eccentricity in welded connection.

## ILLUSTRATIVE PROBLEM

**3-46.** The structural-steel angle shown in Fig. 3-45 is 5 by 3 by $\frac{1}{2}$ in. Its long leg is to be welded to a heavy plate as indicated. (*a*) If a full end weld is used, calculate the additional lengths $L_1$ and $L_2$ of side welds to resist a tensile load of 100 kips, proportioned to eliminate the effect of eccentricity. (*b*) If $L_1$ is limited to approximately 6 in. in length and a full-length end weld is used, what size fillet weld should be used along the heel of the angle? (NOTE: Unless otherwise specified, the same size of weld is to be used all around.)

*Solution:* From Table E-6 we find that $d = 5$ in., $d_1 = 1.75$ in., and $d_2 = 3.25$ in. $R_1$, $R_2$, and $R_3$ are the resistances which must be offered, respectively, by fillet welds 1, 2, and 3. From Table 3-3 we find that a $\frac{7}{16}$-in. fillet weld is recommended along the rounded edge of a $\frac{1}{2}$-in.-thick angle, with an allowable load of 6340 lb per linear inch.

(*a*) Since the length of the end weld is 5 in., $R_3$ may be evaluated. That is,

$$R_3 = (5)(6340) = 31,700 \text{ lb}$$

Since the forces $P$, $R_1$, $R_2$, and $R_3$ are necessarily in equilibrium, $R_1$ and $R_2$ may be determined by moments; thus:

$[\Sigma M_2 = 0]$    $5R_1 + (31,700)(2.5) = (100,000)(3.25)$    and    $R_1 = 49,150$ lb
$[\Sigma M_1 = 0]$    $5R_2 + (31,700)(2.5) = (100,000)(1.75)$    and    $R_2 = 19,150$ lb
$31,700 + 49,150 + 19,150 = 100,000$    *Check.*

The required lengths $L_1$ and $L_2$ then are

$\left[ L = \dfrac{R}{F} \right]$    $L_1 = \dfrac{49,150}{6340} = 7.75$ in.    *Ans.*

$L_2 = \dfrac{19,150}{6340} = 3.02$ in.    *Ans.*

$\dfrac{100,000}{6340} = 15.77$ in. $= 5 + 7.75 + 3.02 = 15.77$ in.    *Check.*

(b) If $L_1$ is limited to 6 in., the strength per linear inch must be

$$\frac{49,150}{6} = 8192 \text{ lb per linear inch}$$

Use $\frac{9}{16}$-in. fillet weld.                                       *Ans.*

## 3-17 DESIGN OF WELDED JOINTS

The four preceding articles together with the problems in them cover only the most elementary phases of welding. For a more complete treatment of the subject, the various publications on welding[1] should be referred to.

**PROBLEMS**

**3-47.** A steel strap $5/16$ in. thick by 4 in. wide is welded to a steel plate by means of $1/4$-in. fillet welds, as indicated in Fig. 3-46. (a) Determine required length $L$ of fillet, if $P = 21,700$ lb. (b) Determine $L$ required to develop the full tensile strength of the strap at 22,000 psi.

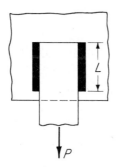

$P$                    **Fig. 3-46   Prob. 3-47.**

**3-48.** Solve Prob. 3-46 when the angle is 4 by $3\frac{1}{2}$ by $1/2$ in. and $1/4$-in. fillet welds are used except in (b), where $L_1$ is limited to 3 in. and the full-length end weld is increased to $3/8$ in. $P = 40$ kips.

**3-49.** In Fig. 3-47, $A$ is a double-angle structural-steel member welded on both sides to the gusset plate. The angles are $102 \times 76 \times 9.5$ mm with their long legs back to back. The connection must resist a total stress $P$ of 200 kN. Using only side fillet welds, determine $L_1$ and $L_2$.

**3-50.** The structural member $B$ in Fig. 3-47 is made of two steel angles 4 by 3 by $1/4$ in. with their long legs welded back to back to the gusset plate. The total stress $T$ in the member is 60,000 lb. The size of the gusset limits the length $L_1$ of the heel fillet to 4.5 in. A full 4-in.-long end weld should be used. Determine $L_1$ and $L_2$ and the fillet sizes, properly proportioned to counteract eccentricity.

[1] S. H. Marcus, *Basics of Structural Steel Design;* 2d Ed., 1980, Reston: American Welding Society, *Welding Handbook.*

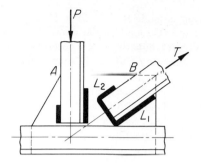

Fig. 3-47    Probs. 3-49 and 3-50.

**3-51.** Figure 3-48 shows a cross-sectional view of the upper chord of a roof truss, in which the chords are T sections (WT 6 × 18) and all web members are of standard pipe sections. The slotted pipe ends fit over the web of the T section and are welded to the web all around. If the pipe shown fits at right angles with the T section, $h = 5.58$ in., and nominal $D = 4.0$ in., what tensile pull $T$ can be developed by a $^1/_4$-in. fillet weld?

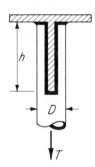

Fig. 3-48    Prob. 3-51.

**SUMMARY** (By Article Number) ————————————————————————

**3-1.**    Whenever two or more pieces are fastened together by tightly driven rivets, the connection is called a **riveted joint.**

**3-3. Failures of riveted joints** are generally in *shear* or *bearing* in the rivets or *intension* in the connected plates or parts. Failure in *end shearing* or *tearing* is prevented when proper end distances are maintained.

**3-4.**    In **structural joints,** rivet holes are generally punched $^1/_{16}$ in. greater in diameter than the rivets. Rivet *shearing and bearing stresses* are based on the original diameter of the undriven rivet. *Tensile stresses* in the connected parts are based on an assumed rivet-hole diameter $^1/_8$ in. greater than the original rivet diameter.

The unit stresses in riveted joints, shearing, bearing, and tensile, are not of uniform intensity, but may vary considerably. Hence, the equation $s = P/A$ gives only an *average stress.* Commonly used allowable stresses have been adjusted to compensate for this nonuniformity.

When, in a riveted joint, two *outside* members enclose a third *inside* member, the rivets are in *double shear.* Bearing stress between the rivets and the outside members is called *outside bearing,* or plain bearing, while bearing stress between the rivets and the inside member is called *inside bearing,* or enclosed bearing.

**3-5.**   An **analysis of a structural joint** is made to determine (1) the stresses in shear, bearing, or tension or (2) the allowable load on the joint.

**3-6.**   The **design of a riveted structural joint** involves determination of the number of rivets required and their diameters and the required thicknesses of plates to resist a given load.

**3-7.**   In **riveted boiler joints,** rivet holes are usually drilled, or are punched and reamed to a diameter $1/16$ in. larger than the rivet. All stresses, shearing, bearing, and tensile, are based on the drilled or reamed size of the hole.

**3-8.**   The **strength of a riveted joint** is the maximum load it can resist without failure. The **efficiency of a riveted joint** is the ratio of its strength to the strength in tension of the unpunched plate.

**3-9.**   **Thin-walled pressure vessels** are generally cylindrical with riveted or welded seams. When internal pressure is caused by a fluid, gas, or liquid, the unit stress in the uniformly thick wall material is

$$\textbf{Unit wall tensile stress } s_t = \frac{pd}{2t} \qquad (3\text{-}2)$$

and

$$\textbf{Force per inch of longitudinal seam } T_L = \frac{pd}{2} \qquad (3\text{-}3)$$

and

$$\textbf{Force per inch of circumferential seam } T_C = \frac{pd}{4} \qquad (3\text{-}5)$$

**3-12.**   A **concentrically loaded joint** is one in which the line of action of the load passes through the centroid of the rivet group. When so loaded, all rivets are presumed to be equally loaded and the joint is efficient.

An **eccentrically loaded joint** results when the line of action bypasses the centroid. The product of the load $P$ and the eccentric distance $e$ is a twisting moment $Pe$ whose center of rotation is the centroid of the rivet group. As a result, twisting forces act on the rivets in addition to the direct force $P/N$ on each rivet.

**3-13.**   **Welding** is a process of joining two or more parts by depositing fused metal from a slender rod along the joint between them and allowing it to cool, the heat of fusion being applied either by an electric arc or by an oxyacetylene torch.

**REVIEW PROBLEMS**

**3-52.**   A typical structural joint of a light roof truss is shown in Fig. 3-49. Each member is made up of two angles placed back to back and riveted to a gusset plate $5/16$ in. thick. The sloping member is made up of two angles $3\frac{1}{2}$ by 3 by $1/4$ in., which are fastened to the gusset plate with four $5/8$-in.-diameter rivets. Calculate the unit shearing and bearing stresses in these rivets caused by a tensile load $T$ of 32,000 lb.

**3-53.**   Calculate the maximum compressive force $C$ which the vertical double-angle member in Fig. 3-49 may safely transmit to the joint if it is fastened with

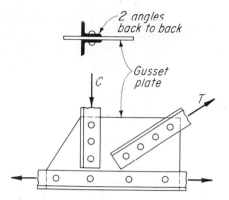

Fig. 3-49   Probs. 3-52 and 3-53.

three 15-mm-diameter rivets to a 12-mm-thick gusset plate. The thickness of each angle is 10 mm.

*Ans.* $C_{max} = 127$ kN

**3-54.** Assume that the beam in Fig. 3-1*a* is a W 16 × 36, the girder is a W 21 × 62, all rivets are of $^3/_4$-in. diameter, and the side angles are $^3/_8$ in. thick. Calculate the value of the maximum load $R$ (beam-end reaction) that may safely be transmitted from the beam to the girder.

**3-55.** In a beam-to-column connection similar to that shown in Fig. 3-14, the beam is a W 21 × 111 and the column is a W 12 × 65. All connecting angles are $^7/_{16}$ in. thick. If $^3/_4$-in.-diameter rivets are used and four rivets connect the seat angle to the column flange, how many additional rivets are required through the web of the beam to transmit a load of 100 kips?

**3-56.** Assume that the beam in Fig. 3-12 is a W 360 × 147, the girder has a 14-mm web, 25-mm-diameter rivets are used, and the load to be transmitted from each beam to the girder is 520 kN. How many rivets must be used to connect the seat angle to the girder? All angles are 15.9 mm thick.

**3-57.** A connection of the type shown in Fig. 3-50 must be designed to support a tensile load $T$ of 30 kips. Calculate (*a*) the required diameter $D$ of the unfinished A307 bolt $B$; (*b*) the required thickness $t$ of bar $A$ and $t'$ of each of the two side plates $C$ (based on bearing); (*c*) the required width $w$ of bar $A$ if allowable tensile stress on net section is 16,000 psi; (*d*) the minimum distances $a$ and $c$ to preclude

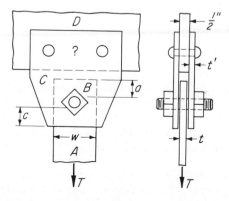

Fig. 3-50   Probs. 3-57 and 3-58.

end shearing in bar $A$ and in plates $C$, if allowable plate shearing stress is 12,000 psi; (*e*) the required number of $3/4$-in. rivets to fasten plates $C$ to plate $D$. Where necessary, round off calculated values to nearest higher multiple of $1/16$ in.

**3-58.** The structural connection shown in Fig. 3-50 is to be designed to support a tensile load $T$ of 18,000 lb. Calculate (*a*) the required diameter $D$ of the unfinished A307 bolt $B$; (*b*) the required thickness $t$ of bar $A$ and $t'$ of each of the two side plates $C$; (*c*) the required width $w$ of bar $A$ if allowable tensile stress on net area is 15,000 psi; (*d*) the minimum distances $a$ and $c$ to preclude end shearing in bar $A$ and in plates $C$ if allowable plate shearing stress is 12,000 psi; (*e*) the required number of $5/8$-in. rivets to fasten side plates $C$ to the main plate $D$. Round off calculated dimensions to nearest higher multiple of $1/16$ in. where necessary.

**3-59.** The pin $A$ of Fig. 3-51 is $1\frac{1}{2}$ in. in diameter and is in double shear. Calculate the shearing stress in the pin if $W = 21$ kips.

*Ans.* $s_s = 10.3$ ksi

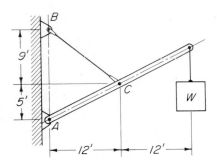

**Fig. 3-51   Prob. 3-59.**

**3-60.** The tubing cutter shown in Fig. 3-52 is used to cut $1\frac{1}{4}$-in.-outside-diameter pipe. If the axle of wheel $A$ is $1/4$ in. in diameter and is supported at both ends, calculate the shearing stress $s_s$ in the axle. The movable wheel $B$ can exert a force of 500 lb on the pipe.

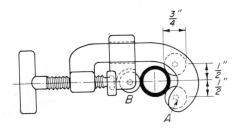

**Fig. 3-52   Prob. 3-60.**

**3-61.** In the continuous triple-riveted lap boiler joint shown in Fig. 3-53, the short pitch $p$ is $2^{13}/16$ in., the long pitch $p_1$ is $5^5/8$ in., the rivet holes are $15/16$ in. in diameter, and the thickness of the plates is $1/2$ in. Calculate the unit stresses in this joint in shear, bearing, and tension due to a load $P$ of 24,000 lb on a repeating section.

*Ans.* $s_s = 8690$ psi; $s_c = 12,800$ psi; $s_t = 10,240$ psi

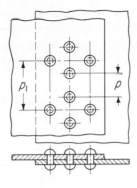

**Fig. 3-53    Probs. 3-61 and 3-62.**

**3-62.** Using the data given in Prob. 3-61, calculate (*a*) the strength of a repeating section of the boiler joint shown in Fig. 3-53 and (*b*) the efficiency of the joint.

**3-63.** At a small hydroelectric power plant, a water pipe with an inside diameter of 30 in. and a wall thickness of 0.375 in. operates under a maximum head *h* of 800 ft. If the ultimate strength of the steel is 50,000 psi, what is the factor of safety?

**3-64.** A spherical gas tank 8 m in diameter is constructed of boiler plate to withstand safely internal gas pressure of 0.4 MPa. Allowable tensile plate stress is 80 MPa. What thickness *t* of plate should be used (*a*) if seams are butt-welded with 100 percent efficiency and (*b*) if seams are riveted and are 80 percent efficient?

*Ans.* (*a*) $t = 10$ mm; (*b*) $t = 12.5$ mm

**3-65.** An open-top water surge tank is 20 ft in diameter and must withstand the pressure caused by a static depth of 98 ft of water. (*a*) What thickness *t* of plate is required if 100 percent efficient butt-welded seams are used? (*b*) Calculate the length of a repeating section and the required plate thickness *t* if 1-in. rivets in $1^{1}/_{16}$-in. holes are used in a triple-riveted lap joint.

**3-66.** Using the eccentrically loaded riveted connection shown in Fig. 3-26*d*, show that the forces exerted on the plate by the three rivets are $P/12$, $4P/12$, and $7P/12$.

**3-67.** The gusset plate shown in Fig. 3-54 will be riveted to a column flange with six equally spaced rivets whose centers lie on a 10-in.-diameter circle. The load *P* is 15,000 lb. (*a*) Calculate the maximum force $F_{max}$ exerted on the plate by the most highly stressed rivet. (*b*) If the rivets are of $^{5}/_{8}$-in. diameter, by what max-

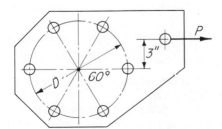

**Fig. 3-54    Prob. 3-67.**

imum percentage may the load $P$ be increased without stressing any rivet beyond its capacity in single shear?

*Ans.* (*a*) $F_{max}$ = 3,870 lb; (*b*) 39 percent

**3-68.** In the eccentrically loaded riveted connection shown in Fig. 3-33, let $a$ = 4 in., $b$ = 5 in., and $e$ = 10 in. and assume that ten $^7/_8$-in.-diameter rivets are used in single shear. Thickness of the column flange is 0.688 in. and of the plate is 0.375 in. Calculate the maximum safe load $P$.

**3-69.** In Fig. 3-55, two $^3/_4$-in. main straps are fastened together by welding $^3/_8$-in. splice plates 3 in. wide on both sides. If $^1/_4$-in. fillet welds and full-width end welds are used, calculate required length $L$ to resist a pull $P$ of 72,000 lb.

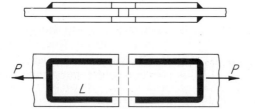

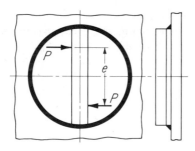

Fig. 3-55   Prob. 3-69.

**3-70.** A 10-mm-thick circular plate 200 mm in diameter is welded to a heavier plate by means of a 6-mm fillet weld around its perimeter, as shown in Fig. 3-56. Calculate the maximum moment $Pe$ that this weld may safely resist. The forces $P$ are exerted against a lug projecting from the plate.

*Ans.* $Pe$ = 37.7 × 10⁶ N·mm

Fig. 3-56   Prob. 3-70.

## REVIEW QUESTIONS

**3-1.** Name some structures in which the separate members are usually riveted together.

**3-2.** Describe an undriven rivet and the manner in which it is usually driven.

**3-3.** What are some of the standard practices in the design of riveted structural joints? Name some examples of structural joints.

**3-4.** Name some of the standard practices in the design of riveted boiler joints. Give some examples and types of boiler joints.

**3-5.** Describe several possible types of failures of riveted joints.

**3-6.** Discuss shearing stresses in riveted joints. Give examples of single shear and double shear.

**3-7.** Explain bearing stresses and give examples of inside bearing and outside bearing.

**3-8.** Discuss tensile plate stresses in a riveted joint.

**3-9.** Are unit stresses in riveted joints fairly uniform or rather nonuniform?

**3-10.** When the equation $s = P/A$ is used in connection with riveted joints, what stress $s$ is thus obtained?

**3-11.** On what bases are allowable rivet unit stresses usually determined? Approximately what factor of safety is already included in the allowable stresses for steel rivets?

**3-12.** What is regarded as the strength of a riveted joint?

**3-13.** Define the term "efficiency of a riveted joint." Estimate a common efficiency of a boiler lap joint.

**3-14.** What is meant by a concentrically loaded riveted joint?

**3-15.** Explain what is meant by an eccentrically loaded riveted joint.

**3-16.** By what method may the centroid of an unsymmetrical group of rivets be located?

**3-17.** By what quantities is an eccentric moment usually measured?

**3-18.** What portion of the total load $P$ is each rivet of a concentrically loaded group of $n$ rivets presumed to carry?

**3-19.** What is the effect of an eccentric moment on the rivets of a group? Describe some characteristics of the force $Q$ and how $Q$ is determined.

**3-20.** Against which rivet of a group of equal-diameter rivets is the maximum force $F$ generally exerted and how is $F$ determined?

**3-21.** Describe generally the process of welding.

**3-22.** For what type of work is welding especially useful?

**3-23.** What are the best assurances of good-quality welding?

**3-24.** Describe briefly fillet welds and butt welds.

**3-25.** What determines the size of a fillet weld?

**3-26.** Where is the critical section in a fillet weld and how are its dimensions determined?

**3-27.** How may eccentric loading of welded joints sometimes be compensated for?

# Chapter 4
# Torsion

## 4-1 TWISTING MOMENTS. TORQUE

Structural members are subjected to various kinds of forces according to the intended function of each member. For example, the wire ropes supporting an elevator cage are acted upon by *tensile forces,* a timber column supporting beams is subjected to *compressive forces,* a beam supporting a footbridge across a stream is subjected to *bending forces,* and a steel shaft transmitting power is subjected to *twisting forces* which produce a **twisting moment,** commonly called a **torque.** A torque is normally applied in a plane perpendicular to the longitudinal axis of the member. When so applied, the member is said to be in **torsion.**

In Fig. 4-1, for example, a piece of shafting is welded to a baseplate $B$, which in turn is fastened to a support by three bolts at equal distances from the shaft center. When the shaft is twisted by the application of the couple $Fa$ to the circular wheel $A$, a resisting torque is formed by the forces $P$ exerted against the plate by the three equispaced bolts. The applied torque is $Fa$, and if $N$ designates the number of bolts, the resisting torque is $NPr$. The axis of rotation is the centroidal axis of the shaft. When two or more bolts are symmetrically spaced about the axis of rotation, each bolt is presumed to resist its proportional part of the total torque.

### ILLUSTRATIVE PROBLEM

**4-1.** If in Fig. 4-1, $F = 1500$ lb, $a = 20$ in., and $r = 5$ in., (a) what force $P$ must be resisted by each bolt, and (b) what diameter bolts should be used if the allowable shearing stress is 12,000 psi?

*Solution:* The applied torque on $A$ is the moment of the couple shown, or $Fa$. This torque is resisted by the moment of the three forces acting on $B$. Equating the acting and resisting torque gives

$$[Fa = NPr] \qquad (1500)(20) = (3)(P)(5) \qquad \text{or} \qquad P = 2000 \text{ lb} \qquad \textit{Ans.}$$

The required bolt diameter can now be found:

$$\left[ A = \frac{F}{s_s} \right] \qquad \frac{\pi D^2}{4} = \frac{2000}{12,000} \qquad \text{or} \qquad D = 0.461 \text{ in. Use } \tfrac{1}{2}\text{-in. bolts.} \qquad Ans.$$

## PROBLEMS

**4-2.** Solve Prob. 4-1 when $F = 3000$ lb, $a = 24$ in., $r = 3$ in., and four equally spaced bolts are used.

**4-3.** Assume in Fig. 4-1 that two 10-mm-diameter bolts fasten the baseplate $B$ to its support and that $a - 200$ mm and $r = 50$ mm. If the ultimate shearing strength of the steel in the bolts is 300 MPa, what value of $F$ is required to shear the bolts off?

*Ans.* $F = 11.78$ kN

**4-4.** If, in Fig. 4-1, $F = 24$ kN, $a = 250$ mm, $r = 60$ mm, and allowable bolt shearing stress is 75 MPa, what number $N$ of 20-mm bolts would be required, equally spaced on a circle of radius $r$?

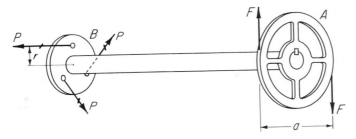

**Fig. 4-1   Twisting moment. Torque.**

## 4-2 STRESSES IN CIRCULAR SHAFTS. THE TORSION FORMULA

If we assume a circular shaft to be made up of a series of thin disks clamped tightly together, it becomes apparent that when the shaft is twisted, a system of forces or stresses is produced on each set of contacting surfaces of adjacent disks which will tend to prevent slippage of any disk with respect to both adjacent disks. The resultants of these stress systems form equal and opposite couples on opposite faces of each disk, one couple being the *torque* and the other the *resisting torque*. Similar stress conditions exist on every pair of cross-sectional planes of a solid shaft.

For the purpose of designing circular shafts to withstand a given torque, we must develop an equation giving the relation between the twisting moment or torque, the maximum stress produced, and a quantity representing the size and shape of the cross-sectional area of the shaft.

Let the shaft shown in Fig. 4-2 be held completely stationary at its left end. When a torque $Pd$ is applied at its right end, that end is rotated until

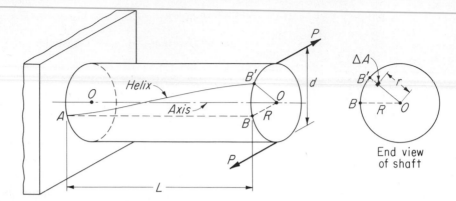

**Fig. 4-2**   Development of torsion formula.

the line $ABO$ assumes the position $AB'O$, of which $AB'$ is a helix and $B'O$ is assumed to be straight. The deformation on a cross-sectional plane (Fig. 4-3) is seen to increase at a uniform rate from zero at the center $O$ to a maximum $BB'$ at the outer surface of the shaft. Consequently the shearing stresses on any cross-sectional area will vary in the same manner, since stress is proportional to deformation.

In Fig. 4-2, consider a small area $\Delta A$ on any cross-sectional plane of the shaft. The relation shown in Fig. 4-3 may now be established. If the maximum unit stress at distance $R$ from $O$ is $s_s$, then by proportion the unit stress at distance $r$ from $O$ is

$$\frac{s_s r}{R} \qquad (a)$$

and the shearing force on area $\Delta A$ will be

$$\frac{s_s r\ \Delta A}{R} \qquad (b)$$

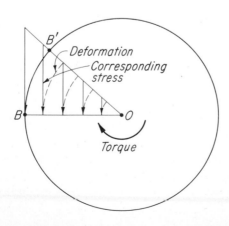

**Fig. 4-3**   Variation of deformations, and corresponding shearing stresses, on cross section of circular shaft.

Taking the moment about $O$ of the shearing force on $\Delta A$ gives

$$\frac{s_s r^2 \, \Delta A}{R} \tag{c}$$

The moment about $O$ of shearing forces on all $\Delta A$ areas is, then

$$\frac{s_s}{R} \Sigma r^2 \, \Delta A = \frac{s_s}{R} \cdot J \tag{d}$$

The quantity $\Sigma r^2 \, \Delta A$ is called the **polar moment of inertia** of the cross-sectional area of the shaft with respect to its centroidal axis $OO$. It is denoted by the symbol $J$.

Now, the twisting moment producing these shearing stresses is the torque $T$, and the moment about $O$ of the shearing forces on all $\Delta A$ areas of the cross section is the resisting torque and equals $s_s J/R$. Since torque and resisting torque constitute action and reaction, they are equal. That is, $T = s_s J/R$ or, since $R = D/2$,

$$\text{\textbf{Twisting moment or torque }} T = \frac{2 s_s J}{D} \tag{4-1}$$

from which

$$\text{\textbf{Maximum unit shearing stress }} s_s = \frac{TD}{2J} \tag{4-2}$$

For a circular area, $J = \frac{1}{2}\pi R^4$, as illustrated in Prob. B-5, Appendix B, or $J = \pi D^4/32$, since $R = D/2$.

For design purposes we use an expression giving the required diameter $D$ of a circular shaft to resist a given torque $T$, using a predetermined allowable shearing stress $s_s$. Since $J = \pi D^4/32$, Eq. (4-1) becomes $T = 2s_s \pi D^4/32D = s_s \pi D^3/16$, and

$$\text{\textbf{Required diameter of solid shaft }} D = \sqrt[3]{\frac{16T}{\pi s_s}} = 1.72 \sqrt[3]{\frac{T}{s_s}} \tag{4-3}$$

where  $T$ = twisting moment or torque, lb·in. or N·mm
$s_s$ = allowable unit shearing stress, psi or MPa (N/mm²)
$D$ = diameter, in. or mm

**Hollow Shafts.** Owing to the fact that the material of a shaft near its center is of comparatively little value in resisting torque, many large-diameter shafts are made hollow. For example, the strength of a hollow shaft whose inside diameter is one-half its outside diameter is about 93 percent of that of a solid shaft.

Hollow shafts are often used in large ships, where a single shaft may transmit more than 10,000 hp (7460 kW) and may exceed 30 in. (760 mm) in diameter. As is more completely stated in Art. 5-6, ''the moment of inertia of an area from which some part has been removed equals the moment of inertia of the total area less the moment of inertia of the part removed.'' Hence the polar moment of inertia $J$ of a hollow shaft of outside diameter $D$ and inside diameter $d$ is

$$\textbf{Polar moment of inertia of hollow shaft } J = \frac{\pi}{32} (D^4 - d^4)$$

$$= 0.0982 \, (D^4 - d^4) \qquad (e)$$

Equation (4-3) for the design of solid shafts cannot be used for hollow shafts. Equation (4-1) should be used. In order to eliminate one of the two unknown diameters in $J$, the inside diameter $d$ is generally expressed in terms of the outside diameter $D$, such as $d = 0.7D$ or $0.8D$, and so forth.

**Noncircular Shafts.** The design of square, rectangular, and other non-circular shafts is a complex problem beyond the scope of this text. By way of comparison, however, it has been found experimentally that the strength of a square shaft of side dimensions $d$ exceeds the strength of a circular shaft of diameter $D$ by a relatively small percentage. Consequently, a square shaft may safely be designed by the circular-shaft formula.

## 4-3 ANALYSIS AND DESIGN OF CIRCULAR SHAFTS

The following problems make use of the basic torsion formula, Eq. (4-1), and of the derived Eqs. (4-2) and (4-3).

ILLUSTRATIVE PROBLEMS

**4-5.** A torque $T$ is applied to a shaft in which the shearing stress must not exceed 60 MPa. Calculate the maximum allowable $T$ ($a$) if the shaft is solid and is 100 mm in diameter and ($b$) if the shaft is hollow and its outside diameter $D = 100$ mm and inside diameter $d = 50$ mm.

*Solution:* The polar moment of inertia $J$ of a solid shaft is $\pi D^4/32$ and of a hollow shaft is $\pi(D^4 - d^4)/32$. Hence for the solid shaft when $D = 4$ in.,

$$\left[ J = \frac{\pi D^4}{32} \right] \qquad J = \frac{\pi(100^4)}{32} = 9.817 \times 10^6 \text{ mm}^4$$

$$\left[ T = \frac{2 s_s J}{D} \right] \qquad T = \frac{2(60)(9.817)(10^6)}{100} = 11.78 \times 10^6 \text{ N·mm} = 11\,780 \text{ N·m} \qquad Ans.$$

For the hollow shaft when $D = 100$ mm and $d = 50$ mm,

$$\left[ J = \frac{\pi(D^4 - d^4)}{32} \right] \qquad J = \frac{\pi(100^4 - 50^4)}{32} = 9.204 \times 10^6 \text{ mm}^4$$

$$\left[ T = \frac{2 s_s J}{D} \right] \qquad T = \frac{2(60)(9.204)(10^6)}{100} = 11.04 \times 10^6 \text{ N·mm} = 11\,040 \text{ N·m} \qquad Ans.$$

**4-6.** A shaft of 3-in. diameter is subjected to a torque $T$ of 40,000 lb·in. Calculate the maximum unit shearing stress in the shaft ($a$) when it is solid and ($b$) when it is hollow with an inside diameter $d$ of 2 in. ($c$) What is the unit stress at the inner surface of the shaft?

*Solution:* For the solid shaft when $D = 3$ in.,

$$\left[ J = \frac{\pi D^4}{32} \right] \qquad J = \frac{\pi(3^4)}{32} = 7.95 \text{ in.}^4$$

$$\left[ S_s = \frac{TD}{2J} \right] \qquad S_s = \frac{(40,000)(3)}{(2)(7.95)} = 7550 \text{ psi} \qquad\qquad Ans.$$

For the hollow shaft when $D = 3$ in. and $d = 2$ in.,

$$\left[ J = \frac{\pi(D^4 - d^4)}{32} \right] \qquad J = \frac{\pi(81 - 16)}{32} = 6.38 \text{ in.}^4$$

$$\left[ S_s = \frac{TD}{2J} \right] \qquad S_s = \frac{(40,000)(3)}{(2)(6.38)} = 9400 \text{ psi} \qquad\qquad Ans.$$

The intensity of stress varies from zero at the center to 9400 psi at the outer surface. Hence at the inner surface

$$S_s = 9400 \frac{1}{1.5} = 6270 \text{ psi} \qquad\qquad Ans.$$

**4-7.** Design a solid steel shaft to resist a torque of 50 kN·m. Allowable shearing stress is 80 MPa.

*Solution:* Equation (4-3) will give directly the required diameter of a *solid* steel shaft. That is,

$$\left[ D = \sqrt[3]{\frac{16T}{\pi S_s}} \right] \qquad D = \sqrt[3]{\frac{(16)(50\ 000\ 000)}{\pi(80)}} = 147 \text{ mm} \qquad\qquad Ans.$$

**4-8.** A hollow shaft must be designed to withstand an estimated torque of 1,708,000 lb·in. It has been predetermined that an inside diameter $d$ equal to $0.6D$ is satisfactory. Maximum allowable unit shearing stress is 10,000 psi.

*Solution:* In terms of $D$, the polar moment of inertia $J$ is

$$\left[ J = \frac{\pi(D^4 - d^4)}{32} \right] \qquad J = \frac{\pi}{32}[D^4 - (0.6D)^4] = (0.0982)(0.870D^4) = 0.0854D^4$$

From Eq. (4-1)

$$\left[ T = \frac{2S_s J}{D} \right] \qquad 1,708,000 = \frac{(2)(10,000)(0.0854D^4)}{D} = 1708D^3$$

from which

$$D^3 = \frac{1,708,000}{1708} = 1000 \qquad D = 10 \text{ in.} \qquad \text{and} \qquad d = 6 \text{ in.} \qquad Ans.$$

**PROBLEMS**

**4-9.** Calculate the polar moment of inertia $J$ of a 120-mm-diameter solid circular shaft.

$$Ans.\ J = 20.36 \times 10^6 \text{ mm}^4$$

**4-10.** A hollow circular shaft has an outside diameter $D$ of 4 in. and an inside diameter $d$ of 2 in. (a) Calculate its polar moment of inertia $J$. (b) Show by comparative equations that the polar moment of inertia of this hollow circular shaft is $^{15}/_{16}$ that of a solid shaft having the same outside diameter.

$$Ans. \ (a) \ J = 23.6 \ \text{in.}^4; \ (b) \ J = (\pi/32)(^{15}/_{16})D^4$$

**4-11.** What maximum torque in pound-inches may safely be applied to a shaft in which the unit shearing stress must not exceed 12,000 psi (a) if the shaft is solid and $D = 5$ in. and (b) if the shaft is hollow, $D = 5$ in., and $d = 4$ in.?

**4-12.** The propeller shaft on a large ship is hollow with an outside diameter $D$ of 24 in. and an inside diameter $d$ of 16 in. If the shearing stress must not exceed 8000 psi, what maximum torque $T$ can the shaft safely withstand?

$$Ans. \ T = 17,400,000 \ \text{lb·in.}$$

**4-13.** Assume in Fig. 4-1 that the left end of the shaft is held stationary and that $F = 1$ kN, $a = 200$ mm, and $D = 40$ mm. Then calculate the maximum unit shearing stress in the shaft. Forces $P$ and $F$ lie in planes perpendicular to the shaft.

**4-14.** Calculate the unit shearing stresses at the outer and inner surfaces of a hollow shaft in which $D = 6$ in. and $d = 4$ in. due to a torque $T$ of 300,000 lb·in.

$$Ans. \ s_s = 8810 \ \text{psi}; \ s_s = 5870 \ \text{psi}$$

**4-15.** Determine the required diameter $D$ of a solid shaft to resist an estimated torque $T$ of 12 kN·m. Allowable unit shearing stress is 100 MPa.

**4-16.** Calculate the required diameter $D$ of a solid shaft to withstand a torque $T$ of 510,000 lb·in. if allowable shearing stress is 12,000 psi.

**4-17.** Design a hollow shaft in which $d = 0.7D$ to withstand a torque $T$ of 9,550,000 lb·in. Allowable shearing stress is 8000 psi.

$$Ans. \ D = 20 \ \text{in.}; \ d = 14 \ \text{in.}$$

**4-18.** Calculate the required inside and outside diameters, $d$ and $D$, of a hollow shaft to withstand a torque $T$ of 100 MPa if $d = ¾D$ and allowable stress is 25 kN·m.

**4-19.** A threaded machine bolt of ¾-in. diameter is screwed up tightly, fastening together two steel parts. To avoid failure of bolt in shear, the unit shearing stress must not exceed the elastic-limit stress of 36,000 psi. What maximum pull may safely be exerted on a wrench with a 20-in. moment arm to avoid failure on the root area if its root diameter is 0.6688 in. and if we assume that friction under the bolt head nullifies 25 percent of the pull?

## 4-4  SHAFT COUPLINGS

Two methods are available for coupling together two sections of a shaft: (1) flanges may be forged on the ends to be connected, which are then bolted together as illustrated in Fig. 4-4, or (2) a separate two-piece coupling may be used, each piece cast with flange and hub. The flanges are then bolted together and the hubs are keyed to the shafts. Since a

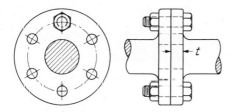

Fig. 4-4    Flange coupling. Prob. 4-20.

keyway weakens a shaft, the allowable torque on a keyed shaft is 25 percent below that on an unkeyed shaft.

When the flange is forged on the shaft, its thickness $t$ must be sufficient to prevent the shaft from shearing out of the flange on the cylindrical surface whose area is the circumference of the shaft $\pi D$ times the thickness $t$, or $\pi Dt$. If the unit shearing stress on this surface is $s_s$, the total resisting force developed on that area is $s_s \pi Dt$ and the resisting moment or resisting torque of this force about the center of the shaft is $s_s \pi DtR$. But any applied torque ($T = 2s_s J/D$) equals the resisting torque ($s_s \pi DtR$). Hence, for equal strength in shear on the cross-sectional area of the shaft and the cylindrical area in the flange, $2s_s J/D = s_s \pi DtR$, from which the required $t$ of flange may be calculated. This reduces to $t = D/8$ for solid shafts. The two allowable stresses $s_s$ are usually, but not always, equal. The resistance offered by the fillet between shaft and flange may be disregarded.

As previously stated, when several bolts in a coupling resist a given torque $T$, each bolt is presumed to resist its proportional part. Also, while the unit shearing stress in a flange bolt is greatest in the portion farthest from the shaft center, it is, for practical purposes, considered to be uniform over the bolt cross section when the bolt diameter is comparatively small in relation to the bolt-circle diameter.

**ILLUSTRATIVE PROBLEM** _____

**4-20.** Two sections of a shaft, connected by a flange coupling similar to that shown in Fig. 4-4, must be capable of transmitting a torque $T$ of 424,000 lb·in. Calculate ($a$) the required diameter $D$ of shaft; ($b$) the minimum thickness $t$ of flange; ($c$) the required diameter of bolts if six are equally spaced on a 9-in.-diameter bolt circle. Allowable $s_s$ in shaft and flange is 10,000 psi; in bolts it is 15,000 psi.

*Solution:* The diameter is determined by Eq. (4-3). That is,

$$\left[ D = \sqrt[3]{\frac{16T}{\pi s_s}} \right] \qquad D = \sqrt[3]{\frac{(16)(424,000)}{\pi(10,000)}} = \sqrt[3]{216} = 6 \text{ in.} \qquad\qquad Ans.$$

Since the torque equals $2s_s J/D$, the resisting torque equals $s_s \pi DtR$, and $J = \pi D^4/32$,

$$\frac{2s_s J}{D} = s_s \pi DtR \qquad \text{or} \qquad t = \frac{2J}{\pi D^2 R}$$

from which

$$t = \frac{2}{\pi D^2 R} \frac{\pi D^4}{32} = \frac{D^2}{16R} = \frac{36}{(16)(3)} = 0.75 \text{ in.}$$    Ans.

Then since $T = NPr$, the force on each bolt is

$$\left[ P = \frac{T}{Nr} \right] \qquad\qquad P = \frac{424,000}{(6)(4.5)} = 15,700 \text{ lb}$$

The required bolt diameter may now be found.

$$\left[ A = \frac{P}{s_s} \right] \qquad \frac{\pi D^2}{4} = \frac{15,700}{15,000} \qquad \text{or} \qquad D = 1.155 \text{ in.} \qquad \text{Use } 1^{1}/_{4}\text{-in. bolts} \qquad Ans.$$

## PROBLEMS

**4-21.** Calculate the thickness $t$ of flange required to develop the full strength of an 80-mm-diameter shaft if allowable shearing stresses $s_s$ in flange and shaft are equal.

Ans. $t = 10$ mm

**4-22.** Two 3-in.-diameter shafts have flanged ends coupled together with eight bolts equally spaced on a 7½-in.-diameter circle. What diameter bolts (to nearest higher ⅛ in.) should be used for coupling and shaft to have the same strength in torque? Allowable shearing stress in shaft and bolts is 12,000 psi.

**4-23.** In Prob. 4-20, let $T = 14\ 726$ N·m, the bolt-circle diameter be 240 mm, and let the allowable shearing stresses be 75 MPa and 100 MPa, respectively. Solve.

## 4-5 ANGLE OF TWIST

When a shaft of length $L$ is subjected to a torque $T$, it will twist about its centroidal axis. If a shaft connects parts whose motions must be closely synchronized, the angle $\theta$ through which the shaft will twist must often be determined. We will now develop an expression giving the angle of twist $\theta$ in terms of torque $T$, length $L$, modulus of elasticity in shear $G$, and polar moment of inertia $J$.

If, in Fig. 4-5, the **total twist** or deformation in length $L$ of an extreme fiber at $B$ is $BB'$, then the **unit twist**, or deformation $\phi$ per inch of length of shaft, is $BB'/L$. But arc $BB' = R\theta$, when $\theta$ is expressed in radians ($360° = 2\pi$ rad). Hence $\phi = R\theta/L$. The modulus of elasticity in shear $G$ is

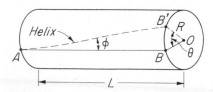

Fig. 4-5    Portion of shaft. Angle of twist.

the ratio of unit stress to unit deformation, or $G = s_s/\phi$, from which $\phi = s_s/G$ [see Eq. (1-5)]. Consequently

$$\phi = \frac{R\theta}{L} = \frac{s_s}{G} \tag{$a$}$$

from which

$$\theta = \frac{s_s L}{RG} \tag{4-4}$$

But from Eq. (4-2)

$$s_s = \frac{TD}{2J} = \frac{TR}{J} \tag{$b$}$$

Hence, substituting Eq. ($b$) in Eq. (4-4), we obtain

$$\textbf{Angle of twist } \theta = \frac{TL}{GJ} \tag{4-5}$$

where $\theta$ = angle of twist, rad (1 rad = 57.3°)

  $T$ = torque, lb·in. or N·mm

  $L$ = length, in. or mm

  $G$ = modulus of elasticity in shear, psi or MPa

  $J$ = polar moment of inertia, in.$^4$ or mm$^4$

## ILLUSTRATIVE PROBLEM

**4-24.** Calculate ($a$) the maximum permissible angle of twist $\theta$ in a 1-in.-diameter shaft 30 in. long if allowable stress is 10,000 psi; ($b$) the angle of twist $\theta$ in a 3-in.-diameter shaft 20 ft long due to a torque of 50,000 lb·in.; ($c$) the required diameter of shaft 4 m long to withstand a torque of 10 kN·m if $\theta$ must not exceed 3° ($G$ = 84 000 MPa).

*Solution:* Part $a$ is solved directly by Eq. (4-4) and parts $b$ and $c$ are solved by Eq. (4-5). That is,

$$\left[\theta = \frac{s_s L}{RG}\right] \qquad \theta = \frac{(10,000)(30)}{(0.5)(12,000,000)} = 0.05 \text{ rad} = 2.87° \qquad Ans.$$

$$\left[\theta = \frac{TL}{GJ}\right] \qquad \theta = \frac{(50,000)(240)(32)}{(12,000,000)(\pi)(3)^4} = 0.1258 \text{ rad} = 7.22° \qquad Ans.$$

$$\left[J = \frac{TL}{\theta G}\right] \qquad J = \frac{10\ 000\ 000(4000)57.3}{(3)(84\ 000)} = 9.095 \times 10^6 \text{ mm}^4$$

$$\left[D^4 = \frac{32J}{\pi}\right] \qquad D^4 = \frac{(32)(9.095 \times 10^6)}{\pi} \qquad D = 98.1 \text{ mm} \qquad Ans.$$

## PROBLEMS

**4-25.** Calculate the angle of twist $\theta$ of a 50-mm-diameter shaft if its length is 3 m and the shearing stress is not to exceed 60 MPa. $G$ = 84 000 MPa.

**4-26.** What unit stress $s_s$ is produced in a 1-in.-diameter shaft 60 in. long when it is twisted through angle $\theta$ of 5°? $G = 12,000,000$ psi.

**4-27.** A solid steel shaft 10 ft long is subjected to a torque $T$ of 2740 lb·in. Because this shaft must closely synchronize two operations, its angle of twist $\theta$ must not exceed 1° in 10 ft. (a) Calculate the minimum required diameter $D$ of this shaft. (b) What unit stress is produced at the maximum angle of twist? $G = 12,000,000$ psi.

> Ans. (a) $D = 2$ in.; (b) $s_s = 1745$ psi

**4-28.** Calculate the angle of twist $\theta$ of a hollow steel ship propeller shaft ($D = 600$ mm and $d = 400$ mm) 30 m long when subjected to a torque of 200 kN·m.

**4-29.** In order to determine the modulus of elasticity $G$ in shear of a certain type of steel, a 1-in.-diameter shaft 60 in. long was tested in a torsion machine. A torque of 1715 lb·in. produced an angle of twist of 5°. (a) Calculate the value of $G$. (b) What maximum twist can be given this shaft without exceeding elastic limit of 36,000 psi in shear?

> Ans. (a) $G = 12,000,000$ psi; (b) $\theta = 20.6°$

## 4-6 POWER TRANSMISSION BY SHAFTS

Power is defined as the time rate of doing work, that is, the amount of work done in a unit of time as, for example, a minute. Work is done when a force displaces a body. The measure of work done by a force is the product of the force and the displacement of its point of application. If the force is in pounds and the displacement in inches, the unit of work done is the inch-pound (in.·lb.). Power, then, being work done per unit of time, may be expressed in inch-pounds per minute (in.·lb/min). But this is seen to be the product of a force $P$ in pounds and the velocity $v$, or rate of displacement in inches per minute, of its point of application. That is, power is $Pv$, the product of force and velocity. The following illustration will show the relation between power, torque $T$, and speed $N$ of pulley or shaft in revolutions per minute (rpm).

Figure 4-6 is the cable drum of a materials hoist. The pull $P$ in the cable turns the drum counterclockwise, the motion being resisted by a friction force $F$.

Let $A$ be a point on the drum where the cable takes off in a straight line. In one revolution of the drum, point $A$ and force $P$ have been displaced a distance equal to the circumference $2\pi R$ of the drum. If the velocity $v$ of the drum is $N$ rpm, then $v = 2\pi RN$ and the power is $Pv = P2\pi RN$. But $PR = T$, the torque. Hence

$$\textbf{Power} = 2\pi NT \qquad (a)$$

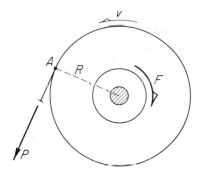

Fig. 4-6    Power transmission by shaft.

**Horsepower.** A commonly used unit of mechanical power is the *horsepower,* equal to 33,000 ft·lb of work per minute or

$$(33,000)(12) = 396,000 \text{ in·lb}$$

of work per minute. Consequently in terms of torque $T$ and shaft speed $N$ in rpm

$$\textbf{Horsepower } H = \frac{2\pi NT}{396,000} \tag{4-6}$$

from which

$$\textbf{Torque } T = \frac{63,025H}{N} \tag{4-7}$$

where $H$ = horsepower
$N$ = number of revolutions per minute
$T$ = torque, in.·lb

Equation (4-6) is used to determine the power-transmitting capacity of a given shaft and Eq. (4-7) to determine the diameter of shaft required to transmit a given horsepower.

When Eq. (4-7) is substituted in Eq. (4-3) and reduced to its simplest form, we obtain

$$D = \sqrt[3]{\frac{16T}{\pi s_s}} = \sqrt[3]{\frac{(16)(63,025H)}{\pi s_s N}} \tag{b}$$

from which, for a solid shaft,

$$D = 68.5 \sqrt[3]{\frac{H}{s_s N}} \tag{4-8}$$

Equation (b) may be put into another useful form, thus:

$$D^3 = \frac{(16)(63,025H)}{\pi s_s N}$$

from which, for a solid shaft,

$$H = \frac{s_s ND^3}{321,000} \qquad (4\text{-}9)$$

For hollow shafts, the basic Eqs. (4-1) and (4-7) should be used.

**Power Using SI Units.** The basic unit of power in SI is the watt. One watt is equivalent to the power expended in moving a one newton force through a distance of one meter in one second. In symbols,

$$1\ W = 1\ N{\cdot}m/s$$

In terms of the shaft torque $T$ and the shaft speed $N$,

$$W = T\left(\frac{2\pi N}{60}\right) \qquad \text{or} \qquad T = \frac{9.549\ W}{N} = \frac{9549\ kW}{N} \qquad (4\text{-}10)$$

where $W$ = watts and $kW$ = kilowatts
  $N$ = revolutions per minute
  $T$ = torque, N·m

When Eq. (4-10) is substituted into Eq. (4-3), with proper regard for units, we obtain

$$D = 365\ \sqrt[3]{\frac{kW}{s_s\ N}}$$

where $D$ = diameter of solid shaft, mm
  $s_s$ = maximum shaft shearing stress, MPa

**Keyed Shafts.** A torque is most commonly transmitted to a shaft through a pulley fastened to the shaft by a key fitted into opposite keyways cut into both pulley and shaft. When the torque is relatively small, the pulley may be fastened to the shaft by a set screw. For higher torques, the pulley may be "shrunk" onto the shaft.

A keyway cut into a shaft naturally weakens it. Customarily, a keyed shaft is permitted to transmit only three-fourths as much torque (or horsepower) as an unkeyed shaft. Conversely, a keyed shaft must be designed to withstand $^4/_3 T$ (or $^4/_3$·power). In the following problems, a shaft will not be considered as keyed unless it is specifically so stated.

**ILLUSTRATIVE PROBLEM**

**4-30.** A motor delivers 45 kW at 200 shaft rpm to pulley $C$ in Fig. 4-7. Of this, 7.5 kW are taken off by pulley $A$, 22.5 kW by pulley $B$, and 15 kW by pulley $D$. (*a*) Determine the required diameter $D$ of the shaft. (*b*) If $a$ = 5 m, $b$ = 2.5 m, and $c$ = 3 m, what is the angle of twist between pulleys $A$ and $C$? (*c*) By how many degrees will pulley $A$ lag behind pulley $D$? Allowable $s_s$ = 58 MPa; $G$ = 84 000 MPa. Appropriately spaced bearings prevent bending of shaft.

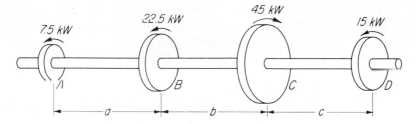

**Fig. 4-7  Power transmission by shaft.**

*Solution:* Close study of Fig. 4-7 will show that 30 kW tends to shear the shaft on some cross-sectional plane between pulleys $B$ and $C$, while only 15 kW tends to shear it on a different cross-sectional plane between pulleys $C$ and $D$. Consequently the shaft need be designed to transmit only 30 kW. Hence from Eq. (4-8)

$$\left[ D = 365 \sqrt[3]{\frac{kW}{s_s N}} \right] \qquad D = 365 \sqrt[3]{\frac{30}{(58)(200)}} = 50.1 \text{ mm} \qquad \qquad Ans.$$

The total twist of the shaft between pulleys $A$ and $C$ is the twist between $A$ and $B$ caused by 7.5 kW plus that between $B$ and $C$ caused by 30 kW. We will obtain $\theta_{AB}$ from Eq. (4-5) after substituting in it the value of $T$ given by Eq. (4-10). That is, since $L = 5$ m, $N = 200$ rpm, $G = 58$ MPa, and $J = \pi D^4/32$,

$$\left[ \theta_{AB} = \frac{TL}{GJ} \right] \qquad \theta_{AB} = \frac{9549(7.5)1000}{200(84\ 000)} \cdot \frac{5000(32)}{\pi(50.1)^4} = 0.034\ 46 \text{ rad} = 1.97° \qquad Ans.$$

The twist between pulleys $B$ and $C$ may be obtained in the same manner, although, since the stress in this section of the shaft is known to be 58 MPa, it may be obtained more easily from Eq. (4-4). That is, since $L = 2.5$ m, $R = 25.05$ mm, and $G = 84\ 000$ MPa,

$$\left[ \theta_{BC} = \frac{s_s L}{RG} \right] \qquad \theta_{BC} = \frac{58(2500)}{25.05(84\ 000)} = 0.068\ 9 \text{ rad} = 3.95°$$

Then

$$\text{Angle of twist } \theta_{AC} = 1.97° + 3.95° = 5.92° \qquad \qquad Ans.$$

The angle of twist between $C$ and $D$ is

$$\left[ \theta_{CD} = \frac{TL}{GJ} \right] \qquad \theta_{CD} = \frac{9549(15)1000}{200(84\ 000)} \frac{3000(32)}{\pi(50.1)^4} = 0.041\ 4 \text{ rad} = 2.37° \qquad Ans.$$

and

$$\text{Lag of } A \text{ behind } D = 5.92° - 2.37° = 3.55° \qquad \qquad Ans.$$

## PROBLEMS

NOTE: In the following problems, no shaft will be considered to be keyed to its pulley unless specifically so stated.

**4-31.** The required diameter of shaft in Prob. 4-30 was found to be 50.1 mm. What diameter will be required if pulleys $C$ and $D$ in Fig. 4-7 are interchanged, all other conditions remaining unchanged?

**4-32.** Calculate the maximum unit shearing stress produced in a 3-in.-diameter shaft transmitting 27 hp at 32 rpm.

**4-33.** What maximum horsepower may be transmitted by a 2-in.-diameter shaft at 320 rpm if allowable shearing stress must not exceed 10,000 psi? What horsepower can safely be transmitted by this shaft if it is keyed to its pulleys?

*Ans.* 79.8 hp; 59.8 hp

**4-34.** Design a solid shaft with keyed pulleys capable of transmitting 60 kW at a shaft speed of 600 rpm, using an allowable stress of 50 MPa. What diameter would be required if no keyways were cut in the shaft?

*Ans.* $D = 50.6$ mm; $D = 46.0$ mm

**4-35.** Pulleys $B$, $C$, and $D$ shown in Fig. 4-8 have diameters of 6, 8, and 10 in., respectively. The shaft $AE$ revolves at 1000 rpm. What horsepower is taken off the shaft at $B$ and $D$? What horsepower is added at $C$?

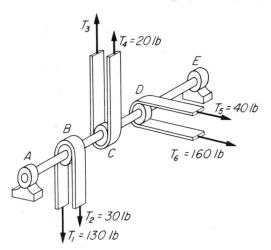

Fig. 4-8 Probs. 4-35 and 4-49.

**4-36.** In Fig. 4-9 a motor transmits 40 hp to gear $B$ and turns it at 600 rpm. Ten horsepower is taken off at gear $C$. All pulleys are adequately attached to the shaft.

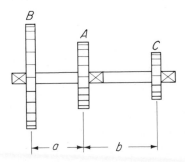

Fig. 4-9 Probs. 4-36 and 4-50.

Calculate the required shaft diameter if the allowable shearing stress is 12,000 psi. By what angle does the gear $C$ lag the driving gear $B$? Assume a constant diameter from $B$ to $C$, $G = 10,000,000$ psi, $a = 20$ in., and $b = 30$ in.

**4-37.** An automobile engine develops a maximum of 96 hp at an engine speed of 3400 rpm, 90 hp of which reaches the rear axle at 240 rpm in low gear and at 725 rpm in high gear. The axle is made of high-strength heat-treated nickel steel having a proportional limit of 96,000 psi. Its minimum diameter is 1¼ in. Calculate the maximum shearing stresses in the shaft in low gear and in high gear.

**4-38.** A 30-kW electric motor drives a centrifugal pump through two gears whose ratio is 5:1. The motor shaft turns at 1750 rpm and the pump at 350 rpm. Shaft bending and gear friction are negligible. One gear is keyed to each shaft. Calculate the minimum required diameters of these shafts, using an allowable stress of 50 MPa.

Ans. Motor $D = 28.1$ mm; pump $D = 48.1$ mm

**4-39.** The engine of a ship delivers 5000 hp to the propeller shaft at 120 rpm. (*a*) Design a hollow shaft in which $d = \frac{2}{3}D$ and allowable shearing stress is 12,000 psi. (*b*) If the shaft is 40 ft long from engine to propeller, through what angle $\theta$ will it twist under full power? $G = 12,000,000$ psi.

Ans. (*a*) $D = 11.16$ in.; $d = 7.44$ in.; (*b*) $\theta = 4.93°$

**4-40.** A certain shaft must transmit 500 hp. The allowable shearing stress is 10,000 psi. Calculate the required diameter $D$ when the shaft speed $N$ equals, respectively, 125, 512, 1000, 1728, 3375, and 8000 rpm. Plot these values and draw a curve in which the vertical scale is the required diameter $D$ and the horizontal scale is speed $N$. [HINT: Use Eq. (4-8). The plot could best be made on log-log paper.]

**SUMMARY** (By Article Number) ─────────────────────────────

**4-1.** When a member is twisted in a plane perpendicular to its longitudinal axis, it is said to be in **torsion.** The **twisting** moment is called a **torque.** Shafts are the most common examples of torsion members.

**4-2. Torsional stresses in circular shafts** caused by a torque are shearing stresses induced on every transverse plane of the shaft. These shearing stresses increase uniformly from zero at the centroidal axis to a maximum at the surface of the shaft. If $T$ = torque, $s_s$ = maximum shearing stress, $J$ = polar moment of inertia, and $D$ = diameter of shaft, the relation of these quantities is

$$\textbf{Twisting moment or torque } T = \frac{2 s_s J}{D} \qquad \textbf{(4-1)}$$

The polar moment of inertia $J$ is

**For solid shaft:** $$J = \frac{\pi D^4}{32}$$

**For hollow shaft:** $$J = \frac{\pi(D^4 - d^4)}{32}$$

The diameter of a solid shaft is obtained from the equation

$$D = \sqrt[3]{\frac{16T}{\pi s_s}} \tag{4-3}$$

Hollow shafts are most easily designed by solving for the required $J$, using Eq. (4-1).

**4-4.** Twisting sections of a shaft may be coupled together end to end (1) by forging flanges on the shaft ends, which are then bolted together, or (2) by use of a two-piece coupling in which the two pieces are bolted together and one piece is keyed to each section of the shaft.

**4-5.** The **angle of twist** is a function of the torque $T$, the length $L$, the shear modulus of elasticity $G$, and the polar moment of inertia $J$. The relation is

$$\text{Angle of twist } \theta = \frac{TL}{GJ} \tag{4-5}$$

**4-6. Transmission of power** is the usual function of shafts. In terms of horsepower $H$ or kilowatts $kW$ and speed $N$ in rpm, the torque $T$ exerted on a shaft is

$$\text{Torque } T = \frac{63{,}025\,H}{N} = \frac{9549\,kW}{N} \tag{4-7) and (4-10}$$

When this value of $T$ is substituted in Eq. (4-3), the required

$$\text{Diameter of solid shaft } D = 68.5 \sqrt[3]{\frac{H}{s_s N}} = 365 \sqrt[3]{\frac{kW}{s_s N}} \tag{4-8) and (4-11}$$

Another convenient form of this equation is

$$H = \frac{s_s N D^3}{321{,}000} \tag{4-9}$$

**Keyways** weaken shafts. A shaft having one or more keyways is permitted to transmit only three-fourths as much torque (or horsepower) as an unkeyed shaft. Therefore, keyed shafts must be designed to withstand $^4/_3 T$ (or $^4/_3 \cdot$power).

## REVIEW PROBLEMS

**4-41.** A threaded bolt of ½-in. diameter is being screwed up tightly, fastening together two steel parts. To avoid failure of the bolt in shear, the unit shearing stress in it must not exceed the elastic limit of 36,000 psi. What maximum pull $P$ on a wrench with a 12-in. moment arm will cause this torsional stress (a) if the gross cross-sectional area resists the stress and (b) if the root area resists the stress?

*Ans.* (a) $P = 73.6$ lb; (b) $P = 37.6$ lb

**4-42.** Calculate the maximum permissible torque $T$ on a shaft if the maximum allowable shearing stress is 100 MPa and if (a) the shaft is solid and $D = 100$ mm and (b) the shaft is hollow, $D = 100$ mm, and $d = 75$ mm.

**4-43.** The elastic limit in shear $s_s$ of the steel in a shaft is determined by twisting the shaft in a torsion machine. This limit was reached when the angle of twist $\theta$

measured 14.8°. Calculate (*a*) the unit shearing stress at the elastic limit and (*b*) the modulus of elasticity *G*. The shaft is 48 in. long and of 1-in. diameter. Torque *T* = 6285 lb·in.

*Ans.* (*a*) $s_s$ = 32,000 psi; (*b*) *G* = 11,900,000 psi

**4-44.** What maximum horsepower may be transmitted by a 125-mm-diameter shaft at 250 rpm if maximum shearing stress is not to exceed 100 MPa and if (*a*) the shaft is not keyed and (*b*) the shaft is keyed?

**4-45.** The engine of an automobile develops a maximum of 100 hp at an engine speed of 3200 rpm. The velocity ratio in low gear between engine and rear axle is 13.68. Calculate the minimum required diameter *D* of axle if allowable stress in the heat-treated nickel steel is 70,000 psi.

**4-46.** A 3-in.-diameter shaft is to transmit 120 hp. (*a*) Calculate the minimum speed *N* (rpm) in order that the maximum shearing stress will not exceed 10,000 psi. What maximum stress will be induced in the shaft when the same horsepower is delivered to the shaft (*b*) at 120 rpm and (*c*) at 160 rpm?

*Ans.* (*a*) *N* = 142 rpm; (*b*) $s_s$ = 11,890 psi; (*c*) $s_s$ = 8930 psi

**4-47.** The engines of a large ship deliver 15,000 hp to each of four hollow propeller shafts at 150 rpm. (*a*) Design one shaft with an inside diameter *d* equal to ¾ of the outside diameter *D*. (*b*) If the shaft is 100 ft long, what will be the angle of twist *θ*? Allowable shearing stress is 10,000 psi. *G* = 12,000,000 psi.

*Ans.* (*a*) *D* = 16.8 in., *d* = 12.6 in.; (*b*) *θ* = 6.82°

**4-48.** Assume that the shaft in Prob. 4-47 is in two 50-ft sections connected by means of a flange coupling. (*a*) How many 1¾-in.-diameter turned bolts are required if they are equally spaced on a 22-in.-diameter bolt circle? (*b*) Calculate the minimum required thickness *t* of flange to prevent the shaft from shearing out of flange. Use $s_s$ = 15,000 psi for bolts and 10,000 psi for flanges.

**4-49.** Pulleys *B* and *D* are driven by a motor connected to the shaft *AE* by means of pulley *C*, as shown in Fig. 4-8. Pulley diameters *B*, *C*, and *D* are 6, 8, and 10 in., respectively. (*a*) Calculate the belt tension $T_3$. (*b*) Calculate the torque in section *CD*. (*c*) If the shaft *CD* is 2 in. in diameter, calculate the maximum shearing stress in this section.

**4-50.** Gear *C*, shown in Fig. 4-9, is driven by an electric motor. Fifteen kW is taken off the 50-mm-diameter steel shaft at *B* and 5 kW is taken off at *A*. Calculate the maximum shearing stress in the shaft and the angle by which the gear *C* leads gear *B*. The shaft rotates at 120 rpm, *a* = 200 mm, and *b* = 300 mm. Assume *G* = 84 000 MPa.

## REVIEW QUESTIONS

**4-1.** What is meant by a torque? In what manner is it usually applied to a member with respect to its longitudinal axis?

**4-2.** What type of stress is induced in a member subjected to a torque?

**4-3.** Describe the variation of deformations and torsional stresses in a circular shaft subjected to a torque.

**4-4.** Give the expression for the polar moment of inertia (*a*) of a solid circular shaft and (*b*) of a hollow circular shaft.

**4-5.** In the design of hollow shafts, is it necessary to predetermine the ratio of *d* to *D*?

**4-6.** What methods are generally used to connect two sections of a shaft end to end?

**4-7.** What determines the minimum thickness *t* of a flange forged at the end of a shaft?

**4-8.** What is meant by the term angle of twist? In what units must the angle of twist be expressed?

**4-9.** How may the shear modulus of elasticity of the steel in a circular shaft be determined?

**4-10.** If 100 hp is to be transmitted by each of two separate shafts, one turning at 200 rpm and the other at 300 rpm, which shaft must have the greater diameter?

**4-11.** How does a keyway affect the strength of a shaft? What is the permissible torque on a keyed shaft compared with that on an unkeyed shaft?

**4-12.** In the design of a keyed shaft, by what fraction must the given torque be increased?

# Chapter 5
# Centroids and
# Moments of
# Inertia of Areas

## 5-1 CENTROIDS

In the analysis and design of beams and columns, covered in Chaps. 7, 8, and 12, we must (a) locate the centroids of their cross-sectional areas, (b) calculate the first, or static, moment of such areas with respect to some axis, and (c) calculate the second moment, or moment of inertia, of such areas with respect to some axis. This necessity accounts for the insertion here of this chapter containing material which may, at first glance, appear to the student to be unrelated to the subject matter of this text.

The concept of the centroid of an area is most easily understood if we consider first the concept of center of gravity of a thin plate of uniform thickness and of homogeneous material. To determine *mathematically* the center of gravity of a flat plate of irregular shape but of uniform thickness and of homogeneous material, the plate may be divided into a number of small elements of equal size, as shown in Fig. 5-1a. If the weight of each element is $w$, concentrated at its center, a parallel force system is formed whose resultant $W$, the weight of the entire plate, passes through the center of gravity of the plate.

Let the weights of these elements be designated $w_1$, $w_2$, and so forth; let their coordinates be $x_1$ and $y_1$, $x_2$ and $y_2$, and so forth; and let the coordinates of the resultant weight $W$ be $\bar{x}$ and $\bar{y}$ (bar $x$ and bar $y$). Now, about any point, the moment of the resultant $W$ equals the algebraic sum of the moments of the separate weights. Consequently, by moments about $O$ in the $XZ$ plane,

$$W\bar{x} = w_1x_1 + w_2x_2 + \cdots = \Sigma wx \qquad (a)$$

and by moments about $O$ in the $YZ$ plane,

$$W\bar{y} = w_1y_1 + w_2y_2 + \cdots = \Sigma wy \qquad (b)$$

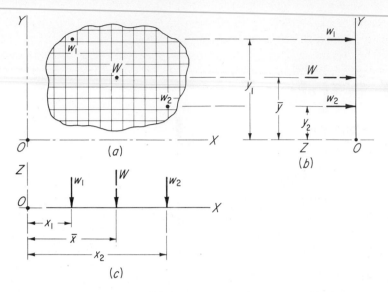

**Fig. 5-1**  Center of gravity of homogenous flat plate. (a) Plate divided into equal-size elements. (b) Forces projected into $YZ$ plane. (c) Forces projected into $XZ$ plane.

or

$$\bar{x} = \frac{\Sigma wx}{W} \quad \text{and} \quad \bar{y} = \frac{\Sigma wy}{W} \tag{5-1}$$

thus locating the center of gravity.

Let us imagine now that the flat plate in Fig. 5-1 gradually reduces in thickness. When its thickness finally becomes zero, it no longer has weight and only a surface or an area remains. The point in the plate which formerly was the center of gravity of the body is now the **centroid** of the area. If $a$ is the surface area of one small element in Fig. 5-1 and $t$ is the thickness of the plate, its volume is $ta$ and its weight $w$ is $\delta ta$, where $\delta$ is the density. If $\delta ta$ is substituted for $w$ in Eq. (a) above and $A$ is the total surface area, $W = \delta tA$, and

$$\delta tA\bar{x} = \delta ta_1 x_1 + \delta ta_2 x_2 + \cdots = \delta t\Sigma ax \tag{c}$$

When the constants $\delta$ and $t$ are canceled from all terms of this equation, we find that $A\bar{x} = \Sigma ax$. Similarly from Eq. (b) we would find that $A\bar{y} = \Sigma ay$. Consequently Eq. (5-1) would become

$$\bar{x} = \frac{\Sigma ax}{A} \quad \text{and} \quad \bar{y} = \frac{\Sigma ay}{A} \tag{5-2}$$

Apparently, then, *about any axis* **the moment of an area equals the algebraic sum of the moments of its component areas,** wherein *the moment of an area is defined as the product of the area multiplied by the perpendic-*

ular distance from the moment axis to the centroid of the area. By means of this principle, we may locate the centroid of any area.

The terms *centroid* and *center of gravity* are often used inter-changeably. Also, a **gravity axis** passing through the center of gravity of a body is often called a **centroidal axis.**

## 5-2 CENTROIDS OF SIMPLE GEOMETRIC AREAS

By geometry and by calculus (see Appendix A), the centroids of some simple geometric areas are found to be located as illustrated in Fig. 5-2.

The centroid of a rectangle or of any parallelogram lies at the intersection $C$ of its two diagonals. This point is also the intersection of the two lines bisecting the two pairs of opposite sides. The truth of this statement can be deduced as follows: Let the rectangle be divided into narrow strips running parallel to the two opposite sides. The centroid of each strip lies at its midpoint, the centroids of all the parallel strips will lie on a line bisecting the two opposite sides, and the centroid $C$ of the entire area will lie at the intersection of the two bisecting lines, which is at their midpoints.

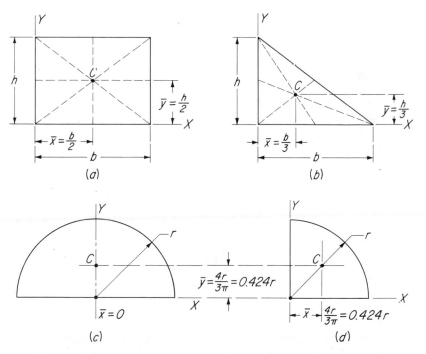

**Fig. 5-2**  Centroids of simple geometric areas. (*a*) Rectangle. (*b*) Triangle. (*c*) Semicircle. (*d*) Quarter circle.

The centroid of a triangle lies at the intersection of its medians. If the triangle is divided into infinitesimally narrow strips running parallel to any base, the centroid of each strip is at its midpoint, the centroids of all the parallel strips will lie on the median of that base, and the centroid $C$ of the entire area lies at the intersection of the three medians, at a distance of one-third the length of the median from its base.

The centroid of a semicircle is most easily determined by integration, as is shown in Prob. A-3, Appendix A. By symmetry, the centroids of a quarter circle and of a semicircle, shown in Fig. 5-2, are seen to lie at equal distances from the $X$ and $Y$ reference axes.

## 5-3  CENTROIDS OF COMPOSITE AREAS

A composite area is one made up of a number of simple areas. To determine its centroid, a composite area is generally divided into two or more simple component areas. The centroid of the composite area is then found by the previously stated principle: about any axis, *the moment of an area equals the algebraic sum of the moments of its component areas.* The following examples illustrate the application of this principle.

ILLUSTRATIVE PROBLEMS

**5-1.** Locate the centroid of the composite area shown in Fig. 5-3 with respect to the $X$ and $Y$ axes.

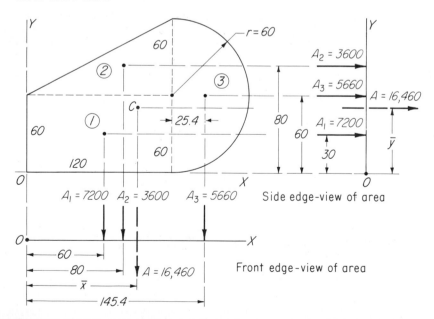

**Fig. 5-3**  Location of centroid of composite area.

*Solution:* Let the X and Y axes intersect at the lower left corner of the area. Then all moments are positive. This composite area divides naturally into three simple geometric areas, a rectangle, a triangle, and a semicircle, whose centroids we obtain from Fig. 5-2. To aid in computing $\bar{x}$ and $\bar{y}$, we represent these areas as forces concentrated at their respective centroids and shown in two edge views of the area similar to a parallel force system in space. The total area is their resultant, which passes through the centroid C of the composite area. From the given dimensions we find that

$$A_1 = (120)(60) = 7200 \qquad A_2 = (\tfrac{1}{2})(120)(60) = 3600$$
$$A_3 = (\tfrac{1}{2})(3.1416)(60)^2 = 5660$$

Hence $\qquad A = 7200 + 3600 + 5660 = 16\ 460\ \text{mm}^2$

Then, in the front view, to find $\bar{x}$,

$[M_o] \qquad\qquad 16\ 460\bar{x} = (7200)(60) + (3600)(80) + (5660)(14)$

or

$16\ 460\bar{x} = 432\ 000 + 288\ 000 + 823\ 000 = 1\ 543\ 000 \qquad$ and $\qquad \bar{x} = 93.7\ \text{mm}$
$$Ans.$$

Likewise, in the side view, to find $\bar{y}$,

$[M_o] \qquad\qquad 16\ 460\bar{y} = (7200)(30) + (5660)(60) + (3600)(80)$

or

$16\ 460\bar{y} = 216\ 000 + 340\ 000 + 288\ 000 = 844\ 000 \qquad$ and $\qquad \bar{y} = 51.3\ \text{mm}$
$$Ans.$$

**5-2.** Figure 5-4 is a cross-sectional area of a composite steel beam made up of an American Standard I beam, 12 by 5¼ in. by 50 lb/ft, with a ½- by 8-in. cover plate riveted to its top flange. Locate the centroid of the composite area.

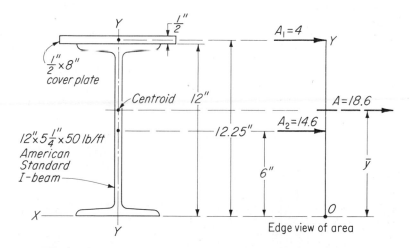

Fig. 5-4   Location of centroid of cross-sectional area of beam.

*Solution:* This area is symmetrical about the *Y* axis. Hence we need determine only the distance $\bar{y}$. The cross-sectional area $A_1$ of the plate is $(\frac{1}{2})(8)$, or 4 in.² From Table E-8 we find the cross-sectional area $A_2$ of the I beam to be 14.6 in.² The total area *A* is then $14.6 + 4$, or 18.6 in.² Then, by moments about *O*, we have

$$[M_o] \qquad 18.6\bar{y} = (4)(12.25) + (14.6)(6) = 136.6 \qquad \text{and} \qquad \bar{y} = 7.35 \text{ in.} \qquad Ans.$$

## PROBLEMS

**5-3.** Locate the centroid with respect to the *X* and *Y* axes of the area shown in Fig. 5-5. (HINT: The moment of an area with a part removed equals the moment of that area *minus* the moment of the part removed. Hence represent the part removed by an arrow applied at its centroid and oppositely directed.)

*Ans.* $\bar{x} = 5.91$ in.; $\bar{y} = 2.85$ in.

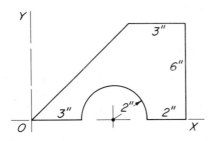

Fig. 5-5   Prob. 5-3.

**5-4.** Determine the location of the centroid of the area shown in Fig. 5-6 (see hint in Prob. 5-3).

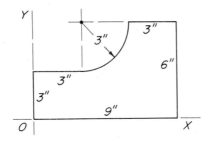

Fig. 5-6   Prob. 5-4.

**5-5.** The irregular body shown in Fig. 5-7 is placed upon two knife-edge sup-

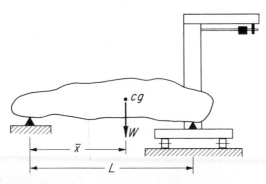

Fig. 5-7   Prob. 5-5.

ports, one resting on a scale that indicates a load of 288 lb. If the weight $W$ of the body is 360 lb and if $L$ is 4 ft, compute the distance $\bar{x}$ from the left support to the vertical plane through the body containing its center of gravity.

**5-6.** The structural-steel angle shown in Fig. 5-8 measures 200 by 150 mm and is 25 mm thick. Compute the distances $\bar{x}$ and $\bar{y}$ from the heel of the angle to its centroid.

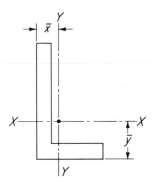

Fig. 5-8    Prob. 5-6.

**5-7.** The timber beam shown in cross section in Fig. 5-9 is made of two planks 20 by 80 mm and one plank 20 × 100 mm, all securely nailed together. Compute the distance $\bar{y}$ from the bottom of the beam to its centroid. $YY$ is an axis of symmetry.

*Ans.* $\bar{y}$ = 59.2 mm

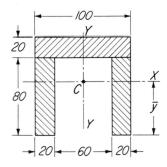

Fig. 5-9    Probs. 5-7 and 5-24.

**5-8.** The cast-iron beam shown in the cross section in Fig. 5-10 is symmetrical

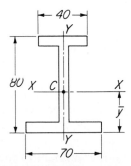

Fig. 5-10    Probs. 5-8 and 5-25.

about the $YY$ axis. The two flanges and the web are 10 mm thick. Compute the distance $\bar{y}$ from the bottom of the beam to its centroid.

**5-9.** Figure 5-11 is a cross section of a steel beam built up of two American Standard channels 10 in. by 20 lb/ft, to whose top flanges are riveted a ½- by 12-in. plate. The area of each channel is 5.86 in.². $YY$ is an axis of symmetry. Compute the distance $\bar{y}$ from the bottom of the beam to its centroid.

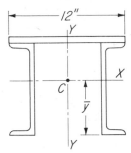

Fig. 5-11    Probs. 5-9 and 5-26.

## 5-4 MOMENTS OF INERTIA OF AREAS

To determine certain *shearing stresses* in beams, a quantity must be determined involving the *first moment of an area Q*, which, with respect to a reference axis, is the product of the area and the perpendicular distance from the axis to the centroid of the area. That is, $Q = A\bar{y}$.

Now, to determine certain *bending stresses* in beams, and also in the design of columns, a quantity must be determined which involves the *second moment of an area,* commonly called the **moment of inertia** and denoted by the symbol $I$. With respect to a reference axis in the plane of the area, **the second moment of an area, or its moment of inertia, is the sum of the products of the elemental areas, each multiplied by the square of the distance from the reference axis to its centroid.** The moment of inertia is therefore the product of the entire area multiplied by the mean value of the squares of the distances from the reference axis to the centroids of the elemental areas. When the number of areas is finite, $I = \Sigma ay^2$, by which equation the *approximate moment of inertia* may be obtained, as outlined in Art. 5-5. When each small area is infinitesimal, or $dA$, and the number of areas is infinite, $I = \int y^2\, dA$, by which equation the *exact moment of inertia* may be obtained, as shown in Appendix B. The exact moment of inertia of a composite area is obtained as illustrated in Art. 5-7.

In practical engineering language, the moment of inertia of the cross-sectional area of a beam or a column of a given material with respect to its neutral axis may be said to represent the relative capacity of the section to resist bending or buckling in a direction perpendicular to the neutral axis.

When bending stresses are involved, the moment of inertia required is with respect to the neutral axis of the cross-sectional area and is called the **rectangular moment of inertia.** In shafts that transmit power and are therefore subjected to twisting or torsional stresses, the moment of inertia requircd is with respect to an axis perpendicular to the plane of the cross-sectional area; it is then called the **polar moment of inertia** of the area. See Appendix B.

**Units ot Moment of Inertia.** Moment of inertia is the product of an area and the square of a distance. If the area is in square inches and the distance in inches, the product will be in *inches to the fourth power*. That is, $(in.^2)(in.^2) = in.^4$. Similarly, $(mm^2)(mm^2) = mm^4$. These units have no physical conception.

## 5-5 APPROXIMATE DETERMINATION OF MOMENT OF INERTIA

The moment of inertia of an area may be obtained approximately by dividing the total area $A$ into a finite number of smaller areas $a$, then multiplying each small area $a$ by the square of the distance $(y^2)$ from the reference axis to its centroid. The moment of inertia is then the sum of these products. Consequently

$$\text{Approximate moment of inertia } I = \Sigma ay^2 \qquad (5\text{-}3)$$

The approximate method is used chiefly with irregular areas. The following example illustrates the method. A rectangular area was chosen here in order that its approximate and exact moments of inertia might be compared and thus show the probable error of the approximate method.

ILLUSTRATIVE PROBLEM

**5-10.** Determine the approximate moment of inertia of the area shown in Fig. 5-12 with respect to axis $XX$ passing through its centroid.

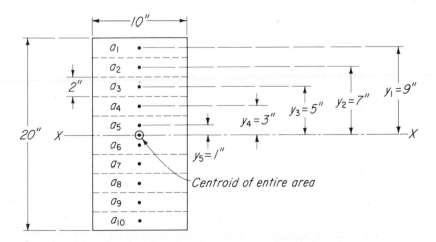

**Fig. 5-12**  Moment of inertia of an area. Approximate method.

*Solution:* Let the area arbitrarily be divided into 10 strips of equal width running parallel with the *XX* axis, each area then containing (2)(10), or 20 in.$^2$. We find that of the distances to the respective centroids, $y_1 = 9$ in., $y_2 = 7$ in., $y_3 = 5$ in., $y_4 = 3$ in., and $y_5 = 1$ in. With respect to axis *XX*, the moment of inertia of the lower half clearly equals that of the upper half; therefore we determine only the latter and then multiply it by 2 to obtain *I* of the entire area. That is,

$$[I = \Sigma ay^2] \qquad I_X = (2)(a_1 y_1^2 + a_2 y_2^2 + a_3 y_3^2 + a_4 y_4^2 + a_5 y_5^2)$$

But each of the areas $a_1$, $a_2$, and so forth, equals 20 in.$^2$. Hence

$$I_X = (2)(20)(81 + 49 + 25 + 9 + 1) \qquad \text{or} \qquad I_X = 6600 \text{ in.}^4 \qquad Ans.$$

The exact moment of inertia of a rectangular area is $^1/_{12}bh^3$, which for the area in this example is 6667 in.$^4$, thus indicating an error in the approximate result of only 1 percent. See Appendix B.

## 5-6 MOMENTS OF INERTIA OF SIMPLE AREAS

The moments of inertia of many simple geometric areas are readily determined by calculus, as is illustrated in Appendix B. These areas and their moments of inertia are generally expressed in terms of simple dimensions which completely define them.

The moments of inertia and the radii of gyration of a number of simple areas are given in Table 5-1. These and many others are usually found in handbooks.

The moments of inertia of other simple areas *from which some parts have been removed,* such as those shown in Fig. 5-13, are easily obtained from the following principle:

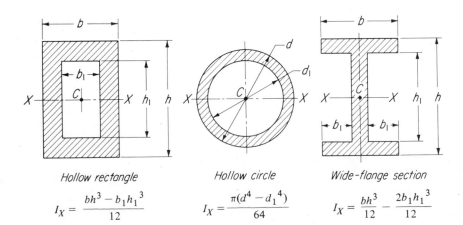

Hollow rectangle
$$I_X = \frac{bh^3 - b_1 h_1^3}{12}$$

Hollow circle
$$I_X = \frac{\pi(d^4 - d_1^4)}{64}$$

Wide-flange section
$$I_X = \frac{bh^3}{12} - \frac{2b_1 h_1^3}{12}$$

**Fig. 5-13**  Special equations for obtaining moments of inertia.

**With respect to any axis, the moment of inertia of an area with a part removed equals the moment of inertia of the area minus the moment of inertia of the part removed.**

When the axes for the area and for the part removed coincide and both are centroidal, as in Fig. 5 13, special simple equations giving the moment of inertia may be written as shown above. Otherwise the general method outlined in the following article must be used.

Table 5-1 MOMENTS OF INERTIA AND RADII
OF GYRATION OF SIMPLE AREAS

| Area | Moment of Inertia | Radius of Gyration |
|---|---|---|
| Rectangle | $I_{Xc} = \dfrac{bh^3}{12}$ <br><br> $I_X = \dfrac{bh^3}{3}$ | $k_{Xc} = \dfrac{h}{\sqrt{12}} = 0.289h$ <br><br> $k_X = \dfrac{h}{\sqrt{3}} = 0.577h$ |
| Triangle | $I_{Xc} = \dfrac{bh^3}{36}$ <br><br> $I_X = \dfrac{bh^3}{12}$ | $k_{Xc} = \dfrac{h}{\sqrt{18}} = 0.236h$ <br><br> $k_X = \dfrac{h}{\sqrt{6}} = 0.408h$ |
| Circle | $I_{Xc} = \dfrac{\pi R^4}{4}$ <br><br> $= 0.7854R^4$ <br><br> $= \dfrac{\pi D^4}{64}$ <br><br> $= 0.0491D^4$ | $k_{Xc} = \dfrac{R}{2} = 0.5R$ |
| Semicircle | $I_{Xc} = \left(\dfrac{\pi}{8} - \dfrac{8}{9\pi}\right) R^4$ <br><br> $= 0.110R^4$ <br><br> $I_X = I_{Yc} = \dfrac{\pi R^4}{8}$ <br><br> $= 0.3927R^4$ | $k_{Xc} = \dfrac{\sqrt{9\pi^2 - 64}}{6\pi} R$ <br><br> $= 0.264R$ <br><br> $k_X = k_{Yc} = \dfrac{R}{2} = 0.5R$ |

## 5-7 MOMENTS OF INERTIA OF COMPOSITE AREAS

In modern structural practice, and especially in aircraft construction, relatively few members used to resist bending are of simple geometric cross-sectional area such as rectangular or circular. Many sections, however, are so shaped that they are readily divided into two or more simple **component areas.** The entire area is then called a **composite area.** The moment of inertia of each component area with respect to a centroidal axis of the composite area is then determined by the following principle, illustrated in Fig. 5-14 and more fully explained in Appendix B:

**The moment of inertia of an area with respect to any axis not through its centroid is equal to the moment of inertia of that area with respect to its own parallel centroidal axis plus the product of the area and the square of the distance between the two axes.**

When this principle is stated in symbols, we have

$$I_X = I_C + Ad^2 \qquad (5\text{-}4)$$

where $I_X$ = moment of inertia of an area with respect to any noncentroidal axis, in.$^4$ or mm$^4$

$I_C$ = moment of inertia of the area with respect to its parallel centroidal axis, in.$^4$ or mm$^4$

$A$ = area, in.$^2$ or mm$^2$

$d$ = perpendicular distance between the parallel axes, in. or mm

Equation (5-4) is called the **transfer formula,** since it "transfers" the moment of inertia of an area from its centroidal axis to any other parallel axis. *The moment of inertia of a composite area* is then the sum of the moments of inertia of the component areas, all with respect to a common axis. Close study of this equation reveals that the moment of inertia of an area is always least with respect to a centroidal axis. Problems 5-11 and 5-12 will illustrate the method.

## 5-8 RADIUS OF GYRATION

In the design of columns a term called the **radius of gyration** and denoted by the symbol $k$ is very useful.

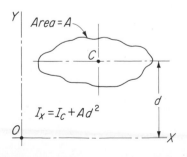

Fig. 5-14   Moment of inertia of an area with respect to a noncentroidal axis.

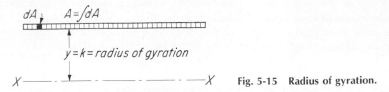

**Fig. 5-15   Radius of gyration.**

The radius of gyration of an area is that distance from its moment-of-inertia axis at which the entire area could be considered as being concentrated without changing its moment of inertia.

That is, if all the elemental areas $dA$ were placed side by side in a strip parallel to the moment-of-inertia axis, and if the thickness of this strip were infinitesimal, the moment of inertia $I_C$ of each elemental area $dA$ would become zero, and that term would therefore drop out of Eq. (5-4). The moment of inertia $I$ of each small area then would be $y^2\,dA$ and of the entire area $I$ would be the sum of these, or $\int y^2\,dA$. But, since $y$ now has the same value for all of the small areas, $y^2 = k^2$, and $\int dA$ (the sum of all the small areas) $= A$. (See Fig. 5-15.) Therefore

$$I_X = Ak^2 \quad \text{or} \quad k = \sqrt{\frac{I}{A}} \tag{5-5}$$

In many instances, the moments of inertia of the cross-sectional area of a column with respect to its two rectangular $X$ and $Y$ axes are not equal. Therefore, one of its two major radii of gyration will be smaller than the other. *A compression member tends to buckle in the direction of its least radius of gyration.*

## ILLUSTRATIVE PROBLEMS

**5-11.** Locate the centroid of the composite T-beam section shown in Fig. 5-16 and determine its moment of inertia $I_X$ with respect to the axis $XX$ and its radius of gyration $k_X$.

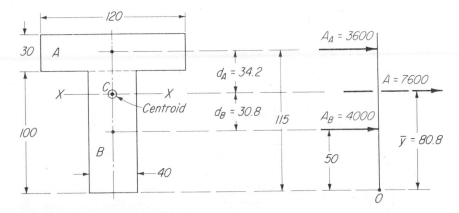

**Fig. 5-16   Centroid and moment of inertia of a composite area.**

*Solution:* The composite area is divided into two rectangular areas *A* and *B*, and the centroid of each area is located by inspection, as in Fig. 5-16. Next we draw the edge view of the entire area and apply the separate areas, represented by arrows, at their respective centroids. Then by moments about *O*

$$7600\bar{y} = (3600)(115) + (4000)(50) = 614\ 000 \quad \text{and} \quad \bar{y} = 80.8 \text{ mm } Ans.$$

Having thus located the centroid of the composite area, we then determine distances $d_A$ and $d_B$ from axis $XX$ to the centroids of areas *A* and *B*; that is, $d_A = 115 - 80.8 = 34.2$ mm, and $d_B = 80.8 - 50 = 30.8$ mm. Now, the moment of inertia of each of areas *A* and *B*, with respect to axis $XX$, is given by Eq. (5-4). That is, $I_X = I_C + Ad^2$. The total moment of inertia of the composite area is then $I_X = \Sigma(I_C + Ad^2)$. Since $A_A = 3600$ mm², $A_B = 4000$ mm², $d_A = 34.2$ mm, and $d_B = 30.8$ mm, we have

$$I_X = \Sigma[I_C + Ad^2] \quad \text{in which} \quad I_C = \frac{bh^3}{12}$$

Area *A*: $\quad I_X = \dfrac{(120)(30^3)}{12} + (3600)(34.2^2) = 4\ 480\ 000 \text{ mm}^4$

Area *B*: $\quad I_X = \dfrac{(40)(100^3)}{12} + (4000)(30.8)^2 = \underline{7\ 130\ 000 \text{ mm}^4}$

Moment of inertia of the composite area = 11 610 000 mm⁴        *Ans.*

From Eq. (5-5), the radius of gyration with respect to axis $XX$ is

$$\left[ k_X = \sqrt{\frac{I_X}{A}} \right] \qquad k_X = \sqrt{\frac{11\ 610\ 000}{7600}} = 39.1 \text{ mm}$$

                                                                        *Ans.*

That is, if the entire area of 7600 mm² of this T section were placed in an exceedingly thin strip parallel to axis $XX$ and at a distance *k* of 39.1 mm from it, the moment of inertia would be $Ak^2$, of $(7600)(39.1)^2 = 11\ 610\ 000$ mm⁴.

**5-12.** Figure 5-17 is a cross-sectional area of a steel beam, built up of one web plate 1 by 14.5 in., four angles 5 by 5 by ¾ in., and two cover plates 1 by 12 in. Compute the moment of inertia of this composite section with respect to axis $XX$ passing through its centroid.

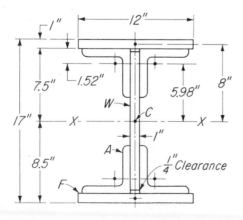

Fig. 5-17   Moment of inertia of a composite area.

*Solution:* This section is symmetrical about the $X$ axis, which, then, is located 8.5 in. above the bottom. From Table E-10 we find that the moment of inertia $I_C$ of each of the four angles about its own centroidal $X$ axis is 15.7 in.⁴, that the area $A$ of each angle is 6.94 in.², and that the distance $y$ from the back of the angle to its centroidal axis is 1.52 in. Considering separately the two flanges $F$, the four angles $A$, and the web $W$, we have

$$I_X = \Sigma[I_C + Ad^2]$$

Two flanges:    $I_X = 2\left[\dfrac{(12)(1^3)}{12} + (12)(8^2)\right] = 2(1 + 768) \quad = 1538 \text{ in.}^4$

Four angles:    $I_X = 4[15.7 + (6.94)(5.98^2)] = 4(15.7 + 248.2) = 1055 \text{ in.}^4$

One web plate:    $I_X = \dfrac{(1)(14.5^3)}{12} = \dfrac{(1)(3048)}{12} \quad\quad = \underline{254 \text{ in.}^4}$

Moment of inertia of the composite area = 2847 in.⁴

PROBLEMS_____

**5-13.** Figure 5-18 is a cross-sectional area of a beam which is symmetrical about its centroidal $X$ axis. Using the transfer formula, compute its moment of inertia $I_X$ and check it by the special equation given in Art. 5-6.

*Ans.* $I_X = 26.56 \times 10^6$ mm⁴

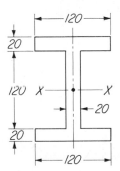

Fig. 5-18    Probs. 5-13 and 7-57.

**5-14.** Locate the centroidal $XX$ axis and compute the moment of inertia $I_X$ of the angle shown in cross section in Fig. 5-19. (Check answers against values given in Table E-6.)

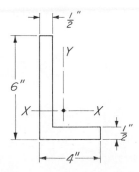

Fig. 5-19    Prob. 5-14.

**5-15.** Determine the location of the centroidal $XX$ axis of the area shown in Fig. 5-20 and compute the moment of inertia $I_x$ of the area.

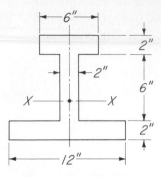

Fig. 5-20   Probs. 5-15 and 7-7.

**5-16.** The timber box beam in Fig. 5-21 is made of four 2- by 10-in. planks firmly nailed together to act as a single unit. Compute the moment of inertia $I_x$ of this section with respect to the centroidal $XX$ axis. Check it by the special equation given in Art. 5-6.

*Ans.* $I_x = 1787$ in.$^4$

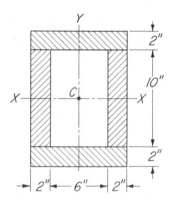

Fig. 5-21   Probs. 5-16, 5-17, and 7-5.

**5-17.** The structural member whose cross-sectional area is shown in Fig. 5-21 is to be used as a column. This column will buckle in the direction of its *least* radius of gyration $k$. Determine the least $k$ for this section. (HINT: Obtain $I_x$ from Prob. 5-16. Then compute $I_Y$, $k_X$, and $k_Y$.)

**5-18.** Figure 5-22 is a cross-sectional area of a steel-pipe column whose outside

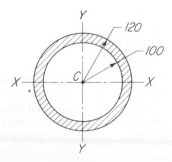

Fig. 5-22   Prob. 5-18.

and inside diameters are 240 and 200 mm, respectively. Using the special equation given in Art. 5-6, compute the moment of inertia and the radius of gyration $k$ of this section with respect to a centroidal axis.

**5-19.** Two sides of the solid wood strut whose cross-sectional area is shown in Fig. 5-23 are of semicylindrical shape. Let $h = 10$ in. and $R = 2$ in. Compute the *least* moment of inertia and radius of gyration of this strut.

$$Ans.\ I_Y = 44.6\ in.^4;\ k_Y = 1.1\ in.$$

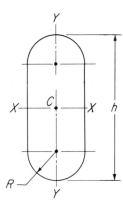

**Fig. 5-23    Probs. 5-19 and 5-39.**

**5-20.** The I beam in Fig. 5-24 is 500 mm deep and its cross-sectional area is 12 500 mm². The moment of inertia with respect to its centroidal $X$ axis is $480 \times 10^6$ mm⁴. A 300- by 20-mm cover plate is riveted to its top flange. Locate the centroidal $X$ axis of the composite area and compute the moment of inertia $I_X$.

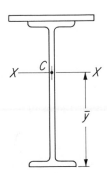

**Fig. 5-24    Prob. 5-20.**

**5-21.** The box girder whose cross-sectional area is shown in Fig. 5-25 is built up of four angles 4 by 4 by ½ in., two web plates 23.5 by ½ in., and two cover plates 25 by ½ in., assembled to provide a section 25 in. deep. The centroidal moment of inertia of each angle is 5.60 in.⁴, its cross-sectional area is 3.75 in.², and the distance $\bar{y}$ from the back of the angle to its own centroid is 1.18 in. Compute the moment of inertia $I_X$ of the entire section.

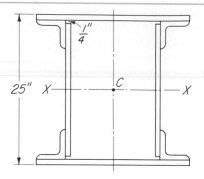

**Fig. 5-25   Prob. 5-21.**

**5-22.** A 2-in.-diameter hole has been bored in the upper half of the 3- by 16-in. timber joist shown in Fig. 5-26. Locate the centroidal $X$ axis. By what percentage has the moment of inertia of the solid joist been weakened by this hole?

> *Ans.* 16.9 percent ($\bar{y} = 7.29$ in.; $I_X = 851$ in.⁴)

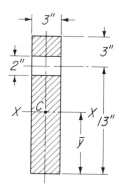

**Fig. 5-26   Prob. 5-22.**

**5-23.** By what percentage will the solid joist in Prob. 5-22 be weakened if the hole is bored through its middle? Compare with answer to Prob. 5-22.

**5-24.** Calculate the moment of inertia of the beam section shown in Fig. 5-9 with respect to both axes $XX$ and $YY$. Calculate also $k_X$ and $k_Y$ ($\bar{y} = 59.2$ mm).

**5-25.** The section shown in Fig. 5-10 has flange and web thicknesses of 10 mm, and $\bar{y}$ is 33.8 mm. Compute the moment of inertia $I_X$ and the radius of gyration $k_X$.

**5-26.** Calculate the moment of inertia $I_X$ and the radius of gyration $k_X$ of the composite beam section shown in Fig. 5-11 ($\bar{y} = 6.78$ in.).

> *Ans.* $I_X = 266.5$ in.⁴; $k_X = 3.88$ in.

**SUMMARY** (By Article Number) _____

**5-1.** The gravitational pull of the earth on a body is its **weight,** also called the **force of gravity.** The point in the body through which this gravity force acts is called its **center of gravity.** The **centroid of an area** occupies the position of the center of gravity of a homogeneous thin plate whose thickness approaches zero.

**5-3.** A **composite area** is one made up of a number of simpler areas. The *centroid of a composite area* is located by use of the following principle: about any axis, *the moment of an area equals the algebraic sum of the moments of its component areas.*

**5-4.** The quantity **moment of inertia,** also called **second moment of area,** is used in the analysis and design of beams and columns. It represents the capacity of a beam or a column to resist bending or buckling. The general expression for the moment of inertia of an area is $I = \int y^2 \, dA$.

**5-5.** To obtain its moment of inertia, an irregular area is divided into narrow strips *running parallel to the reference axis*. The **approximate moment of inertia** is then found by the equation

$$I = \Sigma ay^2 \qquad\qquad (5\text{-}3)$$

in which the area $a$ of each strip is multiplied by the square of the distance from its centroid to the reference axis.

**5-6.** The **exact moment of inertia of a simple geometric area** such as a rectangle, a triangle, and a circle is obtained by integration as described in Appendix B.

**5-7.** The moment of inertia of an area with respect to a noncentroidal axis is given by the following equation, called the **transfer formula:**

$$I_X = I_C + AD^2 \qquad\qquad (5\text{-}4)$$

The **moment of inertia of a composite area** is the sum of the moments of inertia of the component areas, all with respect to a common axis.

**5-8.** The **radius of gyration** $k$ of an area is given by the equation

$$k = \sqrt{\frac{I}{A}} \qquad\qquad (5\text{-}5)$$

This quantity is useful in the design of all compression members, since such members tend to buckle in the direction of their *least radius of gyration*.

### REVIEW PROBLEMS

**5-27.** The timber T beam shown in cross section in Fig. 5-27 is made of two planks 2 by 6 in. securely spiked together. Compute the distance $\bar{y}$ to its centroid $C$.

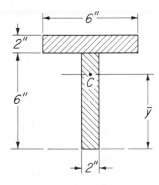

**Fig. 5-27   Prob. 5-27.**

**5-28.** The surface shown in Fig. 5-28 is symmetrical about a vertical centerline through the circular hole whose area is 10 000 mm². Compute the distance $\bar{y}$ to its centroid $C$.

*Ans.* $\bar{y}$ = 225 mm

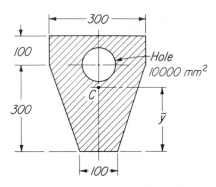

Fig. 5-28   Prob. 5-28.

**5-29.** The cross section of a steel beam built up of a wide-flange section with a channel section riveted to its top flange is shown in Fig. 5-29. The actual depth of the wide-flange section is 18 in., and its area is 14.71 in.². The area of the channel is 6.03 in.², the thickness of its web is 0.28 in., and the distance from the back to its center of gravity is 0.70 in. Compute $\bar{y}$ to the centroid $C$ of the composite section.

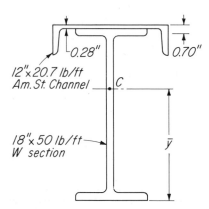

Fig. 5-29   Prob. 5-29.

**5-30.** Figure 5-30 is a cross section of a steel beam, built up of one web plate ¾ by 18 in., one cover plate 1 by 12 in., and four equal-leg angles 5 by 5 by ¾ in., assembled as shown. The distance from the heel of each angle to its centroid is 1.52 in. Compute $\bar{y}$ of the composite section.

**5-31.** In Fig. 5-4, the I beam is 12 in. deep, its cross-sectional area is 14.57 in.², and its moment of inertia with respect to its own centroidal $X$ axis is 301.6 in.⁴. Compute the moment of inertia $I_X$ of the beam with cover plate with respect to the centroidal $X$ axis of the composite section.

*Ans.* $I_X$ = 424 in.⁴

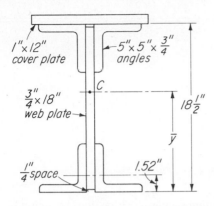

$1''\times 12''$
cover plate

$5''\times 5''\times \frac{3}{4}$
angles

$\frac{3}{4}\times 18''$
web plate

$C$

$18\frac{1}{2}''$

$\bar{y}$

$\frac{1}{4}''$space

$1.52''$

**Fig. 5-30   Probs. 5-30 and 5-35.**

**5-32.** The beam section shown in Fig. 5-31 is built up of a 24-in. wide-flange section with a 15-in. channel riveted to its top flange. The area of the wide-flange section is 23.54 in.², and its centroidal moment of inertia is 2230 in.⁴. The area of the channel is 9.9 in.², the thickness $t$ of its web is 0.4 in., the distance $y$ from its back to its horizontal centroidal axis is 0.79 in., and its moment of inertia about the axis is 8.2 in.⁴. Locate the centroidal $XX$ axis of the composite area, and compute the moment of inertia $I_X$.

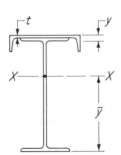

$t$

$y$

$X$————$X$

$\bar{y}$

**Fig. 5-31   Probs. 5-32 and 5-33.**

**5-33.** Refer to Fig. 5-31. The 600-mm-deep wide-flange section has a centroidal moment of inertia of $1000 \times 10^6$ mm⁴ and a cross-sectional area of 16 000 mm². The channel has a centroidal moment of inertia of $4 \times 10^6$ mm⁴ and a cross-sectional area of 9000 mm² with a centroid that is $y = 30$ mm below the top of the web. The web thickness $t = 30$ mm. Compute $I_X$.

**5-34.** Figure 5-32 is a cross-sectional area of a latticed column in which two 15-in. channels are held together by lattice bars riveted to their flanges. The area of each channel is 14.64 in.², the distance $x$ from the back to its centroidal $Y$ axis is 0.8 in., and its moment of inertia about that axis is 11.2 in.⁴. The moment of inertia with respect to the $X$ axis of each channel is 401.4 in.⁴. Compute the distance $d$ required to make $I_X$ and $I_Y$ equal. (The lattice bars, indicated by dashed lines, are neglected, since they serve merely to hold the two channels in position.)

*Ans.  $d = 8.73$ in.*

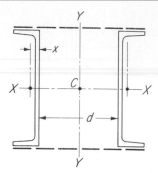

**Fig. 5-32    Prob. 5-34.**

**5-35.** The 4- by 12-in. solid section in Fig. 5-33 has four semicircular grooves cut in its sides as shown, each having a radius of 1 in. Compute the moment of inertia of this section about the $X$ axis.

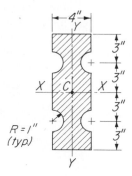

**Fig. 5-33    Probs. 5-35 and 5-38.**

**5-36.** Compute the moment of inertia $I_X$ with respect to the centroidal $X$ axis of the section shown in Fig. 5-30 and described in Prob. 5-30. The area of each angle is 6.94 in.², the distance from the back of the angle to its centroidal axis is 1.52 in., and its moment of inertia with respect to that axis is 15.7 in.⁴. The distance $\bar{y}$ is 11.45 in.

**5-37.** Compute $I_X$ and $I_Y$ for the area shown in Fig. 5-3.

*Ans.* $I_X = 5785$ in.⁴; $I_Y = 18,140$ in.⁴

**5-38.** Compute $I_Y$ for the section of Fig. 5-33.

**5-39.** Compute $I_X$ and $k_X$ for the section of Fig. 5-23. Let $h = 300$ mm and $R = 50$ mm.

## REVIEW QUESTIONS

**5-1.** Explain the meaning of centroid of an area.

**5-2.** By what principle may the centroid of a composite area be located? Explain.

**5-3.** In what phase of engineering practice is the location of the centroid of an area required?

**5-4.** What practical use is made of the quantity moment of inertia?

**5-5.** Define the moment of inertia of an area in terms of its elemental areas and with respect to an axis in the plane of the area.

**5-6.** In practical engineering language, what does the moment of inertia of the cross-sectional area of a beam or a column represent?

**5-7.** In obtaining the rectangular moment of inertia of an area, what is the position of the reference axis with respect to the area?

**5-8.** In obtaining the polar moment of inertia of an area, what is the position of the reference axis with respect to the area?

**5-9.** What is the radius of gyration of an area, and what practical use is made of it?

**5-10.** By what method may the approximate moment in inertia of an area be determined?

**5-11.** How is the exact moment of inertia of a simple geometric area obtained?

**5-12.** State the principle sometimes used to obtain the moment of inertia of an area with some part removed.

**5-13.** Give the transfer formula and a complete word statement of the meaning of its separate terms.

**5-14.** Explain briefly how to obtain the moment of inertia of a composite area.

# Chapter 6
# Shear and
# Moment in
# Beams

## 6-1 TYPES OF BEAMS

In this chapter we consider only horizontal beams supporting vertical loads.

A **beam** is a structural member subjected to loads acting transversely to its longitudinal axis, thus causing it to bend. Various kinds of beams are in common use. The simplest of these is one supported at both outer ends, such as that shown in Fig. 6-1a; it is called a **simple beam.**

If one or both ends of a beam overhang its supports, as shown in Fig. 6-1b and c, it is called an **overhanging beam.** A beam completely supported at one end by being framed into a solid wall or pier, as shown in Fig. 6-1d, is called a **cantilever beam.** The foregoing beams are said to be **statically determinate** because their support reactions may all be determined by the laws of static equilibrium, $\Sigma F = 0$ and $\Sigma M = 0$.

A cantilever beam also supported at its unrestrained end, as in Fig. 6-2a, is called a **supported cantilever.** A single beam extending continuously over more than two supports, as in Fig. 6-2b, is called a **continuous beam.** These two types of beams are of a group said to be **statically indeterminate** because the reacting forces and moments at the supports cannot be determined wholly by the laws of static equilibrium but depend in part upon the elastic properties of the material of which they are made. Such cases are treated in Chap. 10.

## 6-2 LOADS ON BEAMS

A **concentrated load** is one supported on an area of material relatively so small that it may, for convenience of calculation, be assumed to be a point. For example, the force exerted by a chair leg on the supporting floor and the load exerted by a beam on a supporting column are both con-

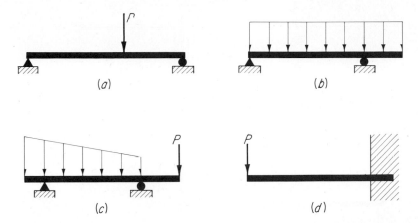

**Fig. 6-1**  Statically determinate beams. (*a*) Simple beam. (*b*) Overhanging beam. (*c*) Overhanging beam. (*d*) Cantilever beam.

sidered as concentrated loads. Another example is shown in Fig. 6-3. Clearly, each of the concentrated loads is actually the resultant of a load distributed over a comparatively small but finite area. For convenience of calculation, an actual load system, such as that shown in Fig. 6-3*a*, is simplified to that shown in Fig. 6-3*b*. A concentrated load, usually expressed in pounds or kips, is conveniently represented by an arrow which indicates the direction and sense of the force.

**Distributed loads,** as the term implies, are loads distributed over larger areas. The usual variety of warehouse contents piled on a floor and pieces of furniture displayed on the floor of a store are examples of distributed floor loads.

If a floor so loaded is supported by joists whose ends in turn are supported on a beam, the load is certainly distributed along the length of each joist, while the beam is subjected to a series of closely spaced concentrated joist loads. Loads distributed over some area are generally expressed in force per unit area, while a load distributed along the length of a beam is usually expressed in force per unit length of beam.

Distributed loads may be *uniform,* as on the beam in Fig. 6-1*b*, or *nonuniform,* as on the beams in Figs. 6-1*c* (trapezoidal), 6-2*a* (triangular), and

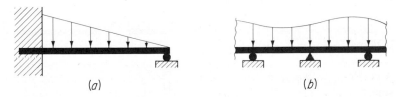

**Fig. 6-2**  Statically indeterminate beams. (*a*) Supported cantilever beam. (*b*) Continuous beam.

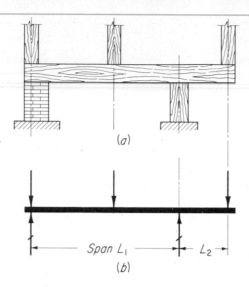

Span $L_1$

$L_2$

(b)

**Fig. 6-3   Concentrated loads.**
(a) Actual load system. (b) Simplified
load system.

6-2b (varying). While distributed loads are often actually uniform, as in the case of the dead weight of a reinforced concrete floor slab of uniform thickness, the trapezoidal and triangular loads are, as a rule, merely the closest approximations of nonuniform loads which may conveniently be dealt with mathematically.

## 6-3  BEAM-SUPPORT REACTIONS

Methods of calculating beam reactions in the most common cases have presumably been studied in previous work. The cantilever beam, however, is a special case, and the nature of the reactions at its support deserves mention.

Consider the cantilever beam $ABC$ in Fig. 6-4a. Assume that the supported end is imbedded solidly in a massive abutment. The probable distribution of the reaction forces on the imbedded portion is as indicated in

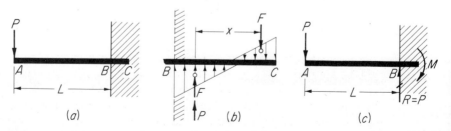

**Fig. 6-4   Reactions at support of cantilever beam.** (a) Cantilever beam. (b) Probable support reactions. (c) Assumed support reactions.

Fig. 6-4*b*. For equilibrium, the algebraic sum of acting and reacting forces must equal zero. That is, $P = P + F - F = P$. The two forces $F$ constitute a couple whose moment is $Fx = M = PL$. The usual simplified manner of indicating the reacting force $R$ and the reacting moment $M$ is shown in Fig. 6-4*c*.

**PROBLEMS**

**6-1.** Calculate the reactions $R_A$ and $R_B$ at supports $A$ and $B$ of the beam shown in Fig. 6-5 if $P_1 = P_2 = P_3 = 12$ kips.

*Ans.* $R_A = 28$ kips; $R_B = 8$ kips

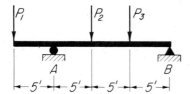

Fig. 6-5    Probs. 6-1 and 6-42.

**6-2.** Compute the support reactions $R_A$ and $R_B$ in Fig. 6-6.

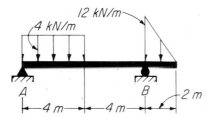

Fig. 6-6    Probs. 6-2 and 6-47.

## 6-4  SHEAR AND MOMENT IN BEAMS

The subjects of stresses in beams and design of beams are discussed in Chaps. 7 and 8. As will be shown in those chapters, the unit shearing stress at any section in a beam is proportional to the transverse shear $V$ at that section and the unit bending stress is proportional to the bending moment.

Beams must be designed to withstand the maximum stresses induced by loads, which then requires determination of the maximum shear $V$ and maximum moment $M$. A thorough understanding of methods of obtaining the shear and moment at any section in a beam is obviously essential to an understanding of the design of beams. Two methods are available: the free-body-diagram method described below and the semigraphical beam-diagram method described in Art. 6-6.

A basic concept of force systems is that when *all* forces acting on a body are accounted for, the force system is complete and is in equilibrium. Hence a loaded beam is in equilibrium under the action of the external loads and the consequent external support reactions. Likewise, *any part of a beam, when isolated from the remainder as a free body, must be and is in equilibrium under the action of the external forces acting on that part and the internal forces at the cut section.*

Consider a small experimental beam made up of three pieces of wood glued together at section $A$ and $B$, as indicated in Fig. 6-7a. Let this beam be supported on two short posts under the two end sections and let it be subjected to a load $W$ resting on a short post placed on the center section. If the glued joints are weaker than the solid material itself on a parallel section, the center part will shear out from between the ends along the transverse planes $AA$ and $BB$ when the failing load is reached. Any load less than that required to cause actual shear failure will produce a *tendency* to shear and will set up transverse resisting shearing stresses in the material of the beam, as indicated by the small arrows. The total *tendency to shear* transversely may readily be visualized and measured.

A simplified free-body diagram of the beam is shown in Fig. 6-7b. Let this beam now be cut at section $BB$, for example, and let the part of it on the left of $B$ be isolated as a free body as in Fig. 6-7c. Only forces $W$ and $R$

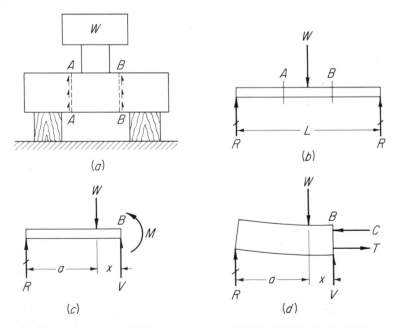

**Fig. 6-7** Transverse shear $V$ and bending moment $M$ in beams. (*a*) Loaded experimental beam. (*b*) Simplified free-body diagram. (*c*) Shear and moment at section $B$. (*d*) Bending-moment couple at section $B$.

act externally on this isolated part. Clearly now, since $R$ is necessarily less than $W$ and the equations of equilibrium, $\Sigma V = 0$ and $\Sigma M = 0$, must be satisfied, the total *internal* vertical force at the cut section is $V = W - R$. The force $V$ represents the sum of all the internal resisting unit shearing stresses and is called the **internal resisting shear.** At any section it is equal to the **external shear,** which is the *algebraic sum of the external forces on either side of the section.*

The existence of the internal resisting moment $M$ at section $B$ may be visualized as follows: Under the influence of the load $W$ the beam will sag between its end supports. Consequently the upper fibers will be shortened as a result of horizontal compressive stresses, the resultant effect of which is denoted by $C$, as shown in Fig. 6-7d. Similarly, the bottom fibers of the beam will be stretched by horizontal tensile stresses, the resultant effect of which is denoted by $T$. These resultants $C$ and $T$ are the only horizontal forces acting on the isolated part of the beam and are therefore equal, since $\Sigma H = 0$. Obviously, $C$ and $T$ form a couple, whose moment $M$ is called the **internal resisting moment,** since it is the moment of the internal stresses resisting bending of the beam. At any section this internal resisting moment is always equal to the **external bending moment,** which is the *algebraic sum of the moments of the external forces on either side of the section.*

To emphasize basic methods rather than details and to simplify solutions, the weight of the beam in the following example and in many of the problems has been disregarded. Later, in Chap. 8, Design of Beams, this weight will be accounted for.

**Signs of Forces, Shears, and Moments.** In most previous work no sign convention was used nor was any found to be essential. However, for the purpose of constructing the diagrams developed later in Art. 6-6 and also later in connection with the design of beams, the following sign convention will be used hereinafter:

**Forces.** Upward-acting forces and forces acting to the right are positive. Conversely, downward-acting forces and forces acting to the left are negative.

**Shears.** The sign of the shear at any section $c$ is the same as the sign of the algebraic sum of the external forces to the *left* of that section, as illustrated in Fig. 6-8a and b.

**Moments.** A positive moment is one which produces tension in the bottom fibers of a beam, as illustrated in Fig. 6-8c. Conversely, a negative moment is one which produces tension in the top fibers, as illustrated in Fig. 6-8d. It is important to realize here that this convention for designating positive and negative moments in beams is purely arbitrary and is

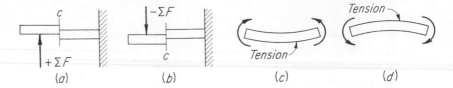

**Fig. 6-8** Signs of forces, shears, and moments. (a) +shear. (b) −shear. (c) +moment.
(d) −moment.

in no way connected with any mathematical convention for the determination of signs.

**In all problems to determine the internal resisting moment $M$ at a cut section of a beam, $M$ will be assumed to be positive (counterclockwise when left end of beam is isolated as a free body, producing tension in bottom fibers). The calculated value of $M$ will then always bear the proper sign.**

A helpful aid in determining the signs of the moments at various sections of a beam is to visualize how it will bend. This is not difficult to do. In Fig. 6-9, for example, it seems likely that the beam will bend as indicated by the dashed line. According to Fig. 6-8c, a "humped" section, as between $A$ and $o$ (Fig. 6-9), indicates tension on the top fibers and, hence, negative moment. A "sagged" section, as between $o$ and $C$, indicates tension on the bottom fibers and hence positive moment. At $o$, the point of contraflexure, there is no bending in the beam and the moment there is zero.

### ILLUSTRATIVE PROBLEM

**6-3.** Calculate the total internal shear $V$ and the internal resisting moment $M$ at each of two sections, respectively, 6 and 16 ft to the right of the left end of the beam shown in Fig. 6-9.

*Solution:* To calculate $V_6$ and $M_6$, we cut the beam 6 ft from the left end and draw a free-body diagram of the *left* portion (always) of the beam, as shown in Fig. 6-10a. This free-body diagram must show *exactly* what remains of Fig. 6-9 if all of it except the left 6 ft is covered with a sheet of paper. In addition, the $V$ and $M$ to be calculated must be shown at the cut section.

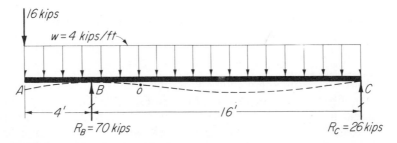

**Fig. 6-9** Calculation of internal shear $V$ and resisting moment $M$ in beam.

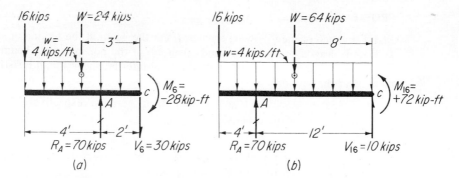

**Fig. 6-10**   Calculation of internal shear $V$ and resisting moment $M$ in beam. (a) Calculation of $V_6$ and $M_6$. (b) Calculation of $V_{16}$ and $M_{16}$.

First, the resultant $W$ of the distributed load is calculated and is shown on the free-body diagram, acting at the centroid of the load area. We now assume $V_6$ to be a positive (upward-acting) force. Then

$$[\Sigma V = 0] \qquad V_6 + 70 = 16 + 24 \qquad \text{and} \qquad V_6 = -30 \text{ kips} \qquad \textit{Ans.}$$

The negative sign indicates that $V_6$ *is* a downward-acting force.

The moment center for calculating $M$ should *always* be a point $c$ on the cut section. We assume $M$ to be positive, that is, producing tension in the bottom fibers of the beam (counterclockwise when *left* portion of beam is isolated). Then, equating the counterclockwise moments to the clockwise moments, we have

$$[\Sigma M_c = 0] \qquad M_6 + (24)(3) + (16)(6) = (70)(2)$$

or $\qquad M_6 = 140 - 72 - 96 \qquad \text{and} \qquad M_6 = -28 \text{ kip·ft} \qquad \textit{Ans.}$

The negative sign indicates $M_6$ to be a negative moment rather than positive, as originally assumed. Consequently the curved moment arrow must act clockwise so as to produce tension in the top fibers.

As soon as they are determined, the values of $V$ and $M$ should be recorded on the free-body diagram, and a check should be made to ensure that $\Sigma V$ and $\Sigma M$ both equal zero. Note here that the calculated $V$ and $M$ are actually the *internal resisting shear* and *internal resisting moment*, respectively.

To determine $V_{16}$ and $M_{16}$ assume $V_{16}$ to be a positive (upward-acting) force. Then

$$[\Sigma V = 0] \qquad V_{16} + 70 = 16 + 64 \qquad \text{and} \qquad V_{16} = 10 \text{ kips} \qquad \textit{Ans.}$$

The positive sign indicates that $V_{16}$ is an upward-acting force, as assumed. We assume that $M_{16}$ is positive, that is, acting counterclockwise so as to produce tension in the bottom fibers.

$$[\Sigma M_c = 0] \qquad M_{16} + (64)(8) + (16)(16) = (70)(12)$$

or $\qquad M_{16} + 512 + 256 = 840 \qquad \text{and} \qquad M_{16} = 72 \text{ kip·ft} \qquad \textit{Ans.}$

The positive sign of $M_{16}$ indicates that it was correctly assumed.

**PROBLEMS**

NOTE: In each of the following problems, cut the beam at the section or sections indicated, draw a *complete* free-body diagram of the portion of the beam on the *left* or the cut section, and calculate the total internal shear $V$ and resisting moment $M$ at the section.

**6-4.** Calculate the shear $V$ and moment $M$ at sections, respectively, 3 and 5 m from the left end of the beam shown in Fig. 6-11.

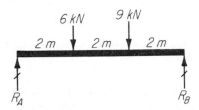

Fig. 6-11   Probs. 6-4, 6-12, and 6-23.

**6-5.** Determine $V$ and $M$ at sections 4 and 10 ft from the left ene of the beam shown in Fig. 6-12.

*Ans.*  $V_4 = 20$ kips $\downarrow$ ; $M_4 = 120$ kip·ft$)$; $V_{10} = 10$ kips $\uparrow$ ; $M_{10} = 150$ kip·ft$)$

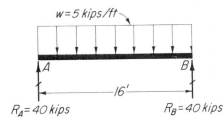

Fig. 6-12   Probs. 6-5, 6-13, and 6-24.

**6-6.** Calculate the reactions $R_A$ and $R_B$ at supports $A$ and $B$ of the beam in Fig. 6-13. Then compute $V$ and $M$ at sections 3 and 7 m from the left end of the beam. $P = 6$ kN.

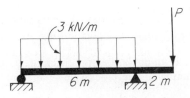

Fig. 6-13   Probs. 6-6, 6-14, and 6-28.

**6-7.** In Fig. 6-14, calculate the support reactions $R_A$ and $R_B$. Then compute $V$ and $M$ at sections 4 and 15 ft from the left end of the beam.

**6-8.** Calculate reactions $R_A$ and $R_B$ of the beam in Fig. 6-15. Then calculate $V$ and $M$ at a section 6 ft from its left end.

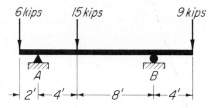

Fig. 6-14   Probs. 6-7 and 6-15.

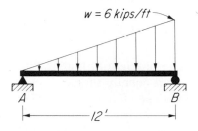

Fig. 6-15   Probs. 6-8, 6-16, and 6-30.

**6-9.** Compute the reactions at supports $A$ and $B$ of the beam shown in Fig. 6-16. Then calculate $V$ and $M$ at a section 6 ft from its left end.

*Ans.* $R_A = 1680$ lb $\uparrow$, $R_B = 2160$ lb $\uparrow$; $V_6 = 120$ lb $\downarrow$, $M_6 = 5760$ lb·ft

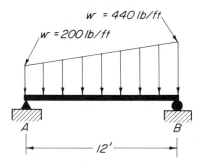

Fig. 6-16   Probs. 6-9, 6-17, and 6-33.

## 6-5 CRITICAL SECTIONS IN BEAMS. SECTIONS OF MAXIMUM MOMENT

A critical section is one at which is found a critical or maximum stress. Two types of critical stresses are generally found in beams: (1) tensile and compressive *bending stresses* and (2) *shearing stresses*. In most ordinary beams and in comparatively long beams, the bending stresses are usually critical and thus determine the design. In relatively short beams, shearing stresses often are critical and thus become the governing factor in design.

As will be shown in Chap. 7, the bending stress at any section in a beam is proportional to the bending moment $M$, and the shearing stress is proportional to the shear $V$. Obviously, then, since beams must be designed to resist the *maximum* probable stresses in bending and shear induced by loads, we must locate the sections of maximum moment $M$ and maximum shear $V$ in order to evaluate those quantities.

**Section of Maximum Shear.** Since the shear $V$ at any transverse section of a beam is the algebraic sum of the transverse forces to the *left* of the section (left, in order to obtain the proper sign), the shear can, in most cases, be evaluated at a glance. To obtain the maximum, it should be evaluated at the beginning and ending of each distributed load and *barely* to the left and right of each concentrated force (including reactions).

**Sections of Maximum Moment.** The problem of finding the section at which the bending moment is greatest is equally simple. It can readily be shown mathematically[1] that *a maximum bending moment occurs at every section where the shear is zero or changes sign.* A change of sign of shear may or may not be found *at a concentrated force.* When a change of sign occurs *between the concentrated forces,* a section of zero shear lies between them which then must be located. Zero shear usually occurs at some section through a distributed load. We must realize here that mathematically speaking, several sections of "maximum moment," positive or negative, may occur in a beam. Only one maximum moment occurs in simple beams. Two may be found in beams overhanging at one end, and three may be found in beams overhanging at both ends. Of these, the moment having the greatest numerical value generally determines the design of the beam.

**ILLUSTRATIVE PROBLEMS** _____

**6-10.** Determine the maximum shear $V$ and moment $M$ in the beam shown in Fig. 6-17.

*Solution:* To determine the maximum shear, we must evaluate the shears at the beginning and end of the distributed load and at the left and right of each concentrated force. Upward-acting forces are positive. Thus we obtain

At $A$ right, $V = 0$

At $B$ left, $V = -(3)(4) = -12$ kN

At $B$ right, $V = 36 - (3)(4) = +24$ kN ← max $V$            *Ans.*

At $C$ left, $V = 36 - (3)(8) = +12$ kN

At $C$ right, $V = 36 - (3)(8) - 6 = +6$ kN

At $D$ left, $V = 36 - 3(16) - 6 = -18$ kN

To obtain the maximum moment in the beam, we must first locate all sections at which the shear is zero or changes sign. In the calculations above, we found a change of sign at $B$, from $-12$ kN at $B$ left to $+24$ kN at $B$ right. We also note that

[1] See Appendix C.

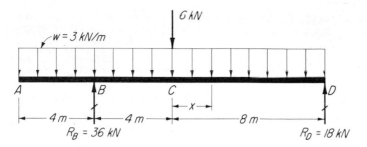

Fig. 6-17   Determination of maximum shear in beam.

between $C$ and $D$ the shear changes from $+6$ to $-18$ kN, indicating a section of zero shear somewhere between $C$ and $D$. Letting $x$ be the unknown distance from $C$, we obtain

$$V = 36 - 6 - (3)(8) - 3x = 0 \quad \text{and} \quad x = 2 \text{ m}$$

Consequently maximum moments occur at sections 4 and 10, and we must calculate both $M_4$ and $M_{10}$ to determine which has the greatest numerical value. The beam is cut at sections 4 and 10, and free-body diagrams are drawn as in Fig. 6-18.

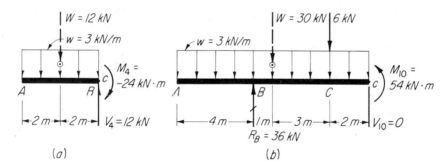

(a)                                    (b)

Fig. 6-18   Determination of maximum moment in beam. Calculation of $M_4$. Calculation of $M_{10}$.

The shears are calculated and shown merely in order to complete the free-body diagrams. At section 4

$$[\Sigma V = 0] \qquad\qquad\qquad V_4 = 12 \text{ kN}$$

As before, we assume $M$ to be positive (producing tension on bottom fibers, acting counterclockwise when *left* portion of beam is isolated). Then

$$[\Sigma M_c = 0] \qquad M_4 + (12)(2) = 0 \quad \text{and} \quad M_4 = -24 \text{ kN·m}$$

The negative moment is now indicated by a clockwise arrow. At section 10, $V$ is known to be zero. Again assuming $M_{10}$ to be positive, we obtain

$$[\Sigma M_c = 0] \quad M_{10} + (6)(2) + (30)(5) = (36)(6) \quad \text{and} \quad M_{10} = 54 \text{ kN·m} \quad \textit{Ans.}$$

**6-11.** Show that the maximum moment $M$ in a simple beam supporting a uniform load occurs at the center and is $M = wL^2/8$, where $w$ is the load per unit of length of beam and $L$ is the span.

*Solution:* At the section of maximum moment the shear $V$ is zero. Hence $W = R$. To locate this section, we solve for $x$. That is, from Fig. 6-19,

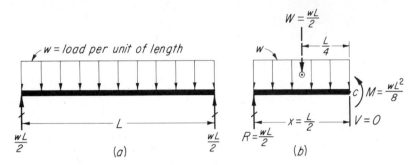

**Fig. 6-19**   Maximum moment in simple beam with uniform load. (a) Loaded beam. (b) Section of maximum moment.

$$[W = R] \qquad\qquad wx = w\frac{L}{2} \quad \text{and} \quad x = \frac{L}{2} \qquad\qquad \textit{Ans.}$$

The maximum moment then is

$$M + \frac{\omega L}{2}\frac{L}{4} = \frac{wL}{2}\frac{L}{2}$$

from which

$$M = \frac{wL^2}{4} - \frac{wL^2}{8} \quad \text{and} \quad M = \frac{wL^2}{8} \qquad\qquad \textit{Ans.}$$

**PROBLEMS**

NOTE: In each of the following problems, calculate the maximum shear $V$ and maximum moment $M$ in the beam in the illustration referred to, using the free-body diagram method illustrated in Fig. 6-18.

**6-12.** See Fig. 6-11.

**6-13.** See Fig. 6-12.  $\qquad\qquad\qquad$ *Ans.* $V_A = 40$ kips; $M_8 = 160$ kip·ft$)$

**6-14.** See Fig. 6-13.  $P = 18$ kN

**6-15.** See Fig. 6-14.

**6-16.** See Fig. 6-15.  $\qquad\qquad\qquad$ *Ans.* $V_{12} = -24$ kips; $M_{5.77} = 55.4$ kip·ft$)$

**6-17.** See Fig. 6-16.

**6-18.** Calculate the reactions at supports $A$, $B$, and $D$ of the beam shown in Fig. 6-30. Then calculate the maximum shear $V$ and bending moment $M$. Let $w = 3$ kN/m, $a = 6$ m, $b = 2$ m, and $c = 4$ m.

**6-19.** Derive an equation giving the maximum moment $M$ in a simple beam of length $L$ supporting a concentrated load $P$ at its center.

**6-20.** Show that in a simple beam of length $L$ supporting a concentrated load $P$ placed at distances $a$ from the left support and $b$ from the right support, the maximum moment $M = Pab/L$.

**6-21.** Prove that in a simple beam of length $L$ supporting a uniformly distributed load $w$ per unit of length, the moment at any distance $x$ from the left support is $M_x = (wx/2)(L - x)$ (see Prob. 6-11).

## 6-6  SHEAR AND MOMENT DIAGRAMS

A semigraphical method of determining shears and moments in beams will now be developed. A distinct, and perhaps the greatest, advantage of these diagrams is that they not only show and evaluate the maximum shear and moment in a beam but also present at a glance a clear and full graphic picture of the *variations* of shear and moment along the entire beam, thus greatly facilitating a clear understanding of the subject. In developing the beam-diagram method, we make use of the free-body-diagram method illustrated in the preceding article.

A simple example is shown in Fig. 6-20. This beam carries a single concentrated load of 18 kips at $B$. By inspection, the shear $V$ at any section

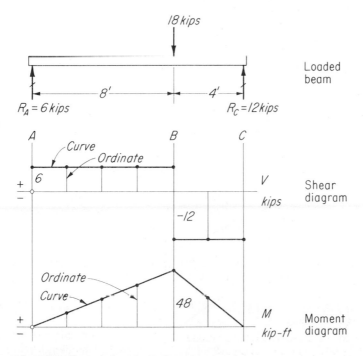

**Fig. 6-20    Shear and moment diagrams.**

between $A$ and $B$ is seen to be $+6$ kips and between $B$ and $C$ is $-12$ kips, which, in each case, is the algebraic sum of the forces to the left of each section. The *shear diagram* is obtained by plotting these values respectively as positive and negative *ordinates* above and below a horizontal reference axis, finally connecting the far ends of all ordinates with a line which in this case is straight.

Similarly, if the beam is cut at 2, 4, 6, and 8 ft from $A$, the bending moments are, respectively, 12, 24, 36, and 48 kip·ft, and all are positive. Note here that the moment $M_8$ at $B$ is $+48$ regardless of whether the beam is cut at $B$ left or $B$ right. If cut at 10 ft from $A$, $M_{10} + (18)(2) = (6)(10)$, or $M_{10} = 60 - 36 = 24$ kip·ft. At $C$ left, $M_{12} + (18)(4) = (6)(12)$, or $M_{12} = 72 - 72 = 0$. The *moment diagram* is now obtained by plotting these values as ordinates, positive and negative, after which their far ends are connected, resulting in straight, sloping lines. Clearly, now, these diagrams give the designer in a single glance not only the maximum shear and moment values but also a complete and graphic picture of the *variations* in shear and moment along the entire beam.

A second example is that of a uniformly loaded simple beam, as illustrated in Fig. 6-21. By inspection, if the beam is cut at $A$ right, the external shear is $+9$ kN. A little study will show that this shear is reduced by 3 kN for every meter to the right of $A$ at which a section might be cut. Then at

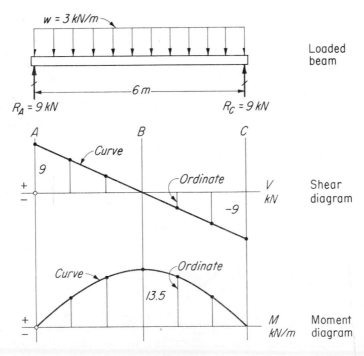

Fig. 6-21    **Shear and moment diagrams.**

1, 2, 3, 4, 5, and 6 m from $A$, the shears will be $+6$, $+3$, $0$, $-3$, $-6$, and $-9$ kN, respectively. When these values are plotted as ordinates and their far ends are connected, the shear diagram is obtained.

For the purpose of obtaining the moments at various sections in this example, the free-body diagrams of the parts to the left should be visualized. At 2 m from $A$, for example, the resultant of the distributed load is 6 kN concentrated 1 m from the cut section. Hence $M_2 = 2(9) - 1(6) = 12$ kN·m. Similarly, at 1, 2, 3, 4, 5, and 6 m from $A$, the moments are, respectively $+7.5$, $+12$, $+13.5$, $+12$, $+7.5$, and 0. When these moment values have then been plotted as ordinates, their far ends are connected by a *smooth curve*, which in this instance is a second-degree parabola.

The process of constructing shear and moment diagrams in this manner is, of course, slow and cumbersome and will not be used hereinafter. But from the two examples thus constructed and shown in Figs. 6-20 and 6-21, certain rules may now be formulated by means of which similar beam diagrams may be constructed rapidly and accurately. These rules, hereinafter referred to as the laws of beam diagrams, are as follows:

**First Law of Beam Diagrams:**

$$\begin{bmatrix} \textbf{Slope of curve at any point} \\ \textbf{in any diagram} \end{bmatrix} = \begin{bmatrix} \textbf{length of ordinate at correspond-} \\ \textbf{ing point in next higher diagram} \end{bmatrix}$$

**Second Law of Beam Diagrams:**

$$\begin{bmatrix} \textbf{Difference in length of any} \\ \textbf{two ordinates in any diagram} \end{bmatrix} = \begin{bmatrix} \textbf{area between corresponding ordi-} \\ \textbf{nates in next higher diagram} \end{bmatrix}$$

The first law is used to determine the *shape* of a curve and the second to determine values of ordinates representing either shear or moment. The load, shear, and moment diagrams are usually constructed in that order from top to bottom of the sheet. The word "higher" in the two laws then means *higher* on the sheet. See Appendix C for mathematical derivations of the laws of beam diagrams.

**Practice Problems.** In order to gain proficiency in the application of the first and second laws of beam diagrams, students should copy and complete the four practice sheets shown in Fig. 6-22. Draw curves proportionately correct, but not to scale.

**Definitions.** Four terms appearing in these two laws will now be defined.

**Curve.** Any line that may be precisely defined by a mathmatical equation is called a **curve.** (Thus, a straight line may, mathematically speaking, be referred to as a curve.)

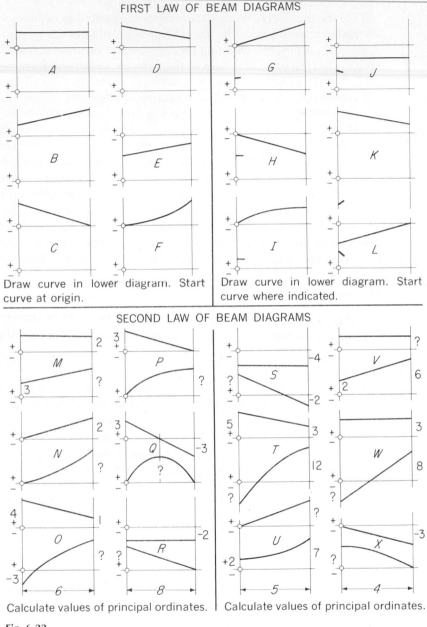

Fig. 6-22

**Slope.** The **slope** of a line is the ratio of $y$ to $x$, as indicated in Fig. 6-23. The slope of a curve at any point is the slope of a tangent to the curve at that point. The slope of a curve may be positive or negative, and it may be uniform, increasing, or decreasing. A horizontal line has a zero slope. These seven possible slopes of curves are shown in Fig. 6-24.

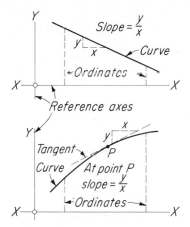

Fig. 6-23   Slope of curves.

**Ordinate.**   With reference to a system of rectangular axes ($X$ and $Y$), an ordinate is a line drawn parallel to the $Y$ axis and extending from the $X$ axis to the curve. **Principal coordinates** are the *initial, final,* and *maximum ordinates* of each segment of a curve.

**Areas under Curves. Area under a curve** means the area lying between the curve and the horizontal $X$ axis and bounded on the left and right by vertical ordinates. Areas under the curves most commonly encountered are shown in Fig. 6-25.

## 6-7   APPLICATIONS OF SHEAR AND MOMENT DIAGRAMS

Before the laws of beam diagrams can successfully be applied with rapidity and accuracy, they, together with the various slopes and areas under curves, must be reasonably well memorized. A load diagram in which positive and negative loads are plotted respectively below and above the horizontal axis may be added to the shear and moment diagrams and

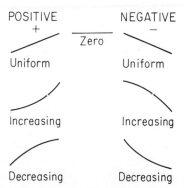

Fig. 6-24   Seven possible slopes of curves.

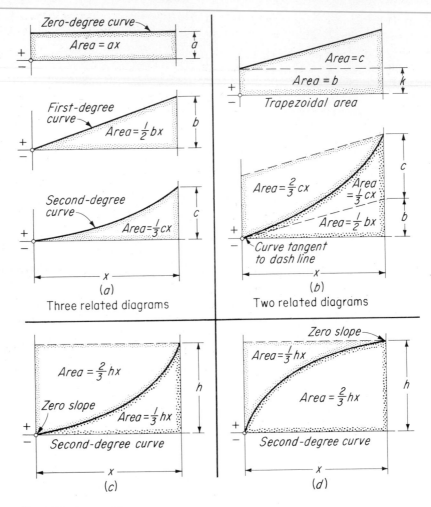

Fig. 6-25 Areas under curves.

should be so added in all problem solutions hereinafter. It will bear the same relation to the shear diagram that the shear diagram bears to the moment diagram; that is, the load diagram conforms in every respect to the laws of beam diagrams. The following illustrative problem presents difficulties somewhat in excess of those normally encountered in practice.

ILLUSTRATIVE PROBLEM

**6-22.** Construct the load, shear, and moment diagrams for the loaded beam shown in Fig. 6-26a. Evaluate all principal ordinates and determine the maximum shear $V$ and moment $M$.

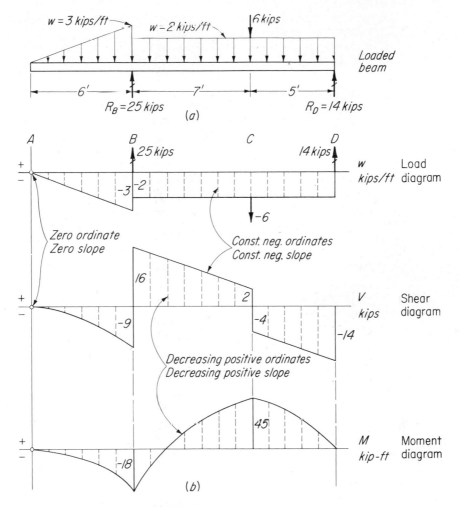

Fig. 6-26   Construction of load, shear, and moment diagrams.

*Solution:* First, a diagram of the loaded beam should always be drawn at the top. Second, the necessary framework of vertical and horizontal axes should be drawn, with + and − indicating the positive and negative sides of the horizontal axes. Third, the load diagram is the constructed. The downward-acting (negative) loads are plotted below the horizontal axis and the upward-acting (positive) reactions above.

Then, the shear diagram for each segment (*AB, BC,* and *CD*) of the beam is constructed and the principal ordinates are evaluated, using the laws of beam diagrams, first in segment *AB*, then in *BC* and *CD*, followed by the three segments of the moment diagram in the same order. The two laws are used simultaneously, the first law to ascertain the shape of the curves and the second law to evaluate the principal ordinates. The explanatory notes on Fig. 6-26*b*, pertaining to the relation between slope of a curve and the ordinate at the corresponding point in the next higher diagram, should be studied closely.

The thought processes while constructing the diagrams might be as follows: Segment $AB$ of shear curve, by first law, increasing negative load ordinates starting with zero ordinate, hence increasing negative *slope* of shear curve starting with zero slope. Shear at $A$ is zero; by second law, *difference* in shear at $A$ right and $B$ left equals load area $AB$, or $(^1/_2)(-3)(6) = -9$. Also, *difference* between shear $B$ left and $B$ right equals concentrated force at $B$, or $-9 + 25 = +16$. (EXPLANATION: Every concentrated force actually is the resultant of a load distributed over a small area; see Art. 6-2.) Next, segment $BC$ of shear curve, first law: constant negative load ordinates and constant negative slope of shear curve; second law, difference between shear $B$ right and $C$ left equals load area $BC$, or $(-2)(7) = -14$; hence shear $C$ left $= 16 - 14 = 2$.

The remaining segments in the shear and moment diagrams are completed in the same manner. Note here that the difference in moments at $A$ and $B$ left equals the shear area $AB$, which is $(^1/_3)(-9)(6) = -18$. Also, the difference in moments at $B$ right and $C$ left equals the shear area $BC$, which is $(^1/_2)(16 + 2)(7) = +63$. Hence the moment at $C$ left is $-18 + 63 = +45$, which also is the moment at $C$ right, since the shear area between $C$ left and $C$ right is zero.

When all diagrams are completed, we see at a glance that $V_{max} = 16$ kips at $B$ right and $M_{max} = 45$ kip·ft at $C$. We note also that in this example, the two maximum moments both occur at sections where the shear changes sign.

The tremendous value of these diagrams now becomes clearly evident when we notice that a single glance at them reveals the variations of shear and moment along the entire length of the beam.

A few hints helpful for constructing beam diagrams may be noted here:

1. The value of the initial shear ordinate is most easily determined by the usual algebraic summation of external forces.
2. An abrupt change in values of adjacent shear ordiinates occurs at every concentrated force, load, or reaction.
3. The moment at the end of a beam is always zero unless the support there is such that it restrains rotation of the beam, as, for example, at the supported end of a cantilever beam.
4. There can never be an abrupt change in value of adjacent moment ordinates. However, an abrupt change in slope of the moment curve occurs at every concentrated force.

## PROBLEMS

NOTE: In each of the following problems, draw a diagram of the beam in the illustration referred to. Then construct the load, shear, and moment diagrams, calculate the values of all principal ordinates, and determine the maximum shear $V$ and moment $M$.

**6-23.** See Fig. 6-11.                    *Ans.*  $V_4 = -8$ kN; $M_4 = 16$ kN·m
**6-24.** See Fig. 6-12.

**6-25.** See Fig. 6-27.          $Ans.$  $V_{15} = -12$ kips; $M_{10} = 80$ kip·ft

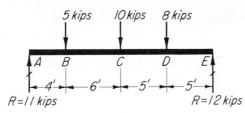

R=11 kips          R=12 kips    **Fig. 6-27    Prob. 6-25.**

**6-26.** See Fig. 6-28.

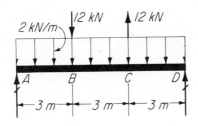

**Fig. 6-28    Prob. 6-26.**

**6-27.** See Fig. 6-29.          $Ans.$  $V_9 = 24$ kips; $M_9 = 48$ kip·ft

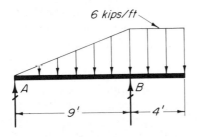

**Fig. 6-29    Prob. 6-27.**

**6-28.** See Fig. 6-13. $P = 9$ kN.
**6-29.** See Fig. 6-30.

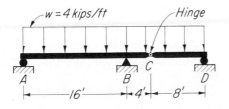

**Fig. 6-30    Prob. 6-29.**

**6-30.** See Fig. 6-15.          $Ans.$  $V_{12} = -24$ kips; $M_{6.93} = 55.4$ kip·ft
**6-31.** See Fig. 6-31. Let $w = 2$ kN/m, $a = 6$ m, $b = 2$ m, and $c = 4$ m.
**6-32.** See Fig. 6-32.
**6-33.** See Fig. 6-16.

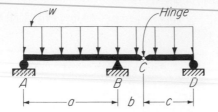

Fig. 6-31    Probs. 6-18 and 6-31.

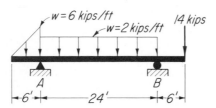

Fig. 6-32    Prob. 6-32.

## 6-8 MOVING LOADS

When a system of concentrated wheel loads, such as those of a truck or locomotive, moves across a simple beam, the problem arises of finding the particular position of the loads at which the bending moment in the beam will be an absolute maximum. Likewise, the position of the loads which will produce the absolute maximum shear must be determined before the beam can be designed. The following rules apply:

1. **The absolute maximum shear $V$ occurs at one of the two end supports and numerically equals the maximum support reaction.**
2. **The absolute maximum shear $V$** can be evaluated only by trial, alternately placing each load over a support and calculating the reactions. The critical position usually occurs when one of the largest loads is over a support while as many as possible of the remaining loads are still on the beam. Remember that the load can move in either direction.
3. **The absolute maximum moment** in a simple span due to a system of concentrated moving loads always occurs under a *large* (critical) load located *near* the resultant of all the loads on the beam. In unusual cases, maximum moment may occur with one or more relatively small end loads off the beam.
4. The critical load for maximum moment must be determined by trial. **The absolute maximum moment will be found when the distance between the critical load and the resultant of all the loads on the beam is bisected by the centerline of the beam.** (The truth of this statement may be proved by the use of calculus. Also, any number of trials using finite values will verify it.)

**ILLUSTRATIVE PROBLEM**_____

**6-34.** Figure 6-33 shows a simple beam subjected to a system of three moving loads, $B = 4$ kN, $C = 8$ kN, and $D = 9$ kN. Position the loads so they will produce the absolute maximum shear and moment. Then calculate the magnitude of each.

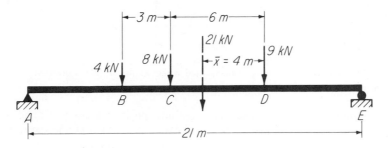

**Fig. 6-33   A system of moving, concentrated wheel loads.**

*Solution:* The resultant $R$ of the loads equals $4 + 8 + 9 = 21$ kN, and may be located by moments about $D$ in Fig. 6-34. Since the moment of the resultant equals the sum of the moments of the separate forces, we obtain

$$[\Sigma M_D = 0] \qquad 21\bar{x} = (8)(6) + (4)(9) = 84 \qquad \text{and} \qquad \bar{x} = 4 \text{ m}$$

According to rule 2, the absolute maximum shear $V$ apparently occurs when the loads are moved to the right until load $D$ is nearly over the right support. The distance from $A$ to the resultant $R$ is then 17 m. By moments about $A$ we obtain

$$[\Sigma M_A = 0] \qquad 21R_E = (21)(17) \qquad \text{and} \qquad R_E = 17 \text{ kN} = V_{max} \qquad \qquad Ans.$$

Study of Fig. 6-33 will quickly show the following results. When load $C$ is over support $E$, $V = 8 + {}^{18}/_{21}(8) = 11.4$ kN; when $B$ is over $A$, $V = 4 + {}^{18}/_{21}(8) + {}^{12}/_{21}(9) = 17$ kN; when $C$ is over $A$, $V = 8 + {}^{15}/_{21}(9) = 14.4$ kN, none of which exceeds 17 kN.

According to rule 3 the maximum moment should be found either under load $C$ or under load $D$. Load $C$ is 8 kN and acts 2 m to the left of the resultant $R$. Load $D$ is 9 kN and acts 4 m to the right of $R$. Two variables determine which is the critical load: (1) *nearness* to the resultant $R$ and (2) *magnitude* of the load. In this case the critical load is apparent. Nevertheless we will compute the moments under both $C$ and $D$.

Assuming first that $C$ is the critical load, we position the three loads $B$, $C$, and $D$ so that the center of the beam will bisect the distance between load $C$ and the resultant $R$, as in Fig. 6-34$a$. By moments about $E$ we evaluate $R_A$:

$$[\Sigma M_E = 0] \qquad 21R_A = (21)(9.5) \qquad \text{and} \qquad R_A = 9.5 \text{ kN}$$

The beam is now cut at $C$, the left portion is isolated as in Fig. 6-34$b$, and the moment at $C$ is calculated as follows:

$$[\Sigma M_C = 0] \quad M_C + (4)(3) = (9.5)(9.5) \qquad \text{or} \qquad M_C = 90.25 - 12 = 78.25 \text{ kN·m}$$

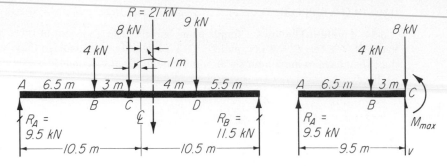

**Fig. 6-34** A system of moving, concentrated wheel loads. (*a*) Loads positioned for max-imum moment at C. (*b*) Calculation of maximum moment at C.

To obtain the moment at section $D$ we now position loads $B$, $C$, and $D$ so that the center of the beam bisects the distance between load $D$ and the resultant $R$, as shown in Fig. 6-35*a*. The reaction $R_E$ is now evaluated by a moment summation about $A$. That is,

$$[\Sigma M_A - 0] \qquad 21R_E = (21)(8.5) \qquad \text{and} \qquad R_E = 8.5 \text{ kN}$$

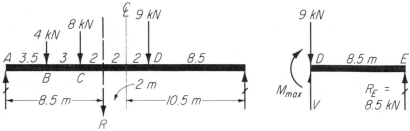

**Fig. 6-35** A system of moving, concentrated wheel loads. (*a*) Loads positioned for max-imum moment at D. (*b*) Calculation of maximum moment at D.

We cut the beam at $D$, isolate the part to the right, as shown in Fig. 6-35*b*, and calculate the moment at $D$ as follows:

$$[\Sigma M_D = 0] \qquad M_D = (8.5)(8.5) = 72.25 \text{ kN·m}$$

The absolute maximum moment for which the beam should be designed is seen to be 78.25 kN·m. It is clearly unnecessary to calculate the moment under load $B$, since $B$ is low in value and is farthest removed from the resultant $R$.

## PROBLEMS

**6-35.** In Prob. 6-34 let the loads at $B$, $C$, and $D$ be 4, 22, and 16 kN, respec-tively, then solve.

**6-36.** Two wheel axles are 18 ft apart. One wheel load is 10 kips and the other is 20 kips. Calculate the absolute maximum shear and moment developed in a 36-ft simple span.

*Ans.* $V_{max} = 25$ kips; $M_{max} = 187.5$ kip·ft

**6-37.** The wheel loads for the truck of Fig. 6-36 are $P_1 - P_2 - 4$ kips, $P_3 - 6$ kips, and $P_4 = 2$ kips. Calculate the absolute maximum moment in the supporting beam directly under the wheels if the span length is 40 ft.

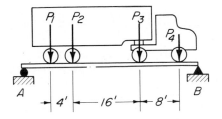

**Fig. 6-36    Probs. 6-37 and 6-50.**

**6-38.** The ore car shown in Fig. 6-37 and its contents weigh 80 kN. Assume that this load is distributed equally to all eight wheels. Calculate the absolute maximum moment in each of the two supporting beams if their span is 12 m.

*Ans.* $M_{max} = 55.2$ kN·m

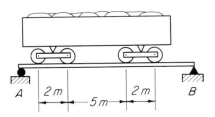

**Fig. 6-37    Probs. 6-38 and 6-51.**

**SUMMARY** (By Article Number)_____

**6-1.** Several **types of beams** are in common use: the simple beam, the overhanging beam, and the cantilever beam, all of which are **statically determinate.** Single beams extending continuously over more than two supports are **statically indeterminate** and are treated in Chap. 10.

**6-2. Loads on beams** may be concentrated or distributed. Distributed loads may be uniform or nonuniform. Nonuniformly distributed loads of considerable variation are generally approximated into triangular or trapezoidal forms of loads.

**6-3. Beam-support reactions** are readily calculated by the principle of moments. At the single support of a cantilever beam are found both a reacting force and a reacting moment.

**6-4.** At any section in a beam, the **transverse shear** $V$ is the algebraic sum of the transverse forces acting on either side of the section. Similarly at any section in a beam, the **bending moment** $M$ is the algebraic sum of the moments of the transverse forces acting on either side of the section.

**6-5.** The **critical section in a beam** is one at which either the transverse shear $V$ or the bending moment $M$ is maximum. A maximum moment is found at all sections at which the shear $V$ is minimum; that is, where the shear is zero or changes sign.

**6-6.** Values of shears and moments at all sections in a beam may be quickly determined from **beam diagrams** drawn in accordance with the following laws:

**First Law of Beam Diagrams:**

$$\begin{bmatrix} \textbf{Slope of curve at any point in} \\ \textbf{any diagram} \end{bmatrix} = \begin{bmatrix} \textbf{length of ordinate at correspond-} \\ \textbf{ing point in next higher diagram} \end{bmatrix}$$

**Second Law of Beam Diagrams:**

$$\begin{bmatrix} \textbf{Difference in length of any} \\ \textbf{two ordinates in any diagram} \end{bmatrix} = \begin{bmatrix} \textbf{area between corresponding ordi-} \\ \textbf{nates in next higher diagram} \end{bmatrix}$$

The diagrams are to be drawn in this order: load, shear, and moment, of which the load diagram is the highest.

**6-8. Movable loads** should be positioned for absolute maximum shear and moment. While some trials must be made, in general, maximum shear will be obtained when a large load is over a support and as many loads as possible on the beam. Maximum moment will be obtained under a large interior load with the loads positioned so that the critical load is as far on one side of the centerline of the beam as the resultant of the loads on the beam is on the other side.

**REVIEW PROBLEMS** _____

**6-39.** Calculate the reactions at supports $A$ and $D$ of the beam shown in Fig. 6-38.

*Ans.* $R_A = 4$ kN ↑ ; $R_D = 4$ kN ↑

**6-40.** Using the free-body-diagram method, calculate $V_3$ and $M_3$ at section 3 and $V_{10}$ and $M_{10}$ at section 10 of the beam shown in Fig. 6-38 (see answer to Prob. 6-39).

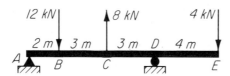

**Fig. 6-38    Probs. 6-39 and 6-40.**

**6-41.** Locate by inspection and calculate the maximum shear $V$ in the beam in Fig. 6-39. Then, using the free-body-diagram method, calculate the maximum moment $M$.

*Ans.* Max $V_{16} = -18.25$ kips; max $M_8 = 62$ kip·ft

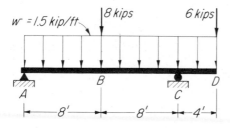

**Fig. 6-39    Probs. 6-41 and 6-45.**

**6-42.** Locate by inspection and calculate the maximum shear $V$ in the beam shown in Fig. 6-5. Them, using the free-body-diagram method, calculate the maximum moment $M$. $P_1 = 12$ kips, $P_2 = P_3 = 18$ kips.

*Ans.* Max $V_5 = 22$ kips; max $M_{15} = 70$ kip·ft)

**6-43.** Calculate the reactions at the supports of the beam in Fig. 6-40. By inspection, locate and calculate the maximum external shear $V$. Then, by the free-body-diagram method, calculate the maximum moment $M$. $P = 9$ kips.

*Ans.* $R_A = 3$ kips ↑; $R_D = 18$ kips ↑; max $V_8 = -10$ kips; max $M_8 = -16$ kip·ft)

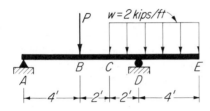

Fig. 6-40   Probs. 6-43 and 6-48.

**6-44.** The two sections of the beam shown in Fig. 6-41 are joined with a frictionless hinge at $B$. Calculate all reactions at supports $A$ and $C$ and determine the maximum shear $V$. Then, using the free-body-diagram method, calculate maximum $M$.

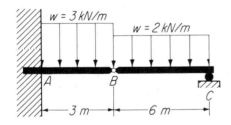

Fig. 6-41   Probs. 6-44 and 6-49.

**6-45.** Draw a complete set of load, shear, and moment diagrams of the beam shown in Fig. 6-39, and determine maximum $V$ and $M$ (see Prob. 6-41 for answers).

**6-46.** For the beam shown in Fig. 6-17, draw a complete set of load, shear, and moment diagrams and verify the maximum shear and moment found in Prob. 6-10.

**6-47.** Draw complete load, shear, and moment diagrams for the beam in Fig. 6-6, and determine maximum $V$ and $M$ ($R_A = 11$ kN; $R_B = 17$ kN).

*Ans.* Max $V_8 = 12$ kN; max $M_{2.75} = 15.125$ kN·m

**6-48.** For the beam in Fig. 6-40 draw complete load, shear, and moment diagrams and determine maximum $V$ and $M$. $P = 15$ kips.

**6-49.** Draw complete load, shear, and moment diagrams for the beam in Fig. 6-41, and determine maximum $V$ and $M$ ($R_A = 15$ kN ↑; $R_C = 6$ kN ↑).

*Ans.* Max $V = +15$ kN; max $M = -31.5$ kN·m)

**6-50.** Refer to Prob. 6-37. Let $P_1 = P_2 = 8$ kips, $P_3 = 5$ kips, $P_4 = 3$ kips, and the span length $L = 60$ ft and solve.

**6-51.** The wheel loads shown in Fig. 6-37 are 12 kN each. Calculate the absolute maximum shear and moment if the span is 16 m.

*Ans.* $M_{max} - 112.7$ kN·m

**6-52.** Two equal vertical loads of 10 kips must remain 12 ft apart. They can be placed in any position along a simple beam of 30-ft span. Calculate the absolute maximum shear and moment.

## REVIEW QUESTIONS

**6-1.** What is meant by (*a*) a simple beam, (*b*) an overhanging beam, and (*c*) a cantilever beam?

**6-2.** Explain the meaning of the terms "statically determinate" and "statically indeterminate." Give examples of beams in each category.

**6-3.** Define the terms "concentrated load" and "distributed load."

**6-4.** Discuss the reactions at the support of a cantilever beam.

**6-5.** Define and discuss the terms "external shear" and "internal shear."

**6-6.** Define and discuss the terms "external bending moment" and "internal resisting moment."

**6-7.** When is the external shear at a section positive? When is the internal resisting moment at a section positive?

**6-8.** What is meant by a critical section in a beam?

**6-9.** At what section in a beam is the shear critical?

**6-10.** At what section in a beam is the moment critical?

**6-11.** At what sections in a beam is the moment a maximum?

**6-12.** Describe the general functions of shear and moment diagrams, also called beam diagrams.

**6-13.** State the first law of beam diagrams. What is its function?

**6-14.** Give the second law of beam diagrams. What is its function?

**6-15.** Where is absolute maximum shear found for beams supporting wheel loads?

**6-16.** How are loads positioned for absolute maximum shear?

**6-17.** Where is absolute maximum moment found for beams supporting wheel loads?

**6-18.** How should loads be positioned on beams for absolute maximum moment?

# Chapter 7 ———————————
# Stresses in Beams ———————————

## 7-1 TENSILE AND COMPRESSIVE STRESSES IN BEAMS

When a beam is bent, the material within it is subjected to both tensile and compressive stresses and, in most cases, also to shearing stresses. In this chapter, we show the distribution of these stresses over a cross-sectional plane of a beam. We also develop two general equations, of which the first expresses the relation at any given section of a prismatic beam between (1) the unit tensile and compressive stresses $s_t$ and $s_c$, (2) the bending moment $M$, and (3) the moment of inertia $I$ of the beam. The second equation expresses the relation at any section of a beam between (1) the unit shearing stress $s_s$, (2) the total transverse shear $V$, and (3) the cross-sectional area $A$ of the beam. These two basic equations are used in various forms in the design of all beams.

As a general rule, beams are bent by transverse loads applied between their supported ends, as illustrated in Fig. 7-1. As a result of such loading, internal tensile, compressive, and shearing stresses are produced. Bending of a beam may also be accomplished by the application at its ends of two equal and opposing moments, as illustrated in Fig. 7-2. When it is so bent, no transverse shearing forces act at any section and the member is said to be in **pure bending.** Hence the material is subjected only to tensile and compressive *bending stresses*.

Let us now consider a straight rectangular experimental beam as shown in Fig. 7-3a. We assume that this beam is made up of a large number of "fibers," all lying parallel to its longitudinal axis. The stresses in these fibers are referred to as **fiber stresses.** We also assume that the material of the beam is homogeneous and is of equal strength and modulus of elasticity in tension and compression. If this experimental beam is made of firm but flexible rubber and is subjected to two equal and oppositely directed bending moments acting in a plane which is also a plane of symmetry of the beam, a condition of pure bending is produced and a number of obser-

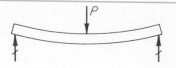

Fig. 7-1 Beam bent by transverse forces.

vations may be made. From these, and from the assumptions stated above, certain conclusions which explain the distribution of internal normal tensile and compressive bending stresses in such a beam may then be made. Some of them are as follows:

| *Observation or assumption* | *Conclusions* |
|---|---|
| 1. The top fibers have been shortened an amount equal to the lengthening of the bottom fibers. | 1. According to Hooke's law, stress is proportional to deformation. Therefore, the compressive stress in the top fibers must equal the tensile stress in the bottom fibers. |
| 2. A plane section (such as $AA$, $BB$, $CC$, and $DD$) before bending remains a plane section after bending. Each plane revolves (from 1 above) about its intersection with central plane $xx$. | 2. The deformations, and consequently the stresses, must then increase uniformly from zero at the central plane to a maximum at the outer fibers. |
| 3. From 1 and 2 it follows that there is no longitudinal deformation of the fibers along the central plane $xx$. Hence line $xx$ equals arc $xx$ in length. | 3. Since no deformation has taken place, there can be no tensile or compressive stress due to bending along this plane, called the "neutral plane." |

These conclusions are illustrated graphically in Fig. 7-4, in which is shown enlarged the portion of the beam between sections $BB$ and $CC$. The intersection of the plane of no stress, the *neutral* plane, with the plane of the cut section is called the **neutral axis of bending** or simply **neutral axis.** The small arrows indicate compressive stresses varying from zero at the neutral axis to a maximum $s_c$ at the top and tensile stresses varying from zero at the neutral axis to a maximum $s_t$ at the bottom.

The forces $C$ and $T$ are the equal and opposite resultants of the compressive and tensile stresses, respectively, and each is applied at the centroid of its stress volume. Consequently $C$ and $T$ form a couple whose resisting moment RM represents the internal resistance offered by the fibers of the beam to the external bending moment $M$. Since, in pure bending, the stresses on the cross-sectional plane $BB$ are equal and opposite to those on plane $CC$, this portion of the beam is in equilibrium as, of course, it must be. That is, $\Sigma V = 0$, $\Sigma H = 0$, and $\Sigma M = 0$. Because this beam is subjected to pure bending, no shearing stresses exist.

Fig. 7-2 Beam bent by end moments. Pure bending.

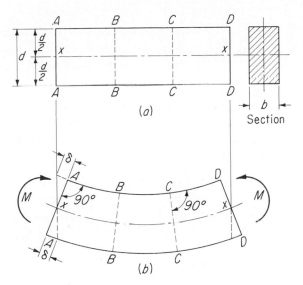

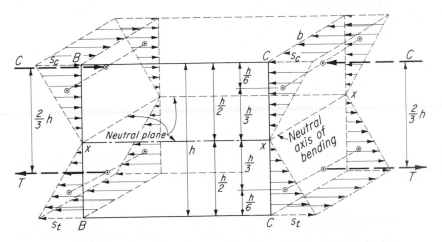

Fig. 7-3  Experimental beam subjected to pure bending. (*a*) Straight beam. (*b*) Bent beam.

## 7-2 RELATION AMONG BENDING MOMENT, FIBER STRESS, AND MOMENT OF INERTIA: THE FLEXURE FORMULA

A relation among the moment $M$, the fiber stress $f$, and the moment of inertia $I$ (Chap. 5) of the cross-sectional area of a beam may now be established.

Consider again the rectangular beam shown in Fig. 7-4 and now shown in two-dimensional form in Fig. 7-5. If the cross-sectional area of this

Fig. 7-4  Distribution of tensile and compressive stresses in bent rectangular beam.

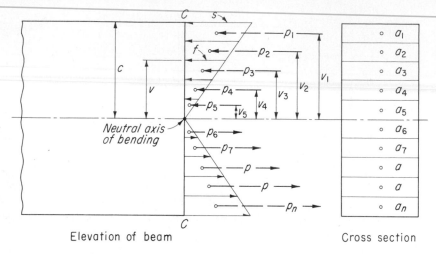

Fig. 7-5   Derivation of the flexture formula.

beam is arbitrarily divided into a number of equal smaller areas, such as $a_1, a_2, a_3, \ldots, a_n$, as shown in Fig. 7-5, and if $p_1, p_2, p_3, \ldots, p_n$ are the resultants of the fiber stresses on those areas, the sum of the moments about the neutral axis of these resultant forces is the total internal resisting moment RM, which at any section equals the bending moment $M$. That is,

$$M = p_1v_1 + p_2v_2 + p_3v_3 + \cdots + p_nv_n \tag{a}$$

But $p$ is the resultant of the stresses acting on area $a$. If $f$ is the unit fiber stress on area $a$, $p = fa$ and

$$M = f_1a_1v_1 + f_2a_2v_2 + f_3a_3v_3 + \cdots + f_na_nv_n \tag{b}$$

These unit stresses $f$, acting at varying distances $v$ from the neutral axis, may now be expressed in terms of the maximum fiber stress $s$ in the extreme fiber acting at a maximum distance $c$ from the neutral axis. That is, from Fig. 7-5, by similar triangles,

$$\frac{f}{v} = \frac{s}{c} \tag{c}$$

or

$$f = \frac{s}{c}v \tag{d}$$

Consequently

$$f_1 = \frac{s}{c}v_1 \qquad f_2 = \frac{s}{c}v_2 \cdots f_n = \frac{s}{c}v_n \tag{e}$$

Substituting these values in Eq. (*b*) gives

$$M = \frac{s}{c}v_1(a_1v_1) + \frac{s}{c}v_2(a_2v_2) + \cdots + \frac{s}{c}v_n(a_nv_n) \qquad (f)$$

or

$$M = \frac{s}{c}(a_1v_1{}^2 + a_2v_2{}^2 + \cdots + a_nv_n{}^2) \qquad (g)$$

Note here that the terms $a_1v_1$, $a_2v_2$, and so forth, represent the *first moment* of these areas about the neutral axis and that the terms $a_1v_1{}^2$, $a_2v_2{}^2$, and so forth, represent the *second moment,* also called the **moment of inertia,** of the areas about that axis. The sum of all the $av^2$ terms in Eq. (*g*) then represents the moment of inertia of the entire cross-sectional area of the beam. If the symbol $I$ designates this moment of inertia (see also Chap. 5),

$$\text{\textbf{Moment of inertia} } I = \Sigma av^2 \qquad (h)$$

Should we desire a practical definition of this term, we might say that *the moment of inertia of the cross-sectional area of a prismatic beam is a measure of the stiffness of the beam, or of its capacity to resist bending.*
When $I$ is substituted for the sum of the $av^2$ terms in Eq. (*g*), we obtain

**The flexure formula:**    $M = s\dfrac{I}{c}$    or    $s = \dfrac{Mc}{I}$    **(7-1)**

where $M$ = bending moment at section, usually in lb·in. or N·mm
$\quad\quad\ \ s$ = fiber stress at distance $c$ from neutral axis, psi or MPa
$\quad\quad\ \ I$ = moment of inertia, in.⁴ or mm⁴
$\quad\quad\ \ c$ = distance from neutral axial to fiber stress $s$, in. or mm

Equation (7-1) is called the **flexure formula.** According to Fig. 7-5 and Eq. (*c*), *s* may, of course, be the stress at any variable distance *c* from the neutral axis. *When c is the distance to the extreme fiber, s is the maximum stress* The flexure formula, Eq. (7-1), was here developed by using a specific rectangular section and by visualizing certain necessary physical relations existing at any cross section of a bent member. The exact mathematical derivation of the flexure formula, using the calculus, is shown in Appendix D.
According to Eq. (*h*), the moment of inertia $I$ of an area may be obtained by dividing it into a finite number of strips all lying parallel to the neutral axis, as illustrated in Fig. 7-5 (see also Prob. 5-10). The moment of inertia of each strip then is $av^2$ and of the entire area comprising all of the strips is $\Sigma av^2$. This method is approximate only and is not generally used except in connection with irregularly shaped areas. The error involved in its use is decreased as the number *n* of areas is increased, and it becomes exact when *n* equals infinity, in which case each area is infinitesimally small, or $dA$. The exact mathematical derivations of the expressions

giving the moments of inertia of a rectangular area, a triangular area, and a circular area, using the calculus, are shown in Appendix B. Appendix B also gives the exact derivation of the transfer formula used for obtaining the moment of inertia of composite areas.

The magnitude of the moment of inertia of an area depends upon two things: the magnitude of the area and its distribution. Of these, the distribution has the greatest effect. For example, let us consider that 18 in.$^2$ of area is available to form the cross section of a beam and that this area is supplied in the form of three $1/2$- by 12-in. steel plates.

If these plates are combined in the three ways illustrated in Fig. 7-6, in each case so that they act as a single unit, striking differences appear in their respective moments of inertia. Clearly, the highest moment of inertia is obtained by placing as much of the material as possible as far away from the neutral axis as is consistent with maintaining a stable cross section. This fact is also disclosed by a study of Eq. ($h$). Consequently, we find most steel and other metal beam sections to be of the wide-flange type shown in Fig. 7-6$c$. Because of its relative weakness in shear, timber cannot generally be used in this manner. Hence timber beams are usually of the rectangular type shown in Fig. 7-6$b$.

Unless otherwise specifically stated or shown, the presumption is that all beams are horizontal and that all loads are vertical and are applied to the narrow face of the beam. Under these conditions, the neutral axis will always be horizontal.

**Location of Neutral Axis.** As previously stated, the neutral axis of bending is the line, or axis, formed by the intersection of the neutral plane with the plane of any cross section of a structural member. Its location is readily determined because *the neutral axis of bending passes through the centroid of the cross-sectional area* and lies perpendicular to the plane of bending. The mathematical proof of this fact is shown in Appendix D.

The centroids of various geometric and symmetrical areas are easily located, of course, without calculation. The methods outlined in Chap. 5 are used to locate centroids of nonsymmetrical areas. Problem 7-1 shows the

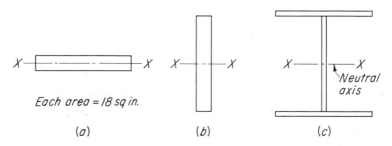

Each area = 18 sq in.

(a)　　　　　(b)　　　　　(c)

Neutral axis

**Fig. 7-6** Effect of distribution of area upon moment of inertia.

calculations involved in the location of the centroid, or neutral axis, of a composite area.

### ILLUSTRATIVE PROBLEM

**7-1.** Figure 7-7 is a cross section of a beam built up of two 2- by 6-in. wood planks, securely spiked together to act as a single unit. It is used as a simple span 10 ft long supporting a load of 200 lb/ft including its own weight. Calculate the maximum bending stress *s* in this beam.

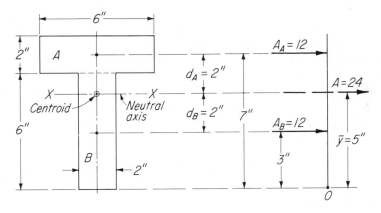

**Fig. 7-7   Location of neutral axis. Calculation of moment of inertia and of bending stress in beam.**

*Solution:* Since this section is unsymmetrical with respect to its horizontal neutral axis, we must first locate that axis. Then the moment of inertia *I* with respect to the neutral axis must be calculated, the dimension *c* must be determined, and finally the bending moment *M* must be computed, after which the maximum bending stress *s* can be calculated.

To determine $\bar{y}$ (see Chap. 5), using the principle that about any point the moment of an area equals the algebraic sum of the moments of its component areas, we take moments about the bottom *O* of the section. That is,

$[M_O] \qquad 24\bar{y} = (12)(7) + (12)(3) = 84 + 36 = 120 \qquad$ and $\qquad \bar{y} = 5$ in.

From Chap. 5, the moment of inertia of a composite area with respect to its neutral (centroidal) axis is $I_X = \Sigma(I_C + Ad^2)$. For a rectangular area, $I_C = bh^3/12$. Then

Area *A*: $\qquad I_X = \dfrac{(6)(2^3)}{12} + (12)(2^2) = 4 + 48 = 52$ in.$^4$

Area *B*: $\qquad I_X = \dfrac{(2)(6^3)}{12} + (12)(2^2) = 36 + 48 = 84$ in.$^4$

Moment of inertia of composite area $= I_X = 136$ in.$^4$

The flexure formula, $s = Mc/I$, clearly shows that $s$ is maximum when $c$ is greatest. Also, Fig. 7-5 shows that the stress is maximum in the extreme fibers lying farthest away from the neutral axis. Consequently, for maximum stress, $c = 5$ in.

Finally, since $w = 200$ lb/ft and $L = 10$ ft, the maximum bending moment is

$$\left[ M = \frac{wL^2}{8} \right] \qquad M = \frac{(200)(10^2)}{8} = 2500 \text{ lb·ft} = 30{,}000 \text{ lb·in.}$$

To be used in the flexure formula, $M$ must be expressed in pound-inches, $c$ in inches, and $I$ in inches⁴. Then the maximum stress in the bottom fibers is

$$\left[ s = \frac{Mc}{I} \right] \qquad s = \frac{(30{,}000)(5)}{136} = 1103 \text{ psi} \qquad\qquad Ans.$$

The maximum stress in the top fibers, using a value of $c = 3$ in., is 662 psi.

## PROBLEMS

NOTE: In solving the following problems, full use should be made of the tables in Appendix E giving properties of sections.

**7-2.** A timber beam of cross section as shown in Fig. 7-8 is full 150 mm wide by 300 mm deep. It is subjected to a bending moment $M$ of 25 kN·m. Calculate the maximum bending stress $s$ in this beam.

$Ans.\ s = 11.1$ MPa

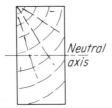

**Fig. 7-8    Probs. 7-2 and 7-3.**

**7-3.** The beam shown in section in Fig. 7-8 is a 6- by 12-in. timber, S4S. It is subjected to a maximum bending moment $M$ of 15 kip·ft. Calculate the maximum bending stress $s$. What is the percentage decrease in stress if the beam is made the full nominal size?

**7-4.** In a problem similar to 7-1, let us assume that the beam is made of a 3- by 12-in. flange and a 4- by 10-in. stem. If the beam is 20 ft long and supports a uniform load of 300 lb/ft including its own weight, what are the maximum tensile and compressive stresses ($\bar{y} = 8.08$ in. and $I_x = 1161$ in.⁴)?

**7-5.** Assume that the box beam mentioned in Prob. 5-16 and shown in Fig. 5-21 is used as a simple span with a length between supports of 20 ft. If the uniformly distributed load is 400 lb/ft including the weight of the beam, what is the value of the maximum fiber stress $s$ ($I_x = 1787$ in.⁴)?

**7-6.** A structural-steel wide-flange section, W 310 × 117, whose cross section is shown in Fig. 7-9, is used as a simple beam 6 m long. It supports a uniform load

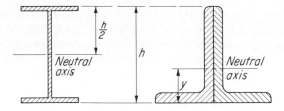

Fig. 7-9   Prob. 7-6.          Fig. 7-10   Prob. 7-8.

of 7 kN/m, which includes the weight of the beam. Calculate the maximum bending stress $s$ in this beam.

**7-7.** A cast-iron beam of the cross section illustrated in Fig. 5-20 and also mentioned in Prob. 5-15 is used as a simple-span beam 8 ft long. It supports a uniform load of 4500 lb/ft including its own weight. Calculate the maximum stresses, $s_t$ in tension and $s_c$ in compression ($\bar{y} = 4$ in.; $I_x = 576$ in.⁴).

*Ans.* $s_t = 3000$ psi; $s_c = 4500$ psi

**7-8.** Two structural-steel angles 5 by 3½ by ½ in. placed with their long legs back to back, as illustrated in Fig. 7-10, are to be used as a lintel beam in a brick wall to span an opening 12 ft wide between supports. The load is triangular and decreases from a maximum of 400 lb per ft at the center of the beam to zero at both ends, giving a total load $W$ of 2400 lb, which includes the weight of the beam. Calculate the maximum bending stress $s$ in this beam.

*Ans.* $s_c = 9600$ psi

**7-9.** A standard steel pipe of 2 in. nominal diameter, used as a clothesline post, extends 72 in. vertically out of a solid concrete base in which it is firmly imbedded. What maximum stress is caused in this pipe by a horizontal pull $P$ at its top of 100 lb?

**7-10.** A footbridge with a clear, simple span of 20 ft uses logs with an average diameter of 12 in. Each log supports a uniform load of 60 lb/ft, which includes its own weight and weight of deck planking, and is also capable of supporting a concentrated load $P$ of 2000 lb at its center. Calculate the maximum bending stress $s$ in one log under this load.

*Ans.* $s = 920$ psi

**7-11.** A 75-mm-diameter steel shaft is 3 m long horizontally between centers of its two supporting bearings. At 1 m from the left end is a downward-acting belt pull of 0.6 kN, and at 2 m from the left is an upward-acting belt pull of 1.2 m, both vertical. Calculate the maximum bending stress in the shaft due to these loads only.

**7-12.** A wood plank 2 in. thick by 12 in. wide and 16 ft long is supported at its outer ends. What stress is produced in this plank by a man weighing 200 lb and standing at its middle when the plank is (*a*) placed on edge and (*b*) laid flat? Assume full dimensions.

*Ans.* (*a*) $s = 200$ psi; (*b*) $s = 1200$ psi

**7-13.** A 1-in.-diameter unfinished bolt is used to fasten together three steel straps, as illustrated in Fig. 3-5. Each outer strap is $^3/_8$ in. thick, and the middle strap is $^5/_8$ in. thick. The bolt holes are $1^1/_8$ in. in diameter. Assuming that the resultants of the bearing stresses against the bolt act along the centerlines of the three straps, calculate the maximum bending stress in the bolt caused by a pull $P$ of 10 kips.

## 7-3 SECTION MODULUS

In general structural practice, the vast majority of beams are selected from available standard stock sizes, the properties of which ($A$, $I$, $c$, etc.) are listed in various easily available handbooks. The properties of non-standard sections and of built-up sections may be calculated by the methods outlined in Chap. 5. The quantity $I/c$ in Eq. (7-1) is called the **section modulus** and is denoted by the symbol $Z$. That is,

$$\text{Section modulus } Z = \frac{I}{c} \tag{7-2}$$

Equation (7-1) then becomes

**The flexure formula:**    $M = sZ$   or   $s = \dfrac{M}{Z}$ $\tag{7-3}$

Since $I$ and $c$ of standard sections are known, their section moduli ($Z$ or $S$ in some literature) are also listed in handbooks. For nonstandard sections and for regular geometric areas, the section modulus may be obtained by calculating the moment of inertia $I$ of the area and then dividing $I$ by $c$, the distance from the neutral axis to the *extreme fiber*. In symmetrical sections, $c$ has only one value, but in unsymmetrical sections $c$ will have two values. In the analysis and design of beams, however, we are usually interested only in the *maximum stress* which occurs in the extreme fiber. In all such problems, the greatest value of $c$ must be used.

Consider, for example, a rectangular area, such as the cross-sectional area of a rectangular timber beam whose width is $b$ and whose depth is $h$. From Table 5-1 its moment of inertia $I$ with respect to its neutral (centroidal) axis is $bh^3/12$. Clearly $c = h/2$ and

**For rectangular areas:**    $Z = \dfrac{bh^3/12}{h/2} = \dfrac{bh^2}{6}$ $\tag{a}$

As an aid in understanding this value of section modulus, an excellent physical concept of the term may be obtained from the following development. From Fig. 7-4, we note that

$$M = C\left(\frac{2}{3}h\right) \tag{b}$$

But $C$ is the resultant of a "stress volume." That is,

$$C = \frac{1}{2}\left(\frac{h}{2}\right)(s)(b) \qquad (c)$$

Consequently

$$M = \frac{1}{2}\left(\frac{h}{2}\right)(s)(b)\left(\frac{2}{3}h\right) \qquad (d)$$

or

$$M = s\,\frac{bh^2}{6} = sZ \qquad \text{(7-4)}$$

from which again we find that the section modulus $Z$ of a rectangular section is $bh^2/6$. We note here also that the section modulus is purely a function of the area of the section. The equation $M = sZ$ is the section-modulus form of the flexure formula which is more generally used by designers and should hereinafter be used in all problems whenever possible and suitable.

## PROBLEMS

**7-14.** Calculate the section modulus $Z$ of a rough, rectangular timber beam whose cross-sectional area is 200 mm wide by 300 mm deep with respect to its horizontal neutral axis.

Ans. $Z = 3.0 \times 10^6$ mm$^3$

**7-15.** If the timber beam in Prob. 7-14 is surfaced on all four sides (S4S), by what percentage has its section modulus, and hence its strength, been reduced from that of the rough beam? Assume each of the given dimensions is reduced by 12 mm, due to surfacing.

**7-16.** A round timber log with a diameter of 10 in. is used as a beam supporting a footbridge which spans a small stream. Calculate its section modulus $Z$ with respect to its neutral axis.

Ans. $Z = 98.17$ in.$^3$

**7-17.** Consider the beam section discussed in Prob. 7-1, for which $I_x = 136$ in.$^4$. Calculate the value of section modulus $Z$ that must be used to obtain the maximum stress ($a$) in tension and ($b$) in compression. ($c$) Which of these two stresses will be the greater?

**7-18.** A beam having the section shown in Fig. 5-16 is to be used on a simple span. Calculate the section modulus $Z$ that would be used in determining the maximum ($a$) tensile stress and ($b$) compressive stress. ($c$) Which of these two stresses will be the greater?

**7-19.** Assume that the beam shown in Fig. 7-7 is built up of two 3- by 8-in. rough planks. Then calculate its section modulus $Z$ with respect to the horizontal neutral axis that must be used to obtain the maximum stress in the beam.

## 7-4 SHEARING STRESSES IN BEAMS

The subject of transverse shearing forces in beams was discussed in Art. 6-4. A discussion of the unit shearing stresses in beams is now in order. In the following development, we will see that both transverse and longitudinal shearing stresses exist and that at any point in a beam the unit transverse and longitudinal shearing stresses are equal. Since beams are normally horizontal and the cross sections upon which bending stresses are investigated are therefore vertical, these shearing stresses in beams are generally referred to as horizontal (longitudinal) and vertical (transverse).

The existence of horizontal shearing stresses in a bent beam can readily be visualized by bending a flexible book or a deck of cards. The sliding of one surface over another, which is plainly visible, is a shearing action, which, if prevented, will set up horizontal shearing stresses on those surfaces.

Consider now a beam made of three identical planks of rectangular cross section, one placed directly above the other, as shown in Fig. 7-11a. Assume the surfaces of contact to be smooth and frictionless. When subjected to a load $P$, as in Fig. 7-11b, all three pieces will bend similarly and independently. Obviously, in this position, the bottom surface of each upper plank is longer than the top surface of the plank below it. Should the movement of these surfaces over one another now be prevented, as, for example, by gluing the three planks together in the unbent position so that they would act as a single beam, as indicated in Fig. 7-11c, longitudinal shearing stresses on the glued planes, and on any other parallel plane within each plank, would result upon reapplication of force $P$.

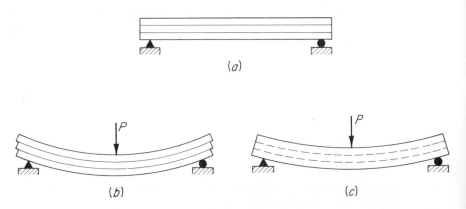

Fig. 7-11 Horizontal shear in beams. (a) Unbent beam of three identical separate planks. (b) Deformation when planks bend separately. (c) Deformation when planks are glued together.

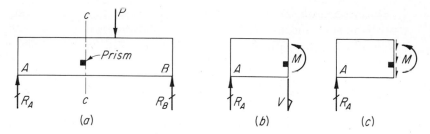

Fig. 7-12   Vertical, or transverse, shearing stresses in beams.

## 7-5 RELATION BETWEEN VERTICAL AND HORIZONTAL SHEARING STRESSES IN BEAMS

From the discussions presented in previous articles it is now seen that in any stressed beam both vertical and horizontal shearing stresses are present as well as the normal tensile and compressive stresses. We will now show that at any point in a stressed beam the vertical and horizontal unit shearing stresses are equal.

Let us consider the loaded beam shown in Fig. 7-12a. When this beam is cut on section $cc$, the shear force $V$ at the cut section (Fig. 7-12b) represents the sum total of all unit vertical shearing stresses on the cut section, as indicated in Fig. 7-12c.

If now we cut out and isolate a small rectangular prism of this beam, as in Fig. 7-13, the following facts become evident. If the dimensions $a$ and $b$ are infinitesimally small, the normal stresses (not shown) on the right and left sides of the prism are equal and opposite and therefore by themselves form a system of forces in equilibrium which may be entirely removed from the prism without altering any remaining forces.

Let now $s_v$ and $s_h$, respectively, be the vertical and horizontal unit shearing stresses on the right and top sides of the prism, the areas of which are $(a)(c)$ and $(b)(c)$. Then the total shear force on the right is

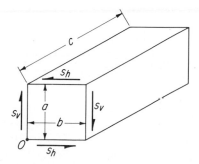

Fig. 7-13   Shearing stresses on elemental prism.

$(s_v)(a)(c)$ and on the top side is $(s_h)(b)(c)$. Since the forces acting on the prism must be and are in equilibrium, it follows that if point $O$ is chosen as a moment center, the moments of the shearing forces on the left and bottom sides of the prism will be zero, and

$[\Sigma M_O]$ $\qquad\qquad s_v ac \cdot b = s_h bc \cdot a$ $\qquad\qquad (a)$

from which

$$s_v = s_h \qquad\qquad (b)$$

*That is, at any point in a stressed beam, the unit vertical and horizontal shearing stresses are equal.* Hence the symbol $s_s$ generally denotes *all* shearing stresses.

### 7-6 THE GENERAL SHEAR FORMULA

The development of the flexure formula in Art. 7-2 provided a means of determining the normal bending stresses, tensile and compressive, at any point in a stressed beam. We will now develop a formula which will give the shearing stress at any point in a stressed beam.

Let us consider a portion of a stressed beam of constant cross section, as shown in Fig. 7-14. From this portion we cut and remove the small piece from the notch in the top surface extending across the beam and part way down to the neutral axis and lying between cross-sectional planes 1 and 2, as shown in Fig. 7-15. We assume that the moment $M_2$ at section 2 is greater than is $M_1$ at section 1. Hence $C_2$, the resultant of the normal stresses on surface 2, is greater than $C_1$, the resultant of the normal stresses on section 1. But since the algebraic sum of the horizontal forces must be zero, an additional horizontal force must be present. Such a force can be exerted only on the bottom surface of the piece and is, of course, the total shear resistance $S$ offered by the fibers on the bottom surface. This shear force equals $C_2 - C_1$.

Consider now the small area $\Delta A$ on the stressed surface 2. According to the flexure formula, Eq. (7-1), the unit normal stress on this small area is $s = My/I$, in which $y$ has replaced $c$ as the distance from the area $\Delta A$ to

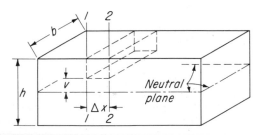

Fig. 7-14    Portion of a stressed beam.

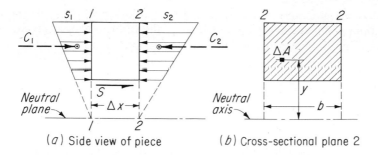

(a) Side view of piece          (b) Cross-sectional plane 2

**Fig. 7-15** Horizontal shear force S equals difference between resultants of opposing normal stresses.

the neutral axis. Then the total stress on area $\Delta A = (My\ \Delta A)/I$ and the total normal force on surface 2 is

$$C_2 = \Sigma \frac{M_2 y\ \Delta A}{I} = \frac{M_2}{I} \Sigma \Delta A\ y \qquad (a)$$

since both $M$ and $I$ are constants at any given section. But $\Sigma \Delta A\ y$ is the sum of all $\Delta A$ areas on surface 2, each multiplied by its distance $y$ to the neutral axis (see Fig. 7-15$b$). This finally becomes $A\bar{y}$, in which $A$ is the stressed area and $\bar{y}$ is the distance from the neutral axis to the centroid of that area. This quantity $A\bar{y}$ is usually denoted by the symbol $Q$ and may be defined as the first or **statical moment** of the stressed area $A$.

It is interesting to note here that the quantity $\Sigma \Delta A\ y$, the *first moment* of the stressed area about the neutral axis, arrived at in the development of the *shear formula,* is analogous to the quantity $\Sigma \Delta A\ y^2$, Eq. ($h$), Art. 7-2, the *second moment* of the stressed area about the neutral axis, arrived at in the development of the *flexure formula.*

When $Q$ is substituted for $\Sigma \Delta A\ y$ in Eq. ($a$), we obtain for both surfaces 1 and 2

$$C_1 = \frac{M_1 Q}{I} \qquad (b)$$

and

$$C_2 = \frac{M_2 Q}{I} \qquad (c)$$

From Fig. 7-15 the difference between $C_2$ and $C_1$ is therefore

$$\frac{M_2 Q}{I} - \frac{M_1 Q}{I} = S \quad \text{or} \quad \frac{(M_2 - M_1)Q}{I} = S \qquad (d)$$

When the beam under consideration is of constant cross section, $Q_1 = Q_2 = Q$ and $I_1 = I_2 = I$.

Before obtaining the final shear formula, we must establish one more relation: the piece shown in Fig. 7-16 is the full top-to-bottom slice of the beam lying between the vertical cross-sectional planes 1 and 2.

On each of the two faces of the slice is shown the shear force $V$ and the moment $M$. Using point $E$ as a moment center, we obtain

$$[\Sigma M_E = 0] \qquad\qquad M_2 - M_1 - V \Delta x = 0 \qquad\qquad (e)$$

or

$$M_2 - M_1 = V \Delta x \qquad\qquad (f)$$

When $V \Delta x$ is substituted for $M_2 - M_1$ in Eq. $(d)$, we obtain

$$\frac{V \Delta x \, Q}{I} = S$$

from which

$$\frac{S}{\Delta x} = \frac{QV}{I} \qquad\qquad (7\text{-}5)$$

where $S/\Delta x$ is the shear flow (usually in pounds per inch of length) on the longitudinal shear surface.

Refer to Fig. 7-15a. The shear force on the bottom surface of the piece is $S$. The unit shearing stress $s_s$ on this surface is the force $S$ divided by the area $(b \, \Delta x)$, or $s_s = S/b \, \Delta x$. Dividing both sides of Eq. (7-5) by $b$ gives

$$\frac{S}{b \, \Delta x} = \frac{QV}{Ib}$$

from which we obtain

**The general shear formula:** $\qquad s_s = \dfrac{QV}{Ib} \qquad\qquad$ **(7-6)**

where $s_s$ = unit shearing stress on horizontal (or vertical) plane at any given cross section of a beam

$V$ = total transverse shear at section

$Q$ = statical moment, about neutral axis, of that part of cross-sectional area lying on either side of horizontal plane at which shearing stress is desired

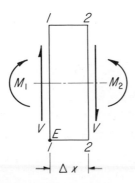

Fig. 7-16  Vertical transverse side of stressed beam.

$I$ = moment of inertia of entire cross-sectional area of beam with respect to its neutral axis

$b$ = width of beam at plane where $s_s$ is calculated

Equation (7-6) is called the **general shear formula** and will give the shearing stress, horizontal or vertical, at any point in a stressed beam.

## ILLUSTRATIVE PROBLEMS

**7-20.** The beam shown in Fig. 7-17 is built up of three 2- by 8-in. planks securely spiked together to act as a single unit. Calculate the unit shearing stresses at plane $aa$ and at the neutral axis caused by a total shear $V$ of 2000 lb. The moment of inertia $I_X$ of the entire section is 895 in.[4].

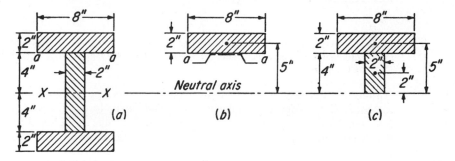

Fig. 7-17  (a) Determination of shearing stresses in a beam. (b) Calculation of Q to determine s, at plane aa. (c) Calculation of Q to determine $s_s$ at neutral axis.

*Solution:* Since $V$ and $I$ are given and $b$ may be obtained by inspection, $Q$ remains the only quantity to be calculated. Now, $Q$ is the first moment about the neutral axis of the area above (or below) the plane on which the shearing stress is desired. The area above plane $aa$ has been isolated and is shown in Fig. 7-17b, and the area above the neutral axis is shown in Fig. 7-17c. At plane $aa$, to illustrate a point, we use both $b = 2$ in. and $b = 8$ in. Then at plane $aa$, $Q = (8)(2)(5) = 80$ in.[3]. Hence, for $b = 2$ in.,

$$\left[ s_s = \frac{QV}{Ib} \right] \qquad s_s = \frac{(80)(2000)}{(895)(2)} = 89.4 \text{ psi} \qquad Ans.$$

When $b = 8$ in.,

$$\left[ s_s = \frac{QV}{Ib} \right] \qquad s_s = \frac{(80)(2000)}{(895)(8)} = 22.4 \text{ psi}$$

These calculations show that at an abrupt change in width of a section there is a correspondingly abrupt change in the unit shearing stress. At the neutral axis, $Q = (8)(2)(5) + (2)(4)(2) = 96$ in.[3]. Hence, for $b = 2$ in.,

$$\left[ s_s = \frac{QV}{Ib} \right] \qquad s_s = \frac{(96)(2000)}{(895)(2)} = 107 \text{ psi} \qquad Ans.$$

**7-21.** The beam shown in Fig. 6-26 has the section shown in Fig. 7-18*a*. The moment of inertia of the entire cross section is 760 in.[4]. Each rivet is capable of transmitting a shear force of 2.44 kips. Find the required rivet spacing at the support *B*.

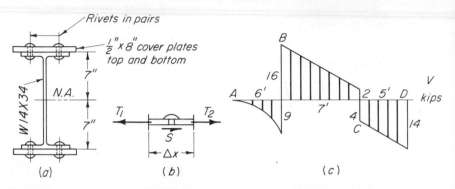

**Fig. 7-18    Determination of longitudinal rivet spacing in cover plate. (a) Beam section. (b) longitudinal forces on cover plate. (c) Shear diagram (see Fig. 6-26).**

*Solution:* Because the moment varies along the beam, the tensile stress in the top cover plate varies in a similar manner. The tensile forces $T_1$ and $T_2$ at half the rivet spacing each side of the rivet will differ by the value of the shear force $S$ (see Fig. 7-18*b*). Since each rivet can transmit a 2.44-kip shear force, a pair of rivets can transmit a force $S = 4.88$ kips.

From the shear diagram of Fig. 6-26 (reproduced in Fig. 7-18*c*), the maximum shear at *B* is $V_B = 16$ kips. Substituting the known values in Eq. (7-5) gives $\Delta x$, the longitudinal spacing of the cover-plate rivets. That is,

$$\left[\frac{S}{\Delta x} = \frac{QV}{I}\right] \qquad \frac{4.88}{\Delta x} = \frac{(8)(0.5)(7.25)(16)}{760} \qquad \text{or} \qquad \Delta x = 8 \text{ in.} \qquad Ans.$$

PROBLEMS_____

**7-22.** Assume in Fig. 7-17 that the web of the I beam is 50 by 300 mm and that each of the two flanges is 50 by 200 mm. The vertical shear force on the section is 12 kN. Calculate $s_s$ at plane *aa* and at the neutral axis.

**7-23.** A rough timber beam 8 in. wide and 12 in. deep, as illustrated in Fig. 7-19,

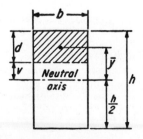

**Fig. 7-19   Prob. 7-23.**

is subjected to a load which causes a maximum transverse shear $V$ of 6400 lb. The maximum unit shearing stress occurs at the neutral axis. Calculate this stress.

*Ans.*  $s_s = 100$ psi

**7-24.** The T beam shown in Fig. 7-20 is made of two 2- by 6-in. planks securely fastened together. Determine the unit shearing stress (*a*) at the top of the flange, (*b*) at the bottom of the flange using both $b = 6$ in. and $b = 2$ in., and (*c*) at the neutral axis, caused by a transverse shear $V$ of 1000 lb ($I_x = 136$ in.⁴).

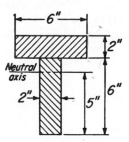

Fig. 7-20  Prob. 7-24.

**7-25.** The timber box beam shown in Fig. 7-21 is made of four planks securely spiked together. Calculate the unit shearing stress at the neutral axis (maximum) caused by a total transverse shear $V$ of 4000 lb ($I_x = 1787$ in.⁴, from Prob. 5-16).

*Ans.*  $s_s = 95.1$ psi

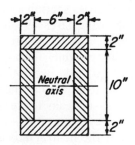

Fig. 7-21  Prob. 7-25.

**7-26.** A structural-steel wide-flange section 360 mm deep by 110 kg/m (W 360 × 110) is used as a simple beam 8 m long. It supports a uniformly distributed load of 90 kN/m, which includes the weight of the beam. Calculate the maximum unit shearing stress $s_s$ in this beam which occurs at the neutral axis.

**7-27.** A structural T section 9 in. deep by 23 lb/ft (WT 9 × 23) is used as a simple lintel beam, flange down, in a brick wall. The span is 12 ft between centers of supports, and the beam carries a load of 480 lb/ft including its own weight. Calculate the unit shearing stress $s_s$ at the neutral axis (maximum) in this beam. Dimensions are those of the lower half of a W 18 × 46.

*Ans.*  Max $s_s = 1245$ psi

**7-28.** The two 2- by 6-in. planks of Prob. 7-24 could be spiked together. If the allowable lateral force on each spike is 120 lb, what spacing would be required? Does this seem to be a practical spacing?

**7-29.** Assume that the T beam in Prob. 7-24 is used on a 10-ft simple span and supports a total uniform load of 50 lb/ft. Determine the maximum permissible spike spacing (*a*) at the end of the beam and (*b*) at a point 2.5 ft from the end of the beam, if the allowable lateral load on each spike is 150 lb.

*Ans.* (*a*) $\Delta x = 3.4$ in.; (*b*) $\Delta x = 6.8$ in.

**7-30.** A W 8 × 21 section is used as a simple beam with a 10-ft span, supporting a total load of 3000 lb/ft. Cover plates $^1/_2$ by 10 in. (similar to those shown in Fig. 7-18*a*) are to be welded to the top and bottom flanges of the wide-flange section with intermittent $^3/_{16}$-in. fillet welds having an effective length of 2 in. Calculate the required longitudinal center-to-center spacing of the welds (*a*) at the supports and (*b*) at the quarter-points of the span.

## 7-7  VARIATION OF SHEARING STRESSES IN BEAMS

Beams must be so designed that they can safely withstand the maximum stresses to which they may be subjected: tensile, compressive, and shearing. The variation of tensile and compressive bending stresses over the cross-sectional area of a beam was discussed and illustrated in Art. 7-2. We now discuss and illustrate the variation of shearing stresses at any section of a beam and show that *with few exceptions, the maximum shearing stress generally occurs at the neutral axis.*

Consider, for example, a rectangular beam of cross section as shown in Fig. 7-22*a*. If we now calculate the unit shearing stresses on several equally spaced horizontal planes in such a beam caused by a total transverse shear *V* and then plot their intensities horizontally from a vertical axis, as illustrated by the graph also shown in Fig. 7-22*a*, we note that in a rectangular beam the shearing stresses increase from zero at the top to a maximum at the neutral axis and then decrease again to zero at the bottom.

Proof that the shearing stress is maximum at the neutral axis of a rectangular beam may be shown as follows. The stress on any random hori-

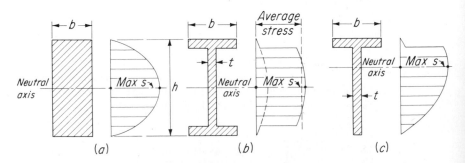

**Fig. 7-22   Variation of shearing stresses in beams. (a) Rectangular beam. (b) I beam. (c) T beam.**

zontal plane, such as that indicated by the dashed line in Fig. 7-19, is $s_s = QV/Ib$, in which $Q$ is the moment of the (shaded) area above the plane with respect to the neutral axis. That is, $Q = A\bar{y}$. Since $A = b(\frac{1}{2}h - v)$ and $\bar{y} = v + \frac{1}{2}(\frac{1}{2}h - v)$, or $\frac{1}{2}(\frac{1}{2}h + v)$, it follows that

$$Q = b\left(\frac{h}{2} - v\right)\frac{1}{2}\left(\frac{h}{2} + v\right) = \frac{b}{2}\left(\frac{h^2}{4} - v^2\right) \tag{g}$$

This equation clearly shows that $Q$, and consequently the unit shearing stress $s_s$, is maximum when $v$ is zero, which it is when the stress desired is on a plane which contains the neutral axis.

The shearing-stress variation over the cross section of an I beam or a wide-flange section is illustrated in Fig. 7-22b. The dashed curve indicates what the stress variation would be if the beam area had remained rectangular with a constant width $b$. This variation, then, would be similar to that shown in Fig. 7-22a. The sudden increase in stress at the bottom of the top flange comes from the change of the width $b$ from $b$ to $t$ in $s_s = QV/Ib$. A similar change occurs at the flange of the T beam in Fig. 7-22c, but here the curve below the neutral axis follows the usual pattern for a rectangular section.

If we multiply the intensity of stress at each of a number of planes in a beam by the width of the beam at that plane and then plot the resulting values horizontally from a vertical reference axis, we obtain a diagram such as shown at the right of Fig. 7-23, in which the central area is a measure of the portion of the total transverse shear resisted by the web of the beam, and the smaller end areas measure the portion resisted by the flanges, which is seen to be relatively small. If the beam shown in Fig. 7-6c is used as an example, we find that the web will resist about 92 percent of the total shear, while the two flanges resist only about 8 percent. As an interesting comparison we find that the web of this beam will resist only about 13 percent of a bending moment to which it may be subjected, while the two flanges will resist about 87 percent.

In some sections, which, however, are rarely used as beams, the maximum shearing stress does not occur at the neutral axis. Consider the H section illustrated in Fig. 7-24a. The centroidal neutral axis of bending lies at $h/2$ from the bottom, while the planes of maximum shearing stress lie at the top and bottom of the horizontal web of the section. In the beam section of the rather unusual shape shown in Fig. 7-24b, the neutral axis

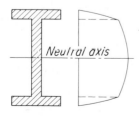

Fig. 7-23    Distribution of shear flow in beam.

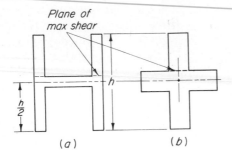

Fig. 7-24  Maximum shearing stress not always at neutral axis. (*a*) H beam. (*b*) Cross.

passes through the centroid, as always, but the maximum shearing stress occurs on the plane between the narrow upper portion and the wider central part.

**PROBLEMS**

**7-31.** Figure 7-19 is a cross section of a rectangular beam 6 in. wide and 12 in. deep. Using the general shear formula, $s_s = QV/Ib$, calculate the unit shearing stresses at the top of the beam and at horizontal planes 2, 4, and 6 in. from the top caused by a total transverse shear $V$ of 5184 lb. Draw a cross-sectional view of the beam. On the right, plot the stresses calculated above to obtain a stress-variation diagram as in Fig. 7-22*a*.

**7-32.** Assume that the beam in Fig. 7-22*b* is a wide-flange section 8 in. deep by 28 lb/ft (W 8 × 28). This beam is subjected to a total transverse shear $V$ of 9800 lb. Calculate the unit shearing stresses at the neutral axis, at 2 in. above the neutral axis, at the bottom of the top flange, using both $b = 0.285$ in. and $b = 6^{1}/_{2}$ in., and at the top of the upper flange. Draw a cross-sectional view of the beam. On the right, plot the stresses obtained above to obtain a stress-variation diagram as in Fig. 7-22*b* ($I_x = 98$ in.$^4$).

*Ans.* Max $s_s = 4408$ psi

**7-33.** A beam has the section shown in Fig. 7-24*a*. All thicknesses are 40 mm, $h = 280$ mm, and the total width of the beam is 280 mm. Calculate and plot the shearing stress at 40-mm intervals from top to bottom of the beam and at the neutral axis. The shear $V$ is 147.4 kN.

## 7-8 SPECIAL SHEAR FORMULAS

The problem of determining the *maximum shearing stresses* in beams is one that constantly confronts designers. The rectangular and circular cross sections occur frequently. For these simple geometric shapes, two greatly simplified maximum-shear formulas may readily be derived. Also, we develop here a very useful formula giving the approximate, but sufficiently accurate, shearing stresses in channels, I beams, and wide-flange sections.

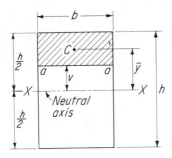

Fig. 7-25   Rectangular-area shear formula.

**Rectangular-Area Shear Formula.** As was shown in Art. 7-7, the maximum shearing stress in a beam of rectangular cross section occurs at the neutral axis. For calculating $s_s$ at the neutral axis, $v$ in Fig. 7-25 becomes zero. The shaded area is then $A = bh/2$, $\bar{y} = h/4$, and

$$Q = A\bar{y} = \frac{bh}{2}\frac{h}{4} = \frac{bh^2}{8}$$

Also, from Table 5-1, $I_X = bh^3/12$. Consequently for rectangular sections the maximum shearing stress is

$$\left[s_s = \frac{QV}{Ib}\right] \qquad s_s = \frac{(bh^2/8)V}{(bh^3/12)b} = \frac{3V}{2bh} \qquad \text{or} \qquad s_s = \frac{3V}{2A} \qquad (7\text{-}7)$$

in which $A$ is the entire cross-sectional area. Equation (7-7) indicates that the maximum stress in a beam of rectangular cross section is 50 percent greater than the average stress, which would be $s_s = V/A$.

**Circular-Area Shear Formula.** The shearing stress in a circular section (see Fig. 7-26) is also maximum at the neutral axis, as can be shown. Hence, in similar manner, with respect to its centroidal neutral axis, the area above the neutral axis is $A = \pi R^2/2$, $\bar{y} = 4R/3\pi$ (from Fig. 5-2), and therefore $A\bar{y} = Q = 2R^3/3$. Also, from Table 5-1, $I_X = \pi R^4/4$ and at the neutral axis $b = 2R$. Consequently for *circular sections the maximum shearing stress is*

$$\left[s_s = \frac{QV}{Ib}\right] \qquad s_s = \frac{(2R^3/3)V}{(\pi R^4/4)2R} = \frac{4V}{3\pi R^2} \qquad \text{or} \qquad s_s = \frac{4V}{3A} \qquad (7\text{-}8)$$

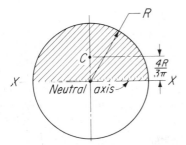

Fig. 7-26   Circular-area shear formula.

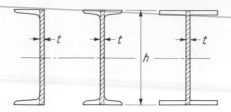

Fig. 7-27   Average web-shearing stress.
The web-shear formula.

in which $A$ is the entire circular area. That is, the maximum shearing stress in a circular beam is one-third greater than the average.

**Web-Shear Formula.** The problem of determining the precise shearing stress in some of the commonly used structural beams such as channels, I beams, and wide-flange sections by use of the general shear formula is a laborious one at best. However, in most cases in practice, the actual shearing stresses in such beams seldom exceed 50 percent of the allowable shearing stress. We know also that the web of such a beam usually resists about 85 to 90 percent of the total transverse shear, as is indicated in Fig. 7-22*b* and *c*. Consequently the following web-shear formula is extremely useful, since, in most cases, it will give an *approximate* shearing stress which is seldom more than 10 to 15 percent *below* that obtained by the precise general shear formula. This web-shear formula, based on the assumption that the web of the beam (see Fig. 7-27), resists all of the transverse shear $V$, is

**The web-shear formula:** $$s_s = \frac{V}{th} \qquad (7\text{-}9)$$

where $V$ = transverse shear, lb or N
$\quad\quad t$ = thickness of web of beam, in. or mm
$\quad\quad h$ = total depth of beam, in. or mm

Clearly, Eq. (7-9) gives only an *average* web-shearing stress. This average stress is generally not more than 10 percent below the actual maximum stress in channels, I beams, and wide-flange sections and is, therefore, satisfactory for investigative purposes. In T beams, the average web-shearing stress may be as high as 33 percent below the maximum.

**PROBLEMS**

**7-34.** Let the beam in Fig. 7-25 be 6 in. wide and 10 in. deep, full size, and assume it to be subjected to a total transverse shear $V$ of 4800 lb. Using the rectangular-area shear formula, calculate the maximum shearing stress $s_s$.

*Ans.* $s_s = 120$ psi

**7-35.** In Prob. 7-34 calculate $s_s$ if the beam is 6 by 10 in. S4S. (NOTE: The maximum shearing stress in this S4S beam is more than 14 percent greater than in the rough beam of the same nominal dimensions.)

**7-36.** A contractor must provide a rectangular timber beam for a simple span of 8 ft to carry a uniform load. He has available three rough beams, 8 by 12 in., 10 by 10 in., and 10 by 12 in., all known to be sufficiently strong in bending. Allowable shearing stress is 100 psi. Which beam should he use for greatest economy (a) if the load is 1600 lb/ft including weight of beam and (b) if the load is 1800 lb/ft?

**7-37.** Assume that Fig. 7-26 represents the cross section of a steel shaft whose diameter is 50 mm and that this shaft is subjected to a transverse shear $V$ of 100 kN. Then calculate the maximum longitudinal shearing stress $s_s$.

*Ans.* $s_s = 67.9$ PMa

**7-38.** In Prob. 7-37 let the diameter of the shaft be $2^1/_2$ in. and let $V = 3750$ lb. Then calculate the maximum $s_s$.

**7-39.** The floor joists in a rustic building are 8-in.-diameter logs 18 ft long carrying a uniform load of 160 lb/ft. Calculate the maximum unit shearing stress $s_s$.

*Ans.* $s_s = 38.2$ psi

**7-40.** A steel channel 10 by $2^5/_8$ in. by 15.3 lb/ft used as a beam is subjected to a maximum transverse shear $V$ of 4800 lb. (a) Using the web-shear formula, calculate the average web-shearing stress $s_s$ in this channel. (b) The precise maximum shearing stress in this beam is 2360 psi. By what percentage is the average web-shearing stress lower than the precise maximum?

**7-41.** A standard steel I beam 310 mm deep by 52 kg/m (S310 × 52) resists a transverse shear $V$ of 100 kN. (a) Calculate the average web-shearing stress and (b) the maximum shearing stress in this beam using Eq. (7-6).

## 7-9 LOAD CAPACITY OF GIVEN BEAMS

Beam problems are generally of three types: to determine (1) the unknown stresses in a given-size beam due to a given load, (2) the unknown load capacity of a given-size beam, using given allowable stresses, or (3) the required size of beam to support a given load, using given allowable stresses. The first type is covered in Arts. 7-2 and 7-6, the third type is covered in Chap. 8, and the second type is discussed here.

### ILLUSTRATIVE PROBLEM

**7-42.** Find the value of the loads $P$ that can safely be supported by a rough 6-by 12-in. Douglas fir beam, as shown in Fig. 7-28a. The maximum allowable stress in bending is 1450 psi and in shear is 120 psi.

*Solution:* The loads $P$ will be determined by the ability of the beam to carry the loads in bending, $s = M/Z$, or in shear, $s_s = 3V/2A$. Both computations should be made, and the least load so computed will be the controlling one. Shear will control in the case of comparatively short beams and bending in the case of long beams.

The shear and bending-moment diagrams are shown in Fig. 7-28b. The maximum bending moment is $PL/3 = P(18)/3 = 6P$ lb·ft. The section modulus $Z$ of

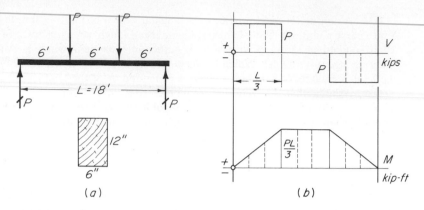

**Fig. 7-28  Prob. 7-42. (a) Load and beam. (b) Shear and moment diagram.**

the beam is $bh^2/6 = (6)(12)(12)/6 = 144$ in.³. From the flexure formula, Eq. (7-3),

$$[M = sZ] \qquad 6P(12) = (1450)(144) \qquad \text{or} \qquad P = 2900 \text{ lb}$$

The special shear formula for rectangular beams, Eq. (7-7), can be used. Since $V_{max} = P$,

$$\left[ s_s = \frac{3V}{2A} \right] \qquad 120 = \frac{3}{2}\frac{P}{72} \qquad \text{or} \qquad P = 5760 \text{ lb}$$

The maximum permissible value of $P$ is the least of these two computed values, or

$$P = 2900 \text{ lb} \qquad\qquad Ans.$$

If the weight of the beam is to be considered, the effect may be approximated by subtracting its weight from the allowable total load. Assuming the beam weighs 20 lb/ft,

$$P = 2900 - (20)(9) = 2720 \text{ lb}$$

**PROBLEMS**

**7-43.** Solve Prob. 7-42 if the maximum allowable stress in bending is 1800 psi and in shear is 140 psi.

*Ans. P* $= 3600$ lb

**7-44.** Solve Prob. 7-42 if the beam is a rough 6- by 18-in. timber.

**7-45.** A 150- by 300-mm rough timber beam and 4 m long between supports must support a concentrated load $P$ at a point 1.5 m from its left end. If allowable stresses are 12 MPa in bending and 1.0 MPa in shear, calculate the maximum permissible value of $P$.

**7-46.** Calculate the maximum load per foot $w$ which may safely be carried by a Ponderosa pine No. 1 grade beam 12 by 18 in. S4S, 20 ft long. The beam weighs 55.9 lb/ft. (See Table 8-1).

*Ans. w* $= 815$ lb/ft

**7-47.** A warehouse floor is supported by 3- by 12-in. joists of Douglas fir dense No. 1 S4S, 16 ft long between centers of end supports and spaced 16 in. on centers. Calculate the allowable uniformly distributed load $w$ in pounds per square foot of floor area if weight of joists is disregarded. (See Table 8-1 and Appendix E-2.)

**7-48.** If the joists in Prob. 7-47 are 100 by 300 mm (S4S) in size and are spaced to 0.5 m on centers, with a 6-m span, what maximum uniformly distributed load can the floor safely support if weight of the joists is disregarded?

*Ans.* $m = 603$ kg/m²

**SUMMARY** (By Article Number)‗‗‗‗‗‗‗‗‗‗‗‗‗‗‗‗‗‗‗‗‗‗‗‗‗‗‗‗‗‗‗‗‗‗

**7-1.** In a bent beam, tensile stresses are produced on the convex side, where the fibers have been stretched, and compressive stresses are produced on the concave side, where the fibers have been compressed. The plane at which the bending stresses are zero, and also change from tensile to compressive, is called the **neutral plane.**

These tensile and compressive stresses increase uniformly from zero at the neutral plane to a maximum at the extreme fiber. The intersection of the neutral plane and any cross-sectional plane of a beam is called the **neutral axis.**

**7-2.** The relation among bending moment $M$, fiber stress $s$, and moment of inertia $I$ at any cross section of a beam is expressed by the **flexure formula,** used in all beam design:

**Flexure formula:** $\qquad M = s\dfrac{I}{c} \qquad \text{or} \qquad s = \dfrac{Mc}{I}$ $\qquad\qquad$ (7-1)

in which $c$ is the distance from the neutral axis to the extreme fiber.

The neutral axis of the cross-sectional area of a beam always passes through the centroid of the area, and it is therefore readily located.

**7-3.** The quantity $I/c$, called the **section modulus,** is denoted by the symbol $Z$. The flexure formula then becomes

**Flexure formula:** $\qquad M = sZ \qquad \text{or} \qquad s = \dfrac{M}{Z}$ $\qquad\qquad$ (7-3)

**7-4.** Shearing stresses, horizontal and vertical, are induced in all beams that are subjected to transverse loads.

**7-5.** At any point in any beam subjected to external shearing forces, the unit vertical and horizontal shearing stresses are equal.

**7-6.** The shear flow $S$ that accumulates in the distance $\Delta x$ can be found from

$$\frac{S}{\Delta x} = \frac{QV}{I} \qquad\qquad (7\text{-}5)$$

The **general shear formula** gives the unit shearing stress at any point in any beam. It is

**General shear formula:** $\qquad s_s = \dfrac{QV}{Ib}$ $\qquad\qquad$ (7-6)

**7-7.** Shearing stresses in beams vary from zero at the top of the beam to a maximum, which generally but not always occurs at the neutral axis, and then to zero again at the bottom.

**7-8. Two special shear formulas** which will give the *maximum unit shearing stresses at the neutral plane* of a rectangular beam and of a circular beam are

Rectangular-area shear formula:  $\qquad s_s = \dfrac{3V}{2A}$  (7-7)

Circular-area shear formula:  $\qquad s_s = \dfrac{4V}{3A}$  (7-8)

In beams with sections such as standard types of channels, I beams, and wide-flange sections, approximately 85 to 90 percent of the total transverse shear is resisted by the web.

Web-shear formula:  $\qquad s_s = \dfrac{V}{th}$  (7-9)

Assuming that the web resists all the shear, we find that the web-shear formula will give an average web-shearing stress which seldom is more than 10 to 15 percent low.

REVIEW PROBLEMS

**7-49.** A round, tapering, straight wood pole, used as a radio aerial mast, extends 27 ft vertically upward from a concrete base into which it is solidly imbedded. Its diameter at the base is 6 in. and at the top is 3 in. If the horizontal pull *P* at the top of the mast is 100 lb, (*a*) what maximum bending stress is produced in the pole and (*b*) what is the maximum shearing stress? (*c*) Why is the shearing stress so low?

**7-50.** A 1-in. round steel reinforcing bar is 20 ft long and weighs 2.67 lb/ft. (*a*) What maximum bending stress is produced in this bar by its own weight if a person holds it horizontally, grasping the bar with one hand at its midpoint, and (*b*) what is the maximum shearing stress? (*c*) What are these same stresses if the bar is held horizontally by two people, one at each end? (*d*) Why is the shearing stress so low?

*Ans.* (*a*) $s = 16{,}000$ psi; (*b*) $s_s = 45.3$ psi; (*c*) same

**7-51.** Calculate the maximum unit bending stress in a simple timber beam 6 by 12 in. S4S caused by its own weight of 17.5 lb/ft when its length between end supports is (*a*) 10 ft, (*b*) 20 ft, and (*c*) 40 ft. If the allowable bending stress in this timber is 1200 psi, what percentage of the capacity of the beam is nullified by its own weight for each given length?

*Ans.* (*a*) $s = 21.7$ psi $= 1.81$ percent; (*b*) $s = 86.8$ psi $= 7.23$ percent; (*c*) $s = 347$ psi $= 28.9$ percent

**7-52.** Calculate the maximum unit horizontal shearing stress in the timber box beam shown in Fig. 7-21 at the junction between the top plank and the side planks caused by a transverse shear *V* of 4000 lb ($I_x = 1787$ in.$^4$).

**7-53.** Assume that the timber T beam mentioned in Prob. 5 11 is used as a simple beam 2 m long and supports a uniform load of 1 kN/m including its own weight. Calculate (*a*) the maximum bending stress in this beam and (*b*) the maximum shearing stress.

**7-54.** A structural-steel wide-flange section 16 by 7 in. by 40 lb/ft (W 16 × 40) is subjected to a maximum transverse shear *V* of 15,000 lb. Its flange is 7 in. wide by 0.503 in. thick. Its web is 0.305 in. thick. Calculate (*a*) the average web-shearing stress; (*b*) the precise maximum shearing stress; (*c*) the percentage by which the average stress is lower than the maximum.

**7-55.** Let the beam of Prob. 7-21 be a W 12 × 40 and the cover plates be $^3/_4$ by 8 in. Solve for the maximum rivet spacing at support *D*.

**7-56.** A 100-mm-OD by 10-mm-thick steel pipe extends 3 m horizontally out of a solid concrete pier. It supports a vertical load of 1 kN at its extreme outer end. Calculate (*a*) the maximum bending stress in this beam and (*b*) the maximum shearing stress. Disregard weight of pipe.

*Ans.* (*a*) $s$ = 51.8 MPa; (*b*) $s_s$ = 0.702 MPa

**7-57.** The beam whose cross section is shown in Fig. 5-18 and is further described in Prob. 5-13 is built up of three 20- by 120-mm wood pieces securely glued together. If the maximum allowable shearing stress in this wood is 0.8 MPa, calculate the greatest value of transverse shear *V* to which this beam may safely be subjected.

**7-58.** A beam of cross section similar to that shown in Fig. 7-24*b* is subjected to a total transverse shear *V* of 1600 lb. Assume that the vertical web is 2 in. wide and 10 in. deep, the horizontal web is 8 in. wide and 4 in. deep, and the section is symmetrical. Calculate the unit shearing stress (*a*) at the horizontal neutral axis and (*b*) at the horizontal plane where the two webs join. (NOTE: In this case the maximum shearing stress does not occur at the neutral axis.)

*Ans.* (*a*) $s_s$ = 37.2 psi; (*b*) $s_s$ = 84.5 psi

**7-59.** The height *h* in Fig. 7-24*b* is 12 in., and the width at the neutral axis is 10 in. All thicknesses are 2 in. If the section is subjected to a vertical shear force *V* of 4 kips, calculate the maximum shear stress $s_s$ and compare with the shear stress at the neutral axis.

*Ans.* Max $s_s$ = 238.7 psi; 438 percent of $s_s$ at neutral axis

**7-60.** Assume that the bolt in Fig. 3-5 is 1 in. in diameter, that the center plate is $^1/_2$ in. thick, each side plate is $^1/_4$ in. thick, and the resultant of the bearing stresses against each plate acts along its centerline. Calculate the maximum bending and shearing stresses in this bolt due to a pull *P* of 12,000 lb.

*Ans.* $s$ = 22,900 psi; $s_s$ = 10,200 psi

**7-61.** Assume that the beam whose cross section is shown in Fig. 7-17 is built up of three 2- by 8-in. rough wood planks securely spiked together and that it is supported at its outer ends. The maximum allowable stresses in this wood are 1200 psi in bending and 120 psi in shear. The length of the beam is 12 ft. What maximum concentrated load *P* may it safely support at its center (*a*) based on bending and (*b*) based on shear ($I_X$ = 896 in.$^4$)?

**7-62.** The wood joists supporting the floor in a small building are of Douglas fir select structural grade, 50- by 300-mm S4S, 6 m long between end supports and spaced $1/3$ m on centers. If the floor itself, without joists, has a mass of 50 kg/m², what additional load in kg/m² may safely be imposed on it? See Table 8-1.

*Ans.* Load = 341 kg/m²

## REVIEW QUESTIONS

**7-1.** Name the types of internal stresses to which a transversely loaded beam is subjected.

**7-2.** What is meant by the term "pure bending"? What stresses are induced in a beam subjected only to pure bending?

**7-3.** Describe in general the deformations and stresses in a beam subjected to pure bending.

**7-4.** Define the terms "neutral plane" and "neutral axis."

**7-5.** The flexure formula defines the relation between which three quantities?

**7-6.** By what other name is the moment of inertia of an area also known?

**7-7.** Give a practical definition of the term "moment of inertia of an area."

**7-8.** Through what point of an area does the neutral axis of bending always pass?

**7-9.** How is the section modulus of an area with respect to its neutral axis of bending most easily obtained?

**7-10.** When the neutral axis of an unsymmetrical beam section lies nearer the top than the bottom, in which of the two section moduli is a designer generally interested?

**7-11.** Describe a demonstration by which the existence of horizontal shearing stresses in beams may easily be visualized.

**7-12.** What relation exists between the unit vertical and horizontal shearing stresses at any point in a beam?

**7-13.** In the general shear formula $s_s = QV/Ib$, what are the quantities $Q$, $V$, $I$, and $b$?

**7-14.** Describe the variation of shearing stresses from top to bottom at any cross section of a rectangular beam.

**7-15.** What causes the sudden change in intensity of shearing stress at the junction of the flange and the web of an I beam?

**7-16.** Name several beam sections in which the shearing stress is maximum at the neutral axis and some section in which it is not.

**7-17.** Name the three special shear formulas and describe the application of each.

**7-18.** By what percentage is the maximum shearing stress greater than the average $(a)$ in a rectangular beam and $(b)$ in a circular beam?

**7-19.** In channels, I beams, and wide-flange sections, by what approximate percentage is the average web-shearing stress lower than the precise maximum?

# Chapter 8
# Design of Beams

## 8-1 CONSIDERATIONS IN BEAM DESIGN

Beams must be so designed that they can safely withstand the maximum stresses to which they are subjected by the imposed loads, both in bending and in shear. The design formulas are the flexure formula $s = Mc/I$ or its section-modulus form $s = M/Z$ and the shear formula $s_s = QV/Ib$ or the appropriate one of its special simplified forms.

Since, in the design of a beam, the bending moment $M$ and allowable stress $s$ are known, its size and shape are the unknown quantities to be determined. Because size and shape are functions of $I/c$, or $Z$, the design forms of the flexure formula become $I/c = M/s$ and $Z = M/s$. The latter of these is generally used.

To accomplish in a single mathematical operation the design of a beam which would meet the strength requirements both in bending and in shear is a laborious process. Instead, the designer usually estimates which of the two stresses will be the critical one, and then proceeds with the design to satisfy that requirement. The chosen beam is thereafter investigated to determine the maximum value of the other stress. If the latter does not exceed the allowable, the design is complete. If it does, a stronger beam must be selected.

In general, the bending stresses are usually critical in "relatively long" beams, while shearing stresses are critical in "relatively short" beams. For example, of four beams all 16 ft in length, a 6- by 8-in. timber beam and a 7-in. steel I beam are relatively long, while a 10- by 20-in. timber beam and a 20-in. I beam are relatively short. The usual practice, however, is to design beams for bending and then to investigate them for maximum shearing stresses. In some cases, the deflection of a beam may be the governing factor. This problem is discussed in Chap. 9.

The **weight of a beam** must be considered in its design. That is, the beam finally selected must be capable of supporting its own weight in addition to the imposed load. In some cases, however, the weight of a beam may be disregarded with insignificant error. In the special case of simple-span timber beams carrying uniformly distributed loads, an estimated weight of beam of 5 percent of the imposed load may be added for design purposes, as illustrated in Prob. 8-1. If the actual weight of the beam selected differs from the estimated weight by more than 50 percent, a revision of the design may be necessary. For all other beams, timber and steel, it is advisable first to select a trial beam based on the moment $M$ and required section modulus $Z$ due to load only and then to calculate the additional moment and additional required section modulus due to the weight of the trial beam. A final selection is then based on the total required section modulus, as will be illustrated in Probs. 8-2, 8-3, 8-22, and 8-23. While not true in all cases, the critical section due to both imposed load and weight of beam may be considered to coincide with the critical section due to load only. Hence the additional moment $M$ due to weight of beam should be that at the section of critical load moment.

The **most economical beam** is the one which has the least cross-sectional area (or the least weight per foot of length) and still satisfies all strength requirements and other specified limitations. In general, it will be found that a deep beam is more economical than a shallower beam of equal strength. Unless otherwise specified, it is taken for granted in general practice that the beam selected will be the most economical (lightest) one. This rule applies also to all problems in this text.

**Lateral Support of Beams.** All beams under load have a tendency to buckle sidewise (laterally) between supports, especially beams that are relatively narrow and deep. In the vast majority of cases, however, such buckling is prevented by furnishing lateral support for the beams. In buildings of the usual type, for example, all floor beams are laterally supported, since the floors they support are fastened to them. Laterally unsupported beams such as crane rails and others are occasionally found in industrial buildings and yards. Such beams require special treatment in their design as illustrated in Art. 8-4. Unless otherwise stated, all beams are presumed to be laterally supported.

## 8-2 DESIGN OF TIMBER BEAMS

Because timber is not a manufactured material, like steel and other metals, wide variances in its strength under different conditions of stress are to be expected. Timber has its greatest usable strength in tension and in bending. In compression parallel to the grain, as in columns, its estimated strength is about 75 percent of that in bending, while in bearing (compression perpendicular to the grain) and in horizontal shear its strength is approximately 20 and 5 percent, respectively, of that in bending. Consequently, shearing stresses in timber beams must be closely watched.

The **span of a timber beam** is considered to be from face to face of girder if the beam is supported on steel hangers fastened to the sides of the supporting girders, or from center to center of girders if the beams rest on top of the supporting girders. Floor girders generally rest on steel column caps especially made for that purpose, and girder spans are then considered to extend from face to face of supporting columns. In general, the span of a simple beam is considered to be the distance between the centers of its two supports. Correct determination of span is important, since the bending moment and hence the required section modulus vary as the square of the span ($M = wL^2/8$). For example, for a span of 20 ft the bending moment is $400w/8$, while for a 21-ft span it is $441w/8$, a difference of 10 percent.

The actual dimensions of rough timbers are presumed to equal their nominal dimensions. Actual dimensions and other properties of dressed timbers (S4S) are given in Table E-2. Three design methods are illustrated below. Methods $A$ and $B$ are basic in concept, while method $C$ makes use of available tabulated information. While experienced designers use such tables extensively, beginners should not use them until they have first mastered the technique of beam design using only fundamental concepts and formulas.

The **depth of a timber beam** is most often about twice its width. In relation to span, a workable rule of thumb is that the depth should be about 1 in. per foot of span (20 in. for a 20-ft span) for uniformly loaded beams and about 1.25 in. for beams and girders subjected to concentrated loads. Narrow, deep timber beams are more economical than are wide, shallow beams, but they are often weak in shear. In general structural practice, however, the width-depth ratio ($b/h$) of structural timber beams is usually not less than ⅓ nor more than ¾. Unless otherwise specified, these limits should be observed in the following timber-beam-design problems. Obviously, the limits apply only to structural beams and not to joists or other "special" beams.

It is important to realize here that in any beam-design problem, the maximum shear $V$ and moment $M$ must always be determined in the usual and most suitable manner. Thereafter, the required section modulus ($Z = M/s$) and cross-sectional area ($A = 1.5V/s_s$) are calculated. From this point on, several methods are available for completing the design, as illustrated in Probs. 8-1 to 8-3.

**Design method A** solves the equation $Z = bh^2/6$ (or $6Z = bh^2$) by the substitution of different values of $b$ (of *available* timber sizes) and uses for convenience a tabulation of the results. This method is especially useful to the beginner in that it shows that several beam sizes are available which satisfy the requirements of strength but that only a few of these are suitable and that other logical restrictions generally reduce the choice to a single beam.

**Design method B,** illustrated in Prob. 8-2, makes use of a predetermined width-depth ratio $(b/h)$, usually ⅓, ½, ⅔, or ¾, whichever best satisfies the requirements as they are known to the designer. This predetermination serves two functions: (1) it represents the designer's estimate of the most desirable ratio for the particular beam being designed and (2) it permits solution of any otherwise insoluble equation containing two unknowns, $b$ and $h$. Any ratio so predetermined and used may, however, in the beam finally selected, vary by one size in either dimension, $b$ or $h$. It must be realized that this predetermination of a ratio of $b$ and $h$ is entirely up to the designer and cannot be fixed by any text or code. When used in problems in this text, it represents, of course, only the judgment of the authors.

**Design method C** makes full use of tables of available beam sizes and their properties, such as area, weight, moment of inertia, and section modulus. Where the required values of $Z$ and $A$ have been calculated, a suitable beam is quickly selected from such a table. See Appendix Table E-2.

### ILLUSTRATIVE PROBLEMS

**8-1. Method A.** Design the most economical rectangular timber beam 6 m long simple span of Douglas fir dense select structural grade S4S to support a uniformly distributed load of 5 kN/m. Weight of beam must be considered. For practical reasons, the ratio of $b/h$ must not be less than ¼ nor more than ¾.

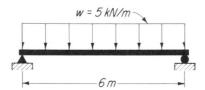

Fig. 8-1  Design of timber beam.

*Solution:* While this method of design is not generally used in practice, its purpose here is to illustrate that several beams may be available which satisfy the strength requirements but that only one, of course, will be the most economical. We first design this beam for bending based only on the given load plus the estimated weight of the beam and later consider its safety in shear.

First we solve for the support reactions $R$ and the maximum bending moment $M$ and transverse shear $V$. Clearly, for the conditions of this problem, $R = wL/2$, $M = wL^2/8$, and $V = R$. The estimated weight of the beam is $5000(0.05) = 250$ N/m. Then, since $w = 5000 + 250 = 5250$ N/m and $L = 6$ m,

$$[R = \tfrac{1}{2}wL] \qquad\qquad R = (\tfrac{1}{2})(5250)(6) = 15\ 750 \text{ N}$$

$$\left[M = \frac{wL^2}{8}\right] \qquad\qquad M = \frac{5250(6)^2}{8} = 23\ 625 \text{ N·m}$$

Table 8-1 ALLOWABLE UNIT STRESSES FOR
VISUALLY GRADED LUMBER, psi (MPa)*

| Species and Commercial Grade | Extreme Fiber in Bending | Tension | Longi-tudinal Shear | Compression† | | | Modulus of Elasticity |
|---|---|---|---|---|---|---|---|
| | | | | Perpen-dicular to Grain | Parallel to Grain | | |
| Douglas fir; larch: | | | | | | | |
| Dense select | 1900 | 1100 | 85 (0.60) | 455 | 1300 | | 1,700,000 |
| structural | (13.1) | (7.6) | | (3.14) | (9.0) | | (12 000) |
| Select structural | 1600 | 950 | 85 (0.60) | 385 | 1100 | | 1,600,000 |
| | (11.0) | (6.5) | | (2.65) | (7.6) | | (11 000) |
| Dense No. 1 | 1550 | 775 | 85 (0.60) | 455 | 1100 | | 1,700,000 |
| | (10.7) | (5.3) | | (3.14) | (7.6) | | (12 000) |
| No. 1 | 1300 | 675 | 85 (0.60) | 385 | 925 | | 1,600,000 |
| | (9.0) | (4.7) | | (2.65) | (6.4) | | (11 000) |
| Hemlock, eastern: | | | | | | | |
| Select structural | 1350 | 925 | 80 (0.55) | 360 | 950 | | 1,200,000 |
| | (9.3) | (6.4) | | (2.50) | (6.5) | | (8 300) |
| No. 1 | 1100 | 775 | 80 (0.55) | 360 | 800 | | 1,200,000 |
| | (7.6) | (5.3) | | (2.50) | (6.5) | | (8 300) |
| Pine, Ponderosa or lodge-pole: | | | | | | | |
| Select structural | 1100 | 725 | 65 (0.45) | 235 | 750 | | 1,100,000 |
| | (7.6) | (5.0) | | (1.60) | (5.2) | | (7 600) |
| No. 1 | 925 | 625 | 65 (0.45) | 235 | 750 | | 1,100,000 |
| | (6.4) | (4.3) | | (1.60) | (5.2) | | (7 600) |

\* Some differences exist among building codes in the matter of allowable unit stresses for timber. Those given in this table are for lumber used under conditions continuously dry, such as in most covered structures, and for "normal" load conditions. Allowable stresses for other conditions are given in "National Design Specification," published by the National Forest Product Association, Washington, D.C.
† Compression parallel to grain in "short" columns. See column formulas, Chap. 12.

From Table 8-1, we find the allowable stress for Douglas fir dense select structural to be 13.1 MPa in bending and 0.6 MPa in horizontal shear. Since $s = M/Z$, $Z = M/s$, and the required section modulus then is

$$\left[ Z = \frac{M}{s} \right] \qquad Z = \frac{23\ 625\ 000}{13.1} = 1.8 \times 10^6 \text{ mm}^3$$

For a rectangular section, $Z = bh^2/6$. Consequently

$$bh^2 = (6)(1.8 \times 10^6) = 10.8 \times 10^6$$

Having found that $bh^2 = 10.8 \times 10^6$, we may then substitute various values for $b$ and solve for $h^2$ and $h$. Table 8-2 gives a clear view of the results. The third column lists the nearest available sizes, obtained from Table E-2. The fifth column gives the cross-sectional area of each beam, also from Table E-2.

Table 8-2 DESIGN OF TIMBER BEAM (Dimensions in mm or mm² for areas)

| $b$ | $h = \sqrt{\dfrac{bh^2}{b}}$ | Beam | Area for Cost Comparison | S4S Area |
|---|---|---|---|---|
| 90 | 347 | 100 × 400 (90 × 388) | 40 000 | 34 920 |
| 140 | 278 | 150 × 300 (140 × 288) | 45 000 | 40 300 |
| 188 | 240 | 200 × 300 (188 × 288) | 60 000 | 54 100 |

Since all three of these beams satisfy the strength requirements and the width-depth ratio limitations, we select the most economical one whose stress in horizontal shear is within the allowable limit of 0.6 MPa.

Using the special shear equation $s_s = 3V/2A$, we now may solve for $s$ in each beam. But it will be simpler here to solve for the required area $A$, which is then compared with the already known areas of the three tabulated beams. That is, $A = 3V/2s_s$. Since $V = 15\ 750$ N and $s_s = 0.6$ MPa,

$$\left[ A = \frac{3V}{2s_s} \right] \qquad \text{Required } A = \frac{(3)(15\ 750)}{(2)(0.6)} = 39\ 375 \text{ mm}^2$$

Obviously, the first beam would be weak in shear, since it has insufficient area. The 150- by 300-mm beam is the logical choice. From Table E-2 its weight is 254 N/m, very slightly more than its estimated weight of 250 N/m. Hence no revision of the design is necessary.

Use 150- by 300-mm S4S Douglas fir beam.      Ans.

**8-2. Method B.** Design the most economical rough timber beam of Douglas fir No. 1 grade 24 ft long to support the load shown in Fig. 8-2a plus its own weight. A ratio of width $b$ to depth $h$ of approximately ½ is considered suitable for this beam.

*Solution:* Because the load is irregular while the weight of the beam is uniformly distributed, we tentatively design this beam for the given load only, leaving its weight for later consideration, In this case, a set of load, shear, and moment diagrams will be helpful, although the maximum moment may, of course, be determined by any other suitable method. The maximum shear $V$ is found to be 5 kips, and the maximum moment $M$ is 25 kip·ft at a point 14 ft from the left end.

From Table 8-1, the allowable stresses for Douglas fir No. 1 are 1300 psi in bending and 85 psi in shear. In the flexure formula, $M$ must always be given in pound-inches. Hence, since $M = 25$ kip·ft, the required section modulus is

$$\left[ Z = \frac{M}{s} \right] \qquad \text{Required } Z = \frac{(25)(1000)(12)}{1300} = 231 \text{ in.}^3$$

Based on a maximum shear $V$ of 5 kips, which must be converted to pounds, the required cross-sectional area is

$$\left[ A = \frac{3V}{2s_s} \right] \qquad \text{Required } A = \frac{(3)(5)(1000)}{(2)(85)} = 88 \text{ in.}^2$$

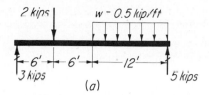

(a)

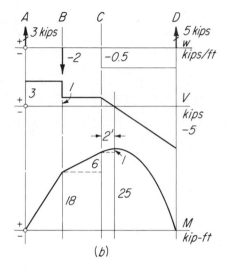

(b)

Fig. 8-2 Design of timber beam.
(a) Loaded beam. (b) Load, shear, and
moment diagrams.

Letting $b = \frac{1}{2}h$, we solve for $h$ and then for $b$. That is,

$$\frac{bh^2}{6} = \frac{h}{2}\frac{h^2}{6} = \frac{h^3}{12} = 231 \text{ in.}^3$$

or $h^3 = (12)(231) = 2772$ and $h = 14.0$ in. Hence $b = 7.0$ in., which suggests the use of a 8- by 14-in. rough beam weighing $(^8/_{12})(^{14}/_{12})40 = 31$ lb/ft. Its section modulus is $8(14)^2/6 = 261$ in.³ and its area is $(8)(14) = 112$ in.².

The additional moment due to the weight of the beam must be evaluated *at the section of maximum moment due to load*, which occurred 14 ft from the left end. When $x = 14$ ft and $L = 24$ ft, this moment (see Prob. 6-21), in a simple beam with a uniformly distributed load, is

$$\left[ M_x = \frac{wx}{2}(L - x) \right] \qquad M_{14} = \frac{(31)(14)}{2}(24 - 14) = 2170 \text{ lb·ft}$$

The additional required section modulus is

$$\left[ Z = \frac{M}{s} \right] \qquad Z = \frac{(2170)(12)}{1300} = 20.0 \text{ in.}^3$$

and the additional required area is

$$\left[ A = \frac{3V}{2s_s} \right] \qquad A = \frac{(3)(31)(12)}{(2)(85)} = 6.6 \text{ in.}^2$$

Hence                     Total required $Z = 231 + 20 = 251$ in.$^3$
                          Total required $A = 88 + 6.6 = 94.6$ in.$^2$

Use 8- by 14-in. beam, Douglas fir rough.                    *Ans.*
                 ($Z = 261$ in.$^3$; $A = 112$ in.$^2$)

**8-3. Method C.** Figure 8-3 is a section of a typical timber beam-and-girder floor system, such as is commonly used in mill-construction buildings for light manufacture. All columns are 12- by 12-in. timbers. Beams and girders are of Douglas fir dense select structural grade S4S. The beams rest on steel hangers fastened to the sides of the girders, and beam spans are from face to face of girders. The girders rest on steel column caps, and girder spans are from face to face of columns. Let $a = 15$ ft, $b = 18$ ft, and $c = 6$ ft.

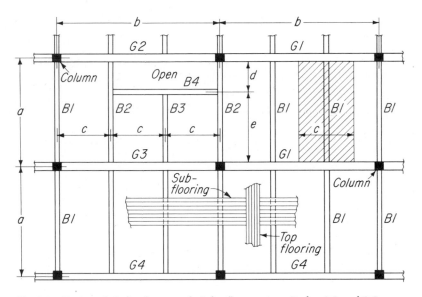

**Fig. 8-3  Design of timber beam-and-girder floor system. Probs. 8-3 and 8-9.**

The floor load to be supported is 125 lb/ft$^2$ uniformly distributed, which includes the weight of the subfloor planking and a finish floor covering only. Design one beam (*B1*) and one girder (*G1*).

*Solution:* In each case, the design load must be calculated from the given floor load. From Table 8-1, the allowable stress in bending is 1900 psi and in shear is 85 psi.

**Design of beam *B*1.** Each beam *B*1 clearly supports all the floor load from face to face of girders and halfway to each adjacent beam. If we assume that each of girders *G*1 is 12 in. wide, the span of the beam is 14 ft. The design load per foot of beam will then be as follows:

Floor load = (125)(6)                               = 750 lb/ft
Estimated weight of beam = (0.05)(750) =    37
Design load                                         = 787 lb/ft

For a uniformly distributed load on a simple beam the maximum moment is

$$\left[ M = \frac{wL^2}{8} \right] \qquad M = \frac{(787)(4^2)}{8} = 19,300 \text{ lb·ft}$$

Since $s = 1900$ psi, the required section modulus is

$$\left[ Z = \frac{M}{s} \right] \qquad \text{Required } Z = \frac{(19,300)(12)}{1900} = 122 \text{ in.}^3$$

Since $s_s = 85$ psi and $V = (787)(7) = 5510$ lb, the required area is

$$\left[ A = \frac{3V}{2s_s} \right] \qquad \text{Required } A = \frac{(3)(5510)}{(2)(85)} = 97.2 \text{ in.}^2$$

From Table E-2, we find that an 8- by 14-in. beam S4S weighing 28 lb/ft meets these requirements. Since the error in estimating its weight is less than 50 percent, no revision is necessary. Hence

Use 8- by 14-in. beam Douglas fir S4S. *Ans.*
($Z = 228$ in.³; $A = 101.3$ in.²)

**Design of girder G1.** The load on the girder is made up of three parts: (1) the concentrated loads from the four beam ends, (2) the floor load directly above the girder, which was not considered in the design of beam B1, and (3) its own weight. We leave item 3, its weight, for later consideration.

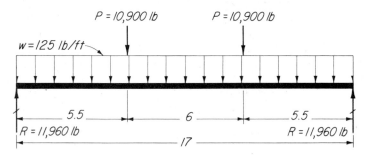

**Fig. 8-4  Girder loads and reactions.**

Each beam transmits half of its load to the girder at the third points of the 18-ft span. Since two beam ends are supported at each third point, $P$ equals the total load on one beam. The previously calculated beam load is 750 lb/ft, and the beam weight is 28 lb/ft. Consequently the concentrated load at each third point is

$$P = (750 + 28)(14) = 10,900 \text{ lb}$$

The girders having been previously assumed to be 12 in. wide, the floor load still unaccounted for and carried directly by the girder is 125 lb/ft. To calculate the maximum bending moment and the required section modulus and area, we then have

Girder end reaction:    $R = 10,900 + (8.5)(125) = 11,960$ lb

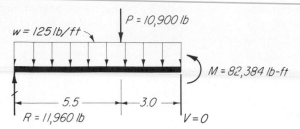

**Fig. 8-5 Calculation of maximum bending moment.**

Maximum bending moment, assumed positive:

$$M + (10,900)(3) + (125)(8.5)(4.25) = (11,960)(8.5) \qquad \text{or} \qquad M = 64,440 \text{ lb·ft}$$

Required section modulus:

$$\left[ Z = \frac{M}{s} \right] \qquad\qquad Z = \frac{(64,440)(12)}{1900} = 407 \text{ in.}^3$$

Required cross-sectional area for $V = 11,960$ lb:

$$\left[ A = \frac{3V}{2s_s} \right] \qquad\qquad A = \frac{(3)(11,960)}{(2)(85)} = 211 \text{ in.}^2$$

From Table E-2, we fine that a 12- by 20-in. beam S4S most nearly satisfies these requirements. Since the beam is just out of the range of our table, we calculate its section modulus to be 729 in.³, its area to be 224 in.², and its weight to be 62 lb/ft. We must determine now the additional required section modulus and area due to its own weight of 62 lb/ft. That is,

$$\left[ M = \frac{wL^2}{8} \right] \qquad \text{Additional } M = \frac{(62)(17^2)}{8} = 2240 \text{ lb·ft}$$

$$\left[ Z = \frac{M}{s} \right] \qquad \text{Additional required } Z = \frac{(2240)(12)}{1900} = 14.1 \text{ in.}^3$$

$$\left[ A = \frac{3V}{2s} \right] \qquad \text{Additional required } A = \frac{(3)(62)(8.5)}{(2)(84)} = 9.3 \text{ in.}^2$$

$$\text{Total required } Z = 407 + 14 = 421 \text{ in.}^3$$
$$\text{Total required } A = 211 + 9 = 220 \text{ in.}^2$$

Use 12- by 20-in. girder, Douglas fir S4S. *Ans.*

$(Z = 729$ in.³; $A = 224$ in.²; less than ½ percent overstressed in shear.)

## PROBLEMS

NOTE: Unless otherwise specifically stated, (*a*) the allowable stresses given in Table 8-1 will be used, (*b*) the weight of the beam will be considered (at 40 pcf), (*c*) the width-depth ratio will not be less than ⅓ nor more than ¾ (joists and "special" beams excepted), (*d*) the most economical (lightest) beam will be selected, and (*e*) a maximum overstress of 2 percent is permitted.

**8-4.** Design a timber beam of Douglas fir dense No. 1 S4S with a simple span of 15 ft to support a uniformly distributed load of 500 lb/ft plus its own weight. Use design method A as illustrated in Prob. 8-1.

*Ans.* 6 by 14 in. S4S

**8-5.** A timber beam of lodgepole pine No. 1 rough with a simple span of 16 ft must support a uniformly distributed load of 700 lb/ft plus its own weight. Design this beam using method A as illustrated in Prob. 8-1.

**8-6.** A timber beam of Douglas fir No. 1 rough in a warehouse floor system spans 18 ft between centers of supports and must carry an estimated uniform load of 670 lb/ft. A width-depth ratio of ½ is considered desirable. Design this beam using method B, as illustrated in Prob. 8-2.

*Ans.* 8 by 14

**8-7.** Design a beam of eastern hemlock No. 1 S4S to support the load shown in Fig. 8-6.

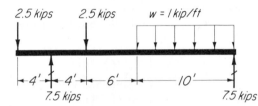

Fig. 8-6    Prob. 8-7.

**8-8.** The beam shown in Fig. 8-7 is to be designed of select-structural eastern hemlock S4S. Weight of beam should be considered.

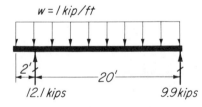

Fig. 8-7    Prob. 8-8.

**8-9.** In Fig. 8-3, let $a = 13$ ft, $b = 16$ ft, and $c = 4$ ft. Note that when $c = 4$ ft, three beams frame into the sides of the interior girders $G1$, instead of two as shown. All columns are assumed to be 12 by 12 in. the floor load is 125 psf, which includes weight of subfloor planking and top finish floor. Design one beam $B1$ and one girder $G1$. Assume width of girder to be 12 in. Use S4S timber with $s = 1,700$ psi and $s_s = 100$ psi.

**8-10.** A section of a common type of floor system is shown in Fig. 8-8. The joist span is 5 m between centers of end supports, and the joist spacing is 0.5 m on centers. What size of 50-mm joist, select-structural lodgepole pine S4S, should be used to support a uniform floor load of 300 kg/m², which includes the estimated mass of the subfloor and top floor?

*Ans.* 50- by 350-mm S4S

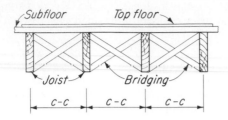

Fig. 8-8   Probs. 8-10 and 8-11.

**8-11.** The joists supporting a warehouse floor are to be of S4S Ponderosa pine No. 1. The joist span is 16 ft and spacing is 2 ft on centers. The maximum practical depth of joist is 14 in. The floor must support a live load of 150 psf plus 17.5 psf for the dead weight of the heavy plank subfloor and the top floor. Design the most economical joist. See Fig. 8-8.

**8-12.** A builder requires 2-in. floor joists to span 15 ft. Only 2- by 10-in. S4S select-structural eastern hemlock joists are available. Calculations show that when spaced the normal 16 in. on centers, the maximum bending stress in them is 1540 psi. (*a*) If a 16-in. spacing is used, by what percentage are the joists overstressed? (*b*) What spacing should be used to bring this maximum stress down to the allowable?

*Ans.* (*a*) 14 percent; (*b*) 14 in.

**8-13.** The springboard shown in Fig. 8-9 is of softwood with $s = 2400$ psi and must be 14 in. wide. Calculate its minimum required thickness $h$ based on bending stress and load only. Then consider weight at 36 pcf and recalculate $h$. Round off to nearest higher multiple of ¼ in.

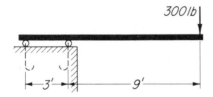

Fig. 8-9   Prob. 8-13.

**8-14.** A pedestrian walk along the side of a timber bridge is supported by timber beams cantilevered out 4 ft, as illustrated in Fig. 8-10. Design one beam of dense select structural Douglas fir S4S.

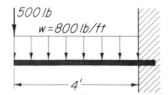

Fig. 8-10   Prob. 8-14.

**8-15.** A round cedar pole used as a radio mast extends vertically 12 m out of a concrete footing in which it is solidly imbedded. The allowable stress in bending is 12 MPa and in shear is 0.6 MPa. Calculate the exact minimum required diameter $D$ of this pole at its base and at its midpoint to withstand a horizontal force of 300 N at the top. Disregard weight of pole.

**8-16.** The temporary earth-retaining wall illustrated in Fig. 8-11 consists of planks driven vertically into the ground. The soft, muddy earth creates a triangular load against the planks with a maximum rate of load at the bottom of $w = 240$ lb/ft against a 12-in.-wide plank. Disregard weight of planks. If allowable stresses are 1000 psi in bending and 100 psi in shear, what minimum thickness $h$ of planks can be used?

*Ans.* $h = 1.47$ in.

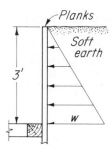

Fig. 8-11   Prob. 8-16.

**8-17.** A semipermanent earth-retaining wall uses horizontal planks supported by vertical timber cantilever posts spaced 1.5 m on centers, as illustrated in Fig. 8-12. The horizontal pressure of the earth produces a triangular load against each post which reaches a maximum $w$ of 10 kN/m at the bottom. Design a square rough post of Douglas fir using $s = 12$ MPa and $s_s = 0.6$ MPa.

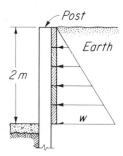

Fig. 8-12   Prob. 8-17.

**8-18.** A timber flume for carrying water is constructed as illustrated in Fig. 8-13. The water pressure produces a triangular hydrostatic load against each vertical post, reaching a maximum $w$ of 1500 lb/ft at the bottom. Each post is supported at the bottom as shown and at the top by a steel rod extending through the opposite

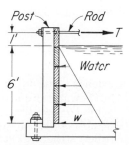

Fig. 8-13   Prob. 8-18.

post. Design the most economical rough timber post. Allowable stress in bending is 800 psi and in shear is 100 psi.

*Ans.* 6 by 8 in. rough

**8-19.** A rustic footbridge in a park is to span 10 m between centers of end supports. Two logs of approximately constant diameter are to be used. Each log must be capable of supporting a uniform load of 6 kN/m plus its own weight. If the allowable stresses are 10 MPa in bending and 0.7 MPa in shear, what diameter logs are required?

**8-20.** A contractor has to provide shoring in a trench excavation which will withstand the pressure of a wet earth mixture. Vertical planks are held in place by wales (beams) and struts as illustrated in Fig. 8-14. The horizontal load against the planks reaches a maximum $w$ of 720 lb/ft against the bottom of a 12-in. plank. If the allowable stresses are 1200 psi in bending and 100 psi in shear, what minimum thickness $h$ of planks can be used?

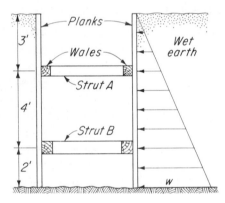

Fig. 8-14   Probs. 8-20 and 8-21.

**8-21.** In order to design the horizontal wales in the trench shoring shown in Fig. 8-14, the reacting forces in struts $A$ and $B$, spaced 6 ft on centers along the trench, must be calculated. Then the total load, uniformly distributed, on a 6-ft section of the upper wale equals $R_A$ and on the lower wale equals $R_B$. Although the wales are continuous beams, assume that maximum $M = wL^2/8$. Using square rough common timbers, design both wales. Refer to Prob. 8-20 for load against vertical planks. Use $s = 1450$ psi and $s_s = 120$ psi.

*Ans.* Upper, 6 by 6 in.; lower, 10 by 10 in.

## 8-3 DESIGN OF STEEL BEAMS

A complete treatment of the design of steel beams is rather extensive, since it includes the design of built-up beams and of plate girders, which is clearly beyond the scope of this text. This article, therefore, is confined to the design or, rather, "selection" of, regular rolled structural-steel beam sections and other beam sections of simple geometric form.

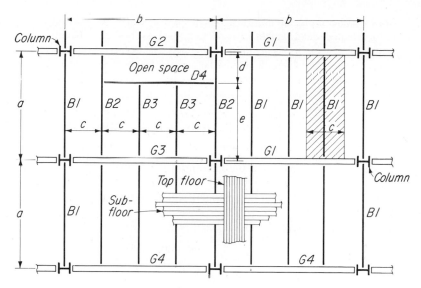

**Fig. 8-15**   Design of steel beam-and-girder floor system.

Some codes allow slightly higher bending stresses for *compact* sections than those permitted for members with thin compression flanges and webs. Most rolled sections qualify as compact sections. Beams that do not meet this requirement are noted in the steel handbooks.

ASTM A36, an all-purpose steel of good strength and weldability, is now generally available. For this steel the allowable bending stresses are usually 22,000 psi (152 MPa) for noncompact sections and 24,000 psi (165 MPa) for compact sections. Higher-strength steels are available for special applications.

The **span of a steel beam** must be carefully determined. In the beam-and-girder floor system shown in Fig. 8-15, for example, the beams $B1$ and $B2$ extend between webs of girders or webs of columns, as the case may be, and their spans are considered to equal the dimension $a$ between centers of columns. The girder spans are from flange to flange of the columns. When so arranged, the beams carry all the distributed floor load. The girders support only the concentrated loads from the beam ends plus their own weight. No floor load comes directly on the girder.

The **depth of steel floor beams,** according to a common specification, should be not less than $1/24$ of the span ($1/30$ for roof beams), which is equal to $1/2$ in. of depth per foot of span. In all cases where beams and girders support plastered ceilings, they must be so proportioned that the deflection will not exceed $1/360$ of the span (1 in. in 30 ft).

## ILLUSTRATIVE PROBLEMS

**8-22.** The roof of a one-story retail-store building 30 by 80 ft is to be supported by wide-flange steel beams spanning the 30 ft and spaced 16 ft on centers longitudinally. The loads, all uniformly distributed, are (a) live load (snow, etc.), 25 psf; (b) estimated dead weight of roof structure exclusive of beam, 15 psf; (c) estimated dead weight of plastered ceiling, 10 psf. Select a suitable W beam. Allowable stresses: bending, 22,000 psi; shear, 14,500 psi. See Table E-1.

*Solution:* Since the beams are spaced 16 ft on centers, the design load per foot of beam is

$$
\begin{aligned}
\text{Live load} &= (25)(16) &&= 400 \text{ lb/ft} \\
\text{Dead load} &= (15 + 10)(16) &&= \underline{400} \\
\text{Beam design load} & &&= 800 \text{ lb/ft}
\end{aligned}
$$

Since this is a simple span, the maximum moment is

$$\left[ M = \frac{wL^2}{8} \right] \qquad M = (0.125)(800)(30)(30) = 90,000 \text{ lb·ft}$$

and the required section modulus is

$$\left[ Z = \frac{M}{s} \right] \qquad \text{Required } Z = \frac{(90,000)(12)}{22,000} = 49.2 \text{ in.}^3$$

Try a W 16 × 36 ($Z = 59.5$ in.$^3$).

The moment and the additional required section modulus due to the weight of 36 lb/ft of this trial beam are

$$\left[ M = \frac{wL^2}{8} \right] \qquad M = (0.125)(36)(30)(30) = 4050 \text{ lb·ft}$$

$$\left[ Z = \frac{M}{s} \right] \qquad \text{Additional required } Z = \frac{(4050)(12)}{22,000} = 2.21 \text{ in.}^3$$

The total required section modulus is then

$$\text{Total required } Z = 49.2 + 2.21 = 51.41 \text{ in.}^3$$

Several other wide-flange sections will carry the load. Trial will show, however, that no lighter wide-flange section will do.

The maximum average web-shearing stress is

$$\left[ s_s = \frac{V}{th} \right] \qquad s_s = \frac{(836)(15)}{(0.295 \times 15.87)} = 2680 \text{ psi} \qquad \text{OK}$$

Use W 16 × 36 steel beam ($Z = 59.5$ in.$^3$).          *Ans.*

**8-23.** Figure 8-15 is a typical structural-steel beam-and-girder floor system. The floor is to be designed for a live load of 100 psf. The estimated dead weight of top and subfloor is 15 psf. The columns are wide-flange sections, W 12 × 65. Design one beam B1 and one girder G1, when $a = 20$ ft, $b = 28$ ft, and $c = 7$ ft.

*Solution:* Since all beams and girders are laterally supported by the subfloor, which fastens to the beams through nailing strips bolted to their top flanges, the allowable bending stress is 22,000 psi. Allowable shearing stress is 14,500 psi.

**Design of beam B1:**

$$
\begin{aligned}
\text{Live load} &= (100)(7) &&= 700 \text{ lb/ft} \\
\text{Dead load} &= (15)(7) &&= \underline{105} \\
\text{Design load} & &&= 805 \text{ lb/ft}
\end{aligned}
$$

$$\left[M = \frac{wL^2}{8}\right] \qquad M = (0.125)(805)(20)(20) = 40,250 \text{ lb·ft}$$

$$\left[Z = \frac{M}{s}\right] \qquad \text{Required } Z = \frac{(40,250)(12)}{22,000} = 21.95 \text{ in.}^3$$

Try a W 10 × 22, $Z = 23.2$ in.$^3$

$$\left[M = \frac{wL^2}{8}\right] \qquad \text{Additional } M = (0.125)(22)(20)(20) = 1100 \text{ lb·ft}$$

$$\left[Z = \frac{M}{s}\right] \qquad \text{Additional required } Z = \frac{(1100)(12)}{22,000} = 0.6 \text{ in.}^3$$

$$\text{Total required } Z = 21.95 + 0.6 = 22.55 \text{ in.}^3 \qquad \text{OK}$$

The maximum average web-shearing stress must now be calculated. That is,

$$\left[s_s = \frac{V}{th}\right] \qquad s_s = \frac{(805 + 22)(10)}{(0.24)(10.125)} = 3403 \text{ psi} \qquad \text{OK}$$

$$\text{Use W 10} \times 22 \ (Z = 23.2 \text{ in.}^3). \qquad \qquad Ans.$$

**Design of girder G1.** Since the columns are 12 in. deep and $b = 28$ ft, the girder span is 27 ft with 7-ft spacing of beams. Each concentrated load equals the load on one beam. That is,

$$P = (827)(20) = 16,540 \text{ lb}$$

and each reaction is one-half the total load, or

$$R = (0.5)(3)(16,540) = 24,810 \text{ lb}$$

Calculating the maximum moment $M$, we obtain

$$M + (16,540)(7) - (24,810)(13.5) \qquad \text{or} \qquad M = 219,000 \text{ lb·ft}$$

The required section modulus is then

$$\left[Z = \frac{M}{s}\right] \qquad \text{Required } Z = \frac{(219,000)(12)}{22,000} = 120 \text{ in.}^3$$

Try W 21 × 62 ($Z = 127$ in.$^3$).

The additional moment due to the weight of the beam is

$$\left[M = \frac{wL^2}{8}\right] \qquad \text{Additional } M = (0.125)(62)(27)(27) = 5650 \text{ lb·ft}$$

and the additional required section modulus is

$$\text{Additional } Z = \frac{(5650)(12)}{22,000} = 3.08 \text{ in.}^3$$

The total required section modulus is

$$Z = 120 + 3.08 = 123.08 \text{ in.}^3 \qquad \text{OK}$$

$$\left[s_s = \frac{V}{th}\right] \text{ Maximum average web } s_s = \frac{24,810 + (62)(13.5)}{(0.40)(21.0)} = 3053 \text{ psi} \qquad \text{OK}$$

$$\text{Use W 21} \times 62 \text{ steel beam } (Z = 127 \text{ in.}^3). \qquad \qquad Ans.$$

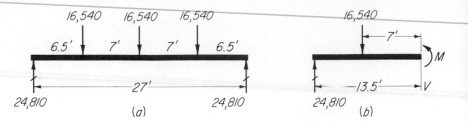

**Fig. 8-16** Design of steel girder. (a) Loaded girder. (b) Maximum moment.

## PROBLEMS

NOTE: Unless otherwise stated, (a) all beams are presumed to be supported against lateral buckling, (b) the allowable stresses are 22,000 psi (152 MPa) in bending and 14,500 psi (100 MPa) in shear, (c) the weight of the beam will be considered, (d) the most economical beam will be selected, and (e) a maximum overstress of 2 percent is permitted.

**8-24.** Assume in Prob. 8-22 that the building is 40 by 100 ft, the beam span is 40 ft, the spacing of beams is 20 ft on centers, and the loads per square foot are the same. Then (a) select a suitable steel wide-flange beam. (b) Calculate the average web-shearing stress.

**8-25.** (a) Select a structural-steel wide-flange beam to carry the load shown in Fig. 6-5 (see and use maximum V and M given with Prob. 6-42). (b) Calculate the average web-shearing stress.

$\qquad$ *Ans.* (a) W 16 × 26; (b) $s_s$ = 5620 psi

**8-26.** A structural-steel beam is to be selected to carry the load shown in Fig. 6-25. Assume that the load includes the weight of the beam. Select (a) the most economical American Standard I beam and (b) the shallowest available wide-flange section. In each case, calculate the average web-shearing stress.

**8-27.** A steel beam is to be selected to carry the load shown in Fig. 8-19. Assume that load includes weight of beam and that the beam is laterally supported. Select (a) the lightest wide-flange section; (b) the lightest American Standard I beam. In each case, calculate the average web-shearing stress.

$\qquad$ *Ans.* (a) W 250 × 17.9, $s_s$ = 19.8 MPa; (b) S 180 × 22.8, $s_s$ = 20.8 MPa

**8-28.** Figure 8-15 is a typical structural-steel beam-and-girder floor system. A similar system for the same building supports the roof, which must be designed for a live load of 40 psf. The estimated dead weight of roof planking and four-ply felt tarred and graveled is 20 psf. (a) Design one beam B1 and one girder G1. (b) Calculate the average web-shearing stresses. Assume a = 20 ft, b = 78 ft, c = 7 ft, and 12-in. columns.

**8-29.** Assume that the beam B3 in Fig. 8-15 supports only its share of the uniformly distributed live floor load of 85 psf plus an estimated dead weight of 15 psf for the floor. Let a = 26 ft, b = 40 ft, c = 10 ft, d = 6 ft, and e = 20 ft. (a) Select a suitable wide-flange beam. (b) Calculate the average web-shearing stress.

**8-30.** Beam B4 in Fig. 8-15 supports only the end-reaction loads from the two beams $B3$. The mass of the uniformly distributed live floor load is 250 kg/m² and the dead mass of the subfloor and top floor is 75 kg/m². Let $a = 6$ m, $b = 10$ m, $c = 2.5$ m, $d = 2$ m, and $e = 4$ m. (a) Select a suitable wide-flange beam for $B4$. (b) Calculate the average web-shearing stress.

**8-31.** To form a wide beam, two American Standard channels are placed back to back but about 16 in. apart. Together they support a uniformly distributed load of 1000 lb/ft over a simple span of 22 ft. (a) What size channels are most suitable? (b) Calculate the average web-shearing stress.

*Ans.* (a) Two 15-in. by 33.9-lb American Standard channels; (b) $s_s = 1550$ psi

**8-32.** The load shown in Fig. 6-15 is to be supported by a beam made up of two American Standard channels placed back to back. Consider also weight of channels. (a) Select the most economical channels to use. (b) Calculate the average web-shearing stress. See Prob. 6-16 for the maximum shear and moment.

**8-33.** Two angles placed with long legs back to back are to be used as a simple lintel (beam) spanning an opening in a brick wall 1.5 m between centers of supports. The estimated uniformly distributed load is 2000 kg/m, which may be assumed to include the mass of the angles. Select the most suitable angles.

**8-34.** A lintel in a brick wall spans 8 ft between centers of supports. The load is considered to be triangular, increasing uniformly from zero at the left end to a maximum of 600 lb/ft at the center, then decreasing uniformly to zero at the right end. Select two angles to carry this load, back to back. The outstanding legs must not be less than 3½ in. wide.

*Ans.* Two 4- by 3½- by ¼-in. angles (4 in. $b - b$)

**8-35.** A solid circular steel shaft 2 m long between centers of supporting bearings must be capable of withstanding a thrust perpendicular to the shaft of 5 kN at any point between the bearings without the maximum stress exceeding 62 MPa. Calculate the appropriate shaft diameter $D$. Neglect weight of shaft.

*Ans.* $D = 75$ mm

**8-36.** Design a solid circular steel shaft, 6 ft long between centers of supporting bearings and acting as a simple beam, to withstand a thrust of 524 lb perpendicular to the shaft at any point along its length. Maximum allowable stress is 12,000 psi. Neglect weight of shaft.

**8-37.** A section of standard steel pipe is to serve as a clothesline pole extending 72 in. vertically out of a solid concrete base. Select a suitable diameter capable of withstanding a horizontal force at the top of 300 lb without exceeding a unit stress of 20,000 psi.

**8-38.** A steel bar is used as a lever to operate a jackscrew. A horizontal force $P$ of 200 lb is applied perpendicular to the bar at point 16 in. from the vertical center line of the jackscrew. If the elastic limit stress of 32,000 psi must not be exceeded, what diameter (a) of solid bar and (b) of standard pipe should be used?

**8-39.** A steel pole supporting several trolley wires extends 21 ft vertically out of a concrete footing into which it is solidly imbedded. The pole is made up of three 7-ft sections of standard pipe and must be capable of withstanding a horizontal force of 650 lb at its top. Determine the required nominal diameter $D$ of (a) the top 7-ft section, (b) the middle 7-ft section, and (c) the bottom 7-ft section. Use $s = 20,000$ psi.

## 8-4 LATERALLY UNSUPPORTED BEAMS

A laterally unsupported beam is one in which the compression side or flange is free to buckle or deflect sidewise under loads, which normally are vertical gravity loads. The vast majority of beams, such as floor and roof beams in buildings, are laterally supported by the floor or roof structures attached to and supported by them. However, some roof beams that support rather light-weight roof coverings are not considered to be laterally supported. Other beams such as crane rails, or beams supporting crane rails, are often laterally unsupported. All beams are presumed to be supported laterally, as well as vertically, at their supports.

Lateral buckling of timber beams has been thoroughly investigated by the Forest Products Laboratory, and is fully and ably discussed in their *Wood Handbook,* obtainable from the U.S. Government Printing Office, Washington, D.C.

In order to visualize what causes a beam to buckle sidewise, under a vertical load, say, let us consider a simple beam, supported at the ends and subjected to a downward-acting vertical load, as illustrated in Fig. 8-17a. Clearly now, the lower portion of the beam is in tension and will

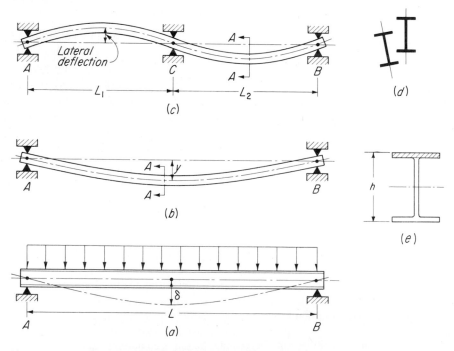

**Fig. 8-17  Laterally unsupported simple steel beams. (a) Side view of simple beam showing vertical deflection. (b) Top view of simple beam showing lateral deflection. (c) Top view of simple beam supported vertically at A and B and laterally at A, B, and C. (d) Section AA. (e) Compressive area used in computing flange area.**

therefore tend to remain straight, as do all tension members. The upper portion, however, is in compression. If this upper portion could now be isolated from the lower portion, the internal compressive stresses in it would tend to make it behave in a manner similar to that of a slender column subjected to axial end loads which would produce similar internal compressive stresses. Under such conditions, this upper portion would buckle sidewise quite readily.

These two opposing tendencies within the same beam, the tendency of the compression side to buckle or deflect sidewise and of the tension side to remain straight, result in a compromise effect: (1) vertical deflection of the beam, due to load, as illustrated in Fig. 8-17a, (2) a certain lateral deflection of the upper portion of the beam, as illustrated in Fig. 8-17b, and (3) a smaller deflection of the lower portion of the beam, caused solely by the sidewise pull of the upper portion upon the lower portion. The combined effect, at section AA, of the three separate effects is a vertical deflection plus a lateral deflection plus a twist, as is illustrated in section in Fig. 8-17d.

If this simple beam is given lateral support at some point C, between the vertical supports A and B, the lateral deflections will be as indicated in Fig. 8-17c, and the laterally unsupported length to be considered in its design will then be the greater of $L_1$ or $L_2$.

## 8-5 DESIGN OF LATERALLY UNSUPPORTED SIMPLE STEEL BEAMS[1]

The tendency of the compression flange of a laterally unsupported steel beam to deflect or buckle sidewise produces in that flange compressive stresses in addition to those produced by normal bending in a laterally supported beam. This fact necessitates the use of allowable design bending stresses somewhat lower than the usual 22,000 psi (152 MPa) in bending. Of the several allowable stress formulas proposed, the ones recommended by the American Institute of Steel Construction are, perhaps, most widely used. The allowable bending stress for laterally unsupported beams of ASTM A36 steel when the compressive flange is solid and approximately rectangular in shape and its area is not less than that of the tension flange is

**Allowable stress in laterally unsupported steel beams;**

$$\left.\begin{array}{l} s = \dfrac{12{,}000{,}000\,C_b}{Ld/A_f} \text{ (but not more than 22,000 psi)} \\[3mm] s = \dfrac{83\,000\,C_b}{Ld/A_f} \text{ (but not more than 152 MPa)} \end{array}\right\} \qquad (8\text{-}1)$$

[1] In the handbook *Steel Construction* are found a series of charts from which certain laterally unsupported steel beams may be selected directly (American Institute of Steel Construction, New York, 1980, pp. 2-51 to 2-77).

where $L$ = laterally unsupported length, in. or mm
    $d$ = depth of beam, in. or mm
    $A_f$ = area of compression flange, in.$^2$ or mm$^2$
    $C_b$ = a coefficient, conservatively taken as unity

The quantity $Ld/A_f$ in Eq. (8-1) is a "slenderness" factor. When $Ld/A_f$ is less than 546, the computed value of $s$ would exceed 22,000 psi (152 MPa), and therefore 22,000 psi (152 MPa) would be used instead of the computed value. If $Ld/A_f$ exceeds 546, the reduced stress computed by Eq. (8-1) would be used. The coefficient $C_b$ may conservatively be taken as unity, or if a more exact value is desired, it may be evaluated from a rather complex formula given in the code.

Since the allowable stress depends upon the dimensions of the section to be used and since these dimensions are obviously unknown, the design of a laterally unsupported beam must be accomplished by successive trials. A suitable procedure is as follows:

# Procedure

1. Calculate the maximum moment $M$ in the beam.
2. Estimate the probable allowable stress, calculate the probable required section modulus, and select a trial section having a comparatively wide flange.
3. Calculate the allowable stress in the trial section, and also the actual stress, and compare the two. If they differ by more than 10 percent, select a more suitable trial section.
4. Consider the effect on the actual stress of the weight of the beam.

NOTE: Failure of laterally unsupported beams is invariably due to buckling of the compression flange. In such beams, the average web-shearing stress is comparatively low and, as a rule, need not be calculated.

ILLUSTRATIVE PROBLEM

**8-40.** A simply supported beam with a span of 28 ft is to carry a uniformly distributed load of 1000 lb/ft. What lightest wide-flange section may be used if the beam is laterally unsupported?

*Solution:* We assume the weight of the beam to be 5 percent of the imposed load. The total load is then $(1.05)(1000) = 1050$ lb/ft and the maximum moment is

$$\left[ M = \frac{wL^2}{8} \right] \qquad M = \frac{(1050)(28)(28)}{8} = 102,900 \text{ lb·ft, or } 1,234,800 \text{ lb·in.}$$

If we now assume the allowable stress to be 18,000 psi, the probable required section modulus is

$$\left[ Z - \frac{M}{s} \right] \qquad \text{Required } Z = \frac{1,234,800}{18,000} = 68.5 \text{ in.}^3$$

The section should be rather compact with a very wide flange to better resist buckling. From Table E-4, we might select a W 18 × 40 with a depth of 17⅞ in., and with flanges 6 by ½ in. Since the length $L = (12)(28) = 336$ in., Eq. (8-1) gives

$$\left[ s = \frac{12,000,000}{Ld/A_f} \right] \qquad \text{Allowable } s = \frac{12,000,000}{[(336)(17.875)]/3} = 5996 \text{ psi}$$

Obviously this is a low allowable stress, so we should seek a section with a better depth-flange area ratio. Possible beams with their dimensions are shown in Table 8-3.

Although calculations indicate that a W 12 × 58 section is adequate from a strength standpoint, its use would violate a common specification that the depth of a beam should be at least ½ in. per foot of span (see Art. 8-3). A W 14 × 61 has an allowable stress of 16,090 psi and a computed stress ($s = M/Z$) of 13,390 psi. Therefore,

Use W 14 × 61 Ans.

## 8-6 DESIGN OF LATERALLY UNSUPPORTED OVERHANGING AND CANTILEVER STEEL BEAMS

Cantilever beams, or the cantilever portions of overhanging beams, present special problems.

Figure 8-18a shows a cantilever beam $AC$ whose actual laterally unsupported length is $L$. Vertical loads have caused its free end $A$ to deflect laterally a distance $y$ from a tangent at the fixed support $C$. If this beam $AC$ is now "reflected," as in a mirror, on the opposite side of the support $C$ and if we then consider the entire length $AB$, we find this imaginary beam $AB$ to be identical to the actual simple beam $AB$ in Fig. 8-18b, which has twice the length of the cantilever beam. We may correctly conclude, then, that the "effective length" of a laterally unsupported cantilever beam, to be used in its design, is twice its actual length, or $2L$.

Table 8-3 DESIGN OF LATERALLY UNSUPPORTED STEEL BEAM

| Beam Section | Depth | Flange Width | Flange Thickness | Allowable Stress, Eq. (8-1) | Section Modulus | Computed Stress, $M/Z$ |
|---|---|---|---|---|---|---|
| W 14 × 48 | 13¾ | 8 | ⅝ | 12,990 | 70.3 | 17,565 |
| W 14 × 61 | 13⅞ | 10 | ⅝ | 16,090 | 92.2 | 13,390 |
| W 16 × 57 | 16⅜ | 7⅛ | 11/16 | 10,700 | 92.2 | 13,390 |
| W 14 × 53 | 13⅞ | 8 | 11/16 | 14,160 | 77.8 | 15,870 |

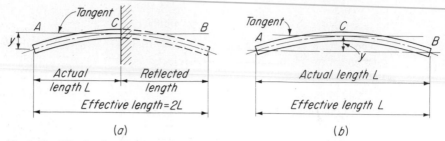

**Fig. 8-18** **Effective length of cantilever beam.** (*a*) **Top view of laterally deflected compression flange of cantilever beam.** (*b*) **Top view of laterally deflected compression flange of simple beam.**

## ILLUSTRATIVE PROBLEM

**8-41.** The overhanging beam shown in Fig. 8-19*a* is laterally unsupported except at the two vertical supports *A* and *B*. End *C* is completely free. Select the lightest available wide-flange section capable of carrying the 20-kN concentrated load and in addition a uniformly distributed load of 5 kN/m, which, we may assume, includes the weight of the beam.

*Solution:* We will find it convenient here to draw the shear and moment diagrams (Fig. 8-19*b* and *c*). The maximum shear is 23.9 kN at *B* left, and the maximum moment, under the concentrated load, is 22.2 kN·m. Study of Eq. (8-1) indi-

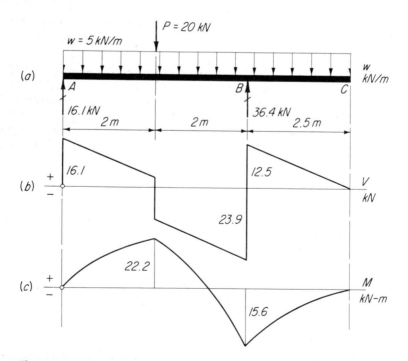

**Fig. 8-19** **Design of laterally unsupported overhanging beam.** (*a*) **Loaded beam.** (*b*) **Shear diagram.** (*c*) **Moment diagram.**

cates that, for any given section, the allowable stress is inversely proportional to the "effective length" of a span. Consequently the *allowable stress* in the 2.5-m cantilever span *BC*, whose effective length is $2L = (2)(2.5) = 5$ m, will be lower than that in the main span *AB*, whose effective length is 4 m. But the *actual stress* in span *BC* will also be lower than that in *AB*, since the bending moment in span *BC* is lower than that in span *AB*. In this case, therefore (though not in all), we must calculate allowable and actual stresses in both spans in order to determine which governs the required size of the beam.

Assuming an allowable stress of 120 MPa and using the higher main-span moment, the probable required section modulus is

$$\left[Z = \frac{M}{s}\right] \quad \text{Probable required } Z = \frac{(22\ 200\ 000)}{120} = 185\ 000 \text{ mm}^3$$

The lightest suitable wide-flange section found in Table E-4 is a W 200 × 22.5 for which $Z = 193\ 000$ mm³, $d = 206$ mm, $b = 102$ mm, and $t = 8.0$ mm.

In the 4-m main span, the allowable stress from Eq. (8-1) is

$$\left[s = \frac{83\ 000}{Ld/A_f}\right] \quad \text{Allowable } s = \frac{83\ 000}{[4\ 000(206)]/[8(102)]} = 82.2 \text{ MPA}$$

The actual stress when the moment $M = 22.2$ kN is

$$\left[s = \frac{M}{Z}\right] \quad \text{Actual } s = \frac{(22\ 200\ 000)}{193\ 000} = 115 \text{ MPa}$$

Clearly the W 200 × 22.5 is inadequate. A W 200 × 26.6 has a wider and thicker flange as well as a greater section modulus. For the W 200 × 26.6 section, $Z = 249\ 000$ mm³, $d = 207$ mm, $b = 133$ mm, and $t = 8.4$ mm. Then for the 4-m main span the allowable stress is

$$\left[s = \frac{83\ 000}{Ld/A_f}\right] \quad s = \frac{83\ 000}{[4000(207)]/[8.4(133)]} = 112 \text{ MPa}$$

and the actual stress is

$$\left[s = \frac{M}{Z}\right] \quad \text{Actual } s = \frac{(22\ 200\ 000)}{249\ 000} = 89.2 \text{ MPa} \quad \text{OK}$$

In the 2.5-m cantilever span, the allowable stress is

$$\left[s = \frac{83\ 000}{Ld/A_f}\right] \quad \text{Allowable } s = \frac{83\ 000}{[2(2500)(207)]/[8.4(133)]} = 89.6 \text{ MPa}$$

and the actual stress when $M = 15.6$ kN·m is

$$\left[s = \frac{M}{Z}\right] \quad \text{Actual } s = \frac{(15\ 600\ 000)}{249\ 000} = 62.7 \text{ MPa} \quad \text{OK}$$

Hence, use a W 200 × 26.6 steel beam.    *Ans.*

## PROBLEMS

NOTE: Unless otherwise specifically stated, the weight of the beam should be considered in the following problems.

**8-42.** A W 14 × 43 section is to be used as a 24-ft simple-span beam. What is the maximum allowable compressive stress if the beam is laterally supported? What is the allowable stress if it is not laterally supported?

**8-43.** A W 12 × 26 section is used as a beam with a simple span of 16 ft. What maximum allowable superimposed uniform load may be placed on this beam if it is not laterally supported?

*Ans.* $w = 1056$ lb/ft

**8-44.** A W 12 × 30 section is used on a simple span of 20 ft. What is the maximum allowable concentrated load $P$ that may be placed at midspan if the beam is not laterally supported?

**8-45.** A W 8 × 40 section is to be used as a laterally unsupported cantilever beam 8 ft long. What concentrated load $P$ may be placed at the outer end without exceeding the allowable stress? Consider also weight of beam.

*Ans.* $P = 7975$ lb

**8-46.** A laterally unsupported simple beam 24 ft long is to carry a superimposed uniform load of 1200 lb/ft. Select the lightest available W section.

**8-47.** Select the lightest available W section with a laterally unsupported simple span of 24 ft to carry a concentrated load $P$ of 20,000 lb at midspan.

*Ans.* W 12 × 58

**8-48.** The W 10 × 26 beam shown in Fig. 8-20 is laterally unsupported except at supports A and B. Calculate the allowable and actual maximum stresses due to the weight of the beam plus a load $P$ of 5320 lb.

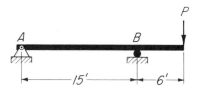

Fig. 8-20   Prob. 8-48.

**8-49.** The beam shown in Fig. 8-21 is laterally unsupported except at supports A and B. Select the lightest available W section when $P_1 = 20,000$ lb and $P_2 = 20,000$ lb.

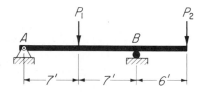

Fig. 8-21   Probs. 8-49, 8-62, and 9-39.

**8-50.** A laterally unsupported simple-span beam 6 m long is to carry a concentrated load of 100 kN at midspan. Select lightest available W section.

**8-51.** A simple beam 24 ft long is laterally supported only at its ends and at midspan in a manner similar to that indicated in Fig. 8-17. Select the lightest W section capable of carrying a uniform load of 1500 lb/ft.

**8-52.** Solve Prob. 8-41 when the main-span length is 6 m instead of 4 m.

SUMMARY

## Design of Beams. Procedure

1. Calculate the maximum shear $V$ in the beam.
2. Locate the section of maximum moment $M$ in the beam and calculate this moment, usually by the most suitable method: (a) the free-body-diagram method, (b) the beam-diagram method, or (c) the method using the appropriate moment formula. (The weight of the beam must be appropriately accounted for, here or in step 5.)
3. Ascertain the appropriate allowable stress, convert the unit of moment to appropriate units, and calculate the required section modulus $Z$.
4. Select a trial beam having a section modulus slightly higher than that required.
5. Calculate the additional moment and the additional required section modulus due to the weight of the trial beam *at the section of maximum load moment.*
6. Ascertain if the section modulus of the trial beam is equal at least to the total required section modulus. If not, select appropriate beam.
7. Determine the safety in shear of the beam selected by calculating the maximum longitudinal shearing stress, if a timber beam, or the average web-shearing stress, if a steel beam, and comparing with the allowable stress.

REVIEW PROBLEMS

**8-53.** A simply supported timber beam 6 m long of Douglas fir dense select structural rough must support two concentrated loads of 19 kN each applied at the third points of the span. Select a suitable and economical beam.

*Ans.* 150 by 350 mm rough

**8-54.** A timber beam of Douglas fir dense No. 1 S4S cantilevers out a distance of 5 ft from its support. It must withstand a concentrated load of 8000 lb at its outer end. Select a suitable and economical beam.

**8-55.** A pedestrian walk alongside a highway bridge is supported by steel beams spaced 15 ft on centers along the bridge and cantilevered out a length of 6 ft from the side of the bridge. The maximum load is estimated to be 200 psf of walk area, which includes the dead weight of the deck, bracing, and beams. Select a suitable W beam.

**8-56.** A steel beam weighing 36 lb/ft is capable of supporting a total uniformly distributed load $W$ of 68,200 lb in addition to its own weight of 396 lb when the span is 11 ft. When the span is 22 ft, $W = 34,100$ lb, and when it is 33 ft, $W = 22,700$ lb. In each case, calculate the percentage of $W$ represented by the weight of the beam and note the importance of considering its weight.

*Ans.* 0.58, 2.32, and 5.23 percent

**8-57.** The second floor of a two-story column-free commercial building 34 by 120 ft is to be supported by steel W girders spanning 32 ft and spaced 12 ft on centers. Smaller beams spanning the 12 ft between girders are spaced 8 ft on centers and attach to the girders at the midpoint and at both quarter points. The floor area supported by each beam is 96 ft². (a) Select a suitable W beam section to withstand a uniform floor load, dead and live, of 150 psf. (b) Each girder then supports three 14.4-kip concentrated loads, one at the center and one at each quarter point. Select a suitable W girder.

*Ans.* (a) W 12 × 14; (b) W 21 × 68

**8-58.** A flagpole 12 m high whose base is solidly imbedded in concrete is made of four equal-length sections of standard pipe of different diameters. If the pole must resist a horizontal load of 200 N/m of height without exceeding a unit stress of 125 MPa, what minimum diameter pipe is required for the bottom section?

**8-59.** A piece of standard steel pipe 66 in. long is to be used as a horizontal lever. The 1000-lb vertical load is 6 in. from the fulcrum, and the applied vertical force of 100 lb is 60 in. from the fulcrum, at the opposite end. What nominal diameter of pipe should be used if the unit stress is not to exceed 20,000 psi?

**8-60.** A rustic footbridge crosses a small stream and is supported by two round logs laid parallel, 6 ft on centers, with a simple span of 20 ft. The deck is of 12-in.-wide planks 12 ft long laid across the logs so that they cantilever 3 ft out beyond the logs at both ends. Deck planks and logs must be capable of supporting a load of 125 psf, which includes weight of deck planks but not of logs. Determine (a) the required thickness $t$ of deck planks, using an allowable bending stress of 500 psi (low because planks are subject to wear), and (b) the required diameter $d$ of logs, using an allowable bending stress of 1000 psi.

*Ans.* (a) $t = 2.6$ in.; (b) $d = 17.1$ in.

**8-61.** In Prob. 8-41, omit the 20-kN concentrated load and solve.

**8-62.** The beam in Fig. 8-21 is laterally unsupported except at supports $A$ and $B$. Select the lightest available W section when $P_1 = 40,000$ lb and $P_2 = 10,000$ lb.

**8-63.** A simply supported beam 8 m long is laterally supported only at its ends and at midspan in a manner similar to that indicated in Fig. 8-17. Select the lightest available W section to carry a concentrated load of 100 kN at midspan.

## REVIEW QUESTIONS

**8-1.** In which type of beam are bending stresses usually critical? In which type are shearing stresses usually critical?

**8-2.** Are shearing stresses often critical (a) in timber beams and (b) in steel beams?

**8-3.** Why should the weight of a beam be considered? Approximately what percentage of the total load is represented by the weight of a simple-span uniformly loaded (a) timber beam and (b) steel beam?

**8-4.** Under what conditions may the estimated weight of a beam be added to the imposed load before design begins? When should it not be so added?

**8-5.** What is meant by the "most economical" beam? What general shape of beam is usually most economical?

**8-6.** Under what conditions may the "most economical" beam not be used?

**8-7.** Define the span of a beam. Discuss spans of timber and steel beams and girders generally used for purpose of design.

**8-8.** By what three methods may the maximum moment in a beam be determined?

**8-9.** Discuss the so-called methods A, B, and C of timber beam design.

**8-10.** What common specification applies to the depth of steel beams?

**8-11.** By what "rule of thumb" may we estimate closely the required depth of (*a*) a timber floor beam and (*b*) a steel floor beam?

**8-12.** What is meant by "lateral support" of a beam? Give some typical illustrations.

**8-13.** How may a beam fail when "laterally unsupported" except at the vertical supports?

**8-14.** Upon what factors does the allowable stress in a laterally unsupported beam depend?

# Chapter 9
# Deflection of
# Beams

## 9-1 RELATION BETWEEN CURVATURE AND STRESS

As previously described in Art. 7-1, a condition of pure bending in a beam may be produced by opposing moments applied at its opposite ends. Pure bending will also exist in any part of a beam in which no shear forces are present, such as length $BC$ of the beam in Fig. 9-1. $EF$ and $GH$ are vertical and parallel planes through the straight, unbent beam.

Consider now the central portion of this beam after the loads have produced bending. The planes $EF$ and $GH$, parallel in the unbent beam, now intersect at $O$. Let $R$ be the radius of curvature of the neutral plane, now called the **elastic curve,** and $c$ the distance from the neutral plane to the extreme fiber. If line $IJ$ is now drawn parallel to $EF$ the distance $JH$ is the deformation in the length $KL$, and $JH/KL$ is the unit deformation $\epsilon$. Clearly, angles $KOL$ and $JLH$ are equal and the similarly designated segments $KOL$ and $JLH$ are similar. Therefore

$$\frac{LH}{OL} = \frac{JH}{KL} \quad \text{or} \quad \frac{c}{R} = \epsilon \qquad (a)$$

Since $\epsilon = s/E$, we obtain

$$\frac{c}{R} = \frac{s}{E} \quad \text{or} \quad s = \frac{cE}{R} \qquad \textbf{(9-1)}$$

In this equation, $c$ and $R$ are expressed in the same units and $s$ and $E$ in units of force per unit area.

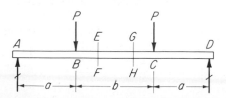

Fig. 9-1   Part $BC$ of beam in pure bending.

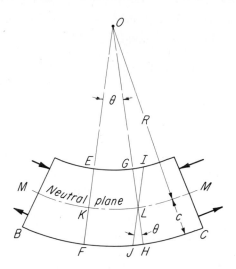

Fig. 9-2   Section of beam in pure bending.

From Eq. (9-1) we may obtain the maximum unit stress $s$ in any portion of a bent member whose radius of curvature $R$ is known or the radius of curvature $R$ corresponding to any given stress $s$.

## PROBLEMS

**9-1.** A band of sheet steel 0.03 in. thick is to be bent around a circular post. (*a*) What bending stress is produced if the post is 40 in. in diameter? (*b*) What minimum radius $R$ may be used if the elastic-limit stress of 36,000 psi is not to be exceeded ($E = 29{,}000{,}000$ psi)?

*Ans.* (*a*) $s = 21{,}700$ psi; (*b*) $R = 12.1$ in.

**9-2.** The main cables of a suspension bridge are made up of strands of $^3/_{16}$-in.-diameter hard-steel wire. Calculate the minimum permissible radius of curvature $R$ of the saddles on which the cables rest at the top of the tower if the unit stress due to bending must not exceed 30,000 psi.

**9-3.** A hard-steel band saw is 1 mm thick and runs on pulleys of 2 m diameter. If the maximum direct tensile stress in the straight portion of the band is 50 MPa, what is the maximum tensile stress in the bent portion?

**9-4.** A 1-m-diameter cylinder is to be covered with thin stainless steel which will bend to the 0.5 m radius without exceeding the elastic-limit stress of 240 MPa. What maximum thickness $t$ of steel may be used ($E = 200\,000$ MPa)?

*Ans.* $t = 1.2$ mm

## 9-2 RELATION BETWEEN CURVATURE AND BENDING MOMENT

According to the flexure formula, Eq. (7-1), $s = Mc/I$. When this latter quantity is substituted for $s$ in Eq. (9-1), we obtain

$$\frac{Mc}{I} = \frac{cE}{R} \qquad \text{or} \qquad R = \frac{EI}{M} \qquad \text{(9-2)}$$

From Eq. (9-2), it is seen that the radius of curvature $R$ at any section in a bent member is equal to the $EI$ of the member divided by the bending moment at that section.

Equations (9-1) and (9-2) were derived to apply to members in pure bending. Since, however, they express only the relations between curvature and stress and between curvature and bending moment, respectively, at any plane section, they are applicable to all bent members.

PROBLEMS

**9-5.** A timber beam, $E = 1,760,000$ psi, is 6 in. wide and 12 in. deep. If the beam is subjected to a constant bending moment of 144,000 lb·in., calculate the radius of curvature of the beam using (a) Eq. (9-1) and (b) Eq. (9-2).

**9-6.** A rough 2- by 12-in. Ponderosa pine plank, laid flat, is supported at its ends. A 160-lb man stands at the midpoint of the 16-ft span. Calculate the minimum radius of curvature by Eq. (9-1) and by Eq. (9-2).

**9-7.** A vertical rough plank 8 m long is supported at the top and bottom ends. The plank is subjected to a horizontal concentrated lateral lead of 800 N at mid-height. Calculate the radius of curvature $R$ of the 40- by 240-mm plank at its top and at points 1, 2, 3, and 4 m from the top end ($E = 12\ 000$ MPa). Give answers in meters. [NOTE: The answers to this problem indicate clearly that the curvature of a simple beam due to a concentrated load at its center varies from zero at either end to a maximum at the center (see Fig. 9-3b).]

*Ans.* $R_0 = \infty$; $R_1 = 38.4$ m; $R_2 = 19.2$ m; $R_3 = 12.8$ m; $R_4 = 9.6$ m

**9-8.** Determine the radius of curvature $R$ of the beam in Prob. 7-1 at the point of maximum stress using (a) Eq. (9-1) and (b) Eq. (9-2). Wood is lodgepole pine.

## 9-3  DEFLECTION OF BEAMS

The deflection of a beam is the deviation of the neutral surface, or elastic curve, of the loaded beam from its original position in the unloaded beam. Since beams are normally horizontal, the deflection is the vertical deviation $\delta$ indicated in Fig. 9-3.

Many reasons exist for determining the deflection of a beam. The most common one is probably the usual limitation that a beam or joist supporting a plastered ceiling must not deflect more than $1/360$ of its span length if cracking of the plaster is to be avoided. Also, in the design of ma-

(a)                                    (b)

Fig. 9-3   Beam deflections. (a) Cantilever beam. (b) Simple beam.

chines and airplanes, the deflection of a member, or of a series of members, is often of vital importance.

Various methods are available for determining beam deflections. Of these, three are presented here: (1) the beam-diagram method, in which slope and deflection diagrams are added to the load, shear, and moment diagrams which were presented in Chap. 6, (2) the moment-area method, in which the two moment-area principles are introduced, and (3) the formula method. In the most commonly occurring problems of the type given herein, the maximum deflection occurs either at the center, as in symmetrically loaded simply supported beams, or at the end, in the case of cantilever beams. Other problems with unsymmetrical loads become more complex, but most of them may be satisfactorily solved by the moment-area method.

## 9-4  DEFLECTION BY THE BEAM-DIAGRAM METHOD

In Chap. 6, a series of three related diagrams was developed which indicated that certain mathematical relations existed between the load and shear diagrams and the shear and moment diagrams. The mathematical proof of these relations, as stated in the first and second laws of beam diagrams, is given in Appendix C. If a fourth and a fifth diagram, drawn in conformity with the first and second laws, are added, ordinates in the fourth diagram will measure the slope of the deflected axis of a beam, and ordinates in the fifth diagram will measure the deflection of that axis from its normal horizontal poisition.

Since the second law necessitates the determination of areas under curves and since such areas in some cases tend to become rather complex, the beam-diagram method is most easily applied only to certain types of problems, among which are the four common types illustrated in Fig. 9-4.

While the values of shear and moment in a beam are not dependent upon its size or upon the elastic properties of its material, it becomes clear, however, that the slope at any point of the deflected axis, and the deflection of that axis, are definitely dependent upon the size and shape $I$ of the beam and upon the stiffness $E$ of its material. Clearly, the deflection must be inversely proportional to $E$ and $I$.

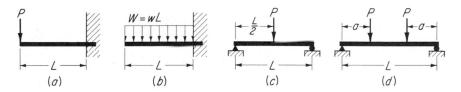

Fig. 9-4   Deflection by the beam-diagram method. Suitable beam types.

Values of ordinates in the slope and deflection diagrams will, therefore, be for a beam whose moment of inertia and modulus of elasticity of material both equal unity. The actual slope (in radians) and deflection (usually in inches or millimeters) may then be obtained by dividing the values of ordinates in these diagrams by $E$ and $I$, since actual slopes and deflections of a beam are *inversely* proportional to its $E$ and $I$. That is, slope $\theta$ = (ordinate value)$/EI$, or ordinate value = $EI\theta$. Also, deflection $\delta$ = (ordinate value)$/EI$, or ordinate value = $EI\delta$.

When the slope and deflection ordinates are thus divided by $E$ and $I$, which express the "elastic properties" of a particular beam, the curves in the slope and deflection diagrams are referred to as the **elastic curves.**

The following examples illustrate the method. For convenience, the laws of beam diagrams are restated:

**First Law of Beam Diagrams:**

$$\begin{bmatrix} \text{Slope of curve at any point} \\ \text{in any diagram} \end{bmatrix} = \begin{bmatrix} \text{length of ordinate at correspond-} \\ \text{ing point in next higher diagram} \end{bmatrix}$$

**Second Law of Beam Diagrams:**

$$\begin{bmatrix} \text{Difference in length of any} \\ \text{two ordinates in any diagram} \end{bmatrix} = \begin{bmatrix} \text{area between corresponding ordi-} \\ \text{nates in next higher diagram} \end{bmatrix}$$

The first law is used to determine the *shape* of a diagram; the second to determine the values of its principal ordinates.

One will usually find it expedient to sketch the $EI\delta$ diagram before drawing the $EI\theta$ diagram. We can usually visualize the exaggerated shape of the deflected beam but cannot as readily visualize the slope diagram.

Table 9-1 SUGGESTED UNITS AND MULTIPLIERS FOR USE IN CALCULATING BEAM SLOPES AND DEFLECTIONS

| English Units | SI Units |
|---|---|
| $P$, kips | $P$, kN |
| $w$, kips/ft | $w$, kN/m |
| $E$, ksi | $E$, MPa |
| $I$, in.$^4$ | $I$, mm$^4$ |
| $L$, ft | $L$, m |
| $\theta$, rad | $\theta$, rad |
| $\delta$, in. | $\delta$, mm |
| $\theta = \dfrac{144}{EI}\left(\begin{matrix}\text{ordinate in}\\ EI\theta \text{ diagram}\end{matrix}\right)$ | $\theta = \dfrac{10^9}{EI}\left(\begin{matrix}\text{ordinate in}\\ EI\theta \text{ diagram}\end{matrix}\right)$ |
| $\delta = \dfrac{1728}{EI}\left(\begin{matrix}\text{ordinate in}\\ EI\delta \text{ diagram}\end{matrix}\right)$ | $\delta = \dfrac{10^{12}}{EI}\left(\begin{matrix}\text{ordinate in}\\ EI\delta \text{ diagram}\end{matrix}\right)$ |
| $\theta = 144$ (formula from Table 9-2) | $\theta = 10^9$ (formula from Table 9-2) |
| $\delta = 1728$ (formula from Table 9-3) | $\delta = 10^{12}$ (formula from Table 9-3) |

Having the deflection curve before us, however, makes it possible to answer the following questions (and hence to sketch the slope curve). Where is the slope a maximum? Is the slope positive or negative at each point along the beam? Where is the slope zero? In addition to the deflection curve, we have the moment diagram to help us draw the slope curve. Keep in mind that the ordinates in the moment diagram tell us the slope of the slope curve at each point along the beam.

Table 9-1 gives suggested units for use in the beam diagrams and the corresponding multipliers to obtain the slope $\theta$ and the deflection $\delta$. Adherence to these suggestions will usually avoid large, and therefore cumbersome, ordinate values.

## ILLUSTRATIVE PROBLEMS

**9-9.** The beam shown in Fig. 9-5 extends 12 ft horizontally out of a solid concrete pier and supports a uniformly distributed load of 0.5 kip·ft. Determine the maximum deflection $\delta_A$ of this beam at its free end $A$ due to ($a$) the load only if a W $10 \times 26$ steel beam is used, ($b$) both load and weight of steel beam, by proportion, ($c$) load only if the beam is a 10- by 16-in. S4S timber with $E = 1760$ ksi, and ($d$) both load and weight of timber beam, by proportion. ($e$) If the maximum permissible deflection is 0.4 in., what size beams should be used?

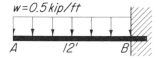

Fig. 9-5    **Maximum deflection by the beam-diagram method.**

*Solution.* The load, shear, and moment diagrams are first drawn in the usual manner, as shown in Fig. 9-6, using the laws of beam diagrams restated above. We note that at $A$ the shear and the moment are both zero. Before the slope and deflection diagrams are drawn, it must be clearly seen and understood from the physical conditions of the problem that the slope of the axis of the beam is zero at $B$, where it extends *horizontally* out of the pier and is maximum at $A$, and that the deflection is zero at $B$ and is maximum at the free end $A$.

The slope of the curve in the slope diagram is negative and increasing, as are the moment ordinates above. Hence the slope diagram must begin with some positive value of ordinate at $A$ in order to end up with zero ordinate at $B$. The slope of the curve in the deflection diagram is positive and decreasing, as are the ordinates in the $EI\theta$ diagram above. But since $\delta$ is zero at $B$, the deflection diagram must start with a negative value of ordinate at $A$.

The principal ordinates may now be evaluated by the second law as follows: The shear ordinate $V$ at $B$ is $(12)(0.5) = 6$ kips, the area under the load curve between $A$ and $B$, and is negative; the moment ordinate $M$ at $B$ is $(\frac{1}{2})(12)(6) = 36$ kip·ft, the area under the shear curve above, and is negative; the $EI\theta$ ordinate at $A = (\frac{1}{3})(12)(36) = 144$ kip·ft², the area under the moment curve, and is positive; and the $EI\delta$ ordinate at $A$ is $(\frac{3}{4})(12)(144) = 1296$ kip·ft³, the area under the $EI\theta$ curve (see Fig. 9-11$e$), and is negative. Before these areas are calculated, it should be noted that the $M$ and $EI\theta$ curves *begin* with zero slope at $A$ because in each

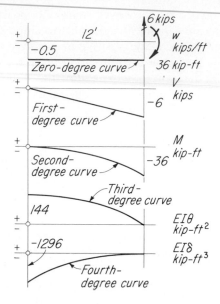

**Fig. 9-6  Beam diagrams.**

case the ordinate at $A$ in the next higher diagram is zero and that the $EI\delta$ curve *ends* with zero slope at $B$, since the $EI\theta$ ordinate at $B$ is zero.

(*a*) For the W 10 × 26 steel beam, $EI\delta_A = -1296$ kip·ft³, $E = 29,000$ ksi, and $I = 144$ in.⁴ Then, $\delta_A = -1296/EI$, or

$$\delta_A = \frac{(1728)(-1296)}{(29,999)(144)} = -0.536 \text{ in.} \qquad Ans.$$

(*b*) By proportion, the deflection at $A$ due both to load, 500 lb/ft, and weight of beam, 26 lb/ft, is

$$\delta_A = -0.536\frac{500 + 26}{500} = -0.564 \text{ in.} \qquad Ans.$$

(*c*) Using a 10- by 16-in. timber beam S4S, for which $I = 2948$ in.⁴ and $E = 1760$ ksi,

$$\delta_A = \frac{(1296)(1728)}{(1760)(2948)} = -0.432 \text{ in.} \qquad Ans.$$

(*d*) Considering also the weight of 40.9 lb/ft of this beam, the total deflection by proportion is

$$\delta_A = -0.432\frac{500 + 40.9}{500} = -0.467 \text{ in.} \qquad Ans.$$

(*e*) If the maximum deflection of the steel beam is limited to $-0.4$ in. and $EI\delta_A = -1296$, its required moment of inertia is $I = -1296/E\delta_A$, or

$$\text{Required } I = \frac{(-1296)(1728)}{(29,000)(-0.4)} = 193 \text{ in.}^4$$

Try W 12 × 26 ($I = 204$ in.⁴).

For both load and weight of beam

$$\text{Required } I = 193 \, \frac{500 + 26}{500} = 203 \text{ in.}^4 \qquad \text{OK}$$

Use W 12 × 26 steel beam *Ans.*

When $\delta_A = -0.4$ for the timber beam,

$$\text{Required } I = \frac{(-1296)(1728)}{(1760)(-0.4)} = 3180 \text{ in.}^4$$

Try 10- by 18-in. timber beam S4S ($I = 4243$ in.$^4$; weight = 46.1 lb/ft). For both load and weight of beam

$$\text{Required } I = 3180 \, \frac{500 + 46.1}{500} = 3475 \text{ in.}^4 \qquad \text{OK}$$

Use 10- by 18-in. S4S ($I = 4243$ in.$^4$) *Ans.*

NOTE: The method of proportion used in this example holds good only for beams with uniformly distributed loads over their entire length.

**9-10.** A simple beam 6 m long supports a single concentrated load of 50 kN at its center, as shown in Fig. 9-7. Calculate the maximum deflection at the center of this beam (*a*) if it is a steel beam with $I = 120 \times 10^6$ mm$^4$ and (*b*) if it is a 200- by 350-mm timber beam S4S, with $E = 12\,000$ MPa.

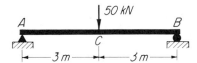

Fig. 9-7 **Maximum deflection by the beam-diagram method.**

*Solution:* The load, shear, and moment diagrams are drawn in the usual manner, as shown in Fig. 9-8. Below them are the slope ($EI\theta$) and deflection ($EI\delta$) diagrams. Because of symmetry of loading, it is clear that the maximum deflection will be at the center of the beam. At this point the slope of the beam axis will be zero, since a tangent to the axis at that point will be horizontal. That is, the $EI\theta$ ordinate at $C$ will be zero. The *slope* of the $EI\theta$ curve between $A$ and $C$ is increasing and positive, as are the ordinates in the moment diagram above. Since the curve must end with zero ordinate at the center $C$, it must necessarily begin at $A$ with a negative ordinate of $EI\theta_A = (\frac{1}{2})(75)(3) = 112.5$, which is the area under the moment curve between $A$ and $C$. The deflections of the beam at supports $A$ and $B$ are clearly zero. Hence $EI\delta_A$ and $EI\delta_B$ are zero. The *slope* of the deflection curve between $A$ and $C$ is decreasing and negative, as are the ordinates in the $EI\theta$ diagram above. The maximum deflection ordinate at the center $C$ of the beam is $EI\delta_C = (\frac{2}{3})(3)(112.5) = 225$ and is negative.

The deflection $\delta_C$ of the steel beam, for which $I = 120 \times 10^6$ mm$^4$ and $E = 200\,000$ MPa, then is

$$\delta_C = \frac{(225)(10^{12})}{200\,000(120)(10^6)} = -9.4 \text{ mm} \qquad \textit{Ans.}$$

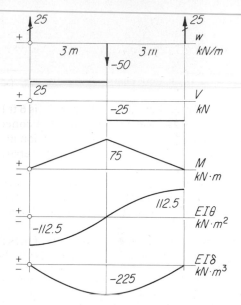

Fig. 9-8   Beam diagrams.

The deflection $\delta_C$ of the timber beam, for which $E = 12\ 000$ MPa and $I = 605 \times 10^6$ mm⁴, is

$$\delta_C = \frac{(225)(10^2)}{(12\ 000)(605)(10^6)} = -31.0 \text{ mm} \qquad Ans.$$

## PROBLEMS

NOTE: Solve the following problems by the beam-diagram method. Disregard weight of beam unless otherwise stated.

**9-11.** Calculate the maximum deflection at the free end of a cantilever beam similar to that shown in Fig. 9-4a when $P = 5$ kips and $L = 6$ ft (a) for an 8- by 16-in. timber beam of dense No. 1 Douglas fir S4S and (b) for a W 10 × 22 steel beam.

**9-12.** Compute the maximum deflection at the free end of a cantilever beam similar to that in Fig. 9-4a when $P = 4$ kips and $L = 12$ ft (a) for an 10- by 18-in. beam of lodgepole pine S4S and (b) for a W 12 × 26 steel beam. If this deflection is limited to 0.5 in., select the appropriate size (c) of timber beam, letting $b/h \approx \frac{1}{2}$, and (d) of steel W beam.

Ans.   (a) $\delta = -0.85$ in.; (b) $\delta = -0.67$ in.; (c) 12- by 22-in.; (d) W 16 × 26

**9-13.** Calculate the maximum deflection at the top of a 120-mm-OD pipe 5 m high due to a 2-kN horizontal load applied at the top. If the deflection must not exceed 50 mm, what diameter of pipe should be used? Steel pipe with 8-mm-thick wall in each case.

**9-14.** A cantilever beam similar to that shown in Fig. 9-4b has a length $L$ of 8 ft. and its load $w$ is 0.6 kip/ft. Calculate the maximum deflection at the free end if the beam is (a) an 8- by 12-in. Ponderosa pine S4S or (b) a steel S 8 × 18.4 section.

**9-15.** Solve all parts of Prob. 9-9 for a beam whose length $L$ is 9 ft and load $w$ is 0.5 kip/ft if the steel beam in ($a$) is an W 8 × 18, the timber beam in ($c$) is an 8- by 14-in. timber S4S, and the maximum permissible deflection in ($e$) is 0.25 in.

> *Ans.* ($a$) $\delta = -0.395$ in.; ($b$) $\delta = -0.409$ in.; ($c$) $= -0.262$ in.;
> ($d$) $\delta = -0.276$ in.; ($e$) W 12 × 16; 6- by 16-in. S4S

**9-16.** A cantilever beam similar to those shown in Fig. 9-4$a$ and $b$ is 6 ft long and carries a uniformly distributed load $w$ of 0.5 kip/ft in addition to a concentrated load $P$ of 2 kips at its left end. Calculate the total maximum deflection at the free end due to both loads if a W 8 × 18 steel beam is used. (HINT: Solve separately for $\delta$ due to each load.)

**9-17.** A 150- by 400-mm (actual dimensions) timber beam, $E = 12\,000$ MPa, S4S 4 m long is supported at its outer ends and carries a concentrated load $P$ of 60 kN at its center, as illustrated in Fig. 9-4$c$. Calculate the maximum deflection $\delta$ at the center.

**9-18.** Calculate the maximum deflection $\delta$ at the center of the shaft described in Prob. 8-35. If $\delta$ must not exceed 0.85 mm, what diameter of shaft should be used?

> *Ans.* $\delta = -2.68$ mm; use 100 mm diameter

**9-19.** A steel girder, W 16 × 45, is 24 ft long and is supported at its outer ends. It carries two concentrated loads of 12 kips each applied at the third points ($a = 8$ ft), similarly to that illustrated in Fig. 9-4$d$. Calculate the maximum deflection $\delta$ at its center.

**9-20.** The loads $P$ in Fig. 9-1 are 12 kN each. The dimensions $a$ and $b$ are 3 m and 2 m, respectively. Calculate the deflection under one of the loads if $E = 200\,000$ MPa and $I = 300 \times 10^6$ mm⁴.

## 9-5 DEFLECTION BY THE MOMENT-AREA METHOD

The determination of the deflections of unsymmetrically loaded simple beams and of overhanging beams by the beam-diagram method becomes rather complex. The closely related **moment-area method,** developed and illustrated in Arts. 9-6 to 9-11, uses the principle embodied in the second law of beam diagrams for the derivation of the first moment-area principle, which in turn is then used to derive the second moment-area principle.

## 9-6 THE FIRST MOMENT-AREA PRINCIPLE

The second law of beam diagrams, as stated in Art. 9-4, is "The difference in length of any two ordinates in any diagram equals the area between corresponding ordinates in the next higher diagram."

In the slope and deflection diagrams of the beam-diagram series, the lengths of ordinates represent, respectively, $EI\theta$ and $EI\delta$. Actual slopes $\theta$

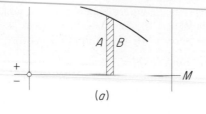

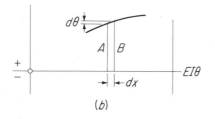

Fig. 9-9   First moment-area principle.
(a) Moment diagram. (b) Slope diagram.

and deflections δ are then obtained by dividing these ordinate values by *EI*.

Figure 9-9 shows portions of a moment diagram (upper) and the corresponding slope ($EI\theta$) diagram (see Art. 9-4). In this slope diagram the length of ordinate at any point, divided by *EI*, expresses the slope-angle $\theta$ (in radians) between a tangent to the elastic curve of the loaded beam at that point (*B*, for example), and the original undeflected horizontal axis of the beam, as illustrated in Fig. 9-10*b*. Therefore, the *difference* in length between slope ordinates *A* and *B* (in Fig. 9-9*b*), divided by *EI*, must equal the *difference* $\phi$ in the slopes of the tangents at *A* and *B*, in Fig. 9-10*b*. But difference, according to the second law, equals the area between corresponding ordinates in the moment diagram. We may therefore state

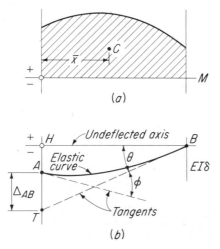

Fig. 9-10   Second moment-area principle.
(a) Moment diagram. (b) Deflection diagram.

### The First Moment-Area Principle:

$$
\begin{bmatrix}
\text{The angle } \phi \text{ between the tan-} \\
\text{gents to the elastic curve of a} \\
\text{beam at any two points } A \text{ and } B
\end{bmatrix}
=
\begin{bmatrix}
\text{the area in the moment dia-} \\
\text{gram, lying between verticals} \\
\text{through } A \text{ and } B, \text{ divided by } EI
\end{bmatrix}
$$

Expressing the first principle in symbols gives

$$\phi_{AB} = \frac{A_{AB}}{EI} \tag{9-3}$$

Clearly, from Fig. 9-10$b$, when either one of the two tangents is horizontal, the angles $\phi$ and $\theta$ are equal.

**Unit of Slope.** In the vast majority of practical problems the actual beam deflections, and hence the slope-angles of tangents to the elastic curve of the beam, are very small, seldom exceeding 1°. A well-known mathematical fact is that, for such small angles, the value of the angle, measured in radians, is very nearly numerically equal to the value of its tangent. The unit of the slope angles $\phi$ and $\theta$ is radians. Conversion to degrees is accomplished by multiplying by $360/2\pi$, the number of degrees in one radian.

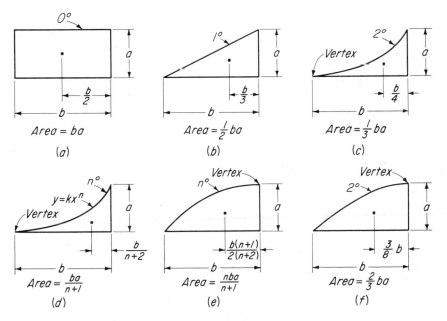

**Fig. 9-11**   Areas under curves. Location of centroids. ($a$) Rectangle. ($b$) Triangle. ($c$) Second-degree parabola. ($d$) $n$th-degree parabola. ($e$) $n$th-degree parabola. ($f$) Second-degree parabola.

The areas under curves shown in Fig. 9-11 are commonly encountered in the application to problems which involve the two moment-area principles.

## ILLUSTRATIVE PROBLEM

**9-21.** The beam shown in Fig. 9-12 is a W 8 × 24 steel section. Calculate the angle $\phi_{AB}$ (in radians and degrees) between the tangents to the elastic curve of this beam at points $A$ and $B$ (left end and center).

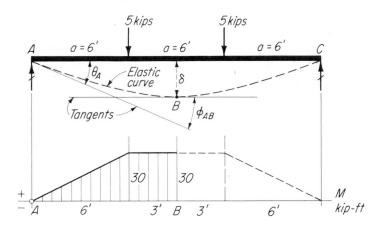

**Fig. 9-12   Calculation of slope angle of beam. Moment diagram.**

*Solution:* The moment of inertia $I$ of this beam is 82.5 in.⁴ (from Table E-4). $E = 29,000$ ksi, the usual value for structural steel. Since the tangent at $B$ is horizontal, $\phi_{AB} = \theta_A$, the slope of the tangent at $A$ with respect to the original horizontal axis of the beam. The bending moment at $B$ is 30 kip·ft. Hence the moment area divided by $EI$ is

$$\phi_{AB} = \theta_A = \frac{[(0.5)(6)(30) + (3)(30)](144)}{(29,000)(82.5)} = 0.01083 \text{ rad}$$

$$= \frac{(0.010\ 83)(360)}{2\pi} = 0.62° \qquad\qquad Ans.$$

The unit of the moment area is (kip·ft)(ft) = kip·ft². The necessary conversion is accomplished by multiplying by 144 (see Table 9-1).

## PROBLEMS

**9-22.** Solve Prob. 9-21 when each of the two loads is 3 kN, $a = 2$ m, and the beam is a 100- by 200-mm rough (full-size) timber of Douglas fir ($E = 12\ 000$ MPa).

**9-23.** A cantilever beam, W 10 × 26, 10 ft long extends horizontally out of a completely fixed support at its right end $B$ and supports a concentrated load of 6 kips at its left end $A$. Calculate the slope angle $\theta_A$, in radians and degrees, of its elastic curve at the left end $A$.

*Ans.* $\theta_A = 0.0103$ rad $= 0.593°$

**9-24.** A steel beam, W 10 × 26, cantilevers 10 ft horizontally out of a completely fixed support at its right end $B$. Calculate the slope angle $\theta_A$ of its elastic curve at the left end $A$ caused by a uniformly distributed load of 3 kip/ft. Check the results by the appropriate formulas found in Tables 9-2 and 9-3.

**9-25.** A man weighing 200 lb stands at the center of a 2- by 12-in. white oak plank S4S laid flat and supported at its outer ends. Calculate the slope angle $\theta_A$ at the left end $A$ of the plank due to the man's weight. (HINT: A tangent to the elastic curve at the center $B$ is horizontal.) Plank is 16 ft long and $E = 1600$ ksi.

**9-26.** A simple beam of length $L$ supports a single concentrated load $P$ at its center. Derive the formula giving the slope angle $\theta_A$ at its left end $A$. Check this against the similar formula shown in Table 9-2.

**9-27.** A 2- by 8-in. (full-size) dense Douglas fir plank 80 in. long and laid flat is supported at its ends and carries a uniform load of 12 lb/in. Calculate the slope angle $\theta_A$ at its left and $A$.

*Ans.* $\theta_A = 0.0283$ rad $= 1.62°$

**9-28.** A simple beam of length $L$ supports two equal concentrated loads $P$, as shown in Fig. 9-4$d$. Derive the formula giving the slope angle $\theta_A$ at the left end $A$ of the beam.

## 9-7 THE SECOND MOMENT-AREA PRINCIPLE

We may now use the first moment-area principle as a means of developing the second moment-area principle. Together, these two principles enable us to calculate actual beam deflections $\delta$.

Figure 9-13 is shows a portion, between two points $A$ and $B$, of the moment diagram for a loaded beam in which all ordinates, for simplicity of development, have been divided by $EI$. This diagram is usually referred to as the $M/EI$ diagram. Below is shown the corresponding portion of the deflected elastic curve of the beam. These two diagrams are similar to those shown in Fig. 9-10. We will now show that the total displacement $\Delta_{AB}$ of point $A$ on the elastic curve from a tangent to the curve at $B$ equals the *moment* about $A$ of the area of the $M/EI$ diagram between $A$ and $B$.

First we select three very closely adjacent ordinates $C$, $D$, and $E$ in the $M/EI$ diagram. The $M/EI$ area between $C$ and $D$ is $A_1$ and that between $D$ and $E$ is $A_2$. Next we project ordinates $C$, $D$, and $E$ downward to intersect the elastic curve and then draw tangents to the curve at the three points of intersection, extending these tangents left to a vertical through $A$, thus obtaining the intercepts $\Delta_{CD}$ and $\Delta_{DE}$. The slopes of the elastic curve and of these tangents are greatly exaggerated in Fig. 9-13, merely for greater clarity of development. Actually, in practical beam problems, these slopes seldom exceed 1° from the original horizontal position of the beam.

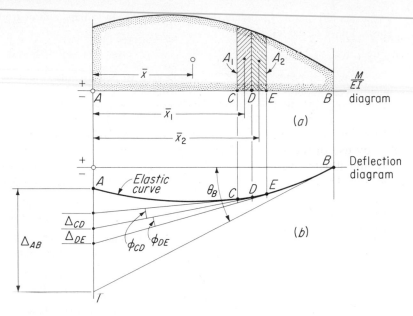

**Fig. 9-13** Derivation of the second moment-area principle.

Under these given circumstances and also under the condition that the distances from $C$ to $D$ and from $D$ to $E$ are infinitesimally small, it is clear that the slope lengths of the tangents at $C$ and $D$ are very nearly equal to the horizontal distance $\bar{x}_1$. As previously explained, the numerical values of the very small slope angles $\phi$ (in radians) and of their tangents are sufficiently close to be considered equal. Consequently $\tan \phi = \phi = \Delta/\bar{x}$ and $\Delta = \phi\bar{x}$. But according to the first moment-area principle, $\phi_{CD} = A_1$ and $\phi_{DE} = A_2$. Therefore $\Delta_{CD} = A_1\bar{x}_1$, the *moment* about $A$ of the area $A_1$ in the $M/EI$ diagram between $C$ and $D$. Similarly, $\Delta_{DE} = A_2\bar{x}_2$. If *all* the intercepts between $A$ and $T$ are now summed up, we obtain $\Delta_{AB}$, the displacement of point $A$ from a tangent at $B$. But $\Delta_{AB}$ would then equal the moment about $A$ of the *entire* area of the $M/EI$ diagram between $A$ and $B$. We may therefore state

**The Second Moment-Area Principle:**

$$
\left[
\begin{array}{l}
\text{The vertical displacement } \Delta_{AB} \\
\text{of any point } A \text{ on the elastic} \\
\text{curve of a beam from a tangent} \\
\text{to the curve at any other point } B
\end{array}
\right]
=
\left[
\begin{array}{l}
\text{the moment about } A \text{ of the area} \\
\text{in the moment diagram lying} \\
\text{between verticals through } A \\
\text{and } B, \text{ divided by } EI
\end{array}
\right]
$$

Expressing the second principle in symbols gives

$$\Delta_{AB} = \frac{(A_{AB})(\bar{x}_A)}{EI} \tag{9-4}$$

Clearly, $\Delta_{AB}$ is *not* the actual deflection $\delta$ of point $A$ from its original position. Methods of calculating actual deflections $\delta$ are outlined and illustrated in the following articles.

## 9-8 DEFLECTION OF CANTILEVER BEAMS

The vertical deflection $\delta$ of any point on a cantilever beam is readily calculated by use of the second moment-area principle, as is illustrated in the following illustrative problem. Unless otherwise stated and shown, all cantilever beams are presumed to be, and to remain, horizontal at the point of fixity. A tangent to the elastic curve of the beam at the point of fixity will then also be horizontal, thus greatly simplifying the solution of this type of problem.

### ILLUSTRATIVE PROBLEMS

**9-29.** The cantilever beam shown in Fig. 9-14 is a 50- by 400-mm rough wood plank laid flat. It is completely fixed at $B$. $E = 12\ 000$ MPa; $I = 4.17 \times 10^6$ mm⁴. Calculate its maximum deflection caused by the load shown.

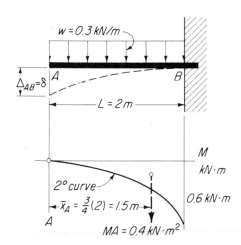

Fig. 9-14    **Deflection of cantilever beam. Moment diagram.**

*Solution:* The upper diagram in Fig. 9-14 shows the originally horizontal beam and also its deflected elastic curve. Clearly, the maximum deflection $\delta$ occurs at $A$. Also, a tangent to the elastic curve at $B$ is parallel to the horizontal axis of the undeflected beam. Therefore, $\delta = \Delta_{AB}$, the vertical displacement of $A$ from a tangent to the elastic curve at $B$. This vertical displacement $\Delta_{AB}$ equals the moment *about* $A$ of the moment area between $A$ and $B$. The moment curve is a second-degree parabola, the moment at $B$ is $(wL)(L/2) = (0.3)(2)(2/2) = 0.6$ kN·m, the moment-area $MA$ equals $(^1/_3)(2)(0.6) = 0.4$ kN·m², and the distance $\bar{x}_A$ from $A$ to the centroid of this moment-area is $(^3/_4)(2) = 1.5$ m (see Fig. 9-11c). Then $\delta_A$

equals the moment *about* $A$ of this moment area divided by $EI$ and is

$$\delta_A = \Delta_{AB} = \frac{1.5(0.4)(10^{12})}{(12\,000)(4.17)(10^6)} = 12 \text{ mm} \qquad Ans.$$

The multiplier $10^{12}$ follows the suggestion of Table 9-1.

**9-30.** Calculate the maximum deflection $\delta$ (at $A$) of the cantilever beam shown in Fig. 9-15. The beam is a W 16 × 67, for which $E$ is 29,000 ksi and $I$ is 954 in.⁴.

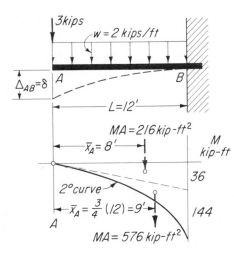

Fig. 9-15  **Deflection of cantilever beam. Moment diagram.**

*Solution:* We find it simplest here to divide the moment diagram into two parts, one for the concentrated load of 3 kips at $A$ forming a triangle and the other for the distributed load forming a parabolic area tangent to the triangle at $A$. Here also, $\delta_A = \Delta_{AB}$, the sum of the moments about $A$ of the moment areas between $A$ and $B$ divided by $EI$. Therefore

$$\delta_A = \Delta_{AB} = \frac{[(216)(8) + (576)(9)](1728)}{(29,000)(954)} = 0.432 \text{ in.} \qquad Ans.$$

Since the moment of the moment area here is expressed in kip-feet³, the necessary conversion to kip-inches³ is made by multiplying by 1728.

**PROBLEMS**

**9-31.** In Prob. 9-29 (Fig. 9-14) let $L = 3$ m and $w = 2.5$ kN/m. Calculate the maximum deflection $\delta_A$ at the left end $A$ if the beam is a structural-steel section with $I = 120 \times 10^6$ mm⁴.

**9-32.** Using the moment-area method, check the deflections obtained in Prob. 9-9.

**9-33.** A 6- by 6-in. rough (full-size) Douglas fir timber beam cantilevers a distance of 6 ft horizontally out of a completely fixed support at its right end $B$. Calculate the deflection $\delta_A$ at its left end $A$ that would be caused by a concentrated load there of 1000 lb.

*Ans.* $\delta_A = 0.678$ in.

**9-34.** The beam shown in Fig. 9-16 is a 3-in.-diameter standard steel pipe. Calculate the maximum deflection $\delta_A$ at the left end $A$. Use $E = 30,000$ ksi.

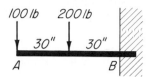

Fig. 9-16   Prob. 9-34.

**9-35.** The cantilever beam in Fig. 9-17 is a 40-mm-diameter solid wood rod, $E = 12\,000$ MPa, completely fixed at $B$. Calculate (*a*) the deflection $\delta_C$ of point $C$ and (*b*) the deflection $\delta_A$ of point $A$. (*c*) As a check on $\delta_A$, calculate $\theta_C$ and multiply $\theta_C$ by 1200 mm to obtain the difference in deflections between $A$ and $C$.

*Ans.*  $\delta_C = 16.1$ mm; $\delta_A = 48.3$ mm

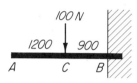

Fig. 9-17   Prob. 9-35.

**9-36.** Figure 9-18 shows a 1- by 4-in. rough board of eastern hemlock 24 in. long fixed at its right end $B$ (1-in. dimension is vertical). Calculate the maximum deflection $\delta_A$ at $A$.

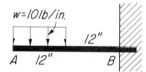

Fig. 9-18   Prob. 9-36.

## 9-9 BENDING-MOMENT DIAGRAMS BY PARTS

In the application of the two moment-area principles, we are concerned with the *area* in the bending-moment diagram between verticals drawn through two points, with the location of the centroid of that area, and with the *moment of that area* about one of these points. It is advantageous, therefore, that we draw these moment diagrams in their simplest possible form. Figure 9-19, for example, shows the three different moment diagrams that might have been drawn for the beam shown at top. All would have given identical results in equally simple manner. Figure 9-20 shows the conventional moment diagram and the moment diagram "by parts," left to right, for the beam shown at the top.

In a moment diagram by parts, a separate moment diagram is drawn for each force or load showing its moment effect along the beam, positive (upward-acting) forces and loads giving positive moment ordinates, and vice versa. As illustrated in Fig. 9-19 and 9-20 and also in Figs. 9-21 and

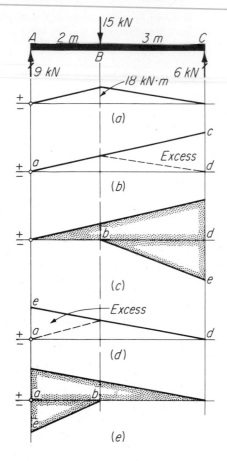

Fig. 9-19  Moment diagram "by parts."
(a) Conventional moment diagram.
(b) Construction. (c) Moment diagram by parts. (d) Construction. (e) Moment diagram by parts.

9-22, these separate diagrams may be drawn from left to right or from right to left, or they may be drawn from left to right up to a certain point and then from right to left back to that same point. The best form to use for any particular problem depends on the type of load and on its position along the beam. Hence only a general *procedure*, such as is outlined in Art. 9-11, can be used as a guide.

## PROBLEMS

**9-37.** Calculate the values of the principal ordinates in the three moment diagrams shown in Fig. 9-19. Show that the algebraic sums of the moment areas of all three diagrams are equal, and also that the algebraic sums of the moments about $A$ of these moment areas are equal.

**9-38.** Refer to Fig. 9-20. Show that the moment areas of $a$ and $b$ are equal and also that the moments about $A$ of these moment areas are also equal.

**9-39.** Draw separate moment diagrams, similar to those in Fig. 9-20, for the beam in Fig. 8-21. Evaluate the principle ordinates if $P_1 = 8$ kips and $P_2 = 7$ kips.

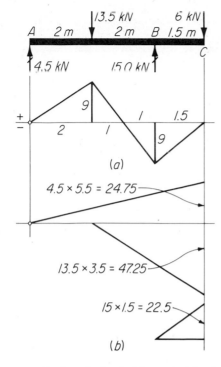

Fig. 9-20   Moment diagram by parts.
(a) Conventional moment diagram.
(b) Moment diagram by parts, left to right.

**9-40.** For the loaded beam in Fig. 9-24, draw the conventional moment diagram and the left-to-right and the right-to-left moment diagrams by parts, and show that the algebraic sums of the moment areas of all are equal and that the moment at $B$ is $-24$ kip·ft in each diagram.

**9-41.** For the beam shown in Fig. 9-26 draw the conventional moment diagram and the left-to-right and right-to-left moment diagrams by parts. Show that the moment areas of the three diagrams are equal and that the moments about $A$ of these areas are also equal.

## 9-10 DEFLECTION OF SYMMETRICALLY LOADED SIMPLE BEAMS

A symmetrically loaded simple beam has complete symmetry of beam, load, and reactions about a vertical centerline. Clearly, then, the maximum deflection of such a beam always occurs at the center. A tangent to the elastic curve at the point of maximum deflection will always be horizontal. This fact greatly simplifies the solution of this type of problem, as is illustrated in Probs. 9-42 and 9-43.

### ILLUSTRATIVE PROBLEMS

**9-42.** A symmetrically loaded simple beam, W 8 × 24, is shown in Fig. 9-21. Calculate (a) the maximum deflection δ at the center of the beam and (b) the deflection at a random point $D$, 3 ft from the left end.

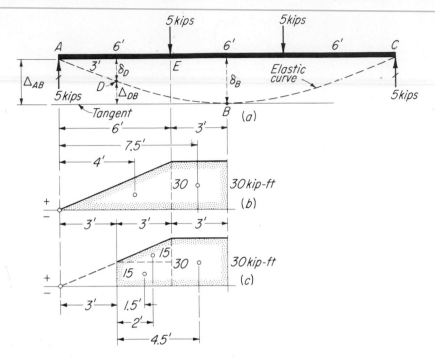

**Fig. 9-21   Deflection of a symmetrically loaded simple beam. (a) Beam and elastic curve. (b) Moment-area between A and B moment diagram. (c) Moment-area between D and B moment diagram.**

*Solution: (a)* When a tangent is drawn to the elastic curve at the center $B$, the point of maximum deflection, we see clearly that this maximum deflection $\delta_B$ is equal to $\Delta_{AB}$, the displacement of point $A$ from the tangent at $B$. According to the second moment-area principle, this displacement $\Delta_{AB}$ equals the moment *about A* of the moment area between $A$ and $B$, divided by $EI$. A conventional moment diagram is most suitable here. Therefore, since $E = 29{,}000$ ksi and $I = 82.8$ in.[4],

$$[M_A] \qquad \Delta_{AB} = \delta_B = \frac{[(90)(4) + (90)(7.5)](1728)}{(29{,}000)(82.8)} = 0.745 \text{ in.} \qquad Ans.$$

*(b)* To obtain the actual deflection $\delta_D$ at $D$, we must first calculate $\Delta_{AB}$ and $\Delta_{DB}$. Then $\delta_D = \Delta_{AB} - \Delta_{DB}$, as is clearly seen in Fig. 9-21. In part $a$, $\Delta_{AB}$ was found to equal 0.745 in. We may now obtain $\Delta_{DB}$ by taking moments *about D* of the moment area between $D$ and $B$, shown in Fig. 9-21c, and dividing the result by $EI$. That is,

$$[M_D] \qquad \Delta_{DB} - \frac{\left[\left(\frac{1}{2}\right)(3)(15)(2) + (3)(15)(1.5) + (3)(30)(4.5)\right](1728)}{(29{,}000)(82.8)} = 0.372 \text{ in.}$$

Then $\qquad\qquad\qquad\qquad \delta_D = 0.745 - 0.372 = 0.373 \text{ in.} \qquad Ans.$

In this example, the conventional moment diagram was used because a moment diagram by parts had no particular advantage. In Prob. 9-43 it will have considerable advantage and will be used.

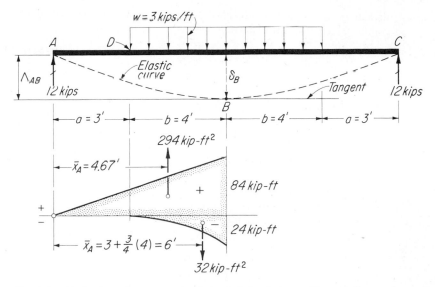

**Fig. 9-22** Deflection of a symmetrically loaded simple beam. Moment diagram by parts. Moment area between $A$ and $B$.

**9-43.** Calculate the maximum deflection $\delta$ of the 10- by 16-in. S4S timber beam shown in Fig. 9-22. Use $E = 1760$ ksi.

*Solution:* The maximum deflection $\delta$ occurs at the center of the beam. A tangent to the elastic curve at the center $B$ will be horizontal. Consequently, the displacement $\Delta_{AB}$ at $A$ equals the deflection $\delta_B$ at $B$. We find it simplest here to draw the moment diagram by parts from left to right *away from* point $A$. The upward-acting (positive) reaction gives a positive triangular moment area between $A$ and $B$. The downward-acting (negative) uniform load gives a negative parabolic moment area between $D$ and $B$. Then $\Delta_{AB} = \delta_B$ and is also equal to the algebraic sum of the moments *about* $A$ of the moment areas between $A$ and $B$. Since $E = 1760$ ksi and $I = 2948$ in.$^4$, we obtain

$$[M_A] \qquad \Delta_{AB} = \delta_b = \frac{[(294)(4.67) - (32)(6)](1728)}{(1760)(2948)} = 0.393 \text{ in.} \qquad Ans.$$

## PROBLEMS

**9-44.** A 3- by 12-in. rough eastern hemlock plank 18 ft long is laid flat on supports at its outer ends. Calculate the maximum deflection $\delta$ caused by a person weighing 176 lb standing at the center of the plank.

**9-45.** In Prob. 9-44 calculate the deflection $\delta$ at a point 3 ft from the left end.

*Ans.* $\delta = 0.549$ in.

**9-46.** A structural-steel beam, W $12 \times 35$, 24 ft long is supported at its ends and carries a uniformly distributed load of 1 kip/ft. Calculate the maximum deflection $\delta$ of this beam.

**9-47.** A simple beam 12 m long of structural steel, W $360 \times 134$, supports three 20-kN concentrated loads applied, respectively, at 3, 6, and 9 m from the left end.

Calculate (a) the maximum deflection $\delta_{max}$ at the center and (b) the deflection $\delta_6$ at a point 6 ft from the left end.

*Ans.* (a) $\delta_{max} = 20.55$ mm; (b) $\delta_6 = 14.6$ mm

**9-48.** A 2- by 12-in. rough (full-size) Ponderosa pine plank 16 ft long laid flat is supported at its outer ends. It carries a uniform load of 150 lb/ft on the extreme left 4 ft and a similar load on the extreme right 4 ft. Calculate the maximum deflection $\delta$.

**9-49.** Calculate the deflection $\delta_3$ of the beam in Prob. 9-43 at a point 3 ft from the left end.

**9-50.** Determine the maximum value of $EI\delta$ in Prob. 9-19.

**9-51.** Calculate the maximum $EI\delta$ in Prob. 9-21.

**9-52.** A steel beam, W 21 × 50, spans 32 ft between the exterior walls of a building. It supports a unform load of 1.5 kips/ft, which includes its own weight. (a) Calculate the maximum deflection $\delta$. (b) Select the most economical W section for which $\delta$ will not exceed $1/360$ of the span. (c) By proportion, what is $\delta$ of the beam selected?

*Ans.* (a) $\delta = -1.24$ in.; (b) W 21 × 57; (c) $\delta = -1.04$ in.

**9-53.** Solve Prob. 9-52 for a W 250 × 17.9 steel beam whose span is 8 m and whose load is 3 kN/m.

**9-54.** Calculate the maximum deflection of a 200- × 400-mm timber beam, $E = 12\,000$ MPa, S4S, whose simple span $L$ is 8 m and uniformly distributed load $w$ is 1.5 kN/m, which includes weight of beam.

*Ans.* $\delta = -7.29$ mm

**9-55.** In Prob. 8-23, adjust the design load on beam $B1$ to include the actual weight of beam. Then (a) calculate the maximum deflection $\delta$ of the beam. (b) If the permissible deflection is $1/360$ of the span, what other W section would be most economical? (c) By proportion, calculate the deflection of the beam selected.

## 9-11  DEFLECTION OF UNSYMMETRICALLY LOADED BEAMS. OVERHANGING BEAMS

Lack of symmetry of either load or reactions (overhang) makes the determination of deflection a somewhat more extensive problem. Determination of *maximum* deflection caused by distributed loads usually requires the solution of a cubic, or higher, equation. Such problems, while not too difficult to solve, are beyond the intended scope of this text. The following procedures and examples will indicate the method of solution in problems with concentrated loads.

### Procedure with English units

1. Draw the beam, its probable elastic curve, and the moment diagram by parts (usually) *from* the support farthest from the load (concen-

trated or resultant). Moments may be in kip-feet, pound-feet, or pound-inches.

2. Draw a tangent to the elastic curve of the beam at the support farthest from load or from the resultant of loads. Using the second principle, calculate desired displacement $\Delta$ from this tangent. To obtain $\Delta$ in inches, the moment must be in pound-inches or must be converted to pound-inches.

3. Calculate the slope angle $\theta$ of this tangent: $\theta = \tan \theta = \Delta/L$, where $\Delta$ and $L$ must be in inches.

4. Using the first principle, solve for distance $m$ from point of maximum deflection $\delta$ to the support *farthest* (usually) from load, which support should also be point of beginning of the moment diagram. The moment *must* be in pound-inches.

5. Draw a second tangent to the elastic curve at point of maximum deflection (this tangent will be horizontal). Then calculate $\delta_{max} = \Delta$, using the second principle. The moment of the area of the moment diagram *must* be converted to pound-inches in this calculation.

NOTE: No set of rules can be established to cover all cases. In general, however, the origin of the moment diagram by parts and the *first* tangent to the elastic curve should be at the same end of the beam. If one scheme does not work out to be simple, try another.

## ILLUSTRATIVE PROBLEMS

**9-56.** Calculate the maximum deflection $\delta$ of the unsymmetrically loaded beam shown in Fig. 9-23. The beam is a 6- by 6-in. (full-size) timber; $E = 1,760,000$ psi and $I = 108$ in.$^4$.

*Solution:* The probable elastic curve of the beam having been drawn, a tangent is drawn to this curve at $B$, farthest from load. Point $B$ is also the starting point of the moment diagram by parts. Positive (upward-acting) forces give positive moment areas, and vice versa. $\Delta_{AB}$ is the displacement of $A$ from a tangent at $B$ and is equal to the moment *about* $A$ of the moment area between $A$ and $B$. That is, using the second principle, we have

$$[M_A] \qquad \Delta_{AB} = \frac{[(67,500)(5) - (27,000)(2)](1728)}{(1,760,000)(108)} = 2.58 \text{ in.}$$

Since the slope angle $\theta_B$ (in radians) is very small, its tangent equals the angle. Therefore

$$\theta_B = \tan \theta_B = \frac{\Delta_{AB}}{L} = \frac{2.58}{(15)(12)} = 0.01435$$

We may now solve for the distance $x$ from $B$ to the point of maximum deflection $C$. At $C$ the moment equals $600x$. The moment area between $B$ and $C$ is

$$\left(\frac{1}{2}\right)(x)(600x) = 300x^2$$

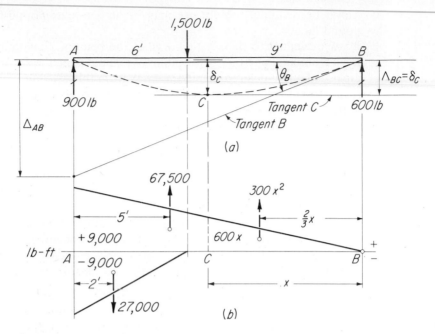

**Fig. 9-23   Maximum deflection of unsymmetrically loaded simple beam. Moment diagram by parts.**

which, according to the first principle, equals $EI\theta_B$. Hence, since moment area $BC = EI\theta_B$,

$$300x^2 = (1,760,000)(108)(0.01435) = 2,720,000$$

from which

$$x^2 = \frac{2,720,000}{300} = 9066 \quad \text{and} \quad x = 95.2 \text{ in.} = 7.93 \text{ ft}$$

A second tangent, drawn to the elastic curve at $C$, the point of maximum deflection, will be horizontal. The displacement $\Delta_{BC}$ of point $B$ from this tangent, according to the second principle, equals the moment about $B$ of the moment area between $B$ and $C$. But $\Delta_{BC}$ equals the required maximum deflection $\delta_C$. Therefore

$$\delta_C = \Delta_{BC} = \frac{(300)(9066)\left(\frac{2}{3}\right)(95.2)}{(1,760,000)(108)} = 0.91 \text{ in.} \qquad Ans.$$

**9-57.** A steel beam, W 8 × 48, overhanging the right support carries two concentrated loads, as shown in Fig. 9-24 ($E = 29,000,000$ psi; $I = 184$ in.$^4$). Calculate (a) the maximum deflection $\delta_D$ between supports $A$ and $B$ and (b) deflection $\delta_C$ of the right end $C$, both from the original horizontal position of the beam.

*Solution:* (a) The probable elastic curve as drawn is undoubtedly correct, at least between $A$ and $D$. In this problem the distance $m$ is most easily obtained if

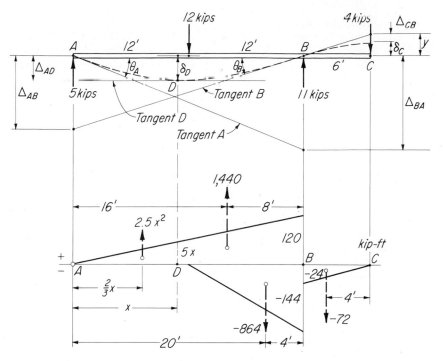

**Fig. 9-24    Maximum deflection of overhanging beam. Moment diagram by parts.**

the moment diagram by parts is drawn *from A to B*. The diagram could be correctly *continued* to $C$, but the part on the overhang may also be drawn from $C$ to $B$, giving a much simpler form.[1] We now draw a tangent to the elastic curve at $A$ so that we may solve for $\Delta_{BA}$, $\theta_A$, and $x$, in that order. Then

$$[M_B] \qquad \Delta_{BA} = \frac{[(1440)(8) - (864)(4)](1728)}{(29,000)(184)} = 2.61 \text{ in.}$$

Also, now,

$$\theta_A = \tan \theta_A = \frac{\Delta_{BA}}{L} = \frac{2.61}{(24)(12)} = 0.00907$$

and, since moment area $AD = EI\theta_A$,

$$(2.5x^2) = (29,000)(184)(0.00907) = 48,400$$

from which

$$x^2 = \frac{48,400}{2.5} = 19,360 \qquad \text{and} \qquad x = 139 \text{ in.} = 11.59 \text{ ft}$$

[1] If a "conventional" moment diagram is also drawn, a little study will show that the moment area between any two points, and also the moment about any point of such moment area, is the same for the conventional diagram and for the diagram by parts in Fig. 9-24.

Had $x$ exceeded 12 ft, the solution would have been incorrect, because the moment area between $A$ and $D$ would not then have equaled $2.5x^2$.

In order now to solve for $\delta_D$, we draw a second tangent $D$ to the elastic curve at $D$. Clearly, now, $\delta_D = \Delta_{AD}$, the displacement of $A$ from tangent $D$. Then, using the second principle, we have

$$[M_A] \qquad \Delta_{AD} = \delta_D = \frac{(2.5)(19{,}360)\left(\frac{2}{3}\right)(139)}{(29{,}000)(184)} = 0.841 \text{ in.} \qquad Ans.$$

(b) To obtain the deflection $\delta_C$, we must draw a third tangent $B$ to the elastic curve at $B$, after which we calculate $\Delta_{AB}$ at $A$ and, by proportion, $y$ at $C$. The slope of the tangent at $B$ will be positive (as it is in *this* problem) if the moment about $A$ of the moment area between $A$ and $B$ is positive and will be negative if this moment is negative. Then

$$[M_A] \qquad \Delta_{AB} = \frac{[(1440)(16) - (864)(20)](1728)}{(29{,}000)(184)} = 1.87 \text{ in.}$$

We now obtain $y$ by proportion thus:

$$\frac{y}{6} = \frac{\Delta_{AB}}{24} \qquad \text{or} \qquad y = \frac{6}{24}\,1.87 = 0.466 \text{ in.}$$

The displacement $\Delta_{CB}$ of point $C$ from the tangent at $B$ is

$$[M_C] \qquad \Delta_{CB} = \frac{(-72)(4)(1728)}{(29{,}000)(184)} = -0.093 \text{ in.}$$

The deflection $\delta_C$ at $C$ is then

$$\delta_C = y - \Delta_{CB} = 0.466 \text{ in.} - 0.093 \text{ in} = 0.373 \text{ in.} \qquad Ans.$$

Since $\Delta_{CB}$ is less than $y$, the deflection $\delta_C$ is *upward* from the original horizontal position of the beam.

## PROBLEMS

**9-58.** A simple beam, W 16 × 45, 30 ft long supports a single concentrated load of 15 kips at a point 12 ft from the left end. Calculate the maximum deflection of this beam.

*Ans.* $\delta = 0.813$ in.

**9-59.** Figure 9-25 shows a wood board 8 in. wide by 1 in. thick, laid flat. Calculate the maximum deflection $\delta$ caused by loads as shown. Use $E = 1760$ ksi.

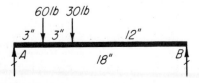

Fig. 9-25   Prob. 9-59.

**9-60.** The beam in Fig. 9-26 is a 2- by 6-in. rough wood plank, $E = 1,760,000$ psi, laid flat. Calculate the deflection $\delta$ at the center (not maximum).

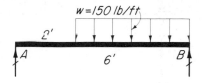

w = 150 lb/ft

2'

A    6'    B

**Fig. 9-26   Probs. 9-14 and 9-60.**

**9-61.** The beam shown in Fig. 9-27 is a full-size 8- by 8-in. timber, for which $E = 1,760,000$ psi. Calculate the deflection $\delta_C$ of its extreme right end $C$.

*Ans.* $\delta_C = 0.207$ in.

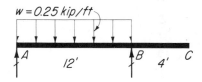

w = 0.25 kip/ft

A    12'    B    4'    C

**Fig. 9-27   Prob. 9-61.**

**9-62.** Calculate the deflection of end $C$ of the beam shown in Fig. 9-28. The beam is a W 200 × 15 structural-steel section.

*Ans.* $\delta_C = 12.5$ mm

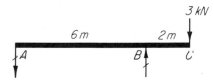

3 kN

6 m    2 m

A    B    C

**Fig. 9-28   Probs. 9-62 and 9-63.**

**9-63.** Determine the maximum deflection between supports $A$ and $B$ of the W 200 × 15 structural-steel beam shown in Fig. 9-28.

*Ans.* $\delta = 5.41$ mm

## 9-12 DERIVATION OF DEFLECTION FORMULAS

In the preceding articles, deflections of beams supporting specified loads were determined by the beam-diagram method and the moment-area method. These methods will now be used to derive a series of simple formulas giving the deflections of various types of beams in terms of their lengths $L$ and loads $P$ or $w$, as the case may be. Formulas for the cases $a$, $b$, and $c$ shown in Fig. 9-4 will be derived by the beam-diagram method. The formula for a simple beam with a uniformly distributed load and that for case $d$ in Fig. 9-4 will be derived by the moment-area method.

## 9-13 CANTILEVER BEAM WITH CONCENTRATED LOAD AT FREE END

The load, shear, and moment diagrams are drawn in the usual manner, as shown in Fig. 9-29, in accordance with the laws of beam diagrams. The shear ordinates have a constant value of $P$ between $A$ and $B$. The moment at the free end $A$ is zero. Hence the moment at $B$ is $PL$, the area under the shear curve between $A$ and $B$.

From the physical conditions of the problem, it is apparent that the beam must remain horizontal at $B$, indicating that the $EI\theta$ ordinate at $B$ is zero. The shape of the curve in the slope diagram must be increasing negative, as are the moment ordinates above. Hence the initial $EI\theta$ ordinate is positive and has a value, according to the second law, of $(^1/_2)(L)(PL) = PL^2/2$, or $PL^2/2EI$ after being divided by $EI$.

Clearly, the deflection at $B$ is zero; that is, $\delta_B = 0$. Hence the maximum deflection ordinate at $A$ is $(^2/_3)(L)(PL^2/2EI) = PL^3/3EI$, the area under the slope curve between $A$ and $B$.

Therefore the maximum deflection at the free end of a cantilever beam due to a load $P$ at that end is

$$\delta = \frac{PL^3}{3EI} \tag{9-5}$$

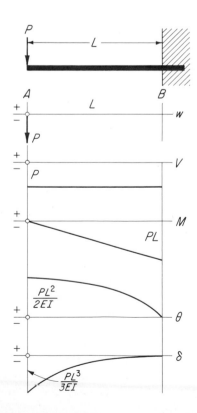

Fig. 9-29  Derivation of deflection formula.

## 9-14 CANTILEVER BEAM WITH UNIFORMLY DISTRIBUTED LOAD

From the physical conditions of this beam we find that the shear $V$ and moment $M$ at $A$ are both zero and that at $B$ the slope $\theta$ and deflection $\delta$ are zero.

The load, shear, and moment diagrams are drawn, as illustrated in Fig. 9-30. The slope of the shear curve is constant and negative, as are the load ordinates above. The shear at $B$ equals $wL$, the area under the load curve between $A$ and $B$.

The slope of the moment curve is increasing and negative, as are the shear ordinates above. The moment at $B$ is $(^1/_2)(L)(wL) = wL^2/2$, the area under the shear curve above.

Clearly, the beam remains horizontal at $B$ and the $\theta$ ordinate at $B$ is therefore zero. The slope of the $\theta$ curve is increasing and negative, as are the moment ordinates above. Hence, since $\theta_B = 0$, the curve starts above the axis and the initial $\theta$ ordinate is $(^1/_3)(L)(wL^2/2) = wL^3/6$, or $wL^3/6EI$ after being divided by $EI$.

Beginning at $A$, the slope of the deflection curve is positive and decreasing, as are the $\theta$ ordinates above. Because $\delta_B$ is zero, the curve must

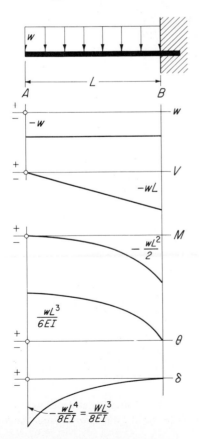

**Fig. 9-30**  Derivation of deflection formula.

start below the axis, and the ordinate $\delta_A$ then equals $(^3/_4)(L)(wL^3/6EI) = wL^4/8EI$ or $WL^3/8EI$, since $W = wL$.

Hence, the maximum deflection at the free end of a cantilever beam due to a uniformly distributed load is

$$\delta = \frac{WL^3}{8EI} \qquad\qquad (9\text{-}6)$$

## 9-15 SIMPLE BEAM WITH CONCENTRATED LOAD AT CENTER

Having drawn the load, shear, and moment diagrams, as in Fig. 9-31, we find the maximum moment to be $(P/2)(L/2) = PL/4$, the area under the shear curve between $A$ and $B$.

Because of symmetry of beam and of loading, it is clear that the deflection $\delta$ will be maximum at the center $B$ of the beam, at which point, therefore, the slope of the beam will necessarily be zero. Also, clearly, the deflections at supports $A$ and $C$ will be zero.

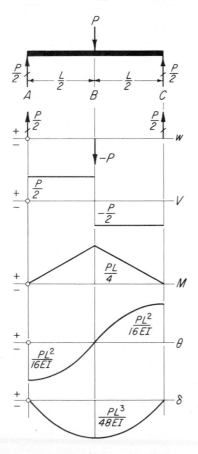

Fig. 9-31  Derivation of deflection formula.

Between $A$ and $B$ the slope of the $\theta$ curve is positive and increasing, and between $B$ and $C$ it is positive and decreasing, as are the moment ordinates. Since $\theta_B = 0$, the curve must start at $A$ with a negative ordinate (according to the second law)

$$\theta_A = \frac{1}{2}\frac{L}{2}\frac{PL}{4}\frac{1}{EI} = \frac{PL^2}{16EI}$$

Starting with zero ordinate at $A$, the slope of the deflection curve is negative and decreasing from $A$ and $B$ and is positive and increasing from $B$ to $C$, as are the $\theta$ ordinates above. The maximum deflection at the center $B$, according to the second law, is then

$$\frac{2}{3}\frac{L}{2}\frac{PL^2}{16EI}$$

or the maximum deflection at the center of a simple beam due to a concentrated load $P$ at the center is

$$\delta = \frac{PL^3}{48EI} \tag{9-7}$$

## 9-16 SIMPLE BEAM WITH UNIFORMLY DISTRIBUTED LOAD

In this derivation we use the moment-area method, which is the simplest for this case.

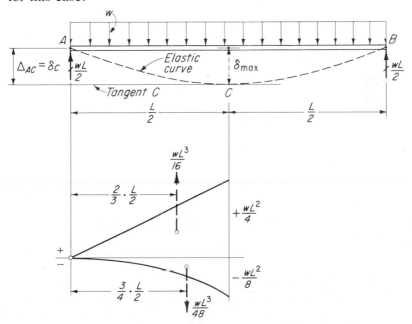

**Fig. 9-32**  Derivation of deflection formula. Moment diagram by parts.

The beam and its deflected elastic curve are shown in Fig. 9-32. Clearly, the maximum deflection occurs at the center $C$, and a tangent to the elastic curve at $C$ will be horizontal, that is, parallel to the undeflected axis of the beam. Hence the maximum deflection $\delta_C$ equals $\Delta_{AC}$, which in turn equals the moment about $A$ of the moment area between $A$ and $C$, divided by $EI$. That is,

$$[M_A] \qquad \delta_{\max} = \Delta_{AC} = \left[\left(\frac{wL^3}{16}\frac{L}{3}\right) - \left(\frac{wL^3}{48}\frac{3L}{8}\right)\right]\frac{1}{EI} = \frac{5wL^4}{384EI} \qquad (9\text{-}8)$$

By the first moment-area principle, the slope at the left end of the beam is (from Fig. 9-32)

$$\theta = \frac{wL^3}{16EI} - \frac{wL^3}{48EI} = \frac{3wL^3}{48EI} - \frac{wL^3}{48EI} = \frac{wL^3}{24EI}$$

### 9-17 SIMPLE BEAM WITH TWO EQUAL CONCENTRATED LOADS SYMMETRICALLY PLACED

Here we also use the moment-area method because it gives us the required formula in the simplest possible manner.

The loaded beam and its deflected elasic curve are shown in Fig. 9-33. Because of symmetry of forces, the maximum deflection $\delta$ occurs at the center. A tangent at this point $C$ will be horizontal. Clearly, therefore, $\delta_{\max} = \Delta_{AC}$, which in turn equals the moment about $A$ of the moment area between $A$ and $C$, divided by $EI$. Hence

$$[M_A] \qquad \delta_{\max} = \Delta_{AC} = \left\{\left(\frac{Pa^2}{2}\frac{2a}{3}\right) + \left[\frac{Pa(L-2a)}{2}\right]\left(a + \frac{L-2a}{4}\right)\right\}\frac{1}{EI}$$

which when simplified becomes

$$\frac{Pa}{24EI}\,(3L^2 - 4a^2) \qquad (9\text{-}9)$$

### 9-18 SUMMARY OF SLOPE AND DEFLECTION FORMULAS

The formulas for the maximum slope of four common beams and loading were found in Arts. 9-13 through 9-16. These slope formulas are summarized in Table 9-2.

The deflection formulas for five commonly used beams and beam loadings were also derived. A score or more additional formulas for other

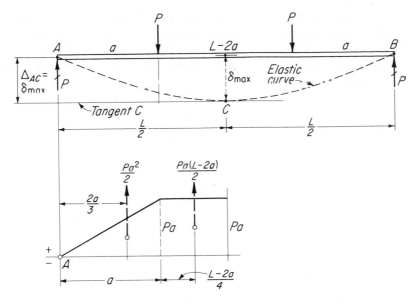

**Fig. 9-33**  Derivation of deflection formula. Conventional moment diagram.

types of beams and loadings and for partial loadings could be derived, some in similar manner and others with the aid of calculus. In practice, the vast majority of deflection problems encountered by engineers and architects are solved by the use of such formulas. Table 9-3 list the five derived deflection formulas plus an additional six, together with the formulas giving the maximum moment in each case.

Table 9-2  BEAM-SLOPE FORMULAS

| Case No. | Beam and load | Maximum slope | Case No. | Beam and load | Maximum slope |
|---|---|---|---|---|---|
| 1 | | $\theta = \dfrac{PL^2}{2\,EI}$ | 3 | | $\theta = \dfrac{PL^3}{16\,EI}$ |
| 2 | | $\theta = \dfrac{wL^3}{6\,EI}$ | 4 | | $\theta = \dfrac{wL^3}{24\,EI}$ |

The calculations of deflections due to irregular or partial loads become rather complex and are rarely performed in practice. For the practical designer, therefore, the following facts are useful: The deflection at the midspan of a simple beam due to a concentrated load applied anywhere on the beam differs from the maximum deflection by less than 3 percent. Similarly, the midspan deflection caused by a partial distributed load, covering not less than half of a simple beam, is less than 5 percent less

Table 9-3 BEAM-DEFLECTION FORMULAS

| Case no. | Beam and load | Maximum moment | Maximum deflection |
|---|---|---|---|
| 1 | | $M = -PL$ | $\delta = -\dfrac{PL^3}{3EI}$ |
| 2 | | $M = -Pa$ | $\delta = -\dfrac{Pa^2}{6EI}(3L - a)$<br><br>($a$ and $L$ must be in inches) |
| 3 | | $M = -\dfrac{wL^2}{2}$<br><br>$= -\dfrac{WL}{2}$ | $\delta = -\dfrac{wL^4}{8EI}$<br><br>$= -\dfrac{WL^3}{8EI}$ |
| 4 | | $M = -\dfrac{wL^2}{6}$<br><br>$= -\dfrac{WL}{3}$ | $\delta = -\dfrac{wL^4}{30EI}$<br><br>$= -\dfrac{WL^3}{15EI}$ |
| 5 | | $M = M_L$ | $\delta = -\dfrac{0.0642ML^2}{EI}$<br><br>at $x = 0.423L$<br><br>($M$ must be in pound-inches) |
| 6 | | $M = \dfrac{PL}{4}$ | $\delta = -\dfrac{PL^3}{48EI}$ |

Table 9-3  BEAM-DEFLECTION FORMULAS (*continued*)

| Case no. | Beam and load | Maximum moment | Maximum deflection |
|---|---|---|---|
| 7 | | $M = \dfrac{Pab}{L}$ | $\delta = -\dfrac{Pab(a + 2b)\sqrt{3a(a + 2b)}}{27EIL}$ <br> at $x = \sqrt{\dfrac{a(a + 2b)}{3}}$ <br> when $x = <a$ and $a - >b$ <br> ($a$, $b$, and $L$ must be in inches) |
| 8 | | $M = Pa$ | $\delta = -\dfrac{Pa}{24EI}(3L^2 - 4a^2)$ <br> ($a$ and $L$ must be in inches) |
| 9 | | $M = \dfrac{wL^2}{8}$ <br> $= \dfrac{WL}{8}$ | $\delta = -\dfrac{5wL^4}{384EI}$ <br> $= -\dfrac{5WL^3}{384EI}$ |
| 10 | | $M = \dfrac{wL^2}{9\sqrt{3}}$ <br> $= 0.1283WL$ <br> at $x = 0.577L$ | $\delta = -\dfrac{0.01304WL^3}{EI}$ <br> at $x = 0.519L$ |
| 11 | | $M = \dfrac{wL^2}{12}$ <br> $= \dfrac{WL}{6}$ | $\delta = -\dfrac{wL^4}{120EI}$ <br> $= -\dfrac{WL^3}{60EI}$ |

than the maximum due to a load of equivalent total magnitude distributed uniformly over the entire span.

## PROBLEMS

NOTE:  Use the appropriate deflection formula listed in Table 9-3.

**9-64.**  Calculate the maximum deflection called for in parts *a* and *c* of Prob. 9-9.

**9-65.** Check, by formulas, the maximum deflections of the two beams described in Prob. 9-10.

**9-66.** Compute the deflections called for in Prob. 9-11.

*Ans.* (a) $\delta = -0.157$ in.; (b) $\delta = -0.182$ in.

**9-67.** Solve Prob. 9-12 by using the appropriate formula listed in Table 9-3.

**9-68.** Using the appropriate formula from Table 9-3, compute the deflection required in Prob. 9-13.

*Ans.* $\delta = 93.9$ mm

**9-69.** Calculate the deflections called for in Prob. 9-14.

**9-70.** Compute by formulas the deflections called for in Prob. 9-16.

*Ans.* $\delta = -0.139$ in.; $-0.078$ in.; $-0.217$ in.

**9-71.** Using the appropriate formula, calculate the deflection called for in Prob. 9-17.

*Ans.* $\delta = 8.33$ mm

**9-72.** Calculate the maximum deflection of the shaft of Prob. 8-35 using the appropriate formula listed in Table 9-3.

**9-73.** Calculate by formula the deflection called for in Prob. 9-19.

**9-74.** Using the appropriate formulas from Table 9-3, calculate the deflection at the center of the beam of Prob. 9-20.

**9-75.** Calculate, using the proper formulas, the maximum deflection $\delta$ of girder $G1$ in Prob. 8-23 that results from the three concentrated loads and the actual weight of the beam.

[NOTE: Calculate the deflection in three separate parts: (1) $\delta$ due to weight of beam, (2) $\delta$ due to the middle concentrated load, and (3) $\delta$ due to the two outer concentrated loads.]

*Ans.* $\delta = -0.019$ in. $-0.304$ in. $-0.405$ in. $= -0.728$ in.

**SUMMARY** (By Article Number)

**9-1.** The **relation between radius of curvature $R$ and stress $s$** in a member subjected to pure bending is

$$s = \frac{cE}{R} \tag{9-1}$$

in which $c$ is the distance from the neutral plane to the extreme fiber and $E$ is the modulus of elasticity of the material.

**9-2.** The **relation between radius of curvature $R$ and bending moment $M$** is

$$R = \frac{EI}{M} \tag{9-2}$$

in which $E$ is the modulus of elasticity and $I$ is the moment of inertia of the section.

**9-4. Deflections by the beam-diagram method** are readily obtained only for cantilever beams and for simple beams with concentrated loads symmetrically placed. The five beam diagrams are drawn in accordance with the first and second laws of beam diagrams. Ordinates in the fourth (slope) and fifth (deflection) diagrams must be divided by $E$ and $I$.

**9-5.** The **moment-area method** is slightly more complex than the other methods but it is also more useful in that it is readily applicable to beams with unsymmetrical loads. The two moment-area principles should be memorized.

**9-12. Deflection formulas** are most commonly used in practice. Table 9-3 gives the formulas for 11 typical cases. In more difficult problems, approximate results, usually sufficiently close, may readily be obtained by calculating the midspan deflection in the manner outlined in the second paragraph of Art. 9-18.

## REVIEW PROBLEMS

**9-76.** A horizontal cantilever beam 9 ft long supports a uniformly distributed load of 800 lb/ft, which includes its own weight. Using the beam-diagram method, calculate the deflection $\delta$ at the free end if the beam is (a) a W 10 × 22 steel section and (b) a 6- by 18-in. dense Douglas fir timber S4S.

$$\text{Ans. } (a)\ \delta = -0.331 \text{ in.; } (b)\ \delta = -0.272 \text{ in.}$$

**9-77.** Assume that the beam described in Prob. 9-76 is a steel W 12 × 26. Calculate the actual maximum deflection $\delta$ (a) by proportion, using the answer to Prob. 9-76, and (b) using the appropriate formula.

**9-78.** A simple beam 18 ft long supports two equal concentrated loads of 8 kips each at the third points of the beam. Calculate the maximum value of $EI\delta$ using (a) the beam-diagram method, and (b) the moment-area method.

$$\text{Ans. } EI\delta = -1656 \text{ kip·ft}^3$$

**9-79.** Using the moment-area method, calculate the maximum deflection $\delta$ of girder $G1$ in Prob. 8-3. Adjust the load to include actual weight of beam. Check $\delta$ by the appropriate formulas.

**9-80.** A cantilever beam, W 200 × 26.6, extends 2 m horizontally out from a completely solid abutment. It supports a single concentrated load of 5 kN at its free end. Using the beam-diagram method, calculate (a) the slope angle $\theta$ at the free end in radians and degrees and (b) the maximum deflection $\delta$. Neglect weight of beam.

**9-81.** Consider a simple beam of length $L$ and with uniform load $w$ per foot. Using the moment-area method, show that the maximum slope angle at either end is $wL^3/24EI$ and that the maximum deflection at the center is $5wL^4/384EI$.

**9-82.** Calculate the deflection $\delta$ of the beam mentioned in Prob. 9-42 and illustrated in Fig. 9-21 at the point of application $E$ of the left 5-kip load. Use the moment-area method.

**9-83.** Using the moment-area method, calculate the maximum deflection $\delta$ of the structural-steel beam ($I = 1752.4$ in.$^4$) shown in Fig. 9-34.

$$\text{Ans. } \delta = 0.501 \text{ in.}$$

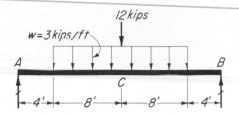

Fig. 9-34   Prob. 9-83.

**9-84.** The cantilever beam shown in Fig. 9-35 is a 2- by 6-in. wood plank (full size). Using the moment-area method, calculate the deflections at $C$ and $A$. Check the difference in the deflections at $C$ and $A$ by calculating $\theta_A$ and evaluating this difference in terms of $\theta_A$ and the distance $AC$. Use $E = 1,760,000$ psi.

*Ans.* $\delta_C = 0.943$ in.; $\delta_A = 1.47$ in.

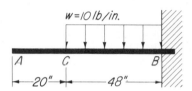

Fig. 9-35   Prob. 9-84.

**9-85.** A symmetrical simple beam is loaded as shown in Fig. 9-22. Let $w = 20$ kN/m, $a = 2$ m, $b = 4$ m, $L = 12$ m, $E = 200\ 000$ MPa, and $I = 620 \times 10^6$ mm⁴. Calculate the maximum deflection.

**9-86.** Refer to Fig. 9-20. Calculate the slope $\theta_B$ over the support $B$. Use the second moment-area principle and solve by parts. Also calculate the deflection at $C$. Assume $E = 200\ 000$ MPa and $I = 10 \times 10^6$ mm⁴.

## REVIEW QUESTIONS

**9-1.** What is meant by pure bending?

**9-2.** What is the elastic curve?

**9-3.** If a section of a bent member of prismatic cross section and of homogeneous material is in pure bending, is the radius of curvature of the elastic curve necessarily constant?

**9-4.** Express in equation form the mathematical relation between unit stress $s$ and radius of curvature $R$.

**9-5.** Give the mathematical relation between radius of curvature $R$ and bending moment $M$.

**9-6.** What is meant by the deflection of a beam?

**9-7.** What objectionable results can be caused by the excessive deflection (*a*) of a beam in a building and (*b*) of a part in a precision machine?

**9-8.** What is the commonly accepted maximum deflection of beams and joists to avoid cracking of plastered ceilings?

**9-9.** Describe the beam-diagram method of determining deflections.

**9-10.** Explain the first moment-area principle.

**9-11.** If a moment area is expressed in kip-feet², how may a conversion to pound-inches² be accomplished, in order to obtain $\phi$ of $\theta$ in radians?

**9-12.** For very small angles (say, less than 3°), what relation exists between the angle (in radians) and its tangent?

**9-13.** Give the essence of the second moment-area principle.

**9-14.** When is the displacement Δ of a point equal to the actual deflection δ? Cite some cases.

**9-15.** When the moment of a moment area is expressed in kip-feet$^3$, how may a conversion to pound-inches$^3$ be accomplished in order to obtain δ in inches?

**9-16.** Explain how a moment diagram by parts is constructed.

**9-17.** If in a beam-deflection problem $W$ is in pounds and $L$ in feet, what will be the units of $EI\delta$?

**9-18.** In the equation $\delta = (EI\delta)/EI$, in what units must each quantity be expressed to give δ in inches?

**9-19.** Give the units in which each of the following quantities in formulas 1 to 11, Table 9-3, must be expressed in order to give the deflections δ in millimeters: $P$, $W$, $w$, $L$, $a$ and $b$, $E$, and $I$.

# Chapter 10
# Statically
# Indeterminate
# Beams

## 10-1 STATICALLY INDETERMINATE BEAMS

Any loaded beam for which the simple laws of statics ($\Sigma V = 0$, $\Sigma H = 0$, and $\Sigma M = 0$) are insufficient for the determination of *all* support reactions is said to be **statically indeterminate.** Such a condition exists whenever the number of unknown reactions (forces or moments) exceeds the number of available *independent* statics equations.

When a beam extends over several supports, as illustrated in Fig. 10-1*a*, it is said to be a **continuous beam.** The four unknown vertical support reactions make this beam statically indeterminate, since only two *independent* statics equations, $\Sigma V = 0$ and $\Sigma M = 0$, are available. A similar situation exists in the ''supported cantilever'' beam in Fig. 10-1*b*. Here there are three unknowns, the support reactions $R_1$ and $R_2$ and the support reacting moment $M_2$, and only two independent equations can be written ($\Sigma H = 0$ is trivial here).

To solve problems of this type, therefore, we must introduce certain other relations that are known to exist. According to the second moment-area principle, stated in Chap. 9, and using a slight modification of the subscripts, we have the following relation:

**Second Moment-Area Principle:**

$$\begin{bmatrix} \text{The vertical displacement of any} \\ \text{point 1 on the elastic curve of a} \\ \text{beam from a tangent to the curve} \\ \text{at any other point 2} \end{bmatrix} = \begin{bmatrix} \text{the moment about } A \text{ of the} \\ \text{area in the moment diagram} \\ \text{lying between verticals through} \\ \text{1 and 2, divided by } EI \end{bmatrix}$$

We will find this relation useful in the following derivation.

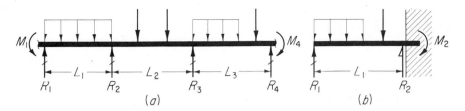

**Fig. 10-1   Statically indeterminate beams.**

## 10-2 DERIVATION OF THE THREE-MOMENT EQUATION

We now show that the second moment-area principle may be used to derive an equation expressing a relation existing among the bending moments $M$ at any three consecutively adjacent supports of a continuous beam, such as supports 1, 2, and 3 of the beam shown in Fig. 10-2a.

Let us assume, for example, that $L_1$ and $L_2$ are any two adjacent spans cut out of a multiple-span continuous beam and that the loads shown merely represent any loads that may be applied on these two spans. If each span were a simple beam, the moments produced by the loads would be *positive*, and the simple-beam moment diagrams could then be drawn on the positive side of the axis, as shown in Fig. 10-2b. Figure 10-2c shows the probable deflected "elastic curve" of this two-span section. The curvature over each of the three supports indicates tension in the top fibers, which in turn indicates negative moments at supports 1, 2, and 3. We may designate these moments $M_1$, $M_2$, and $M_3$ and may indicate them as ordinates on the *negative* side of the axis in Fig. 10-2b. When we now join the lower ends of these ordinates with straight lines, we have the complete moment diagrams for spans $L_1$ and $L_2$. For simplicity, the trapezoidal areas are divided into two triangles.

While it is true that the bending moments at interior supports of continuous beams are negative in the vast majority of problems, such is not always so. We thus consider them to be positive in this derivation. By so doing, each moment solved for by the three-moment equation will bear its true sign, positive or negative.

When this beam is considered to be weightless and is not otherwise loaded, a tangent to its elastic curve at support 2 would coincide with the undeflected horizontal axis of the beam and would, of course, be a horizontal line passing through support points 1, 2, and 3. But a tangent to the elastic curve at 2 of the *loaded* beam would slope, as shown in Fig. 10-2c, and the displacements at 1 and 3 of this tangent from the horizontal would be $\Delta_1$ and $\Delta_3$, positive and negative, respectively.

If we now take moments about point 1 of the moment-diagram areas in span $L_1$, we will, according to the second moment-area principle, obtain $EI\Delta_1$. Similarly, moments about 3 of the moment-diagram areas in span $L_2$

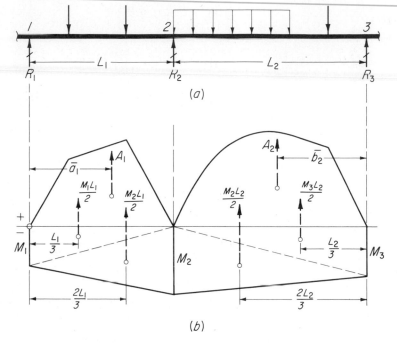

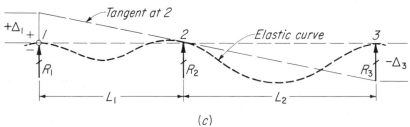

**Fig. 10-2** Derivation of the three-moment equation. (a) Any two spans of continuous beam. (b) Moment diagram. (c) Deflection diagram.

would yield $EI\Delta_3$. All moment-areas are assumed to be positive, or upward-acting. If $A_1$ and $A_2$ are the simple-beam moment areas in spans $L_1$ and $L_2$, and $\bar{a}_1$ and $\bar{b}_2$ are the distances from 1 and 3, respectively, to the centroids of those areas, we obtain, by moments about point 1,

$$[M_1] \qquad EI\Delta_1 = A_1\bar{a}_1 + M_1\frac{L_1}{2}\left(\frac{L_1}{3}\right) + M_2\frac{L_1}{2}\left(\frac{2L_1}{3}\right)$$

or

$$\Delta_1 = \left(A_1\bar{a}_1 + \frac{M_1L_1^2}{6} + \frac{M_2L_1^2}{3}\right)\frac{1}{EI} \qquad (a)$$

Similarly, by moments about point 3,

$$[M_3] \qquad EI\Delta_3 = A_2\bar{b}_2 + M_2\frac{L_2}{2}\left(\frac{2L_2}{3}\right) + M_3\frac{L_2}{2}\left(\frac{L_2}{3}\right)$$

or
$$\Delta_3 = \left( A_2 b_2 + \frac{M_2 L_2^2}{3} + \frac{M_3 L_2^2}{6} \right) \frac{1}{EI} \qquad (b)$$

A simple relation may now be established: From Fig. 10-2c by similar triangles, if $\Delta_1$ is positive and $\Delta_3$ is negative,

$$\frac{\Delta_1}{L_1} = -\frac{\Delta_3}{L_2} \qquad (c)$$

If we now divide each term in Eq. (a) by $L_1$ and each term in Eq. (b) by $L_2$ and then equate the right sides of Eqs. (a) and (b) using the relation established in Eq. (c), the $\Delta$ and $1/EI$ terms drop out and we obtain

$$\frac{A_1 \bar{a}_1}{L_1} + \frac{M_1 L_1}{6} + \frac{M_2 L_1}{3} = -\left( \frac{A_2 \bar{b}_2}{L_2} + \frac{M_2 L_2}{3} + \frac{M_3 L_2}{6} \right) \qquad (d)$$

If we then multiply each term in Eq. (d) by 6 and rearrange the terms, we obtain

$$M_1 L_1 + 2M_2(L_1 + L_2) + M_3 L_2 = -\frac{6A_1 \bar{a}_1}{L_1} - \frac{6A_2 \bar{b}_2}{L_2} \qquad \textbf{(10-1)}$$

Equation (10-1) is called the **three-moment equation.** It expresses, in terms of the loads on any two adjacent spans (such as 1–2 and 2–3), the relation among the bending moments at the three consecutive supports (1, 2, and 3) of those two spans. The presumptions are (1) that the beam is initially straight before loading, (2) that its $E$ and $I$ are constant throughout its entire length, and (3) that the supports remain in a straight line after loading.

The three-moment equation [Eq. (10-1)] enables us to calculate the bending moments at *all* supports of continuous and other statically indeterminate beams. When these moments have been evaluated, the support reactions are easily calculated.

## 10-3 LOAD TERMS FOR THE THREE-MOMENT EQUATION

The two terms on the right side of Eq. (10-1) are called **load terms.** For any given type of load, the value of a load term may be expressed in terms of the load and span.

Each type of load, except those that are symmetrically distributed about the vertical centerline of a span, will have two slightly differing load terms, depending on whether the load is on the *left* or the *right* span of the two-span section under consideration.

For example, to determine the two load terms for a single off-center concentrated load $P$, as illustrated in Fig. 10-3, we draw the moment diagrams shown in Fig. 10-3b and f. To obtain the left-span load term, the $A\bar{a}$ portion of the term $6A\bar{a}/L$ is obtained by calculating the moment of the moment-diagram area about point 1. For the right-span load term, the $A\bar{b}$ portion of $6A\bar{b}/L$ is obtained by moments about point 3.

This moment of the moment area may be obtained in one of several ways. Perhaps the simplest of all, for the left span, is to construct the two triangles 1–2–$n$ and 1–$m$–$n$, as in Fig. 10-3$c$. Clearly, now, according to the principles of moments of areas outlined in Chap. 5, the moment about point 1 of the moment-area 1–2–$m$ equals the moment of area 1–2–$n$ less the moment of area 1–$m$–$n$. Triangles $n$–1–2 and $m$–$o$–2 are similar. Hence, by proportion,

$$\frac{n-1}{L} = \frac{Pab/L}{b} \quad \text{or} \quad \frac{n-1}{L} = \frac{Pa}{L} \quad \text{and} \quad n-1 = Pa$$

Then by moments about point 1

$$[M_1] \qquad A_1\bar{a}_1 = \left(\frac{PaL}{2}\frac{L}{3}\right) - \left(\frac{Pa^2}{2}\frac{a}{3}\right) = \frac{Pa}{6}(L^2 - a^2) \qquad (a)$$

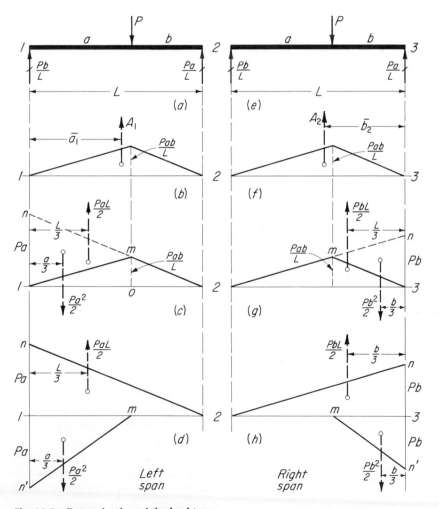

Fig. 10-3   Determination of the load terms.

To obtain the complete left-span load term for this case, this expression must be multiplied by $6/L$ [see Eq. (10-1)]. We then have

**Case 1, Left-Span Load Term:**

$$\frac{6Aa}{L} = \frac{Pa}{L}(L^2 - a^2) \tag{10-2}$$

In similar manner, the right-span load term for case 1 is

**Case 1, Right-Span Load Term:**

$$\frac{6Ab}{L} = \frac{Pb}{L}(L^2 - b^2) \tag{10-3}$$

Table 10-1  LOAD TERMS FOR USE
WITH THE THREE-MOMENT EQUATION
$w$ = kips/ft (or lb/ft) or kN/m (or N/m) of span; $W$ = total load on span.

| Left span $\dfrac{6A\bar{a}}{L}$ | Load no. | Type of loading | $\dfrac{6A\bar{b}}{L}$ right span |
|---|---|---|---|
| $\dfrac{Pa}{L}(L^2 - a^2)$ | 1 | | $\dfrac{Pb}{L}(L^2 - b^2)$ |
| $\dfrac{wL^3}{4} = \dfrac{WL^2}{4}$ | 2 | | $\dfrac{wL^3}{4} = \dfrac{WL^2}{4}$ |
| $\frac{8}{60}wL^3 = \frac{8}{30}WL^2$ | 3 | | $\frac{7}{60}wL^3 = \frac{7}{30}WL^2$ |
| $\frac{7}{60}wL^3 = \frac{7}{30}WL^2$ | 4 | | $\frac{8}{60}wL^3 = \frac{8}{30}WL^2$ |
| $\frac{5}{32}wL^3 = \frac{5}{16}WL^2$ | 5 | | $\frac{5}{32}wL^3 = \frac{5}{16}WL^2$ |
| $\dfrac{w(b^2 - a^2)}{4L}[2L^2 - (b^2 + a^2)]$ | 6 | | $\dfrac{w(d^2 - c^2)}{4L}[2L^2 - (d^2 + c^2)]$ |

But note here that the two triangles making up the moment diagram in Fig. 10-3c could also be shown separately, as in Fig. 10-3d, without changing the value of $A\bar{a}$. In this illustration, the ordinate at any point from line 1–2 to curve $n$–2 represents the moment about point 1 of the reaction $Pa/L$; similarly, the ordinate at any point from line 1–$m$ to curve $n'$–$m$ represents the moment at that point of load $P$. From this it is clear that **a moment diagram may be constructed by parts.** This concept was used in deriving the three-moment equation (Fig. 10-2b) and will be of great advantage in determining the $A\bar{a}$ and $A\bar{b}$ portions of the other load terms listed in Table 10-1.

Because of symmetry of load, the load terms for a uniformly distributed load (Fig. 10-4a) are identical for left and right spans. Figure 10-4b shows the moment diagram (by parts) for this load on a left span. Ordinate $n$–1 represents the moment about 1 of the reaction at 2, which is $(wL/2)(L) = wL^2/2$. Area $n$–1–2 then equals $(\frac{1}{2})(L)(wL^2/2) = wL^3/4$. Similarly, ordinate $n'$–1 represents the moment about 1 of the load on the beam, which is $(wL)(L/2) = wL^2/2$. Area $n'$–1–2 then equals $(\frac{1}{3})(L)(wL^2/2) = wL^3/6$. Now, $A\bar{a}$ equals the moment about 1 of these two areas, or

$$[M_1] \qquad A\bar{a} = \left(\frac{wL^3}{4}\frac{L}{3}\right) - \left(\frac{wL^3}{6}\frac{L}{4}\right) = \frac{wL^4}{24} \qquad (b)$$

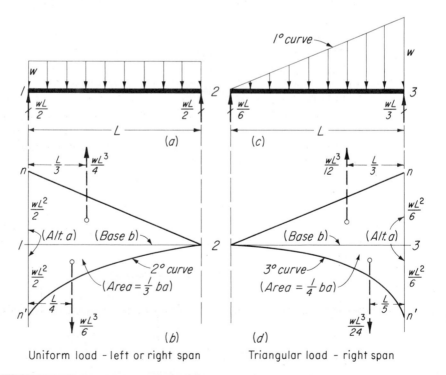

Fig. 10-4   Determination of the load terms.

The load term is now obtained by multiplying Eq. (*b*) by 6/*L*. That is,

**Case 2, Uniform Load, Any-Span Load Term:**

$$\frac{6A\bar{a}}{L} = \frac{6A\bar{b}}{L} = \frac{wL^3}{4} = \frac{WL^2}{4} \tag{10-4}$$

The moment diagram (by parts) for a triangular load (Fig. 10-4*c*) on a *right span* is shown in Fig. 10-4*d*. When moments are taken about point 3, we obtain

$$[M_3] \qquad A\bar{b} = \left(\frac{wL^3}{12}\frac{L}{3}\right) - \left(\frac{wL^3}{24}\frac{L}{5}\right) = \frac{wL^4}{36} - \frac{wL^4}{120} \tag{c}$$

or

$$A\bar{b} = \frac{10wL^4}{360} - \frac{3wL^4}{360} = \frac{7wL^4}{360} = \frac{7WL^3}{180} \tag{d}$$

since $W = wL/2$. Multiplying these terms by 6/*L* gives

**Case 3, Triangular Load, Right-Span Load Term:**

$$\frac{6A\bar{b}}{L} = \frac{7wL^3}{60} = \frac{7WL^2}{30} \tag{10-5}$$

The left-span load term for case 3 is $8wL^3/60 = 8WL^2/30$.

## 10-4 APPLICATIONS OF THE THREE-MOMENT EQUATION

When applied to a two-span beam with only one unknown moment at the middle support, a single application and solution of the three-moment equation will give this moment, as is illustrated in Prob. 10-1. For all other multiple-span beams, one three-moment equation must be written for every unknown moment. These equations are then solved simultaneously, as is illustrated in Prob. 10-2.

ILLUSTRATIVE PROBLEMS

**10-1.** Using the three-moment equation, solve for the unknown moment $M_B$ at support *B* of the two-span beam shown in Fig. 10-5. Disregard weight of beam.

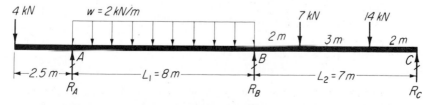

Fig. 10-5 Application of the three-moment equation.

*Solution:* A restatement here of the three-moment equation is convenient:

$$M_1 L_1 + 2M_2(L_1 + L_2) + M_3 L_2 = -\frac{6A_1\bar{a}_1}{L_1} - \frac{6A_2\bar{b}_2}{L_2}$$

We note that $M_1 = M_A = -(4)(2.5) = -10$ kN·m (negative because of tension in top fibers), $M_2 = M_B$ (unknown), and $M_3 = M_C = 0$. From Table 10-1 the load term (No. 2) for the left span $(AB)$ is

$$-\frac{6A_1\bar{a}_1}{L_1} = -\frac{wL_1^3}{4} = -\frac{(2)(8)(8)(8)}{4} = -256$$

and for the right span $(BC)$ the two load terms (No. 1) are

$$\frac{6A_2\bar{b}_2}{L_2} = -\frac{Pb}{L}(L^2 - b^2) = -\frac{(7)(5)}{7}[(7)(7) - (5)(5)] = -120$$

and

$$-\frac{6A_2\bar{b}_2}{L_2} = -\frac{Pb}{L}(L^2 - b^2) = -\frac{(14)(2)}{7}[(7)(7) - (2)(2)] = -180$$

The subscripts 1 and 2 indicate that the quantities refer to spans 1 and 2, respectively.

Substituting the appropriate values[1] in the three-moment equation, we obtain, since $M_1 = -10$,

$$(-10)(8) + 2M_B(8 + 7) + 0 = -256 - 120 - 180$$

or    $30M_B = -556 + 80 = -476$    and    $M_B = -15.9$ kN·m    *Ans.*

**10-2.** Calculate the moments at supports $B$ and $C$ of the continuous beam shown in Fig. 10-6. Neglect weight of beam.

*Solution:* We note that $M_A$ and $M_D$ are both zero. Since there are two unknown support moments, $M_B$ and $M_C$, two three-moment equations must be written, one involving points $A$, $B$, and $C$ and one involving points $B$, $C$, and $D$. These equations are then solved simultaneously. They are

$$M_A L_1 + 2M_B(L_1 + L_2) + M_C L_2 = -\frac{6A_1\bar{a}_1}{L_1} - \frac{6A_2\bar{b}_2}{L_2} \qquad (a)$$

$$M_B L_2 + 2M_C(L_2 + L_3) + M_D L_3 = -\frac{6A_2\bar{a}_2}{L_2} - \frac{6A_3\bar{b}_3}{L_3} \qquad (b)$$

Evaluating the load terms, we obtain

Left span 1:    $-\dfrac{6A_1\bar{a}_1}{L_1} = -\dfrac{8}{60}wL^3 = -\dfrac{8}{60}(3)(10)(10)(10) = -400$

Right span 2:    $-\dfrac{6A_2\bar{b}_2}{L_2} = -\dfrac{Pb}{L}(L^2 - b^2) = -\dfrac{(10)(8)}{20}(400 - 64) = -1344$

---

[1] Two types of errors are commonly made where a beam overhangs a support. It is frequently forgotten (1) that the moment due to load on overhanging portion of beam is usually *negative* and (2) that this moment must be multiplied by the adjacent full-span length.

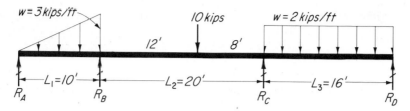

**Fig. 10-6**   Application of the three-moment equation.

Left span 2:   $-\dfrac{6A_2\bar{a}_2}{L_2} = -\dfrac{Pa}{L}(L^2 - a^2) = -\dfrac{(10)(12)}{20}(400 - 144) = -1536$

Right span 3:   $-\dfrac{6A_3\bar{b}_3}{L_3} = -\dfrac{wL^3}{4} = -\dfrac{(2)(16)(16)(16)}{4} = -2048$

Substituting the appropriate values in Eqs. (*a*) and (*b*), respectively, we obtain

$$0 + 2M_B(10 + 20) + 20M_C = -400 - 1344$$

and   $\qquad 20M_B + 2M_C(20 + 16) + 0 = -1536 - 2048$

or   $\qquad\qquad\qquad 60M_B + 20M_C = -1744$   $\qquad\qquad$ (*c*)

and   $\qquad\qquad\qquad 20M_B + 72M_C = -3584$   $\qquad\qquad$ (*d*)

Multiplying Eq. (*c*) by 3.6 and subtracting Eq. (*d*) gives

$$216M_B + 72M_C = -6278 \qquad\qquad\qquad (e)$$
$$\underline{\phantom{2}20M_B + 72M_C = -3584} \qquad\qquad\qquad (d)$$
$$196M_B \qquad\quad = -2694 \quad \text{and} \quad M_B = -13.74 \text{ kip·ft} \qquad Ans.$$

and, multiplying Eq. (*d*) by 3 and subtracting Eq. (*c*) gives

$$60M_B + 216M_C = -10,752 \qquad\qquad\qquad (f)$$
$$\underline{60M_B + \phantom{2}20M_C = -\phantom{0}1,744}$$
$$196M_C = -\phantom{0}9,008 \quad \text{and} \quad M_C = -45.96 \text{ kip·ft} \qquad Ans.$$

Substituting the values of $M_B$ and $M_C$ in either Eq. (*c*) or Eq. (*d*) will check only the correctness of the simultaneous solution. Such a check, however, will not disclose possible errors in Eqs. (*c*) and (*d*). Substituting in Eq. (*c*), we have

$$(60)(-13.74) + (20)(-45.96) = -1744$$

or   $\quad -824.4 - 919.2 = -1744 \quad$ and $\quad -1743.6 = -1744 \qquad$ *Check.*

## PROBLEMS

NOTE: In each of the following problems, determine all unknown bending moments at the supports.

**10-3.**

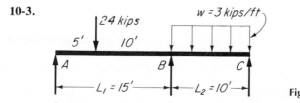

**Fig. 10-7**

**10-4.**

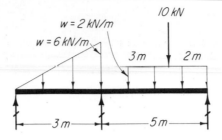

Fig. 10-8

*Ans.* $M_B = -10.5$ kN·m

**10-5.**

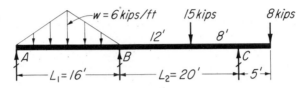

Fig. 10-9

**10-6.**

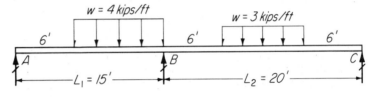

Fig. 10-10

*Ans.* $M_B = -82.7$ kip·ft

**10-7.**

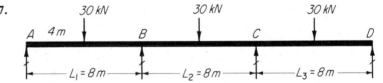

Fig. 10-11

*Ans.* $M_B = M_C = -36.0$ kN·m

**10-8.**

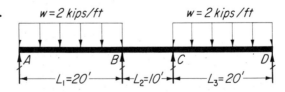

Fig. 10-12

**10-9.**

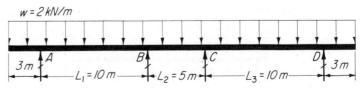

Fig. 10-13

**10-10.**

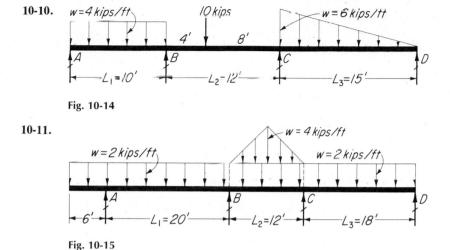

Fig. 10-14

**10-11.**

Fig. 10-15

## 10-5 INDETERMINATE BEAMS WITH FIXED ENDS

The application of the three-moment equation to indeterminate beams with fixed ends is made possible by the addition, at the fixed end, of an "imaginary span." The following example will establish proof of this statement.

Let Fig. 10-16 represent the right-end span of any multiple-span beam, terminating in a fixed-end support at the right end $D$. At $C$, the left end of that span, the beam has been cut barely to the right of support reaction $R_C$, and $V_C$ and $M_C$ are, respectively, the shear and moment at the cut section.

We assume always that any fixed end of a beam is horizontal and that a fixed end means that a tangent to the elastic curve at the fixed end must also be horizontal. We now remove the fixed-end support and add an imaginary "reflected span" at the right, in such manner that the two spans with loads are perfectly symmetrical about $D$. When this symmetry exists, a tangent to the elastic curve at $D$ must necessarily be horizontal, and we have, therefore, created a condition that is mathematically and

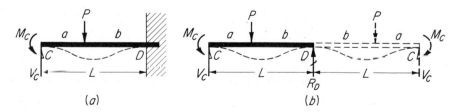

Fig. 10-16 Indeterminate beams with fixed ends.

physically equivalent to that of a fixed end, insofar as the original beam to the left of $D$ is concerned.

The original form of the three-moment equation (10-1) is

$$M_1 L_1 + 2M_2(L_1 + L_2) + M_3 L_2 = -\frac{6A_1\bar{a}_1}{L_1} - \frac{6A_2\bar{b}_2}{L_2} \qquad (a)$$

When this equation is applied to the two spans in Fig. 10-16b, with the proper substitution of subscripts, we obtain

$$M_C L + 2M_D(L + L) + M_C L = -\frac{6A\bar{a}}{L} - \frac{6A\bar{a}}{L} \qquad (b)$$

or

$$2M_C L + 4M_D L = 2\left[-\frac{6A\bar{a}}{L}\right] \qquad (c)$$

Dividing each term by 2 gives

$$M_C L + 2M_D L = -\frac{6A\bar{a}}{L} \qquad (d)$$

But note now that Eq. ($d$) could have been obtained directly from Eq. ($a$), if the load on the imaginary span had been considered to be zero and if the length approached zero. Equation ($a$) would then become

$$M_C L + 2M_D(L + 0) + M_C(0) = -\frac{6A\bar{a}}{L} - 0 \qquad (e)$$

or

$$M_C L + 2M_D L = -\frac{6A\bar{a}}{L} \qquad (f)$$

which is the same as that previously obtained by Eq. ($d$). Correct results will therefore be obtained by the three-moment equation when at any fixed-end support an imaginary span is added whose length is short and load is zero.

## ILLUSTRATIVE PROBLEM

**10-12.** Using the three-moment equation, calculate the moment $M_B$ at the fixed-end support $B$ of the beam shown in Fig. 10-17. Weight of beam is not to be considered.

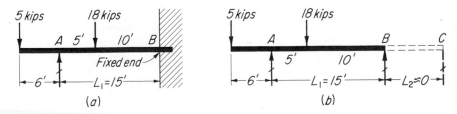

Fig. 10-17 Moment at fixed end of indeterminate beam.

*Solution:* An imaginary span $L_2$ of extremely short length and with zero load is added at the fixed end as shown. For the resulting two-span beam, the three-moment equation is

$$M_A L_1 + 2M_B(L_1 + L_2) + M_C L_2 - \frac{6A_1\bar{a}_1}{L_1} \quad \frac{6A_2\bar{b}_2}{L_2} \quad (a)$$

The moment $M_A$ at $A$ can readily be evaluated and is $(-5)(6) = -30$ kip·ft. Since $L_1 = 15$ ft, $L_2 \approx 0$, and $A_2 = 0$, Eq. (*a*) reduces to

$$(-30)(15) + 2M_B(15 + 0) + 0 = -\frac{(18)(5)}{15}(225 - 25) - 0 \quad (b)$$

and $\qquad 30M_B = 450 - 1200 = -750 \qquad$ and $\qquad M_B = -25$ kip·ft $\qquad$ *Ans.*

## PROBLEMS

In each of the following problems, determine all unknown bending moments at the supports.

**10-13.** Refer to Fig. 10-18. $P = 3$ kips, $w = 4$ kips/ft, $L_1 = 4$ ft, and $L_2 = 12$

$$\text{Ans. } M_A = -12 \text{ kip·ft}; M_B = -66 \text{ kip·ft}$$

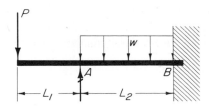

Fig. 10-18   Probs. 10-13 and 10-14.

**10-14.** Refer to Fig. 10-18. $P = 6$ kN, $w = 8$ kN/m, $L_1 = 2$ m, and $L_2 = 4$ m.
**10-15.** Refer to Fig. 10-19. $P_1 = P_2 = 0$, $w = 2$ kips/ft, $L_1 = 4$ ft, and $L_2 = 10$ ft.

$$\text{Ans. } M_A = -16 \text{ kip·ft}; M_B = -17 \text{ kip·ft}$$

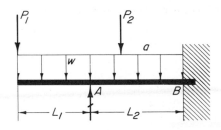

Fig. 10-19   Probs. 10-15 and 10-16.

**10-16.** Refer to Fig. 10-19. $P_1 = 6$ kips, $P_2 = 10$ kips, $w = 4$ kips/ft, $a = 10$ ft, $L_1 = 5$ ft, and $L_2 = 15$ ft.
**10-17.** Refer to Fig. 10-20. $L_1 = 6$ ft, $L_2 = 12$ ft, and $w - 6$ kips/ft.

$$\text{Ans. } M_A = -12 \text{ kip·ft}; M_B = -68.4 \text{ kip·ft}$$

**10-18.** Refer to Fig. 10-20. $L_1 = 1$ m, $L_2 = 3$ m, and $w = 12$ kN/m
**10-19.** Refer to Fig. 10-21. $L = 20$ ft, $w_1 = 2$ kips/ft, and $w_2 = 5$ kips/ft.

$$\text{Ans. } M_B = -180 \text{ kip·ft}$$

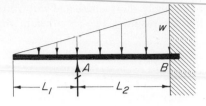

Fig. 10-20 Probs. 10-17 and 10-18.

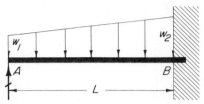

Fig. 10-21 Probs. 10-19 and 10-20.

**10-20.** Refer to Fig. 10-21. $L = 20$ ft, $w_1 = 3$ kips/ft, and $w_2 = 5$ kips/ft.

**10-21.** Refer to Fig. 10-22. Use symbols.

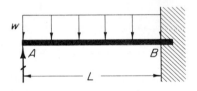

Fig. 10-22 Probs. 10-21 and 10-22.

**10-22.** Refer to Fig. 10-22. $L = 16$ ft and $w = 4$ kips/ft.

**10-23.** Refer to Fig. 10-23. Use symbols.

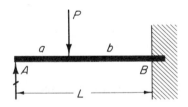

Fig. 10-23 Probs. 10-23 and 10-24.

**10-24.** Refer to Fig. 10-23. $P = 10$ kN, $a = 2$ m, $b = 4$ m, and $L = 6$ m.

**10-25.** Refer to Fig. 10-24. $L_1 = 10$ ft, $L_2 = 12$ ft, and $w = 3$ kips/ft.

*Ans.* $M_B = -27.6$ kip·ft; $M_C = -40.2$ kip·ft

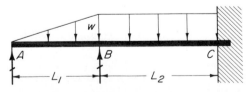

Fig. 10-24 Prob. 10-25.

**10-26.** Refer to Fig. 10-25. Use symbols.

**10-27.** Refer to Fig. 10-25. $L = 6$ m and $w = 12$ kN/m

**10-28.** Refer to Fig. 10-26. $P_1 = 5$ kips, $P_2 = 9$ kips, $w = 2$ kips/ft, $L_1 = 6$ ft, $L_2 = 18$ ft, $b = 6$ ft, and $L_3 = 12$ ft.

*Ans.* $M_B = -18.0$ kip·ft; $M_C = -27$ kip·ft

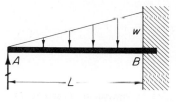

**Fig. 10-25   Probs. 10-26 and 10-27.**

**Fig. 10-26   Prob. 10-28.**

**10-29.** Refer to Fig. 10-27. Use symbols.

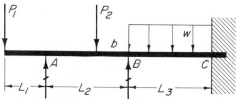

**Fig. 10-27   Probs. 10-29 and 10-30.**

**10-30.** Refer to Fig. 10-27. $L = 24$ ft and $w = 3$ kips/ft.

**10-31.** Refer to Fig. 10-28. $P = 12$ kips, $w = 2$ kips/ft, $L_1 = 6$ ft, $L_2 = 20$ ft, $a = 5$ ft, and $L_3 = 15$ ft.

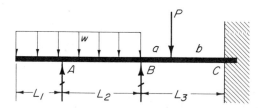

**Fig. 10-28   Prob. 10-31.**

**10-32.** Refer to Fig. 10-29. $P = 18$ kips, $a = 10$ ft, and $b = 5$ ft.

*Ans.* $M_A = -20$ kip·ft; $M_B = -40$ kip·ft

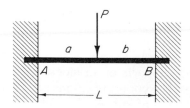

**Fig. 10-29   Probs. 10-32 and 10-33.**

**10-33.** Refer to Fig. 10-29. $P = 10$ kips, $a = 9$ ft, and $b = 16$ ft.

**10-34.** Refer to Fig. 10-30. $P = 10$ kips, $w = 3$ kips/ft, $L_1 = 6$ ft, $L_2 = 16$ ft, and $L_3 = 12$ ft.

**10-35.** Refer to Fig. 10-31. $L = 12$ ft and $w = 6$ kips/ft.

*Ans.* $M_A = M_B = -45$ kip·ft

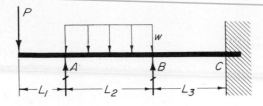

Fig. 10-30   Probs. 10-34.

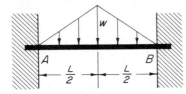

Fig. 10-31   Probs. 10-35 and 10-36.

**10-36.** Refer to Fig. 10-31. $L = 4$ m and $w = 4$ kN/m.
**10-37.** Refer to Fig. 10-32. $L_1 = 10$ ft, $L_2 = 12$ ft, and $w = 6$ kips/ft.

Ans. $M_A = -44$ kip·ft; $M_B = -62$ kip·ft; $M_C = -77$ kip·ft

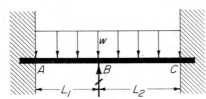

Fig. 10-32   Prob. 10-37.

**10-38.** Refer to Fig. 10-33. $L_1 = 20$ ft, $L_2 = 15$ ft, and $w = 3$ kips/ft.

Ans. $M_A = -121.5$ kip·ft; $M_B = -57.1$ kip·ft; $M_C = +28.5$ kip·ft

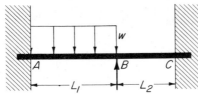

Fig. 10-33   Prob. 10-38.

## 10-6 DETERMINATION OF SUPPORT REACTIONS

After the unknown moments at the supports of an indeterminate beam have been determined, the values of the support reactions are readily calculated, as is illustrated in the following problems.

### ILLUSTRATIVE PROBLEMS

**10-39.** Calculate the reactions at supports $A$, $B$, and $C$ of the beam shown in Fig. 10-34 (see Prob. 10-1); $M_B = -15.9$ kN·m.

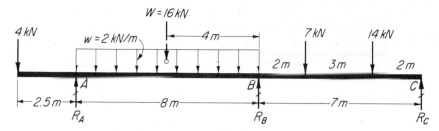

**Fig. 10-34** Calculation of support reactions of continuous beams.

*Solution:* The beam in Fig. 10-35 is shown cut at $B$ and the *negative* moment there ($M_B = -15.9$ kN·m) is shown so as to put tension in the top fibers. The forces $V_{B_L}$ and $V_{B_R}$ are the shears barely at the left and right, respectively, of reaction $R_B$. Their algebraic sum equals $R_B$.

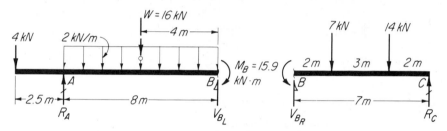

**Fig. 10-35** Calculation of support reactions of continuous beams.

For the left part, by moments about $B$, we obtain

$[\Sigma M_B = 0]$
$$8R_A = (10.5)(4) + 16(4) - 15.9 = 90.1 \qquad \text{and} \qquad R_A = 11.26 \text{ kN} \qquad Ans.$$

For the right part, also by moments about $B$, we obtain

$[\Sigma M_B = 0]$
$$7R_C = (7)(2) + (14)(5) - 15.9 = 68.1 \qquad \text{and} \qquad R_C = 9.37 \text{ kN} \qquad Ans.$$

Using the value of $R_A$, $R_B$ may now be found by moments about $C$, in Fig. 10-34, after which a check by $\Sigma V$ should be made. That is,

$[\Sigma M_C = 0] \qquad 7R_B + (15)(11.26) = (14)(2) + (7)(5) + (16)(11) + (4)(17.5)$

or $\qquad\qquad\qquad 7R_B = 140.07 \qquad$ and $\qquad R_B = 20.01$ kN $\qquad\qquad\qquad Ans.$
$[\Sigma V = 0]$
$$11.26 + 9.73 + 20.01 = 4 + 16 + 7 + 14 \qquad \text{or} \qquad 41 = 41 \qquad Check.$$

**10-40.** Calculate the reactions $R_A$ and $R_B$ at supports $A$ and $B$ of the fixed-end beam shown in Fig. 10-36 ($M_A = -12$ kip·ft and $M_B = -24$ kip·ft).

*Discussion:* The actual distribution of reacting forces around the imbedded end of a fixed-end beam, the resultant of which forces would be the single-force reaction, is rather complex. It is customary, therefore, to show a reaction at a fixed end as a force acting at the surface where the beam enters the supporting wall or abutment, as illustrated in Fig. 10-36. If the beam were cut at the wall, the shear $V$

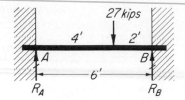

Fig. 10-36 Reactions at fixed ends.

at the cut section would equal that reaction. Hence, when a reaction at a fixed end is called for, it may be thought of as actually being the shear $V$, or vice versa.

*Solution:* The beam is cut at the supporting walls and is isolated as a free body, as shown in Fig. 10-37. The known negative moments at $A$ and $B$ are applied so as to produce the required tension on the top fibers. Solving for $V_A$ and $V_B$, which, respectively, equal $R_A$ and $R_B$, we obtain

$[\Sigma M_B = 0]$ $\quad 6V_A + 24 = (27)(2) + 12$

$\qquad 6V_A = 54 + 12 - 24 = 42 \qquad$ and $\qquad V_A = 7$ kips $\qquad$ *Ans.*

$[\Sigma M_A = 0]$ $\quad 6V_B + 12 = (27)(4) + 24$

$\qquad 6V_B = 108 + 24 - 12 = 120 \qquad$ and $\qquad V_B = 20$ kips $\qquad$ *Ans.*

$[\Sigma V = 0]$ $\qquad\qquad\qquad\qquad 7 + 20 = 27 \qquad\qquad\qquad\qquad$ *Check.*

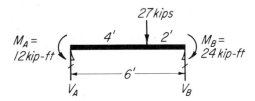

Fig. 10-37 Calculation of reactions.

**PROBLEMS**

**10-41.** Calculate the support reactions of the beam referred to in Prob. 10-3 ($M_B = -47.0$ kip·ft).

$\qquad$ *Ans.* $R_A = 12.9$ kips; $R_B = 30.8$ kips; $R_C = 10.3$ kips

**10-42.** Determine all support reactions of the beam referred to in Prob. 10-4 ($M_B = -10.5$ kN·m).

**10-43.** Evaluate the support reactions of the beam referred to in Prob. 10-5 ($M_B = -70.22$ kip·ft).

$\qquad$ *Ans.* $R_A = 19.6$ kips; $R_B = 35.9$ kips; $R_C = 15.5$ kips

**10-44.** Calculate the four reactions at the supports of the beam referred to in Prob. 10-7.

$\qquad$ *Ans.* $R_A = R_D = 10.5$ kN; $R_B = R_C = 34.5$ kN

**10-45.** Evaluate the support reactions of the beam referred to in Prob. 10-8.

**10-46.** Determine the reactions at all supports of the beam referred to in Prob. 10-9 ($M_B = M_C = -13.5$ kN·m).

$\qquad$ *Ans.* $R_A = R_D = 15.55$ kN; $R_B = R_C = 15.45$ kN

**10-47.** Calculate the reactions at all supports of the beam referred to in Prob. 10-11 ($M_B = -69.03$ kip·ft; $M_C = -67.19$ kip·ft).

**10-48.** Evaluate reactions $R_A$ and $R_B$ of the beam in Prob. 10-12 ($M_B = -25$ kip·ft)

$$Ans. \quad R_A = 17.33 \text{ kips}; \quad R_B = 5.67 \text{ kips}$$

**10-49.** Determine the reactions at supports $A$ and $B$ of the beam referred to in Prob. 10-27 ($M_B = -28.8$ kN·m).

**10-50.** Calculate the support reactions at $A$ and $B$ of the beam referred to in Prob. 10-19.

**10-51.** Show that in Prob. 10-21 $R_A = 3wL/8$ and $R_B = 5wL/8$ ($M_B = -wL^2/8$).

**10-52.** Evaluate the reactions at all supports of the beam referred to in Prob. 10-25.

**10-53.** Calculate support reactions $R_A$, $R_B$, and $R_C$ of the beam referred to in Prob. 10-28.

$$Ans. \quad R_A = 9.63 \text{ kips}; \quad R_B = 15.71 \text{ kips}; \quad R_C = 12.66 \text{ kips}$$

**10-54.** Determine the support reactions of the beam of Prob. 10-32.

**10-55.** In Prob. 10-37, calculate the reactions $R_A$, $R_B$, and $R_C$.

**10-56.** Calculate the support reactions $R_A$, $R_B$, and $R_C$ of the beam referred to in Prob. 10-38.

**10-57.** Determine the reactions $R_A$ and $R_B$ of the beam in Prob. 10-17.


## 10-7 SHEAR AND MOMENT DIAGRAMS

While shear and moment diagrams are very useful for the determination of shears and moments in statically *determinate* beams, they cannot be used to determine shears and moments in statically *indeterminate* beams. After the support moments and reactions have been determined, however, as outlined in the preceding articles, the load, shear, and moment diagrams may readily be drawn; they are very helpful to the designer, since they show so clearly the variation in shear and moment over the full length of the beam and also indicate clearly the value and location of maximum shear and maximum moment.

The maximum moment in an indeterminate beam generally occurs at a support and is negative. Under certain load conditions, however, the maximum moment could occur at some point between two supports and might then be positive. A complete set of diagrams is the best safeguard against overlooking the *actual* maximum moment.

**ILLUSTRATIVE PROBLEM**——————————————————————

**10-58.** Draw a complete set of load, shear, and moment diagrams for the beam shown in Fig. 10-34. Calculate and show the maximum positive moments between

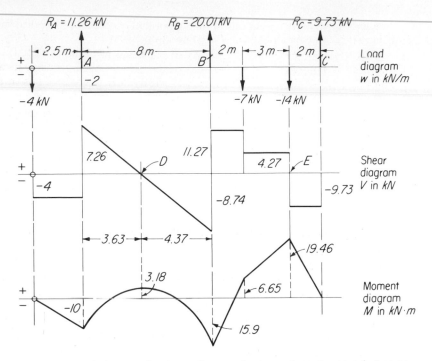

**Fig. 10-38**   Load, shear, and moment diagrams for beams in Probs. 10-1 and 10-39.

supports; $M_B = -15.9$ kN·m, $R_A = 11.26$ kN, $R_B = 20.01$ kN, $R_C = 9.73$ kN (see Probs. 10-1 and 10-39).

*Solution:* The shear diagram clearly reveals a section of zero shear at $D$ and a change in sign of shear at $E$, with consequent maximum positive moments at these points. The moment at $A$ is $-10$ kip·ft. The *difference* between this moment and the moment at $D$ is $(\frac{1}{2})(3.63)(7.26) = 13.18$, and $M_D = -10 + 13.18 = +3.18$. Since $M_C = 0$ and the difference between $M_E$ and $M_C = (-9.73)(2) = -19.46$, $M_E$ must equal $+19.46$.

Clearly now, the maximum moment in the beam is $+19.46$ kip·ft under the 14-kN load. If these diagrams are completed using the given loads and the calculated reactions, and the moments $M_A = -10$, $M_B = -15.9$, and $M_C = 0$ are obtained, it will be a check on the correctness of the support reactions only. These diagrams cannot check the correctness of $M_B$.

## PROBLEMS

NOTE: Instructors will find it convenient to have students draw the load, shear, and moment diagrams for Probs. 10-40, 10-41, 10-44, 10-46, and 10-53, for which the support moments and reactions have already been calculated.

## SUMMARY (By Article Number)

**10-1. Statically indeterminate beams** constitute that group of beams having more than two unknown reacting forces or moments at the supports, thus making the simple laws of statics, $\Sigma F = 0$ and $\Sigma M = 0$, insufficient for the solution. Beams having three or more supports, called **continuous** beams, fall in this category.

**10-2.** The **three-moment equation** establishes a mathematical relation among the moments at any three consecutive supports of a "continuous" beam. Its application makes possible the determination of the moments and reactions at all supports of any continuous beam having more than two supports. Its simplest form is

$$M_1L_1 + 2M_2(L_1 + L_2) + M_3L_2 = -\frac{6A_1\bar{a}_1}{L_1} - \frac{6A_2\bar{b}_2}{L_2} \qquad (10\text{-}1)$$

in which $L_1$ and $L_2$ are any two consecutive spans of a continuous beam and $M_1$, $M_2$, and $M_3$ are the bending moments at the three consecutive supports of those two spans. The numerical values of the two *load terms* in the right side of Eq. (10-1) are determined by the type and magnitude of the load and also upon whether the load is on the *left* or the *right* span of a given two-span section of a beam. The six types of loading, for which load terms are given in Table 10-1, will cover satisfactorily all situations commonly met in practice.

Other methods are available for "solving" statically indeterminate beams. Of these, the most popular, and perhaps the most useful, is the so-called moment-distribution method. Further study along this line, however, is usually reserved for books on structural analysis and design.

**10-5. Indeterminate beams with fixed ends** are solved simply by the three-moment equation by adding, at the fixed end, an additional "imaginary" span whose length and load both are zero. Consequently, the terms containing the length of and load on this imaginary span become zero and therefore drop out of the three-moment equation.

**10-6. Determination of support reactions** is a simple matter after the bending moments at the supports have been calculated. The beam may be cut successively at the different supports in such manner that only one unknown, a reaction, needs to be calculated.

**10-7. Shear and moment diagrams** cannot in themselves solve for the bending moments and the reactions at the supports of indeterminate beams. They are extremely useful, however, as a check upon calculated reactions and as a way to determine the maximum positive moments *between* the supports that are not disclosed by the application of the three-moment equation.

**REVIEW PROBLEMS**

**10-59.** Using the three-moment equation, calculate the bending moment at support $B$ of the beam shown in Fig. 10-39.

*Ans.* $M_B = -27.55$ kip·ft

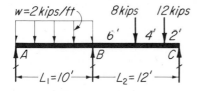

Fig. 10-39   Probs. 10-59 and 10-69.

**10-60.** Determine the moment at support $B$ of the beam in Fig. 10-40.

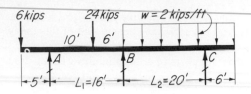

Fig. 10-40   Probs. 10-60 and 10-70.

**10-61.** Calculate the moment at support $B$ of the beam shown in Fig. 10-41.

*Ans.* $M_B = -60$ kip·ft

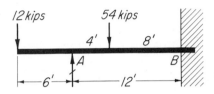

Fig. 10-41   Probs. 10-61 and 10-71.

**10-62.** Evaluate the moment at support $B$ of the beam shown in Fig. 10-42. $L = 9$ m and $w = 3$ kN/m.

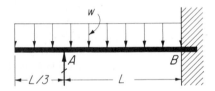

Fig. 10-42   Probs. 10-62 and 10-72.

**10-63.** Determine the moment at support $B$ of the beam shown in Fig. 10-43.

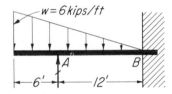

Fig. 10-43   Probs. 10-63 and 10-73.

**10-64.** Calculate the moment at supports $A$ and $B$ of the beam shown in Fig. 10-44.

*Ans.* $M_A = -120$ kip·ft; $M_B = -80$ kip·ft

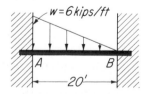

Fig. 10-44   Probs. 10-64 and 10-74.

**10-65.** Calculate the support moments in the beam of Fig. 10-45 if $a = 3$ m, $b = 2$ m, and $P = 10$ kN.

**10-66.** In Fig. 10-45 let $P = 12$ kN, $a = 4$ m, and $b = 2$ m. Determine the moment at support $A$.

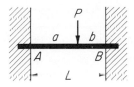

Fig. 10-45    Probs. 10-65 and 10-66.

**10-67.** Determine the moments at supports $A$, $B$, and $C$ of the beam shown in Fig. 10-46.

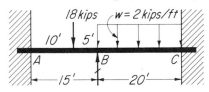

Fig. 10-46    Probs. 10-67 and 10-75.

**10-68.** Calculate the moments at supports $A$, $B$, and $C$ of the beam shown in Fig. 10-47.

Ans. $M_A = -12.38$ kip·ft; $M_B = -55.24$ kip·ft; $M_C = -72.38$ kip·ft

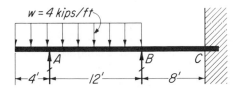

Fig. 10-47    Prob. 10-68.

NOTE: In the following problems, instructors may find it convenient to have students draw the load, shear, and moment diagrams and calculate the maximum positive moment in each span, in addition to the reactions called for.

**10-69.** Calculate the reactions $R_A$, $R_B$, and $R_C$ of the beam shown in Fig. 10-39 ($M_B = -27.55$ kip·ft).

Ans. $R_A = 7.25$ kips; $R_B = 21.05$ kips; $R_C = 11.70$ kips

**10-70.** Determine the reactions at supports $A$, $B$, and $C$ of the beam shown in Fig. 10-40 ($M_B = -71.4$ kip·ft).

**10-71.** Evaluate the reactions at supports $A$ and $B$ of the beam shown in Fig. 10-41 ($M_B = -60$ kip·ft).

Ans. $R_A = 49$ kips; $R_B = 17$ kips

**10-72.** Calculate the support reactions $R_A$ and $R_B$ of the beam shown in Fig. 10-42. $L = 9$ m and $w = 3$ kN/m ($M_b = -23.63$ kN·m).

**10-73.** Compute the reactions at supports $A$ and $B$ of the beam shown in Fig. 10-43 ($M_B = +14.4$ kip·ft).

**10-74.** Determine the support reactions $R_A$ and $R_B$ of the beam shown in Fig. 10-44 ($M_A - -120$ kip·ft; $M_B = -80$ kip·ft).

Ans. $R_A = 43$ kips; $R_B = 17$ kips

**10-75.** Calculate the reactions at supports $A$, $B$, and $C$ of the beam shown in Fig. 10-46 ($M_B = -37.33$ kip·ft; $M_C = +18.67$ kip·ft).

**10-76.** Evaluate the reactions at $A$, $B$, and $C$ of the beam referred to in Prob. 10-68.

**10-77.** Compute the four reactions of the beam referred to in Prob. 10-2.

*Ans.* $R_A = 3.63$ kips; $R_B = 13.76$ kips; $R_C = 26.49$ kips; $R_D = 13.12$ kips

## REVIEW QUESTIONS

**10-1.** When is a beam statically indeterminate?

**10-2.** What is meant by a "continuous" beam?

**10-3.** Discuss briefly the essence of the second moment-area principle.

**10-4.** What relation is expressed by the three-moment equation?

**10-5.** Is the three-moment equation applicable to a continuous beam with different $I$ values in each span?

**10-6.** What manner of loading will give the same load terms for both left and right spans?

**10-7.** What errors are commonly made in the application of the three-moment equation?

**10-8.** How many applications of the three-moment equation are required to solve for four unknown support moments in a beam?

**10-9.** What is done to make the three-moment equation applicable to a fixed-end support?

**10-10.** Can support moments and reactions of indeterminate beams be found by use of beam diagrams?

**10-11.** In what manner are beam diagrams useful in connection with indeterminate beams?

**10-12.** How may positive moments between supports readily be evaluated?

# Chapter 11 _____
# Combined _____
# Stresses _____

## 11-1 COMBINED AXIAL AND BENDING STRESSES

The problems of the determination of direct (axial) tensile and compressive stresses in structural members and of the design of simple members to withstand direct (axial) tensile and compressive loads were discussed in Chap. 1. Bending stresses and the design of members to withstand bending loads were discussed in Chaps. 7 and 8.

In many instances, however, structural members are subjected both to axial and bending loads, as, for example, the cantilevered bar shown in Fig. 11-1a. The load $P$ is clearly a direct axial load producing a unit axial stress of $P/A$ at any cross section $AA$ of the bar and $Q$ is a bending load producing a unit bending stress of $Mc/I$, which is tensile at the top of the bar and compressive at the bottom. The combined stress, axial and bending, is then $s = s_d \pm s_b$, or

$$s = \frac{P}{A} \pm \frac{Mc}{I} \qquad (11\text{-}1)$$

If the axial stress is tensile, it will add to the tensile bending stress; if it is compressive, it will add to the compressive bending stress. In either case the total, or maximum, stress is always the sum of the axial and the bending stresses. The bending-stress term is here kept in the form $s = Mc/I$ in order to make it possible to calculate either the tensile or compressive bending stress, especially in sections that are unsymmetrical about the neutral axis, where the distances $c$ to the most highly stressed tensile and compressive fibers are not equal.

The bending load $Q$ will cause a deflection $\delta$ of the bar, as indicated, and greatly exaggerated, in Fig. 11-1b. The axial load $P$ will, however, reduce this deflection, since the moment $P\delta$ is opposite in sense to the bending moment $QL$ and hence will actually reduce the bending stress. Had $P$ been a compressive force, however, its moment $P\delta$ would have

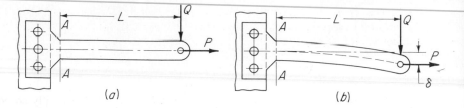

**Fig. 11-1   Combined axial and bending stress. (a) Deflection disregarded. (b) Deflection exaggerated.**

added to the moment $QL$ and hence would have increased the bending stress and the lateral deflection.

When the deflection $\delta$ is relatively small, it may be, and customarily is, disregarded without appreciable error.

In textbook problems, ideal conditions and absolute precision are necessarily presumed. In practice, however, such presumed precision rarely exists. In most ordinary problems, the effect of unknown errors due to ordinary lack of precision is often far greater than the effect of some known secondary condition which may purposely be disregarded. Secondary effects, such as the change in stress due to the deflection $\delta$, may be disregarded in "relatively short" members but not in "relatively long" members, which, of course, is somewhat ambiguous. Only study and experience will enable a designer to detect the line of demarcation. As a guide, however, the structural members mentioned in all problems in this chapter are of the relatively short type.

Since three basic stresses exist—axial, bending, and torsional (shear, as in shafts)—four combinations of stress are possible: (1) axial and bending; (2) axial and torsional; (3) axial, bending, and torsional; and (4) bending and torsional. Only cases 1 and 4, the most commonly occurring ones, are discussed here. The others are usually considered in books on machine design.

### ILLUSTRATIVE PROBLEM

**11-1.** Assume that the bar in Fig. 11-1 is of structural steel 4 in. deep (vertical) and $3/4$ in. thick, that $P = 6000$ lb, $Q = 1000$ lb, and $L = 30$ in. Then calculate (a) the maximum combined bending and direct stress at section $A$ and (b) the deflection $\delta$ and percentage of error in stress involved in disregarding $\delta$.

*Solution:* The cross-sectional area of the bar $A = (0.75)(4) = 3$ in.$^2$, the moment of inertia $I = bd^3/12 = (0.75)(4^3)/12 = 4$ in.$^4$, and $c = 4/2 = 2$ in. Hence

$$\left[ s = \frac{P}{A} + \frac{Mc}{I} \right] \qquad s = \frac{6000}{3} + \frac{(1000)(30)(2)}{4} = 17{,}000 \text{ psi} \qquad Ans.$$

The maximum deflection (case 1, Table 9-3) is

$$\left[ \delta = \frac{QL^3}{3EI} \right] \qquad \delta = \frac{(1000)(27{,}000)}{(3)(29{,}000{,}000)(4)} = 0.0775 \text{ in.} \qquad Ans.$$

The decrease in moment due to this deflection is $P\delta = (6000)(0.0775) =$ 465 lb·in., and the decrease in bending stress is $Mc/I = (465)(2)/4 = 233$ psi. The percentage of error is then

$$100 - \left(\frac{16,767}{17,000}\right)(100) = 100 - 98.7 = 1.3 \text{ percent} \qquad Ans.$$

So small an error may well be disregarded in this case, especially so because bending accounts for about 88 percent of the total stress.

### PROBLEMS

**11-2.** In Fig. 11-1, let $P = 40$ kN, $Q = 10$ kN, and $L = 500$ mm and let the bar be of structural steel 20 by 80 mm, which latter dimension is measured parallel to force $Q$. Then calculate (*a*) the maximum combined bending and direct stress at section $A$, and (*b*) the deflection $\delta$ and percentage of error involved in disregarding $\delta$.

**11-3.** The steel plate $A$ in Fig. 11-2 is 4 in. wide and $^5/_8$ in. thick. Calculate the weight $W$ which will produce a maximum combined stress of 10,000 psi in the plate.

*Ans.* $W = 2940$ lb

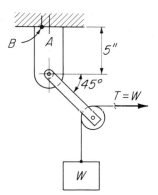

**Fig. 11-2    Probs. 11-3 and 11-33.**

**11-4.** A $1^1/_2$-in.-diameter standard steel pipe extends vertically out of a solid support as shown in Fig. 11-3. Calculate the value of $P$ which will produce a combined stress of 12,000 psi when $h = 20$ in. and $\theta = 30°$.

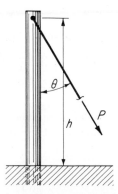

**Fig. 11-3    Prob. 11-4.**

**11-5.** Figure 11-4 illustrates the cross section of an earth-retaining wall which is supported by round timber piles. The portion $h$ of each pile extending above the ground level is assumed to act as a cantilever beam with complete fixity at ground level. Each pile is subjected to a lateral earth load whose resultant is $P$ and a vertical axial load $R$ from a beam resting on top of the pile. Calculate the maximum combined stress in one 200-mm-diameter pile when $P = 2$ kN, $R = 100$ kN, and $h = 3$ m.

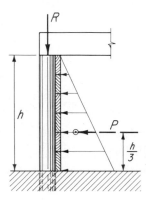

**Fig. 11-4   Probs. 11-5 and 11-35.**

**11-6.** In the type of timber truss shown in Fig. 11-5, closely spaced joists are often laid directly on the upper chord, thus subjecting the chord to bending as well as to direct stress, as illustrated by the chord section $AB$, shown isolated above the truss. If the chord measures $9^1/_2$ by $11^1/_2$ in. (vertical) in cross section and if $P = 25,000$ lb, $w = 1000$ lb/ft, and $L = 14$ ft, what is the maximum stress $(a)$ in the top fibers and $(b)$ in the bottom fibers of the chord?

*Ans.* $(a)$ 1634 psi; $(b)$ 1176 psi

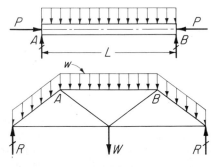

**Fig. 11-5   Probs. 11-6 and 11-7.**

**11-7.** Solve Prob. 11-6 when the chord is a 150- by 200-mm (vertical) timber, $P = 120$ kN, $w = 15$ kN/m, and $L = 3$ m.

**11-8.** Member $AB$ of the crane shown in Fig. 11-6 is a steel wide-flange section, W 6 × 15, and its length $L$ is 8 ft. Calculate $(a)$ the maximum combined stress in $AB$ when a load of 5000 lb is applied at the midpoint and weight of beam and its deflection are neglected and $(b)$ the maximum deflection $\delta$ and the increase in bending stress caused by the moment $P\delta$.

*Ans.* $(a)$ $s = 12,400$ psi; $(b)$ $\delta = 0.109$ in.; $s_h = 67.4$ psi

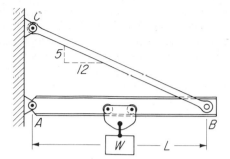

Fig. 11-6    Probs. 11-8 and 11-9.

**11-9.** The horizontal member $AB$ of the crane shown in Fig. 11-6 is 12 ft long and must be capable of supporting a load $W$ of 15,000 lb at its midpoint. What is the most economical 10-in. wide-flnage section to use if the allowable combined stress is 18,000 psi? (NOTE: A trial selection may be based on bending only.)

**11-10.** The bracket $BEC$, attached to the vertical member $AD$ in Fig. 11-7, must be capable of supporting a load $P$ of 600 lb. $AD$ is a 2- by 4-in. timber S4S whose greatest dimension is in the plane of bending. Calculate the maximum combined stress in $AD$ when $a = 4$ ft, $b = 4$ ft, $c = 4$ ft, and $d = 2$ ft ($a$) barely above $C$ and ($b$) barely below $C$.

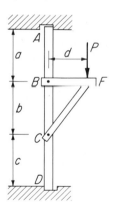

Fig. 11-7    Probs. 11-10 and 11-43.

**11-11.** The ladder shown in Fig. 11-8 leans against what may be considered to be a frictionless edge support at $C$. Hence $R_C$ is perpendicular to $AC$. Each of the two identical sidepieces measures 1 by 3 in. Holes 1 in. in diameter for the rungs are bored partway through both sidepieces, as indicated in the cross section, but should be considered to extend all through for the purpose of calculating the moment of inertia of the sidepieces. What maximum combined stress is produced in one sidepiece when $P = 260$ lb and $L = 10$ ft and load is centered on ladder rung?

$$Ans. \quad s_c = -60 - 1038 = -1098 \text{ psi}$$

**11-12.** What vertical load $P$ can safely be supported at the midpoint $B$ of the ladder described in Prob. 11-11 if $L = 13$ ft and safe allowable combined stress is 1200 psi? See Fig. 11-8.

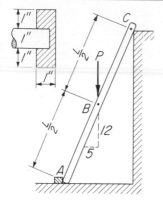

**Fig. 11-8   Probs. 11-11 and 11-12.**

**11-13.** The loading-yard boom shown in Fig. 11-9 is a 200- by 200-mm full-size timber. Assume that $a = 5$ m, $b = 1$ m and $W = 50$ kN. Calculate the maximum combined stress in this boom (*a*) barely above $B$ and (*b*) barely below $B$.

*Ans.* (*a*) $s = 1.0 + 22.5 = 23.5$ MPa; (*b*) $s = 1.675 + 22.5 = 24.175$ MPa

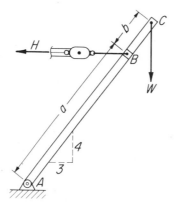

**Fig. 11-9   Probs. 11-13 and 11-44.**

## 11-2 COMBINED STRESSES CAUSED BY ECCENTRIC AXIAL LOAD

In short tension and compression members subjected to axial (concentric) loads, the resulting stress is assumed to be uniformly distributed over the entire cross-sectional area. When the action line of the load is parallel to, but does not coincide with, the centroidal axis of the supporting member, as, for example, in Fig. 11-10*a* and *b*, $P$ is referred to as an **eccentric load** and the distance $e$ is called the **eccentricity** of the load or the **eccentric distance.** In such instances, then, the product $Pe$ becomes an eccentric moment which will bend the post about the axis $XX$, thus adding tensile and compressive bending stresses to the already existing compressive direct stresses.

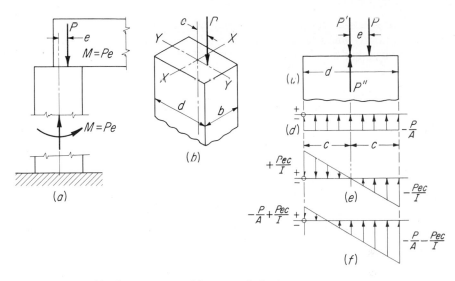

Fig. 11-10   Combined stresses caused by eccentric load.

If dimension $d$ of the post is greater than $b$, the moment of inertia of the post will be greatest with respect to axis $XX$, which is therefore called the **major axis,** while $YY$ is called the **minor axis.** We might then locate load $P$ by stating that it is applied on the minor axis at a point a distance $e$ from the major axis.

A clearer understanding of these combined stresses may be obtained by a study of Fig. 11-10c, $d$, $e$, and $f$. The upper portion of the post is shown in $c$. As an aid in this development, we will now apply two equal and opposite forces, $P'$ and $P''$, along the axis of the post. These forces have the magnitude of $P$. This may be done without in any way changing the resultant force system or its effect. Clearly, now, the post is subjected to a direct or axial force $P'$ and to a bending moment $Pe$ ($P$ and $P''$ form a couple whose moment $M = Pe$). This bending moment $M$ acting at the top of the post is counteracted by an equal and opposite resisting moment $M$ at any horizontal plane between the top and bottom of the post, thus indicating that the changes in unit stresses caused by the eccentric moment are constant at all transverse sections in the post.

If $-$ indicates compression and $+$ tension, the uniformly distributed direct stress is then $-P/A$, as shown in diagram $d$. The bending stress caused by the moment $Pe$ is $\pm Mc/I$, or $Pec/I$, and is clearly compressive $(-)$ on the right side of the post and tensile $(+)$ on the left side, as indicated in diagram $e$. The combined axial and bending stress on the right side is then $-P/A - Pec/I$ and on the left side is $-P/A + Pec/I$, as indicated in diagram $f$. When, as in this illustration, the unit tensile stress caused by bending exceeds the unit compressive stress caused by the axial load, the resultant will actually be a tensile stress in the post, which in general appearance is subjected only to compressive stresses. Consequently such eccentricity of load should be avoided wherever possible, but must be taken into account when it is known to exist.

### ILLUSTRATIVE PROBLEM

**11-14.** In Fig. 11-10, let the rectangular post be of timber, let $b = 100$ mm, $d = 180$ mm, and $e = 50$ mm, and let the load $P$ transmitted from the beam to the post be 100 kN. Then calculate the combined stress in the extreme fibers on the right and left sides of the post, respectively.

*Solution:* The moment of inertia $I$ with respect to the axis of bending $XX$ is $bd^3/12 = 100(180^3)/12 = 48.6 \times 10^6$ mm$^4$, and distance $c$ to extreme fiber is 90 mm. Hence

$$\frac{P}{A} = -\frac{100\ 000}{(100)(180)} = -5.56 \text{ MPa} \quad \text{and} \quad \frac{Mc}{I} = \pm\frac{(100\ 000)(50)(90)}{48.6 \times 10^6} = \pm 9.26 \text{ MPa}$$

On the right side

$$s = -5.56 - 9.26 = -14.82 \text{ MPa compression} \qquad\qquad Ans.$$

and on the left side

$$s = -5.56 + 9.26 = +3.70 \text{ MPa tension} \qquad\qquad Ans.$$

### PROBLEMS

**11-15.** The timber post shown in Fig. 11-10 measures 8 in. ($b$) by 10 in. ($d$) and is 4 ft long. A load $P$ of 40,000 lb is applied at a point located by measuring 2 in. along the $Y$ axis from the centroid and then 1 in. at right angles to that axis. Calculate the maximum unit stress caused by this doubly eccentric load. (NOTE: Add one additional bending-stress diagram to the three shown in Fig. 11-10.)

**11-16.** A section of $1^1/_2$-in. standard steel pipe is solidly imbedded in a concrete footing, as shown in Fig. 11-11. Calculate the unit stresses on the left and right side of the vertical portion of the pipe when $P = 60$ lb and $d = 20$ in.

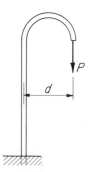

**Fig. 11-11   Probs. 11-16 and 11-39.**

**11-17.** A 6- by 6-in. timber post 5 ft long is subjected to an axial load $P$, as illustrated in Fig. 11-12. It is found that the axis is 0.2 in. out of straight line at the mid-

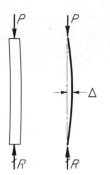

**Fig. 11-12  Probs. 11-17 and 11-40.**

point under a load of 36,000 lb. Calculate the maximum combined stress and the percentage increase in stress due to deviation of axis.

*Ans.* 1000 psi + 200 psi; 20 percent

**11-18.** The bearing bracket shown in Fig. 11-13 has a symmetrical cross section and is subjected to a maximum load $P$ of 10 kN. Let $b = 100$ mm and, at section $AA$, $c = 25$ mm, $d = 50$ mm, $I_A = 0.5 \times 10^6$ mm$^4$, and $A = 1000$ mm$^2$. Calculate the combined unit stresses at the top and bottom of the bracket at section $AA$.

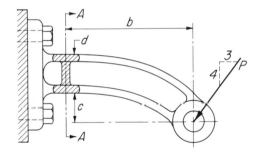

**Fig. 11-13  Probs. 11-18 and 11-19.**

**11-19.** The cast-iron bracket bearing in Fig. 11-13 is subjected to a maximum load of 500 lb. Assume that $b = 8$ in. and that, at section $AA$, $c = 2$ in., $d = 4$ in., $I_A = 8$ in.$^4$, and $A = 6$ in.$^2$. Calculate the unit combined stresses at the top and bottom of the bracket at section $AA$.

*Ans.* $s = +1050$ psi; $s = -1150$ psi

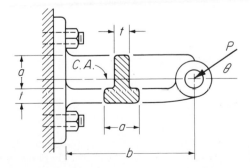

**Fig. 11-14  Prob. 11-20.**

**11-20.** In Fig. 11-14 let $a = 3$ in., $b = 6$ in., $t = \frac{1}{2}$ in., $P = 2000$ lb, and $\theta = 30°$. Find the maximum unit combined stresses in the horizontal member.

**11-21.** Calculate the maximum tensile and compressive unit combined stresses at section $AA$ of the clamp shown in Fig. 11-15 when $e = 2$ in. and $P = 100$ lb (area $= 0.375$ in.$^2$; $I = 0.0332$ in.$^4$).

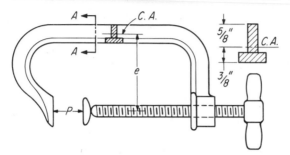

Fig. 11-15   Probs. 11-21 and 11-41.

**11-22.** In the cast-iron press frame shown in Fig. 11-16, $a = 6$ in., $b = 6$ in., $t = 2$ in., and $d = 17$ in. Calculate the maximum tensile and compressive unit combined stresses in the vertical member of this frame when $P = 10,000$ lb.

*Ans.* +4837 psi, −6933 psi

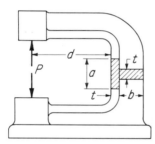

Fig. 11-16   Probs. 11-22 and 11-23.

**11-23.** If in Fig. 11-16 $a = 200$ mm, $b = 300$ mm, $t = 50$ mm, and $d = 490$ mm, what maximum value of $P$ may the press be subjected to without exceeding the following unit stresses: 20 MPa in tension and 70 MPa in compression?

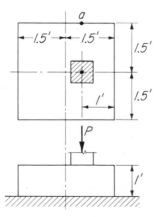

Fig. 11-17   Prob. 11-24.

**11-24.** The concrete footing shown in Fig. 11-17 supports a timber post carrying an axial compressive load $P$ of 16,650 lb. The concrete weighs 150 pcf; weight of the post may be disregarded. The load is centered on one of the two centroidal axes at a point $1/2$ ft away from the other. (*a*) Calculate the maximum unit soil pressure under footing in psf. (NOTE: Calculate $I$ in feet$^4$.) (*b*) Considering the solution to part *a*, what will be the maximum soil pressure if the post is moved $1/2$ ft toward point *a*, thereby giving the load double eccentricity? A simple addition will give the answer.

**11-25.** A small concrete diversion dam has the cross-sectional dimensions shown in Fig. 11-18. Assume that concrete weighs 150 pcf. Consider a 1-ft length of dam and calculate the unit combined stresses at *a* and *b* due to (*a*) weight of dam only, if $h = 12$ ft, and (*b*) weight of dam plus the effect of 9 ft of water against right side.

*Ans.* (*a*) $-0.80$ psi at *a*, $-14.8$ psi at *b*; (*b*) $-20.6$ psi at *a*, $+4.9$ psi at *b*

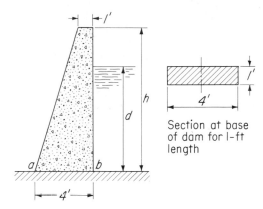

Section at base
of dam for 1-ft
length

**Fig. 11-18   Prob. 11-25.**

## 11-3  STRESSES ON INCLINED CUTTING PLANES

While the principal stress in a bar subjected to an axial pull $P$ is tension, one should be aware of the fact that shearing stresses exist on diagonal planes. As a matter of fact, many materials weak in shear fail along surfaces inclined at 45° to the geometric axis when tested in axial compression or tension.

The forces acting on an inclined plane through an axially loaded member are shown in Fig. 11-19*c*. The area over which these forces are distributed is $A/\cos\theta$. Therefore the diagonal shearing stress is $P/A \sin\theta \cos\theta$, and the diagonal tension is $P/A \cos^2\theta$, as shown in Fig. 11-19*d*. It may be of interest to note that the diagonal shearing stress is maximum when $\theta$ equals 45°.

Now the diagonal shearing stresses can arise from bending stresses as well as from stresses due to axial loads. Furthermore, diagonal tensile

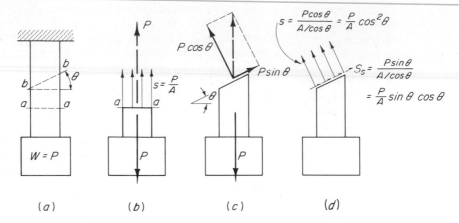

**Fig. 11-19** Diagonal tension and shear due to axial load. (*a*) Loaded member. (*b*) Axial forces. (*c*) Forces on inclined plane. (*d*) Stresses on inclined plane.

stresses are caused by shear. This explains the diagonal tension cracks sometimes observed in regions of high shear in concrete beams. Steel stirrups are usually placed in concrete beams to control the diagonal tension cracking.

Consider the body shown in Fig. 11-20*a*. The stresses at point *O* in the body depend upon the orientation of the cutting planes used to expose them. If cuts are made in the *X* and *Y* directions, the stresses are as shown in Fig. 11-20*b*. Suppose we know these stresses but wish to compute the stresses found on a cutting plane making an angle $\theta$ to the *XZ* plane. We may cut out a wedge-shaped element (see Fig. 11-20*c*) from the cube shown in Fig. 11-20*b*. Assume that the sloping face on the wedge has an area $\Delta A$. The areas of the right face and the bottom face will be $\Delta A \sin \theta$ and $\Delta A \cos \theta$, respectively. The **forces** acting on each face of the wedge are shown in Fig. 11-20*c*.

To find the force $s_N \Delta A$, let us sum forces in the *N* direction.

$$s_N \Delta A = (s_X \Delta A \sin \theta) \sin \theta + (s_S \Delta A \sin \theta) \cos \theta$$
$$+ (s_Y \Delta A \cos \theta) \cos \theta + (s_S \Delta A \cos \theta) \sin \theta$$

Dividing all terms by $\Delta A$ gives

$$s_N = s_X \sin^2 \theta + s_Y \cos^2 \theta + 2s_S \sin \theta \cos \theta \qquad (11\text{-}2)$$

Equation (11-2) gives the normal stress on the inclined cutting plane in terms of the stresses on the *X* and *Y* planes. A summation of forces in the $\theta$ direction will yield

$$(s_S)_\theta = s_S(\cos^2 \theta - \sin^2 \theta) + (s_X - s_Y) \sin \theta \cos \theta \qquad (11\text{-}3)$$

Equation (11-3) gives the shearing stress on the inclined cutting plane.

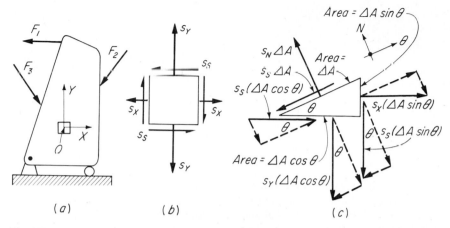

**Fig. 11-20   Stresses on inclined cutting planes. (a) Body. (b) Stresses on XY cutting planes. (c) Forces on wedge-shaped element cut from cube shown in b.**

By means of calculus, it can be shown that the maximum normal stress at point $O$ is

$$\textbf{Max } s = \frac{s_Y + s_X}{2} + \sqrt{s_S^2 + \left(\frac{s_X - s_Y}{2}\right)^2} \qquad \textbf{(11-4)}$$

and the maximum shearing stress is

$$\textbf{Max } s_S = \sqrt{s_S^2 + \left(\frac{s_X - s_Y}{2}\right)^2} \qquad \textbf{(11-5)}$$

The principal (maximum) tensile and compressive stresses in a shaft, beam, or other structural element are given by Eq. (11-4) and the principal shearing stresses are given by Eq. (11-5). The cutting planes that expose these stresses are called the principal planes.

One must be careful about signs when using the equations above. Positive stresses have been indicated in Fig. 11-20$b$. If any stress is opposite in direction to that shown in the sketch, it should be assigned a negative sign. Also recall that the square root of a number can be either plus or minus, as required to fit the physical problem.

The principal tension in a beam may be greater than the maximum tensile bending stress (see, for example, Prob. 11-29). In beams of the usual cross section and span, however, bending stresses and shearing stresses on horizontal and vertical planes are usually maximum stresses.

## ILLUSTRATIVE PROBLEMS

**11-26.** The bending and shearing stresses at a certain point in a beam are as shown in Fig. 11-21$a$. (a) Calculate the normal and shearing stresses at this same

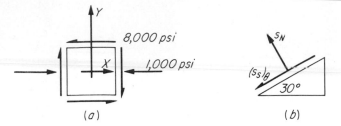

**Fig. 11-21**   (*a*) Stresses on *XY* planes. (*b*) Stresses on $\theta$ plane.

point on a cutting plane making an angle of 30° with the *XZ* plane. (*b*) Calculate the maximum tensile, compressive, and shearing stresses at this point in the beam.

*Solution:* The normal and shearing stresses on the 30° plane are

$$[s_N = s_X \sin^2 \theta + s_Y \cos^2 \theta + 2s_S \sin \theta \cos \theta]$$
$$s_N = (-1000)(0.5)^2 + 0(0.866)^2 + 2(8000)(0.5)(0.866)$$
$$s_N = -250 + 6930 = +6680 \text{ psi (tension)} \qquad \textit{Ans.}$$

$$[(s_S)_\theta = s_S(\cos^2 \theta - \sin^2 \theta) + (s_X - s_Y)\sin \theta \cos \theta]$$
$$(s_S)_\theta = 8000(0.866^2 - 0.5^2) + (-1000)(0.5)(0.866)$$
$$(s_S)_\theta = +4000 - 433 = +3567 \text{ psi} \qquad \textit{Ans.}$$

The plus sign on the shearing stress indicates that the direction of the computed stress on the inclined plane is that shown on the derivation sketch, that is, downward to the left.

The maximum tensile and compressive stresses are

$$\left[ \text{Max } s = \frac{s_X + s_Y}{2} + \sqrt{s_S^2 + \left( \frac{s_Y - s_X}{2} \right)^2} \right]$$

$$\text{Max } s = \frac{-1000 + 0}{2} + \sqrt{(8000^2) + \left[ \frac{0 - (-1000)}{2} \right]^2}$$

$$\text{Max } s = -500 + \sqrt{64,250,000} = -500 + 8016$$

$$= 7516 \text{ psi tension} \qquad \textit{Ans.}$$

$$\text{Min } s = \frac{-1000 + 0}{2} + \sqrt{(8000^2) + \left[ \frac{0 - (-1000)}{2} \right]^2}$$

$$= -500 - \sqrt{64,250,000} = -8516 \text{ psi compression} \qquad \textit{Ans.}$$

The maximum shear stress is

$$\left[ \text{Max } s_S = \sqrt{s_S^2 + \left( \frac{s_Y - s_X}{2} \right)^2} \right]$$

$$\text{Max } s_S = \sqrt{(8000^2) + \left[ \frac{0 - (-1000)}{2} \right]^2} = \sqrt{64,250,000} = 8016 \text{ psi} \qquad \textit{Ans.}$$

Note that all signs must be handled with care.

**11-27.** The bracket in Fig. 11-22*a* is subjected to the force $F = 5000$ lb at an angle $\theta = 45°$. Calculate the principal tensile stress at *O*.

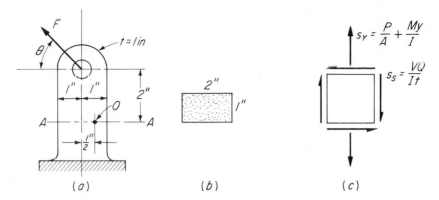

**Fig. 11-22**  (a) Bracket. (b) Cross section AA. (c) Vertical and horizontal stresses at O.

*Solution:* The bending moment on section *AA* is the horizontal component of the force *F* multiplied by 2 in. That is, $M = (0.707)(5000)(2) = 7070$ lb·in. The vertical tensile stress at point *O* is (see Fig. 11-22c)

$$\left[ s_Y = \frac{P}{A} + \frac{My}{I} \right] \qquad s_Y = \frac{0.707(5000)}{2} + \frac{7070\left(\frac{1}{2}\right)}{\frac{(1)(2^3)}{12}} = 1768 + 5302$$
$$= 7070 \text{ psi}$$

The horizontal and vertical shearing stresses at *O* are

$$\left[ s_S = \frac{VQ}{Ib} \right] \qquad s_S = \frac{0.707(5000)\frac{1}{2}(1)\left(\frac{3}{4}\right)}{\frac{(1)(2^3)}{12}(1)} = 1988 \text{ psi}$$

The maximum tensile stress at *O* is

$$\left[ \text{Max } s = \frac{s_Y + s_X}{2} + \sqrt{s_S^2 + \left(\frac{s_Y - s_X}{2}\right)^2} \right]$$
$$\text{Max } s = \frac{7070}{2} + \sqrt{1988^2 + \left(\frac{7070 - 0}{2}\right)^2}$$
$$= +3535 + \sqrt{16,400,000} = 3535 + 4055 = 7590 \text{ psi} \qquad Ans.$$

**PROBLEMS**————————————————————————————————

**11-28.** Solve Prob. 11-27 if $F = 8000$ lb and $\theta = 0°$.

*Ans.* Max $s = 13,500$ psi

**11-29.** Calculate the magnitude of the uniform load $w$ pounds per foot that will cause a maximum bending stress of 10,000 psi in the beam shown in Fig. 11-23. Then calculate the principal tension at point $A$ (in the web just below the flange).

*Ans.* $w = 23,100$ lb/ft; $s = 29,000$ psi

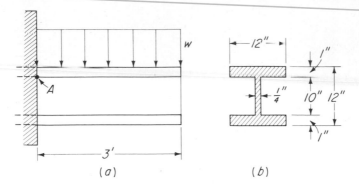

**Fig. 11-23**   (a) Loaded beam. (b) Cross section.

**11-30.** Solve Prob. 11-27 for the principal tension at a point $\frac{1}{2}$ in. to the left of $O$, that is, on the centerline of the bracket.

## 11-4 MOHR'S CIRCLE

Equations were developed in Art. 11-3 for calculating the maximum and minimum normal and shearing stresses at a point in a body subjected to plane stresses. Mohr's circle provides an alternate way of accomplishing these calculations. The advantage of the Mohr's circle technique is that it simplifies the handling of signs. If $s_N$ from Eq. (11-2) is plotted against $(s_S)_\theta$ from Eq. (11-3) for all possible values of the angle $\theta$, a circle will result. This circle is commonly called **Mohr's circle.**

Suppose that the stresses on the $X$- and $Y$-cutting planes at a point in a stressed body are as shown in Fig. 11-24a. We see that $s_X = 1000$ psi ($T$), $s_Y = 3000$ psi ($T$), and $s_S = 1000$ psi. By comparison with the definition sketch, Fig. 11-20b, we see that all stresses happen to be positive. Mohr's circle is a plot of $s_S$ against $s_N$, so we first draw the mutually perpendicular $s_N - s_S$ axes (see Fig. 11-24b). Next we draw a vertical line 1000 psi ($s_X$) to the right of the $s_S$ axis (to the *right* because $s_X$ is positive). Point $X$ lies on this line either 1000 psi (the shearing stress $s_S$ on the right-hand face of the element) above or below the $s_N$ axis. In this example, $X$ lies *above* the $s_N$ axis because the shearing stress was positive, that is, agreed in direction with the definition sketch. Point $Y$ lies on a vertical line 3000 psi to the right of the $s_S$ axis ($s_Y = +3000$ psi). Point $Y$ will always be on the opposite side of the horizontal $s_N$ axis from point $X$.

The center of Mohr's circle will always be midway between points $X$ and $Y$. One does not ordinarily draw the Mohr's circle sketch to scale, but all dimensions should be kept to at least approximate size. We now have Mohr's circle, Fig. 11-24b.

The maximum normal stress corresponds to point $M$ on the diagram. Since $CP - 1000$ psi and $PY = 1000$ psi, the radius of the circle is

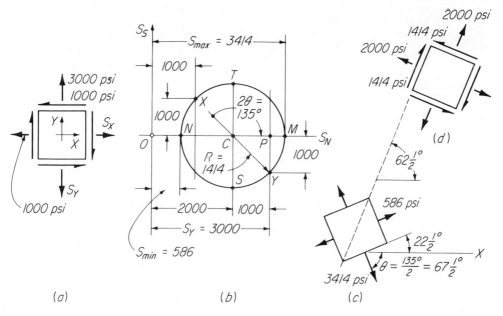

**Fig. 11-24   Mohr's circle. (a) Stressed element showing stresses on X and Y planes at point A. (b) Mohr's circle. (c) Principal normal stresses. (d) Principal showing stresses.**

1414 psi. We see, then, that the maximum normal stress (distance $OM$) is 3414 psi. The minimum normal stress (distance $ON$) is 586 psi. These stresses are shown on the element drawn in Fig. 11-24c.

Point $N$, representing the *minimum* normal stress, and point $M$, representing the *maximum* normal stress, are 180° apart on Mohr's circle. Actually, the two stresses are only 90° apart in space. We conclude, then, that all angles obtained from Mohr's circle must be divided by 2 to get the correct spatial relation. We note that angle $PCY$ is 45° and therefore that angle $XCM$ is 135°. The true angle between the maximum normal stress and the $X$ axis is, then, $135°/2 = 67^1/_2°$ (see Fig. 11-24c). The cutting planes exposing the maximum and minimum normal stresses, shown in Fig. 12-24c, are called the principal planes.

The maximum shearing stress is represented on Mohr's circle by the dimension $CT$, which is equal to $R$ or 1414 psi. Point $T$ lies 90° from $M$ on the circle, and therefore the planes exposing the maximum shearing stresses are oriented 45° from the principal planes (see Fig. 11-24d).

**ILLUSTRATIVE PROBLEMS**

**11-31.** A body is subjected to stresses in the $X$ and $Y$ planes (plane stress). At point $A$ these stresses are as shown in Fig. 11-25a. Using Mohr's circle, calculate the maximum and minimum normal stresses and the maximum shearing stresses at point $A$.

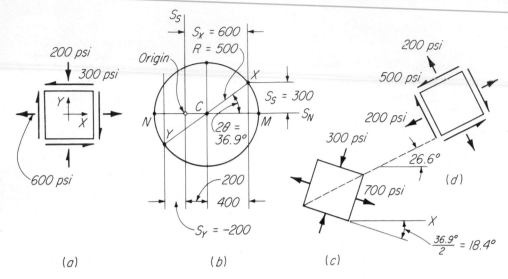

**Fig. 11-25   Mohr's circle. (a) Stresses on X and Y planes. (b) Mohr's circle. (c) Maximum and minimum normal stresses. (d) Maximum shearing stresses.**

*Solution:* Note that $s_Y$ is negative and that the other stresses are positive. Draw vertical lines 200 psi to the left of the $s_S$ axis and 600 psi to the right. Point $X$ lies on this latter line since $s_X = +600$ psi. Point $X$ must be 300 psi *above* the $s_N$ axis since the shearing stresses on the $X$ and $Y$ planes are positive, that is, are as shown in Fig. 12-20$b$. Point $Y$ must lie 300 psi *below* the $s_N$ axis (on the opposite side from $X$) and 200 psi to the left of the $s_S$ axis.

The center of the circle lies midway between $X$ and $Y$, that is, 400 psi to the left of $X$ and 400 psi to the right of $Y$. The radius of the circle may now be found. It is

$$R = \sqrt{300^2 + 400^2} = 500 \text{ psi}$$

Point $M$ lies $500 + 200 = 700$ psi to the right of the $s_S$ axis. Therefore the maximum normal stress is

$$\text{Max } s = 700 \text{ psi } (T) \qquad\qquad Ans.$$

The minimum normal stress is

$$\text{Min } s = 500 - 200 = 300 \text{ psi } (C) \qquad\qquad Ans.$$

The maximum shearing stress is equal to the radius of the circle or

$$\text{Max } s_S = 500 \text{ psi} \qquad\qquad Ans.$$

These stresses are shown on properly oriented elements in Fig. 11-25$c$ and $d$.

**11-32.** The stresses on the $X$ and $Y$ planes at point $A$ in a body are as shown in Fig. 11-26$a$. Using Mohr's circle, calculate the principal normal and shearing stresses at that point.

*Solution:* Note that all stresses are opposite in direction to those shown in the definition sketch, Fig. 11-20$b$, and are therefore negative. Point $X$ (associated with

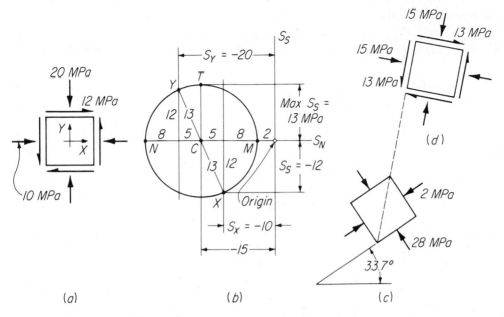

(a)                          (b)                          (c)

**Fig. 11-26   Problem 11-32. (a) Stress on X and Y planes. (b) Mohr's circle. (c) Principal normal stresses. (d) Principal shearing stresses.**

$s_X$) is at $(-10, -12)$ and $Y$ is at $(-20, +12)$. The center of the circle is midway between these two points at $(-15, 0)$ (see Fig. 11-26b). The radius of Mohr's circle is

$$R = \sqrt{5^2 + 12^2} = 13 \text{ MPa}$$

The maximum and minimum normal stresses are

$$\text{Max } s = 15 - 13 = 2 \text{ MPa } (C) \qquad \qquad Ans.$$
and
$$\text{Min } s = 15 + 13 = 28 \text{ MPa } (C) \qquad \qquad Ans.$$

By custom, a positive stress is assumed to be the maximum stress compared with a negative stress even though the numerical value of the negative stress is larger than that of the positive stress. In a similar manner, $-2$ MPa of our illustrative problem is considered to be the maximum normal stress and $-29$ MPa is the minimum normal stress.

Note that angle $MCX$ is $\tan^{-1} (^{12}/_5)$ or $67.4°$. Half of $67.4°$ is $33.7°$ (see Fig. 11-26c). The principal planes, then, make angles of $33.7°$ and $-56.3°$ with the $XZ$ (horizontal) plane.

The maximum shearing stress is represented by the distance $TC$, or 13 MPa. These stresses are shown on properly oriented shearing planes in Fig. 11-26d. Note that in *all* cases the normal stresses on the principal shearing planes are equal. Furthermore, if we compare the algebraic sum of the two mutually perpendicular normal stresses in (a), (c), and (d) of *each* figure, we see that the sum remains constant.

## PROBLEMS

**11-33.** Refer to Prob. 11-3. Using Mohr's circle, calculate the maximum normal stress in the steel plate $A$ at point $B$. Point $B$ is 1 in. from the plate centerline. Assume $W = 2940$ lb.

**11-34.** Refer to Prob. 11-26. Solve for the maximum and minimum normal stresses and the maximum shearing stress using Mohr's circle.

**11-35.** Refer to Prob. 11-27. Solve for the maximum and minimum normal stresses and the maximum shearing stress using Mohr's circle. Also find the angles $\theta_H$ that expose each of the above stresses.

**11-36.** Using Mohr's circle, calculate the maximum normal and maximum shearing stresses at points $A$ and $C$ in Fig. 11-27.

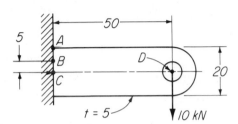

**Fig. 11-27   Probs. 11-36, 11-37, and 11-52.**

**11-37.** Using Mohr's circle, calculate the absolute maximum tensile and shearing stress at point $B$ in Fig. 11-27.

*Ans.* Max tension = 767 MPa; max shear = 392 MPa

## 11-5 COMBINED SHEARING AND BENDING STRESSES IN CIRCULAR SHAFTS

Shearing stresses in circular shafts caused by torque were discussed in Chap. 4. However, horizontal shafts as well as other nonvertical shafts are also subjected to bending due to their own weight and the weight of pulleys. Belt tensions, which most often act in directions other than the vertical, are the greatest single cause of bending in vertical shafts. Shafts supported by more than two bearings are continuous beams. Only shafts supported by two bearings will be dealt with here.

The maximum shearing stress in a circular shaft due to a combined torque $T$ and a bending moment $M$ will be found on an inclined plane within the shaft. It can be shown that this maximum shearing stress is equal in magnitude to that on planes normal to the axis caused by a single hypothetical resultant torque $T_R$ equal to

$$T_R = \sqrt{T^2 + M^2} \tag{11-6}$$

This simple relation enables us to use Eqs. (4-2) and (4-3) in the following forms:

### Maximum Resultant Shearing Stress:

$$s_R = \frac{T_R D}{2J} = \frac{5.09 T_R}{D^3} \qquad \text{(11-7)}$$

### Required Diameter of Solid Shaft:

$$D = \sqrt{\frac{16 T_R}{\pi s_R}} = \sqrt{\frac{5.09 T_R}{s_R}} \qquad \text{(11-8)}$$

In any given situation the point of maximum stress remains fixed in space but moves around the surface of the shaft as different points on the surface pass through that fixed point. This continual reversal of stress (due to bending) calls for lower allowable design stresses.

### ILLUSTRATIVE PROBLEM

**11-38.** The shaft shown in Fig. 11-28$a$ is subjected to the following loads: $W_1 = 60$ lb, $W_2 = 40$ lb, $T_1 = 50$ lb, $T_2 = 10$ lb, $T_3 = 50$ lb, and $T_4 = 110$ lb. The principal dimensions are $a = 6$ in., $b = 12$ in., $c = 18$ in., $r_1 = 9$ in., and $r_2 = 6$ in. Neglect the weight of the shaft itself. Calculate the required diameter $D$ of shaft if the combined shearing stress $s_R$ must not exceed 10,000 psi. Small bearing friction may be disregarded.

*Solution:* The shear and bending moment diagrams are shown in Fig. 11-28$b$. Note that the torque $T$ between the pulleys is

$$T = (50 - 10)(9) = (110 - 50)(6) = 360 \text{ lb·in.}$$

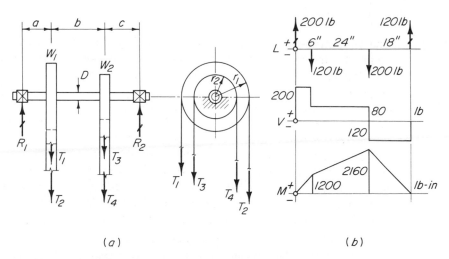

(a)  (b)

**Fig. 11-28   Shaft subjected to combined bending and torsion (Prob. 11-38). (a) Shaft and pulleys. (b) Shear and moment diagrams.**

The resultant torque $T_R$ that will give a stress by Eq. (4-2) equivalent to that caused by the combined torque $T$ and the bending moment is

$$[T_R = \sqrt{T^2 + M^2}] \qquad T_R = \sqrt{(360^2) + (2160^2)} = 2190 \text{ lb·in.}$$

The required diameter $D$ of shaft is, then,

$$\left[D = \sqrt[3]{\frac{5.09 T_R}{s_R}}\right] \qquad D = \sqrt[3]{\frac{(5.09)(2190)}{10,000}} = 1.04 \text{ in.} \qquad\qquad Ans.$$

## PROBLEMS

**11-39.** In Prob. 11-38 let $a = 6$ in., $b = 12$ in., $c = 15$ in., $r_1 = 8$ in., $r_2 = 4$ in., $W_1 = 60$ lb, $W_2 = 50$ lb, $T_1 = 30$ lb, $T_2 = 10$ lb, $T_3 = 30$ lb, and $T_4 = 70$ lb. Calculate the required shaft diameter $D$ if the allowable stress $s_R$ is 8000 psi.

*Ans.* $D = 0.986$ in.

**11-40.** In Fig. 11-28a let $a = 4$ in., $b = 8$ in., $c = 5$ in., $r_1 = 20$ in., $r_2 = 10$ in., $W_1 = 30$ lb, $W_2 = 20$ lb, $T_1 = 55$ lb, $T_2 = 5$ lb, $T_3 = 10$ lb, $T_4 = 110$ lb, and $D = 1$ in. Calculate the absolute maximum shearing stress $s_R$ due to the combined bending and torsion.

*Ans.* $s_R = 5936$ psi

**11-41.** A 50-mm diameter shaft is subjected to a bending moment $M$ of 1.5 kN·m. If the allowable absolute maximum shearing stress $s_R$ cannot exceed 100 MPa, what torque $T$ may be applied to the shaft in addition to the above bending moment $M$?

## SUMMARY (By Article Number)

**11-1.** When relatively short members are subjected to loads which produce both direct (axial) and bending unit stresses, the combined unit stress at any point is the algebraic sum of these stresses. That is,

$$\textbf{Combined unit stress} = \frac{P}{A} \pm \frac{Mc}{I} \qquad\qquad \textbf{(11-1)}$$

In such members, deflections due to loads give rise to secondary bending stresses which are generally small and may usually be disregarded.

**11-2.** An eccentric load $P$ applied along an axis parallel to the centroidal axis of a member will produce bending stresses equal to $\pm Mc/I$ in addition to the direct stress $P/A$, where the moment $M = Pe$ and $e$ is the distance between the two axes.

**11-3.** The stresses at a point in a loaded structure depend upon the orientation of the cutting planes that expose the stresses. Such cuts are usually made in the vertical or horizontal direction. The maximum stresses (principal stresses) may occur on inclined cutting planes (the principal planes).

**11-4.** Mohr's circle provides a convenient way to calculate the principal normal and shearing stresses at any point in a body undergoing plane stress.

**11-5.** A shaft subjected to both bending and torque may be treated as though it were subjected to a single torque of

$$T_R = \sqrt{T^2 + M^2} \qquad (11\text{-}6)$$

**REVIEW PROBLEMS**

**11-42.** If, in Prob. 11-5, the pile is 10 in. in diameter and $P = 900$ lb, $R = 12,000$ lb, and $h = 9$ ft, (a) what is the maximum combined unit stress? (b) If the allowable combined stress is 900 psi, can an 8-in. pile be used? Show all work. See Fig. 11-4.

*Ans.* (a) $-483$ psi; (b) yes, $s_c = 884$ psi

**11-43.** Assume in Fig. 11-7 that $a = b = c = 1$ m and $d = 0.75$ m. What size of square timber S4S will be required to support a load of 60 kN? Assume maximum allowable combined stress $= 12$ MPa.

**11-44.** In Prob. 11-13, assume that $W = 10$ kN and that the boom is a round timber of 150 mm diameter. Then solve for the maximum combined unit stress (a) barely above $B$ and (b) barely below $B$ when boom is in position shown. See Fig. 11-9.

**11-45.** A vertical 6- by 10-in. rough timber post 4 ft long supports a vertical load $P$ of 18,000 lb similar to that shown in Fig. 11-10. $P$ is applied on the $Y$ axis at a point 2 in. from the $X$ axis. Calculate the unit combined stresses on the left and right sides of the post.

*Ans.* Left $s_t = 60$ psi; right $s_c = 660$ psi

**11-46.** If, in Fig. 11-11, $P = 140$ lb and $d = 24$ in., what size standard steel pipe should be used to keep the maximum combined unit stress less than 5000 psi?

**11-47.** The longitudinal axis of a short structural-steel post, W 8 × 31, is found to deviate 0.1 in. from straight in the plane of the major axis under a load of 150,000 lb. The longitudinal axis is straight in the plane of the minor axis. Calculate the maximum combined stress in the post. See Fig. 11-12.

*Ans.* $s_c = 18,000$ psi

**11-48.** If, in Fig. 11-15, $e = 3$ in., what maximum force $P$ can be exerted by the clamp without exceeding a combined stress of 10,000 psi at section $AA$? The cross-sectional area at $AA$ is 0.375 in.², and its moment of inertia about the neutral axis is 0.0332 in.⁴

*Ans.* $P = 186$ lb

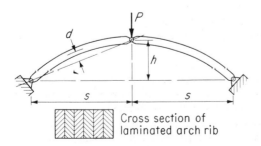

Cross section of laminated arch rib

**Fig. 11-29   Prob. 11-49.**

**11-49.** The three-hinged arch shown in Fig. 11-29 is made of two identical sections of laminated wood with constant cross-sectional areas measuring 5 by 12 in., the latter dimension lying in the vertical plane. The maximum lateral deviation $d$ of the longitudinal axis is 2 ft at the midpoint under a load $P$ of 6000 lb. Calculate the maximum combined stress in one section when $s = 20$ ft and $h = 8$ ft.

**11-50.** The cast-iron frame of a jigsaw is shown in Fig. 11-30. The exterior dimensions of the hollow section at $AA$ are 50 by 100 mm, the walls are 7 mm thick, and $d = 400$ mm. What maximum load $P$ may be exerted if the compressive stress at section $AA$ is not to exceed 20 MPa?

*Ans.* $P = 1.91$ kN

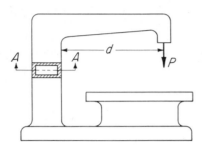

**Fig. 11-30   Prob. 11-50.**

**11-51.** Refer to Fig. 11-4. Assume a solid round pile with an 8-in. diameter, $R = 50$ kips, $P = 10$ kips, and $h = 9$ ft. Calculate the maximum normal and shearing stresses at the center of the pile at the lower ground level. Use Mohr's circle and check answers using Eqs. (11-4) and (11-5).

**11-52.** Find the maximum normal and shearing stresses at point $B$ in Fig. 11-27 if an additional 12-kN horizontal force is added at point $D$. Assume this additional force acts toward the left. Use Mohr's circle.

**11-53.** A $1\frac{1}{2}$-in.-diameter vertical shaft 6 ft long is supported at its ends by bearings. Belt tensions cause a lateral force of 200 lb at midheight. The shaft transmits 60 hp to the belt when turning at 1800 rpm. The shaft is driven from the top. Compute the absolute maximum shearing stress $s_R$ due to the combined torque and bending.

**11-54.** If the load $w$ in Fig. 11-23 is 10,000 lb/ft, calculate the principal tension at a point 6 in. above the bottom of the beam at a section adjacent to the wall.

*Ans.* $s = 11,080$ psi.

## REVIEW QUESTIONS

**11-1.** In general, what is meant by the expression "combined stress"?

**11-2.** Name the four possible combinations of stress and cite examples of each.

**11-3.** Give the formula for the combination of direct and bending stresses.

**11-4.** Explain the combination of stress due to an eccentric load.

**11-5.** In a section symmetrical about each of two rectangular axes, which is referred to as the "major axis" and which the "minor"?

**11-6.** If an 8- by 12-in. timber post is to support an eccentric load $P$ applied on one of the axes at a distance 2 in. away from the other, on which axis should the load be applied to give the least combined stress?

**11-7.** In Fig. 11-16, should the direct and bending stresses be added to give the maximum compressive combined stress?

**11-8.** In Fig. 11-28 should the direct and bending stresses be added to give the maximum tensile combined stress?

**11-9.** Consider the cast-iron frame in Fig. 11-16, and discuss some reasons why the flange of the T section is on the inside instead of the outside.

**11-10.** If the clamp in Fig. 11-15 is of malleable iron, would the reasons advanced for the cast-iron frame mentioned in question 11-9 hold also for the clamp?

**11-11.** Show by means of an appropriate free-body diagram that shearing stresses exist on inclined planes through short columns under axial compressive loads.

**11-12.** How can we take into account the combined effects of torque and bending on solid circular shafts?

# Chapter 12
# Columns

## 12-1 INTRODUCTION

A column is a structural member subjected to an end load whose action line parallels that of the member and whose length is generally ten or more times its least lateral dimension. When the action line of the end load coincides with the axis of the member, it is said to be **axially** or **concentrically loaded;** when this action line and the axis do not coincide, the member is said to be **eccentrically loaded.** A column is presumed to be axially loaded unless it is known not to be.

The **ideal column** is perfectly straight, its material is perfectly homogeneous, it is entirely free from any flaws of manufacture, and it is perfectly axially loaded. This ideal column would fail by direct crushing of the material in a manner similar to that of a short block. It would never fail by buckling sidewise. But the ideal column does not exist. The columns that we actually use are never perfectly straight, their material is not perfectly homogeneous, they contain unavoidable flaws of manufacture, however small, and they are rarely, if ever, perfectly axially loaded. Consequently columns used in practice fail by buckling sidewise in the direction of their least strength and stiffness. **Buckling** of a column is here defined as elastic instability resulting in lateral buckling due to an axial load.

## 12-2 TYPES OF COLUMN CROSS SECTIONS

The type of section most suitable for a given column depends largely upon its particular function. The piston rod of an engine, the leg of a chair, and the boom of a derrick are all compression members coming under the classification of columns. The columns with which we are perhaps most familiar are those commonly used in buildings of timber, steel, or reinforced concrete. A group of commonly used structural-steel sections is

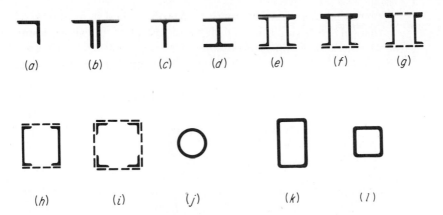

Fig. 12-1  Typical column cross sections.

shown in Fig. 12-1. The solid lines indicate sections that extend the full length of the member. The dashed lines indicate lacing bars or tie plates whose sole function is to hold rigid the load-carrying elements of the column, as indicated in Fig. 12-2.

The single angle shown in $a$ is often used for web members in light trusses and for bracing in general, the double-angle section in $b$ is generally used as a chord member in riveted trusses, and the section in $c$ as a chord member in welded trusses. The wide-flange section in $d$ is the standard structural column in steel-frame buildings, but it is also used in heavier types of trusses. The types in $e, f, g,$ and $h$ may be used in buildings or bridges. Type $g$ is frequently used for long columns supporting moderate loads. Type $i$ is very suitable for derrick booms and for light towers. The hollow tube in $j$ is extensively used in aircraft work and is the familiar pipe column so widely used in light buildings. Structural tubing, both rectangular ($k$) and square ($l$), is available in many sizes and thicknesses. The dimensions and properties of these tubing sections are now listed in the steel handbooks. Many other sections have been developed to meet requirements in specialized work.

Channels are very suitable for built-up columns of the latticed type. When they are so used, the spacing back to back is generally such that the column has equal moment of inertia with respect to the two major axes. The use of lattice bars saves material, gives columns of relatively great stiffness, and also provides access for painting inside surfaces.

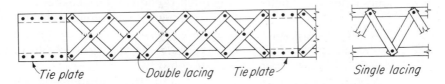

Fig. 12-2  Latticed columns.

The design of a column cannot be accomplished in as simple a manner as that of an axially loaded tension member or short compression member, where $A = P/s$, or of a beam, where $Z = M/s$, in both of which cases intensity of stress is directly proportional to the magnitude of the load. Long columns exhibit no such linearity. Short columns fail by yielding while long columns fail by elastic instability, that is, by lateral buckling. The load-carrying ability of intermediate length columns is influenced by both the strength and elastic properties of the column material. Empirical design formulas, based on much authoritative study and numerous actual tests, have been devised which satisfactorily cover the design of columns within the limits of each group. These formulas differ also with the material used, and thus we find separate sets of formulas for wood, steel, and aluminum. To complicate matters still further, various authorities have devised special formulas for use within their own particular fields of work. While such a multiplicity of column formulas may temporarily confuse students, designers experience little difficulty, since they soon limit themselves to those formulas that are most applicable to their particular field of work.

## 12-3 SLENDERNESS RATIO AND COLUMN GROUPS

While the three groups of columns are referred to above as (1) short, (2) intermediate, and (3) long, the length of a column alone cannot determine the group in which it belongs; rather, its slenderness is the determining factor.

The slenderness ratios of metal columns, such as pipe sections, rolled structural-steel or aluminum sections, and others, is measured by $L/k$, where $k$ is the radius of gyration (see Art. 5-8), often given in handbooks under the symbol $r$. The slenderness ratio of timber columns is measured by $L/d$, where $d$ is the least dimension of a square or rectangular cross section.

## 12-4 CRITICAL OR BUCKLING LOADS ON COLUMNS

The critical or buckling load is here defined as the maximum load a column can support. The buckling load for any given column depends upon many items: (1) whether the column is short, intermediate, or long, and (2) whether the ends are free to rotate or are completely or partially restrained against rotation. For the purpose of this discussion, we select an "ideal" column, perfect in every detail, with ends free to rotate and falling within the "long" range of columns.

We now place this slender column in a vertical position and apply a horizontal force $H$ at its midpoint until a lateral deflection $\Delta$ has been produced, as shown in Fig. 12-3a. In Chap. 6 we learned that the resisting

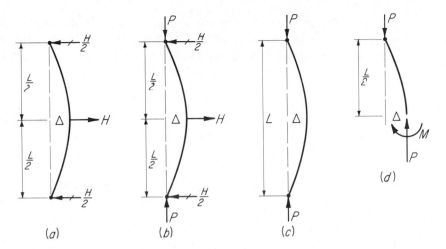

**Fig. 12-3    Critical, or buckling, load on columns.**

moment $M$ was proportional to the load $H$ and in Chap. 9 that the deflection $\Delta$ (case 6, Table 9-3) is also proportional to the load $H$. Hence in Fig. 12-3 the deflection $\Delta$ at the midpoint is proportional to the moment $M$ at that point.

Let us now gradually apply two equal and opposite forces $P$ at the ends of the column, as in Fig. 12-3$b$, while simultaneously we decrease $H$ so as to maintain a constant deflection $\Delta$. When $H$ finally becomes zero, the forces $P$ alone maintain the column in its bent position, as in Fig. 12-3$c$, and the resisting moment $M$ at the midpoint (Fig. 12-3$d$) is then

$$M = P\Delta \tag{12-1}$$

But we concluded above that $M$ was proportional to $\Delta$. Hence, for *all* values of $M$ and $\Delta$, $P$ will remain constant. If $P$ is now increased, $M$ and $\Delta$ will both increase fairly rapidly. It is evident, therefore, that $P$ is the critical or buckling load, that is, the maximum load this slender column can support without buckling. Interpreted further, it means that any load $P$ sufficient to cause the axis of a column (with ends that are not restrained against rotation) to bend or buckle even slightly is the maximum load the column can support. In practice, columns are usually restrained to some degree against rotation of the ends. The effect of such restraint diminishes as the length of the column increases and is usually quite small in columns falling within the "long-column" range.

## 12-5 EULER'S FORMULA FOR ANALYSIS AND DESIGN OF SLENDER COLUMNS

This formula, derived in 1757 by Leonard Euler, a Swiss mathematician, will give the critical or buckling load on columns with ends free to rotate that fall within the long-column range. The mathematics involved is some-what complex and is beyond the intended scope of this book. The formula is therefore given without derivation. It is

$$\frac{P}{A} = \frac{\pi^2 E}{(L/k)^2} \tag{12-2}$$

where $P/A$ = average unit stress on cross-sectional area $A$ of column when supporting buckling load $P$, also referred to as average unit load, same units as $E$ below

$E$ = modulus of elasticity of column material

$L/k$ = slenderness ratio in which $L$ is *effective* column length and $k$ is *least* radius of gyration of cross section. $L$ and $k$ should be in same units.

It is interesting to note here that in Euler's formula the buckling load $P$ is dependent solely upon the stiffness $E$ (modulus of elasticity) of the material and not upon its strength. This is evident because the formula contains no factor measuring that strength. The explanation is that in the long, slender columns to which Euler's formula applies, buckling failure occurs *before* the actual maximum unit stress reaches the yield strength of the material. It should be noted here also that Euler's formula contains no design factor of safety, since $P$ is the greatest load a column can support without failure. A suitable factor of safety must, however, be applied to the design.

## 12-6 EFFECT OF END RESTRAINT ON COLUMN STRENGTH. EFFECTIVE LENGTH

Figure 12-4 shows examples of typical column-end restraints. When both ends are free to rotate (ends rounded, hinged, pinned), as in case $a$, the column may be referred to as a **simple column.** The *actual length* of each of the four columns is $L$, but the **effective length,** or *simple-column length,* of each is the distance between adjacent points of inflection $I$. A point of inflection is found at every column end that is free to rotate and at every inflection point $I$ *between* the ends where there is a change of curvature of the axis.

Hence in case $a$, both ends hinged, the effective length equals the actual length $L$; in $b$, one end hinged and other end fixed, the effective length, when calculated, is found to be $0.7L$ between the top of the column and inflection point $I$; in $c$, both ends fixed, the effective length is $L/2$, the distance between the two inflection points; and in $d$, one end free and laterally unsupported and the other end fixed (flagpole), the free end will sway sideways, and the curvature in the length $L$ will be similar to that of

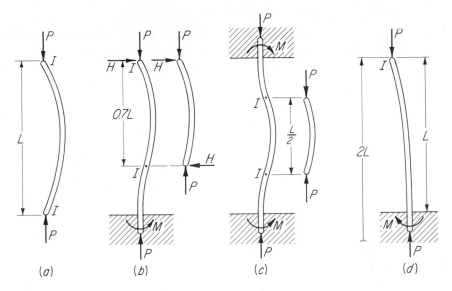

**Fig. 12-4    Effect of end restraint on column strength.**

the upper half of the simple column in $a$, which gives type $d$ an effective length of $2L$.

When the effective length for each of the four cases is substituted in Euler's formula, we obtain

Case $a$:
$$\frac{P}{A} = \frac{\pi^2 E}{(L/k)^2} = \frac{(1)\pi^2 E}{(L/k)^2} \qquad (12\text{-}3)$$

Case $b$:
$$\frac{P}{A} = \frac{\pi^2 E}{(0.7L/k)^2} \approx \frac{2.05\pi^2 E}{(L/k)^2} \qquad (12\text{-}4)$$

Case $c$:
$$\frac{P}{A} = \frac{\pi^2 E}{(0.5L/k)^2} = \frac{4\pi^2 E}{(L/k)^2} \qquad (12\text{-}5)$$

Case $d$:
$$\frac{P}{A} = \frac{\pi^2 E}{(2L/k)^2} = \frac{\frac{1}{4}\pi^2 E}{(L/k)^2} \qquad (12\text{-}6)$$

These equations show that if a column restrained as in case $a$ will support a load $P$, the same column will support twice that load, or $2P$, when restrained as in $b$, four times $P$, or $4P$, when restrained as in $c$, but only ¼ of $P$ when restrained as in $d$; *provided, however, that in no case must $P/A$ exceed the value permissible on the "simple-column length" of the column in question.* For various conditions of partial end restraint, the effective length may be satisfactorily estimated to have values between $L$ and $0.5L$. All such estimates should be conservative. Table 12-1 provides a simple summary.

Table 12-1 EFFECTIVE LENGTH OF COLUMN

| End Condition | Effective Length | Number of Times Stronger that Hinged-End Column |
|---|---|---|
| Both ends hinged | $L$ | 1 |
| One hinged, other fixed | $0.7L$ | 2 |
| Both ends fixed | $0.5L$ | 4 |
| One free, other fixed | $2L$ | $1/4$ |
| Either or both ends partially restrained | Between $L$ and $0.5L$ | Between 1 and 4 |

While these values of effective length for the four cases shown in Fig. 12-4 are mathematically correct for the end conditions shown, they are seldom found in practice. In buildings, bridges, roof trusses, and other similar structures, columns and all other compression members are nearly always fastened to other members of the structure with rivets or bolts, or, in the case of steel members, they may be welded to a connection with other members, thus causing some degree of end restraint. An experienced designer will disregard this restraint if it appears to be small, in which case the factor of safety will be increased, or will estimate its value.

In case $a$, for the end of a compression member to be completely free to rotate, the pin must be lubricated so as to offer no resistance to rotation. This condition is found only in machines, for example, heavy earth-moving and road-grading machines, where extensive use is made of hydraulic jacks to manipulate certain moving parts. Connecting rods in automobile engines and the side bars connecting the driving wheels of a locomotive are other examples of lubricated pinned-end compression members. Such cases are not dealt with in this text.

Case $d$ is sometimes referred to as the "flagpole type," but it occurs only rarely. The column can, relatively speaking, be completely fixed at the bottom. The axial load $P$ at its top would be an unusual situation, such as might be created by the "flagpole sitters" of a few decades ago. However, should such a load exist, the effective length would equal $2L$.

Obviously, therefore, when the degree of end restraint is indeterminate, the designer must rely on accumulated, available experience.

## 12-7 LIMITATIONS OF EULER'S FORMULA

A column will always buckle in the direction of its *least strength*, which means its greatest $L/k$ ratio. As previously stated, the magnitude of the critical load depends solely upon the slenderness ratio of the column and the stiffness $E$ of the material. But, since $E$ is a constant only up to the proportional limit, Euler's formula holds only for values of $P/A$ up to that limit. For example, for structural steel, with $E$ equal to 29,000,000 psi and

Table 12-2 LIMITATIONS OF EULER'S FORMULA

| Material | Modulus of Elasticity, ksi (MPa) | Proportional Limit, ksi (MPa) | Lowest Value of $L/k$ for which Euler's Formula Applies |
|---|---|---|---|
| Structural steel, low alloy | 29,000 (200 000) | 50 (342) | 76.0 |
| Structural steel, carbon | 29,000 (200 000) | 31 (214) | 96.0 |
| Douglas fir, select structural | 1600 (11 000) | 4.5 (31.0) | 59.2 |

a proportional limit of 31,000 psi, the lowest value of $L/k$ for which Euler's formula applies is

$$\frac{L}{k} = \sqrt{\frac{\pi^2 E}{P/A}} = \sqrt{\frac{\pi^2(29,000,000)}{31,000}} = 96$$

Other limits are given in Table 12-2.

The upper limits of $L/k$ depend entirely upon good judgment and safe design and are usually fixed by code or sound professional practice. For timber columns the upper limit of length divided by least-side dimension ($L/d$) is generally 50, which corresponds to a slenderness ratio ($L/k$) of approximately 170. For structural-steel columns the upper limit of $L/k$ is usually 200.

## 12-8 ANALYSIS OF LONG COLUMNS BY EULER'S FORMULA

A column of given size and length may be analyzed to determine its load-carrying capacity $P$, the average unit stress $P/A$ due to a given load, and other factors. For this purpose, the following form, to which a factor of safety ($FS$) must be added, seems most suitable:

$$\frac{P}{A} = \frac{\pi^2 E}{(L/k)^2} \quad \text{or} \quad P = \frac{\pi^2 E A k^2}{L^2} = \frac{\pi^2 E I}{L^2} \tag{12-7}$$

since $I = Ak^2$ (Art. 5-8).

### ILLUSTRATIVE PROBLEM

**12-1.** A 3- by 5-in. wood column of select-structural Douglas fir is 12 ft long and is supported laterally at midlength against buckling in its weakest direction. Assume ends to be hinged. Calculate the maximum safe load $P$ if a factor of safety of 3 is used. Check to make sure that Euler's formula holds for this case.

*Solution:* Support of this column laterally at midlength has the effect of replacing the 12-ft simple column with two 6-ft simple columns in the weaker direction, while in the stronger direction it remains a 12-ft column. It is possible, therefore, that the 12-ft column may buckle in its strongest direction (parallel to the

5-in. dimension) before the 6-ft column buckles in its weakest direction (parallel to the 3-in. dimension). This may be determined by calculating the safe load on each of the two columns. For Douglas fir, $E = 1,600,000$ psi (Table 8-1). Factor of safety is 3. In the stronger direction, $L = 144$ in. Hence

$$\left[I = \frac{bd^3}{12}\right] \qquad I = \frac{(3)(5^3)}{12} = 31.2 \text{ in.}^4$$

$$k = \sqrt{\frac{I}{A}} = \sqrt{\frac{31.2}{15}} = 1.44 \quad \text{and} \quad \frac{L}{k} = \frac{144}{1.44} = 100 \quad \text{OK}$$

$$\left[P = \frac{\pi^2 EI}{L^2(FS)}\right] \qquad P = \frac{(3.1416^2)(1,600,000)(31.2)}{(144)(144)(3)} = 7920 \text{ lb} \qquad Ans.$$

In the weaker direction, $L = 72$ in. and

$$\left[I = \frac{bd^3}{12}\right] \qquad I = \frac{(5)(3^3)}{12} = 11.25 \text{ in.}^4$$

$$k = \sqrt{\frac{I}{A}} = \sqrt{\frac{11.25}{15}} = 0.866 \quad \text{and} \quad \frac{L}{k} = \frac{72}{0.866} = 83.1 \quad \text{OK}$$

$$\left[P = \frac{\pi^2 EI}{L^2(FS)}\right] \qquad P = \frac{(3.1416^2)(1,600,000)(11.25)}{(72)(72)(3)} = 11,400 \text{ lb}$$

showing the column is now actually weakest in the direction parallel to its 5-in. dimension and that the safe load is 7920 lb. The values of $L/k$ also show that Euler's formula applies.

PROBLEMS

NOTE: Column ends shall be presumed to be "hinged" or "pinned" unless otherwise stated. Steel columns are formed from carbon structural steel.

**12-2.** In Prob. 12-1, assume the column to be 2 by 6 in. and solve for the safe load $P$.

**12-3.** Standard steel pipe is used making sections of scaffolding commonly used by the building industry. Calculate the safe axial load on a 1½-in. pipe column 78 in. long if a factor of safety of 4 is used.

Ans. $P = 3650$ lb

**12-4.** In the preceding problem, what will be the answer if 2- by 4-in. wood uprights (S4S) are used instead of steel pipe? Assume $E = 1,760,000$ psi.

**12-5.** A yardstick is 1 in. wide and ¼ in. thick, of pine ($E = 1,200,000$ psi), and exactly 36 in. long. Assuming it to be straight, what axial load $P$ will cause it to buckle if the end restraint is similar (a) to case a of Fig. 12-4; (b) to case b; (c) to case c?

**12-6.** Calculate the safe axial load $P$ on a 75-mm-OD steel pipe column 6 m long if a factor of safety of 2 is used. Wall thickness of pipe = 7.5 mm.

**12-7.** What is the safe axial load $P$ on a rough 125- by 125-mm lodgepole pine column 5 m long if a factor of safety of 3 is used?

Ans. $P = 20.3$ kN

**12-8.** Calculate the safe axial load $P$ on a steel column, W 8 × 31, 25 ft long if a factor of safety of 3 is used.

*Ans. P* = 39,300 lb

## 12-9 DESIGN OF LONG COLUMNS BY EULER'S FORMULA

To design a column means to determine its required size. With Euler's formula that is accomplished by solving for the required moment of inertia $I$ in Eq. (12-7). For design, then, the following form is most suitable:

$$ I = \frac{PL^2}{\pi^2 E} \tag{12-8} $$

in which $L$ is the *effective* length of the column. $I$ must be multiplied by a suitable factor of safety before a corresponding size is selected, or $P$ may be multiplied by the factor of safety, as in Prob. 12-9.

To make sure that Euler's formula applies to the problem, the slenderness ratio $L/k$ for the member selected should be calculated and checked against the lowest value (Art. 12-7) for which the formula applies.

ILLUSTRATIVE PROBLEM ————————————————

**12-9.** A structural-steel column 25 ft long must support an axial load of 75,000 lb. Select the lightest available W section, using a factor of safety of 2.5, and calculate its slenderness ratio $L/k$. The proportional limit of this steel is 31,000 psi.

*Solution:* The design load $P$ now is (2.5)(75,000), and

$$ \left[ I = \frac{PL^2}{\pi^2 E} \right] \qquad \text{Required } I = \frac{(2.5)(75,000)(300^2)}{\pi^2 (29,000,000)} = 59 \text{ in.}^4 $$

Use W 8 × 48 ($I_Y$ = 60.9 in.⁴).          *Ans.*

For this section, $k$ = 2.08. Hence

$$ \frac{L}{k} = \frac{300}{2.08} = 144 \qquad (>96, \text{ hence OK}) $$

PROBLEMS————————————————

NOTE: Column ends will be presumed to be "hinged" or "pinned" unless otherwise stated.

**12-10.** A copper rod 600 mm long with hinged ends is subjected to an axial compressive load $P$ of 2 kN. Calculate its required diameter $d$, if a factor of safety of 3 is used, and its slenderness ratio $L/k$.

*Ans. d* = 14.2 mm; $L/k$ = 170

**12-11.** A round wood column of select-structural Douglas fir 8 ft 4 in. long must support an axial compressive load $P$ of 2090 lb. Calculate (*a*) its required diameter $d$ if a factor of safety of 3 is used and (*b*) its slenderness ratio $L/k$.

**12-12.** A square wood column of select-structural Douglas fir S4S 12 ft long must support an axial compressive load $P$ of 19,000 lb. Determine the approximate size of column if a factor of safety of 3 is used. Calculate its slenderness ratio $L/k$.

**12-13.** An axial load of 14 kN must be supported by a square select-structural Douglas fir column 4 m long. What least-dimension $d$ is appropriate if a factor of safety of 4 is used?

*Ans.* $d = 100$ mm; $L/k = 138$

**12-14.** A main member in a large truss is 16 ft long and is to be made of two structural-steel angles placed back to back with a ⅜-in. gusset plate between. It is to support an axial compressive load of 20,000 lb. If a factor of safety of 2 is appropriate, what size angles will be most economical? What is $L/k$?

**12-15.** A structural-steel equal-leg angle 10 ft long is to be used as a brace capable of withstanding an axial compressive load of 6950 lb. Using a factor of safety of 3.5, select the lightest available angle. Calculate $L/k$.

*Ans.* Use L 4 × 4 × ¼; $L/k = 151$

## 12-10 INTERMEDIATE COLUMNS

As stated in Art. 12-5, no single column-design formula has yet been devised which has proved to be satisfactory for all lengths of columns. For hinged-end columns in the "long" range, where only the *stiffness* $E$ of the material, and not its *strength*, determines the buckling load, Euler's formula will predict closely the actual buckling load. In the "intermediate" range, however, the *strength* of the material as well as its stiffness determines the buckling load.

Since many different materials are used, such as steel, brass, bronze, aluminum, wood, and others, with working strengths varying perhaps from 1000 to 30,000 psi, it is not surprising that literally dozens of formulas have been devised. One edition of the "Structural Aluminum Handbook" lists 19 different column formulas for as many different aluminum alloys in the intermediate range. Despite this, the problem of column design is not difficult, since each industry usually confines itself to a few well-tested and proved formulas.

In order to simplify matters, only the most commonly used column formulas are presented here, grouped according to kind of material. Because all are of empirical origin, arrived at after much authoritative experimental testing, they are merely presented here along with explanations about their proper use. Unlike Euler's formula, in which the required moment of inertia may be solved for directly, all formulas in the intermediate range, with $L/k$ up to 126.1 for A36 steel, give merely the allowable average unit load $P/A$ on a given column; this load in turn is dependent upon the slenderness ratio $L/k$. Because there are two variables, $A$ and $k$, in each formula, an appropriate column section can be selected only by the trial-and-error method. Two or three trials at the most are usually sufficient.

Since all structural design, whether it applies to a building, a bridge, a machine, or an airplane, demands absolute safety with respect to strength, the column formulas to be used in different geographical areas are specified by the code-making authorities within those areas, such as municipal, state, federal, Army, Navy, and others, who are then guided by the special conditions within their areas.

Rolled steel sections are now available in several grades, designated according to yield strength and other properties as A7, A36, A242, A373, A440, and A441 by the American Society for Testing and Materials (ASTM). Only A36 steel, the most commonly used, will be considered here.

## 12-11 COLUMN FORMULAS, STRUCTURAL STEEL

Analysis here implies the determination of the allowable unit load $P/A$ on a given column, or of its allowable total load $P$.

The **Gordon-Rankine formula**, developed more than a century ago, is

$$\frac{P}{A} = \frac{18,000}{1 + (1/18,000)(L/k)^2} \tag{12-9}$$

The **American Institute of Steel Construction (AISC)** recommends the following formula for main and secondary compression members (columns, etc.) where the slenderness ratio $KL/k$ equals $C_c$ or less ($C_c$ = 126.1 for A36 steel).

**For main and secondary members ($KL/k$ not over 126.1):**

$$\frac{P}{A} = \frac{\left[1 - \dfrac{(KL/k)^2}{2C_c^2}\right] s_y}{FS} \tag{12-10}$$

where 
$$C_c = \sqrt{\frac{2\pi^2 E}{s_y}}$$

and    Factor of safety $FS = \dfrac{5}{3} + \dfrac{3(KL/k)}{8C_c} - \dfrac{(KL/k)^3}{8C_c^3}$

(For A36 steel, $FS = 1.92$ when $C_c = KL/k$, or 126.1)

For main members with slenderness ratios greater than $C_c$ (the so-called long columns), the AISC recommends

**For main members when** $KL/k$ **exceeds** $C_c$:

$$\frac{P}{A} = \frac{12\pi^2 E}{23(KL/k)^2} \tag{12-11}$$

For braces and secondary members, subjected only to temporary wind loads or other temporary forces, when $L/k$ exceeds $C_c$ (126.1 for A36 steel), the AISC recommends

**For braces and secondary members when** $L/k$ **exceeds 120:**

$$\frac{P}{A} = \frac{P/A \text{ by Eq. (12-10) or (12-11)}}{1.6 - L/200k} \tag{12-12}$$

For Eq. (12-12), $K$ is always taken as unity. The modification (in the denominator) of Eq. (12-12) has the effect of increasing the allowable $P/A$ for braces and secondary members.

Allowable $P/A$ values for all three formulas are tabulated in Table E-8 in Appendix E. The maximum allowable $L/k$ for any member is 200.

These empirical AISC column formulas (12-10) to (12-12) apply to compression members in structures that are braced against sidesway (lateral displacement) by diagonal bracing, shear walls, or other adequate methods of preventing sidesway. When these conditions are met, as they are in the great majority of buildings, the effective-length factor $K$ is taken as unity, that is, $K = 1$. However, in such structures, further analysis may indicate that some degree of end restraint is present which may permit a value of $K$ less than unity and which then would permit a higher value of unit load $P/A$ to be used. In other design problems, such as building frames in which the rigidity of the joints connecting beams and columns offers the only resistance to sidesway, the end-restraint factor $K$ will lie between 1 and 2, which then has the effect of decreasing the allowable $P/A$.

Obviously, therefore, only structural designers fully aware of all the requirements of the complete design can determine a safe and appropriate value of $K$, either by some rational analysis or by exercise of their design experience and sound engineering judgment.

For the reasons stated above, the effective-length factor $K$, as it appears in Eqs. (12-10) to (12-12), is taken as unity ($K = 1$) in all problems, unless otherwise specified.

When values of $P/A$ are calculated for various ratios of $L/k$ up to 200 and are then plotted, as in Fig. 12-5, formula (12-10) gives the $AB$ portion of the curve $ABC$ and formula (12-11) the $BC$ portion. The curve $ABC$ is for *main* members and curve $BD$ for *secondary* members. Allowable AISC stresses may then be obtained by solving the applicable formulas, from the curves in Fig. 12-5, or from Table E-14. These allowable stresses are based on a factor of safety of about 2 based on the elastic limit and of about 4 based on ultimate strength.

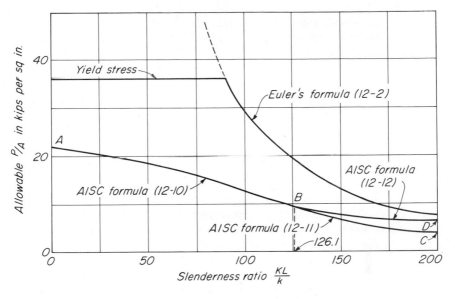

**Fig. 12-5** Allowable unit loads *P/A* on axially loaded column of ASTM A36 steel.

The **American Railway Engineering Association (AREA)** has adopted the following formulas for main compression members in railway bridges:

$$\textbf{AREA:} \qquad \frac{P}{A} = 15{,}000 - \frac{1}{3}\left(\frac{L}{k}\right)^2 \qquad \text{for pinned ends} \qquad \textbf{(12-13)}$$

$$\text{and} \qquad \frac{P}{A} = 15{,}000 - \frac{1}{4}\left(\frac{L}{k}\right)^2 \qquad \text{for riveted ends} \qquad \textbf{(12-14)}$$

for *L/k* values not greater than 140.

The **American Association of State Highway Officials (AASHO)** has adopted the same formulas for compression members in highway bridges, with a limit of *L/k* of 120 for main members and 140 for secondary members.

## 12-12 SOME PRACTICAL ASPECTS OF COLUMN DESIGN

In most cases, column ends are not hinged or pinned, as has been presumed in all formulas presented in this article with the exception of the AREA formula (12-14), which allows a slightly higher stress when the column ends are riveted instead of hinged. Most cases of hinged-end columns occur in machine work; connecting rods in automobile engines and the side bars connecting the driving wheels of a locomotive are examples of hinged-end columns.

In structural work, the ends of the steel columns are usually restrained to some degree at the bottom by being welded to a baseplate, which in turn is bolted to a concrete footing, and at the various floor levels by the usual beam connections. Steel-pipe columns generally have plates welded to their ends which, when bolted to other parts of a structure, also restrain the ends to some degree. Such restraints, however, vary greatly and are difficult to evaluate. They are generally not used to permit higher stresses and therefore serve to add to the factor of safety. On the other hand, such restraints also serve to offset some possible lack of straightness of the column or some small unintentional and often unnoticed eccentricity of loading which, in some degree, is almost certain to exist.

**Main members** are those that resist *permanent,* known, calculated, or closely estimated loads. **Secondary members** are those that resist *temporary* loads, such as braces that may become loaded only temporarily because of wind, earthquake, or other temporary conditions.

Tables E-6 and E-7 list a wide selection of *structural-steel shapes* and their properties, sufficient for use in solving problems given in this text. Acquisition of "Steel Construction," the manual of the American Institute of Steel Construction, is strongly recommended for all who wish to extend their work beyond the problems of this text.

## 12-13 ANALYSIS OF STRUCTURAL-STEEL COLUMNS

Analysis here implies the determination of the allowable unit load $P/A$ on a given column, or of its allowable total load $P$, as illustrated in the following problems.

**ILLUSTRATIVE PROBLEM** _____

**12-16.** Calculate the allowable load $P$ on an A36 structural-steel column, W 8 × 31, when its unsupported length is (*a*) 8 ft, (*b*) 16 ft, and (*c*) 24 ft. Use AISC formulas or Table E-8. (Since $K = 1$, it may be omitted.)

*Solution:* To obtain the allowable load $P$, we first determine the allowable unit load $P/A$, which in turn depends upon the value of $L/k$. From Table E-4, $A = 9.13$ in.², and the *least* $k_y = 2.02$ in. Then

$$\frac{L}{k} = \frac{96}{2.02} = 47.5 \tag{a}$$

Since $L/k < 126.1$, use Eq. (12-10) or see Table E-8. From Table E-8.

$$\text{Allowable } \frac{P}{A} = 18{,}570 \text{ psi}$$

and

$$P = (18{,}570)(9.13) = 169{,}500 \text{ lb} \qquad \textit{Ans.}$$

$$\frac{L}{k} = \frac{192}{2.02} = 95.0 \tag{b}$$

Since $L/k < 126.1$, use Eq. (12-10) or see Table E-8. From Table E-8.

$$\text{Allowable } \frac{P}{A} = 13,600 \text{ psi}$$

and
$$P = (13,600)(9.13) = 124,000 \text{ lb} \qquad\qquad Ans.$$

$$\frac{L}{k} = \frac{288}{2.02} = 142.6 \qquad\qquad (c)$$

Since $L/k > 126.1$, use Eq. (12-11) or see Table E-8.

$$\left[\frac{P}{A} = \frac{12\pi^2 E}{23(KL/k)^2}\right] \qquad \frac{P}{A} = \frac{12\pi^2(29,000,000)}{23(142.6^2)} = 7344 \text{ psi}$$

and
$$P = (7344)(9.13) = 67,000 \text{ lb}. \qquad\qquad Ans.$$

Interpolating in Table E-8 will give $P/A = 7344$ psi. By studying the answers obtained, we see clearly that the allowable load $P$ on this column decreases as its slenderness ratio increases.

## PROBLEMS

NOTE: The allowable unit load $P/A$ may be obtained from Table E-8. All members are main members unless specifically designated as secondary or braces. Steel is A36. $K = 1$.

**12-17.** Determine the allowable unit load $P/A$ and the allowable load $P$ for a structural-steel column, W 10 × 49, when its unsupported length is (a) 14 ft, (b) 21 ft, and (c) 28 ft.

**12-18.** Determine the allowable unit load $P/A$ and allowable load $P$ for a 3-in. standard steel-pipe column when its length is (a) 8 ft, (b) 10 ft, and (c) 12 ft.

**12-19.** Determine the allowable unit load $P/A$ and allowable load $P$ for a 4-in. standard steel-pipe column when its length is (a) 8 ft, (b) 12 ft, and (c) 18 ft.

*Ans.* (a) $P/A = 17,100$ psi, $P = 54,300$ lb; (b) $P/A = 13,430$ psi, $P = 42,650$ lb; (c) $P/A = 7300$ psi, $P = 23,200$ lb

**12-20.** A double-angle truss compression member 8 ft long is made up of two angles 4 by 3 by ¼ in. riveted together long legs back to back with a ⅜-in. gusset plate between. Determine the allowable load $P$ for this member.

**12-21.** A brace in a structural-steel roof system is 14 ft long and is made up of two angles 3½ by 2½ by ¼ in. riveted together long legs back to back with a ⅜-in. gusset plate between. Determine the allowable unit load $P/A$ and the allowable load $P$ for this member.

*Partial Ans.* $P = 21.9$ kips

**12-22.** The top-chord member in a flat structural-steel roof truss is made up of two angles 4 by 3 by ⅜ in. short legs riveted back to back with a ⅜-in. gusset plate between. The 3-in. legs lie in the vertical plane in which the chord is braced by web members at 6-ft intervals. Diagonal steel rods brace the chord at 12-ft intervals in a horizontal plane. Determine the allowable load $P$ on this member.

*Ans.* $P = 75,100$ lb

**12-23.** A structural-steel column, W 10 × 33, has a laterally unsupported length of 24 ft in its strongest direction but is laterally supported at midlength in its weakest direction. Determine the allowable unit load $P/A$ and the allowable load $P$ for this column.

**12-24.** Two C 250 × 37 channels are welded together to form a 150- by 250-mm box column, with an unsupported length of 7 m. Calculate the allowable load $P$ on this column.

*Ans.* $P = 733$ kN

**12-25.** A box column is built up of two 4- by 4- by ¼-in. angles to form a square section. If its unsupported length is 18 ft, what is the allowable load $P$ for this column?

**12-26.** A construction-derrick boom is built up of four 3- by 3- by ⅜-in. steel angles latticed together to form a section such as shown in Figs. 12-1*i* and 12-2. The boom will be 60 ft long and will measure 24 in. from back to back of corner angles across each of the four sides. Calculate the allowable axial load $P$.

## 12-14 DESIGN OF STRUCTURAL-STEEL COLUMNS

As previously mentioned, the design of columns is not a straightforward process but must be accomplished by successive trials. If we assume that $L/k$ lies between 30 and 120, we find from Table E-8 that the corresponding allowable unit loads (stresses) range from 19,940 psi for $L/k = 30$ down to 10,280 psi for $L/k = 120$. The midpoint would be $L/k = 75$, for which value the allowable stress is 15,900 psi. These stresses were obtained by solving formula (12-10). If conditions of a design problem do not suggest a more appropriate allowable stress for the first trial, 16,000 psi, say, is probably as good as any to use. For braces and secondary members with $L/k$ between 120 and 200, an estimated allowable $P/A$ of 8000 psi is recommended. A five-step procedure is outlined below.

## Design Procedure. Structural-Steel Columns

1. Using an estimated allowable stress $s'$ (from 10,000 to 20,000 psi), calculate the probable required area $A' = P/s'$.
2. Select trial section having approximately the probable required area and a high *least* radius of gyration $k$ ($r$ in some handbooks). (If a *"most economical"* section is required, it will invariably be found to be one of the *lightest* three of the "square" groups, that is, 8 by 8, 10 by 10, 12 by 12, 14 by 14.)
3. Compute $L/k$ and determine actual allowable stress $s$ from formula (12-10) or (12-11) or use Table E-8.

4. Calculate allowable load $P' = sA$ on trial section and percent to which it is stressed:

$$\text{Percent stressed} = \frac{\text{actual load}}{\text{allowable load}} \, 100 = \frac{P}{P'} \, 100$$

5. Repeat steps 2, 3, and 4 if section is stressed less than 95 percent or more than 102 percent (if stressed less than 95 percent, a more economical section can usually be found; a maximum overstress of 2 percent is usually permitted in practice).

The fact that a column is 100 percent stressed is not in itself sufficient evidence that it is the *lightest* available. For example, under a load of 200 kips, a W $14 \times 68$ is 100 percent stressed when $L$ is 20 ft, while a W $12 \times 64$ is only 80 percent stressed.

## ILLUSTRATIVE PROBLEMS

**12-27.** A six-story building has a structural-steel beam-and-column frame appropriately fireproofed. The columns are spaced 20 ft on centers in one direction and 25 ft on centers in the perpendicular direction. Hence an interior column supports a floor area of 500 ft². The governing building code specifies that the frame must be designed to withstand the actual dead weight of the structure plus a possible roof snow load of 40 psf and a live load on each floor of 125 psf. The dead weight of the roof is estimated to be 80 psf and of each floor 100 psf. The unsupported length of the ground-floor column is 20 ft and of the others 16 ft. Design the third-floor column, using the most economical W section.

*Solution:* The third-floor column supports the fourth, fifth, and sixth floors and the roof. The load on this column is therefore

$$\text{Roof dead load } (500)(80) = 40{,}000 \text{ lb}$$
$$\text{Snow load} \qquad (500)(40) = 20{,}000$$

Sixth-, fifth-, and fourth-floor loads:

$$\text{Dead load } (500)(100)(3) = 150{,}000$$
$$\text{Live load } (500)(125)(3) = \underline{187{,}500}$$
$$\text{Design load, third-floor column} = 397{,}500 \text{ lb}$$

Proceeding through the five recommended steps, we have

1. Let $s' = 16{,}000$ psi; then required $A = 397{,}500/16{,}000 = 24.85$ in.². Consulting Table E-4 we find that a W section 14 by 14½ in. by 90 lb/ft comes nearest to having the required area. Then

2. *First Trial:*

$$\text{W } 14 \times 14\frac{1}{2} \times 90 \qquad A = 26.5 \text{ in.}^2 \qquad \text{least } k_y = 3.70$$

3. $\dfrac{L}{k} = \dfrac{192}{3.70} = 51.9 \qquad$ From Table E-8, allowable $s = 18{,}180$ psi.

4. Allowable load $P' = sA = (18,180)(26.5) = 482,000$ lb.

$$\text{Percent stressed} = \frac{P}{P'}\,100 = \frac{397,500}{482,000}\,100 = 82.5 \text{ percent} \qquad \text{NG}$$

The capacity of this column is more than 17.5 percent in excess of that required. A more economical section can undoubtedly be found.

We will try a lighter section, W 12 × 79, which has about 9 percent less area. Hence

2. *Second Trial:*

$$\text{W } 12 \times 12 \times 79 \qquad A = 23.2 \text{ in.}^2 \qquad \text{least } k_y = 3.05$$

3. $\dfrac{L}{k} = \dfrac{192}{3.05} = 62.9$ \qquad From Table E-8, allowable $s = 17,150$ psi.

4. Allowable load $P' = sA = (17,150)(23.2) = 398,000$ lb.

$$\text{Percent stressed} = \frac{397,500}{398,000}\,100 = 99.9 \text{ percent} \qquad \text{OK}$$

$$\text{Use W } 12 \times 79. \qquad\qquad\qquad\qquad\qquad Ans.$$

Since this W 12 × 79 is stressed to its capacity, a lighter 12 by 12 wide-flange section will not carry the load. (A W 12 × 72 can carry only 362,000 lb.)

**12-28.** The top-chord member of a flat roof truss is supported vertically at 8-ft intervals by web members and at 16-ft intervals horizontally by a system of bracing rods. Select the most economical double-angle section to withstand a compressive load of 83,000 lb. Gusset plates are ⅜ in. thick. Use AISC specifications.

*Solution:* We will proceed with an estimated allowable unit stress of 16,000 psi, since the problem data give little inkling about what might be a better estimate.

Owing to the method of bracing this member, the effective column length in the vertical plane is 8 ft and in the horizontal plane is 16 ft. Clearly, therefore, the resistance of the member to buckling in the horizontal plane must be about twice its resistance in the vertical plane. This ratio can be obtained by selecting two angles with *short* legs back to back. Hence

1. Let $s' = 16,000$ psi; then required $A' = P/s' = 83,000/16,000 = 5.19$ in.$^2$. From Table E-7 we find that the area of two 4- by 3½- by ⅜-in. angles is 5.34 in.$^2$.

2. *First Trial:* \qquad Two 4- by 3½- by ⅜-in. angles \qquad $A = 5.34$ in.$^2$

3. Vertical plane,

$$\frac{L}{k_X} = \frac{96}{1.06} = 90.5 \qquad \text{and} \qquad s = 14,150 \text{ psi} \qquad \text{(Table E-8)}$$

Horizontal plane,

$$\frac{L}{k_Y} = \frac{192}{1.88} = 102.1 \qquad \text{and} \qquad s = 12,700 \text{ psi} \qquad \text{(governs)}$$

4. Allowable load $P' = sA = (12,700)(5.34) = 67,900$ lb \qquad NG

This allowable load is 18 percent short of the actual load, so another choice must be made. The most economical section is usually one of the two *lightest* sections of a group. Hence we should look to the 5- by 3-in. and the 5- by 3½-in. groups. But we discover immediately that all the 5 by 3s have $k_Y$ values lower than that used in the first trial, which means that all have lower allowable stresses. The 5 by 3½ by ⅜ has about the same $k_X$ value and about 14 percent higher area. Hence

2. *Second Trial:*     Two 5- by 3½- by ⅜-in. angles     $A = 6.10$ in.²
3. Vertical plane,

$$\frac{L}{k_X} = \frac{96}{1.02} = 94 \quad \text{and} \quad s = 13{,}720 \text{ psi} \quad \text{(governs)}$$

Horizontal plane,

$$\frac{L}{k_Y} = \frac{192}{2.40} = 80 \quad \text{and} \quad s = 15{,}360 \text{ psi}$$

4. Allowable load $P' = sA = (13{,}720)(6.10) = 83{,}700$ lb

$$\text{Percent stressed} = \frac{P}{P'} = \frac{83{,}000}{83{,}700} \, 100 = 99.2 \text{ percent} \quad \text{OK}$$

Use two 5- by 3½- by ⅜-in. angles.     *Ans.*

## PROBLEMS

NOTE: Unless otherwise stated, the AISC formulas will be used. Allowable stresses may be obtained from Table E-14. All members are main members unless specifically designated as secondary or braces. $K = 1$.

**12-29.** Select the most economical structural-steel W column 16 ft long to support an axial compressive load of 225,000 lb.

*Ans.* W 10 × 49

**12-30.** A structural-steel column 24 ft long must support an axial compressive load of 360,000 lb. Select the most economical W section to use.

**12-31.** A steel column with an unsupported length of 24 ft must support an axial load of 66,000 lb. Select the most economical W section to use.

**12-32.** Select the most economical structural-steel W column 20 ft long to support an axial load of 36,000 lb.

*Ans.* W 8 × 24

**12-33.** Design the fourth-floor column in Prob. 12-27.

**12-34.** Design the most economical double-angle main truss member 12 ft long to resist a compressive force of 66,000 lb. The long legs are to be riveted together back to back with a ⅜-in. gusset plate between.

**12-35.** A main truss member 10 ft long is made up of two angles riveted together back to back with a ⅜-in. gusset plate between. Design the most economical section to withstand a compressive load of 60,000 lb.

**12-36.** A riveted double-angle truss chord member is to be supported vertically at 6-ft intervals by web members and horizontally at 18-ft intervals by a system of braces. Gusset plates are ½ in. thick. Select the most economical angles to use to withstand a load of 120 kips.

**12-37.** A riveted double-angle main truss member 9 ft long must resist a compressive load of 50,000 lb. What size angles will give the most economical section? Gusset plates are ⅜ in. thick.

**12-38.** Design a riveted double-angle brace (secondry member) 14 ft long capable of withstanding a maximum compressive load of 25,000 lb; ⅜-in. gusset plates are used.

**12-39.** A standard steel-pipe column 14 ft long must support an axial compressive load of 80,000 lb. Select the lightest available suitable pipe section.

**12-40.** Select a steel-pipe column 4 m long to support an axial load of 200 kN.

**12-41.** A standard steel-pipe column 17 ft long must support an axial compressive load of 25,000 lb. Select a suitable size.

**12-42.** A piece of standard steel pipe 12 ft long is used as a brace and must be capable of resisting a maximum compressive load of 10 kips. What minimum diameter of pipe can be used?

**12-43.** A latticed column of the type shown in Figs. 12-1h and 12-2 is to be 30 ft long and must support an axial compressive load of 260,000 lb. (a) Select the most economical American Standard channels to use. (b) Calculate the required distance b from back to back of these channels to give equal moments of inertia with respect to the two major axes.

$$Ans. \quad (a) \text{ C } 12 \times 30; \ (b) \ b = 9.78 \text{ in.}$$

**12-44.** Solve Prob. 12-43 for a load of 90 kips and a length of 40 ft.

**12-45.** A square structural tube (see Fig. 12-1l) with a wall thickness of 6 mm is to be used as a column. What minimum outside dimension is required to support a 100-kN axial load safely if the column is 4 m long?

**12-46.** A construction-derrick boom is to be built up of four equal-leg angles latticed together to form a section such as is illustrated in Figs. 12-1i and 12-2. The boom will be 80 ft long and will measure 20 in. from back to back of corner angles along each of the four sides. It must be capable of withstanding an axial load of 115,600 lb. Select the most economical angles to use. (NOTE: For preliminary design purposes, assume that the radius of gyration $k$ of the boom equals 45 percent of its side dimension of 20 in. Then calculate $L/k$ of boom, solve for required area $A$, select trial angles, and determine actual $k$, $L/k$, $P/A$, and allowable load $P'$.)

**12-47.** A compression member in a railway bridge is 24 ft long and must withstand an axial load of 240 kips. Using the AREA formula for riveted ends, select the most economical W section to use.

## 12-15 ANALYSIS AND DESIGN OF STRUCTURAL-ALUMINUM COLUMNS

Aluminum alloys are widely used in aircraft construction and in stationary structures where lightweight materials are required. Many alloys and tempers are available. Yield strengths vary from 11,000 to 53,000 psi. Be-

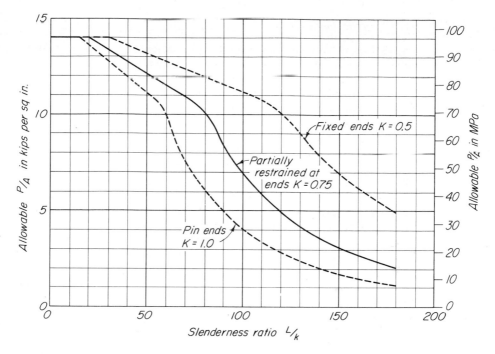

**Fig. 12-6** Allowable unit loads *P/A* on axially loaded columns of 6061-T6 aluminum.

cause the strength and other properties vary so greatly, we find many formulas for the design of aluminum columns. Some of these formulas give the *ultimate* unit load, so an appropriate factor of safety, usually from 2 to 4 in stationary structures, must be used in calculating the allowable load. Other formulas or graphs, such as those suggested by the American Society of Civil Engineers, give allowable average compressive stresses. The **ASCE** graph[1] for allowable compressive stresses in axially loaded columns of aluminum alloy 6061-T6 is shown in Fig. 12-6.

The **ASCE** specification suggests that the curve for partial end restraint (*K* = 0.75) should ordinarily be used. The curves for pin-ended and fixed-ended columns are to be used as a guide in the selection of the allowable compressive stress for those cases in which the degree of end restraint is known to be different from that represented by partial restraint. The factor *K* corresponds to the *effective-length* coefficients shown in Fig. 12-4 and given in Table 12-1.

NOTE: The cross sections of structural aluminum angles have properties similar to those of structural-steel angles. Properties of angles mentioned in the following problems may therefore be obtained from Tables E-6 and E-7. Aluminum pipe may be obtained with the same dimensions as steel pipe as given in Table E-9.

[1] Paper 970, *Journal of the Structural Division of the American Society of Civil Engineers*, vol. 82, no. ST3.

## ILLUSTRATIVE PROBLEM

**12-48.** Select an aluminum (6061-T6) pipe column to carry a 21,000-lb axial load. The ends of the 12-ft-long member are partially restrained against rotation.

*Solution:* An estimated allowable stress of 7000 psi would require a cross-sectional area of 3 in.². From Table E-9 choose for the first trial a 4-in. pipe with ANSI Schedule 40 dimensions. Note that $A = 3.174$ in.² and $k = 1.51$ in. Therefore, $L/k = 144/1.51$, or 95.4. From Fig. 12-6, with $L/k = 95.4$ and using the curve for partially restrained ends, the allowable $P/A = 7,600$ psi. Hence,

$$P = (7600)(3.174) = 24,100 \text{ lb} \qquad \text{OK}$$

Use 4-in. Schedule 40 aluminum pipe.                                     *Ans.*

## PROBLEMS

**12-49.** A round, hollow tube of aluminum alloy, 6061-T6, with outside diameter of 1.5 in. and inside diameter of 1.25 in., is subjected to an axial compressive load $P$. The section has the following properties: $A = 0.54$ in.²; $I = 0.129$ in.⁴; $k = 0.488$ in. If the length of this tube is 30 in. and its ends are assumed to be hinged, what maximum load $P$ may it safely support?

**12-50.** A rectangular hollow tube (see Fig. 12-1k) is formed from 6061-T6 aluminum. The outside dimensions are 150 × 100 mm, and the wall thickness is 5 mm. Compute the allowable axial load if the column is 4 m long and the ends are partially restrained against rotation.

*Ans.* $P = 115$ kN

**12-51.** Solve Prob. 12-50 if outside dimensions are 6 × 4 in., wall thickness is ¼ in., and $L = 10$ ft.

**12-52.** Four 3- by 3- by ¼-in. structural-aluminum angles are riveted together back to back to form a symmetrical cross. Calculate the safe allowable axial load $P$ for this column if its unsupported length is 100 in. and the end-restraint factor $K$ is 0.75.

**12-53.** Select the most economical double-angle main truss member 10 ft long to resist a compressive force of 25,000 lb. The angles are attached to a ⅜-in. gusset plate, they have an end-restraint factor $K = 0.75$, and they are of 6061-T6 aluminum.

**12-54.** A round aluminum-alloy tube, 6061-T6, 2 m long, must withstand a calculated axial compressive force of 10 kN. Its ends are assumed to be hinged. If its inside diameter $d$ is to be 0.8 of its outside diameter $D$ ($d = 0.8D$), what should be the value of $D$?

## 12-16 ANALYSIS AND DESIGN OF AXIALLY LOADED TIMBER COLUMNS

The most widely used formulas for the design of timber columns are those of the National Forest Products Association. This article will follow their recommendations. Solid wood columns can be classified into three length categories, each characterized by the mode of failure under ultimate load.

**Short columns** have a length-to-least-cross-sectional-dimension $(L/d)$ ratio of 11.0 or less. Such columns will fail by crushing if tested at sufficient load. The allowable unit stress to be used in the design of short columns is

$$s_C = \left[\begin{array}{l}\text{Allowable compressive stress parallel}\\ \text{to the grain (see Table 8-1)}\end{array}\right] \qquad \text{(12-15)}$$

**Long columns** will fail by elastic instability, that is, by buckling. A long column is defined as one whose $L/d$ ratio is equal to or greater than $K$, calculated from $K = 0.671\sqrt{E/s}$. The modulus of elasticity $E$ and the allowable compressive stress parallel to the grain $s$ may be taken from Table 8-1. In no case should the $L/d$ ratio exceed 50. The average compressive stress in a long column $L/d$ greater than $K$ (but less than 50) should not exceed

$$s_C = 0.3 \frac{E}{(L/d)^2} \qquad \text{(12-16)}$$

Equation (12-16) is Euler's equation for column instability with a factor of safety of 2.74. Euler's equation, Eq. (12-2), is

$$\frac{P}{A} = \frac{\pi^2 E}{(L/k)^2}$$

Since $k = \sqrt{I/A}$, for a rectangular column section $k = \sqrt{bd^3/[12(bd)]} = d/\sqrt{12}$. Substitution into Euler's equation and adding a factor of safety of 2.74 yields

$$s_C = \frac{1}{2.74} \frac{\pi^2 E}{(L\sqrt{12}/d)^2} = 0.3 \frac{E}{(L/d)^2}$$

as given by Eq. (12-16)

Columns with $L/d$ ratios greater than 11.0 but less than $K$ are said to be of **intermediate length** and the average compressive stress for such columns should not exceed

$$s_C = \left[1 - \frac{1}{3}\left(\frac{L/d}{K}\right)^4\right]\left[\begin{array}{l}\text{Allowable compressive stress parallel}\\ \text{to the grain (see Table 8-1)}\end{array}\right] \qquad \text{(12-17)}$$

**Square timber columns** are generally used under normal design conditions. When axially loaded, the side dimensions *may* differ by 2 in. or, occasionally, 4 in. When eccentrically loaded, the side dimensions may differ by 2 in., 4 in., or even more. **Round-** and other-shaped columns may be analyzed by substituting $k\sqrt{12}$ for $d$ in each of the design formulas. Here $k$ is the least radius of gyration of the column section.

The formulas presented in this article are for normal conditions of loading for columns in reasonably dry locations. For columns having loads of unusually long or short duration, or ones used under wet conditions, refer to the National Design Specifications for Wood Construction as published by the National Forest Products Association.

## ILLUSTRATIVE PROBLEMS

**12-55.** Design a timber column 15 ft long of select-structural Douglas fir S4S to carry an axial load of 20,000 lb.

*Solution:* From Table 8-1, the allowable compressive stress parallel to the grain is 1100 psi and the modulus of elasticity for this species is 1,600,000 psi. Therefore,

$$\left[ K = 0.671 \sqrt{\frac{E}{s}} \right] \qquad K = 0.671 \sqrt{\frac{1,600,000}{1100}} = 25.59$$

*First trial:* A 10-ton axial load on a 15-ft-long column should suggest a 6-in. or 8-in. column for the first trial. Try $d = 7\frac{1}{2}$ in. with $L/d = [15(12)]/7.5 = 24$. Since the $L/d$ is more than 11 but less than 25.59, the trial column is intermediate in length and Eq. (12-17) applies.

$$\left[ s_C = \left\{ 1 - \frac{1}{3} \left( \frac{L/d}{K} \right)^4 \right\} s \right] \qquad s_C = \left[ 1 - \frac{1}{3} \left( \frac{24}{25.59} \right)^4 \right] 1100 = 816 \text{ psi}$$

The required column area is

$$\left[ A = \frac{P}{s_C} \right] \qquad A = \frac{20,000}{816} = 24.5 \text{ in.}^2$$

An 8- by 8-in. (S4S) column has an area of 56.25 in.² and is therefore adequate. Perhaps a 6- by 8-in. or a 6- by 6-in. column would do. In this case a second trial with $d = 5.5$ in. will be necessary.

*Second trial:* Assume $d = 5.5$ in. and therefore $L/d = [15(12)]/5.5 = 32.73$. A 5.5-in.-wide column will be a long column since $L/d$ exceeds $K = 25.59$. Equation (12-16) applies to long columns.

$$\left[ s_C = 0.3 \frac{E}{(L/d)^2} \right] \qquad s_C = 0.3 \frac{1,600,000}{(32.72^2)} = 448 \text{ psi}$$

The required column area is then

$$\left[ A = \frac{P}{s_C} \right] \qquad A = \frac{20,000}{448} = 44.64 \text{ in.}^2$$

A 6- by 8-in. column has an area of 41.25 in.² and is therefore inadequate. A 6- by 10-in. column has an area of 52.25 in.² and can therefore be used. A 6- by 10-in. column would contain slightly less board measures of lumber than an 8- by 8-in. column.

Use 6- by 10-in. S4S *Ans.*

**12-56.** What concentric axial load may be supported by a No. 1 grade 150- by 150-mm Ponderosa pine column 3 m long?

*Solution:* From Table 8-1, $E = 7600$ MPa and $s = 5.2$ MPa. We calculate $L/d = 3000/150 = 20$ and $K = 0.671\sqrt{7200/5.2} = 25.65$. Since $L/d = 20$ is less than $K$ but more than 11, the column is of intermediate length. Equation (12-17) is applicable. The allowable average compressive stress is

$$\left[ s_C = \left\{ 1 - \frac{1}{3}\left(\frac{L/d}{K}\right)^4 \right\} (s) \right] \qquad s_C = \left[ 1 - \frac{1}{3}\left(\frac{20}{25.65}\right)^4 \right] (5.2) = 4.56 \text{ MPa}$$

The allowable axial load is then

$$[P = A(s_C)] \qquad\qquad P = 150(150)(4.56) = 102\ 600 \text{ N} = 102.6 \text{ kN} \qquad\qquad Ans.$$

**12-57.** Design a rough-size timber column 8 ft long of eastern hemlock No. 1 grade to support an axial load of 25 kips.

*Solution:* From Table 8-1, $E = 1,200,000$ psi and $s = 800$ psi. Try a column with $d = 6$ in. and $L/d = [8(12)]/6 = 16$. Then $K = 0.671\sqrt{E/s} = 0.671\sqrt{1,200,000/800} = 26.0$. Since $L/d$ of 16 is more than 11 but less than $K$, the column is intermediate in length and the allowable compressive stress is

$$\left[ s_C = \left\{ 1 - \frac{1}{3}\left(\frac{L/d}{K}\right)^4 \right\} s \right] \qquad s_C = \left[ 1 - \frac{1}{3}\left(\frac{16}{26}\right)^4 \right] 800 = 762 \text{ psi}$$

The required area is then

$$\left[ A = \frac{P}{s_C} \right] \qquad\qquad A = \frac{25,000}{762} = 32.8 \text{ in.}^2$$

A 6- by 6-in. column has an area of 36 in.² and is therefore adequate.

Use a 6- by 6-in. rough No. 1 eastern hemlock column        Ans.

## PROBLEMS

NOTE: Obtain values of $c$ and $E$ from Table 8-1. Specify only the most economical commercially available timber sizes.

**12-58.** Calculate the allowable average stress and the allowable load $P$ on each of three 6- by 6-in. S4S timber columns of Douglas fir select structural, respectively 5, 10, and 15 ft long.

Ans. 1100 psi, 33,300 lb; 907 psi, 27,400 lb; 448 psi, 13,550 lb

**12-59.** Three 200- by 200-mm timber columns of lodgepole pine No. 1 rough are, respectively, 1.5, 4.0, and 6.0 m long. Calculate the allowable unit load $P/A$ and the allowable load $P$ for each column.

**12-60.** (a) Design a timber column 16 ft long of Douglas fir No. 1 S4S to support an axial load of 12,000 lb. (b) What size should be used if load is increased to 17,000 lb?

**12-61.** A timber column 16 ft long of Ponderosa pine No. 1 rough must support an axial load of 70,000 lb. Select the most economical available size.

**12-62.** Design a round timber column 5 m long of dense No. 1 Douglas fir to support an axial load of 400 kN.

**12-63.** Determine the required size of timber column of Douglas fir select structural rough 17 ft long to support an axial load of 100 kips.

**12-64.** (a) Design a timber column 12 ft long of Douglas fir No. 1 S4S to support an axial load of 48,000 lb. (b) What size should be used for a load of 60,000 lb?

**12-65.** Determine the required size of a timber column 16 ft long of Ponderosa pine No. 1 S4S to support an axial load of 100 kips.

**12-66.** (*a*) Design a timber column 5 m long of select-structural eastern hemlock rough to support an axial load of 145 kN. (*b*) What size would be required to support 180 kN?

**12-67.** Design a round timber column 10 ft long of Douglas fir No. 1 to support an axial load of 35,500 lb.

## 12-17 ECCENTRICALLY LOADED COLUMNS

When the action line of the load on a column does not coincide with its longitudinal centroidal axis, the load is said to be **eccentric** and the column is **eccentrically loaded.** A complete treatment of the analysis (investigation) and design of eccentrically loaded columns would be very extensive and is beyond the intended scope of this text. Only the most commonly occurring cases can be discussed here, such as some eccentrically loaded structural-steel and timber columns that are usually found in structural frames of buildings.

The effects of eccentricity of load on the stresses in *short* members were discussed and illustrated in Chap. 11. The corresponding effects on the stresses in longer members, such as columns, are similar. However, a "long" column will necessarily *bend* slightly when subjected to an eccentric load, causing its longitudinal axis to deviate an amount $\Delta$ from a straight line, as illustrated in Fig. 11-12. This deviation $\Delta$ (a lateral deflection) will cause an additional moment $M = P\Delta$ and consequent internal stresses in addition to those obtained by Eq. (11-1). In comparatively long columns, these additional stresses due to $P\Delta$ may reach a sizable fraction of the total stress.

When it is necessary to consider the additional moment $P\Delta$, the maximum deflection $\Delta$ caused by the eccentric moment $M = P\Delta$ may be calculated, using the deflection equation in case 5, Table 9-3. The total combined stress in the column then becomes

$$s = \frac{P}{A} + \frac{Pe}{Z} + \frac{P\Delta}{Z} \tag{12-18}$$

Eccentricity of loading should be avoided whenever possible because it leads to uneconomical design. It does, however, occur in several ways in normally designed structures.

Consider the stresses produced at a column cross section when an eccentric load is applied at the point of lateral support. As shown in Fig. 12-7*a*, a steel beam is fastened to the inside flange of an exterior steel column. We would assume the resultant load $P$ to be concentrated at the face of the inner column flange. The eccentric distance is then one-half the depth of the column section, and the eccentric moment is $Pe$. In Fig. 12-7*b*, a crane rail is supported by a bracket welded onto the flange of a

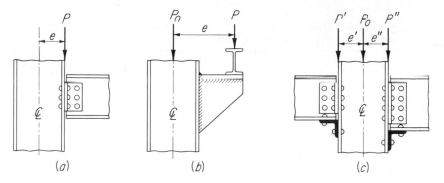

**Fig. 12-7   Eccentric loads on steel columns.**

steel column. The axial load $P_O$ may come from a roof load. The total direct load is then $P_O + P$, and the eccentric moment is $Pe$. In Fig. 12-7c, the total direct load is $P_O + P' + P''$, and the eccentric moment is $P''e'' - P'e'$ (assuming that $P''$ is greater than $P'$). The loads $P'$ and $P''$ may be assumed to act at the outer face of the column flange.

For the three cases in Fig. 12-7, the combined-stress formula becomes

Case *a*:
$$s = \frac{P}{A} \quad \frac{Pe}{Z} \tag{12-19}$$

Case *b*:
$$s = \frac{P_O + P}{A} + \frac{Pe}{Z} \tag{12-20}$$

Case *c*:
$$s = \frac{P_O + P' + P''}{A} + \frac{P''e'' - P'e'}{Z} \tag{12-21}$$

In all cases, $Z$ is the section modulus in the direction of bending.

When a beam-to-column connection is sufficiently rigid to transmit fully a moment from the beam to the column, the designer in practice will usually assume that the beam loads are carried in to the axis of the column and that no eccentricity of load exists. Completely welded connections would be in this category, as well as some types of riveted connections.

## 12-18  DESIGN OF ECCENTRICALLY LOADED STEEL COLUMNS

The possible lateral deflection $\Delta$ of structural-steel columns in building frames is usually small and is often disregarded. The total actual combined unit stress is then considered to be

$$s = \frac{\Sigma P}{A} + \frac{M}{Z} \tag{12-22}$$

If the actual axial stress $\Sigma P/A$ is designated by $f_a$ and the actual bending stress $M/Z$ by $f_b$, Eq. (12-22) becomes

$$s = f_a + f_b \qquad\qquad (a)$$

If we now divide each term by $s$, we obtain

$$1 = \frac{f_a}{s} + \frac{f_b}{s} \qquad\qquad \textbf{(12-23)}$$

For the design of structural-steel columns, the American Institute of Steel Construction (AISC) offers the following similar, but slightly changed, version of Eq. (12-23), which is now commonly used:

$$\frac{f_a}{F_a} + \frac{f_b}{F_b} \leqq 1 \qquad\qquad \textbf{(12-24)}$$

in which $f_a$ = actual axial stress

$F_a$ = axial stress allowable if only axial column stress exists

$f_b$ = actual bending stress in column

$F_b$ = bending stress allowable if only bending stress exists

Equation (12-24) states, in effect, that the ratio of the actual direct axial stress to the allowable axial stress plus the ratio of the actual bending stress to the allowable bending stress must not exceed unity.

Equation (12-24) may be referred to as a **stress factor** SF. A stress factor of 1.000 would indicate that the member is 100 percent stressed; 0.952 would indicate 95.2 percent stressed.

The quantities $f_a$, $F_a$, $f_b$, and $F_b$ in Eq. (12-24) obviously cannot be evaluated until a specific trial section has been selected. Clearly, therefore, the design of eccentrically loaded steel columns must be accomplished by the trial-and-error method.

To provide for the additional stress resulting from the lateral deflection $\Delta$ between points where a column is laterally supported, represented by the term $P\Delta/Z$ in Eq. (12-18), the latest AISC specification magnifies the computed bending stress $f_b$ given in Eq. (12-24) by an amplification factor.

The detailed design of an eccentrically loaded steel column is beyond the intended scope of this book. The interested reader is referred to the latest edition of the *Manual of Steel Construction* of the AISC for a worked-out design of a building column subjected to both an axial load and a bending moment.

## 12-19 DESIGN OF ECCENTRICALLY LOADED TIMBER COLUMNS

Timber columns subjected to eccentric loads are usually designed by a trial-and-error proceedure. Since relatively few timber sections are commercially available (6 by 6 in., 8 by 8 in., 8 by 10 in., etc), the selection of an appropriate section can usually be accomplished in one or two trials. The formulas presented here are those of the Forest Products Laboratory, U.S. Department of Agriculture. The formulas are intended for the design of square or rectangular eccentrically loaded timber columns whose ends can be assumed to be pin-connected. The eccentricity of the load is assumed to be relatively small but large lateral loads may be present on the column. The general formula for the stress factor is

$$\frac{P/A}{s_c} + \frac{M/Z + P/A(6 + 1.5J)(e/d)}{s_B - J(P/A)} \leq 1.0 \qquad (12\text{-}25)$$

where $s_C$ = allowable compressive stress determined from Eq. (12-15), (12-16), or (12-17)

$s_B$ = allowable bending stress from Table 8-1

$J$ = a unitless convenience factor defined by $J = [(L/d) - 11.0]/[K - 11.0]$ (however, $J$ cannot be less than zero or more than unity)

$Z$ = section modulus about bending axis

$M$ = maximum bending moment due to the lateral load; $M$ will be zero if no lateral load is applied

$e$ = eccentricity of load

$d$ = side dimension of rectangular column in direction of eccentricity

A simple column with an eccentrically applied load is shown in Fig. 12-8a. An eccentrically loaded column with an additional lateral load is shown in Fig. 12-8b. If the eccentric load is applied on a bracket at some height $L'$ less than the full height $L$, the column is analyzed for the loaded condition shown at the right in Fig. 12-8c. The equivalent lateral load $P'$ is

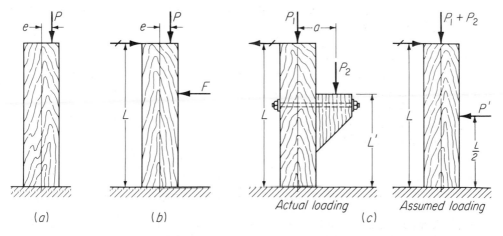

(a)  (b)  Actual loading  Assumed loading  (c)

**Fig. 12-8**  Eccentric loads on timber columns. (a) Eccentric load. (b) Eccentric load and lateral load. (c) Column with bracket and its equivalent.

calculated from

$$P' = \frac{3aL'P_2}{L^2} \tag{12-26}$$

A real lateral load $F$ or the equivalent lateral load $P'$ is used to calculate the bending moment $M$ in Eq. (12-25). Note that for a short column loaded as shown in Fig. 12-8a, Eq. (12-25) becomes (with $M = 0$ and $J = 0$)

$$\frac{P/A}{s_C} + \frac{6Pe/Ad}{s_B} = \frac{P/A}{s_C} + \frac{Pe/Z}{s_B} \leq 1.0$$

## ILLUSTRATIVE PROBLEMS

**12-68.** Design a rough timber column 8 ft long of select structural eastern hemlock to support an eccentric load $P = 15,000$ lb in a manner similar to that shown in Fig. 12-8a. Assume the eccentricity $e = 2$ in.

*Solution:* From Table 8-1, the allowable compressive stress parallel to the grain is 950 psi, the allowable bending stress is 1350 psi, and the modulus of elasticity $E$ is 1,200,000 psi. In order to obtain an approximate column size, assume an average stress of half the short column allowable stress, or 475 psi. The required column area would then be $A = P/s = 15,000/475 = 31.6$ in.$^2$. A 6-in.$^2$ column has approximately this area.

*First trial:* Assume a 6- by 6-in. column, $A = 36$ in.$^2$, $Z = bd^2/6 = 36$ in.$^3$, $K = 0.671\sqrt{E/s} = 0.671\sqrt{1,200,000/950} = 23.85$, $L/d = [(8)(12)]/6 = 16$, and $J = (16 - 11)/(23.85 - 11) = 0.389$. Since $L/d$ of 16 is greater than 11 but less than 25.85, the column is intermediate in length. The allowable compressive stress from Eq. (12-17) is

$$\left[ s_c = \left\{ 1 - \frac{1}{3}\left(\frac{L/d}{K}\right)^4 \right\} s \right] \qquad s_C = \left[ 1 - \frac{1}{3}\left(\frac{16}{23.85}\right)^4 \right] (950) = 886 \text{ psi}$$

Next we evaluate the stress factor using Eq. (12-25). Note that since there is no lateral load, $M = 0$. Also, $P/A = 15,000/36 = 416.7$ psi.

$$\left[ \frac{P/A}{s_C} + \frac{P/A(6 + 1.5J)(e/d)}{s_B - J(P/A)} \leq 1 \right]$$

$$\frac{416.7}{886} + \frac{416.7[6 + 1.5(0.389)](2/6)}{1350 - 0.389(416.7)} = 0.47 + 0.77 > 1$$

Since the stress factor is greater than 1.0, the section is inadequate, so we must look for a larger one.

*Second trial:* Try a 6- by 8-in. section with the 8-in. dimension in the direction of the eccentricity. We calculate $A = 48$ in., $Z = 6(8^2)/6 = 64$ in.$^3$, $L/d = 8(12)/8 = 12$, and the convenience factor $J = (12 - 11)/(23.85 - 11) = 0.0778$. Note that $K$ is still 23.85 because of possible buckling in the 6-in. direction. The allowable compressive stress for this intermediate length column with $L/d = 96/6 = 16$ is

$$\left[ s_c = \left\{ 1 - \frac{1}{3}\left(\frac{L/d}{K}\right)^4 \right\} s \right] \qquad s_C = \left[ 1 - \frac{1}{3}\left(\frac{16}{23.85}\right)^4 \right] (950) = 886 \text{ psi}$$

We next calculate the stress factor using Eq. (12-25) with $M = 0$ and $P/A = 15,000/48 = 312.5$,

$$\left[\frac{P/A}{s_C} + \frac{P/A(6 + 1.5J)(e/d)}{s_B - J(P/A)} \leqq 1\right]$$

$$\frac{312.5}{886} + \frac{312.5[6 + 1.5(0.0778)](2/8)}{1350 - 0.0778(312.5)} = 0.35 + 0.36 = 0.71$$

Since the stress factor is now less than 1.0, the second trial section is satisfactory.

Use 6- by 8-in. rough select-structural grade eastern hemlock column *Ans.*

**12-69.** A 12-ft-long column of eastern hemlock No. 1 grade must support a 30-kip eccentric load. The eccentricity is 3 in. Design a suitable S4S column for this condition.

*Solution:* From Table 8-1, the allowable bending stress is 1100 psi, the allowable compressive stress parallel to the grain is 800 psi, and the modulus of elasticity is 1,200,000 psi. Assume a 10- by 12-in. column. Then $L/d = [12(12)]/9.5 = 15.16$ and $K = 0.671\sqrt{1,200,000/800} = 25.99$. The column is, then, of intermediate length and

$$\left[s_c = \left\{1 - \frac{1}{3}\left(\frac{L/d}{K}\right)^4\right\} s\right] \qquad s_C = \left[1 - \frac{1}{3}\left(\frac{15.16}{25.99}\right)^4\right] (800) = 769 \text{ psi}$$

When calculating the convenience factor $J$, we use the $L/d$ ratio in the direction of the eccentricity or bending, $144/11.5 = 12.52$. Therefore

$$\left[J = \frac{L/d - 11}{K - 11}\right] \qquad J = \frac{12.52 - 11}{25.99 - 11} = 0.101$$

The stress factor is (with $M = 0$ and $P/A - 30,000/109.25 - 274.6$ psi)

$$\left[\frac{P/A}{s_C} + \frac{P/A(6 + 1.5J)(e/d)}{s_B - J(P/A)} \leqq 1\right]$$

$$\frac{274.6}{769} + \frac{274.6[6 + 1.5(0.101)](3/11.5)}{1100 - 0.101(274.6)} = 0.36 + 0.41 = 0.77 \quad \text{OK}$$

Use 10- by 12-in. S4S eastern hemlock column *Ans.*

**12-70.** What lateral load $F$ may be applied to the eccentrically loaded column of Fig. 12-9 in addition to the 40-kip load shown? The wood is dense select-structural grade Douglas fir.

*Solution:* From Table 8-1, the allowable bending stress is 1900 psi, the allowable compressive stress parallel to the grain is 1300 psi, and the modulus of elasticity is 1,700,000 psi. The section modulus is $Z = bd^2/6 = [(8)(12^2)/6 = 192 \text{ in.}^3$, $K = 0.671\sqrt{E/s} = 0.671\sqrt{1,700,000/1300} = 24.26$, and $L/d = [(14)(12)]/8 = 21$. Since 21 is between 11 and 24.26, the column is of intermediate length and $s_C$ is given by Eq. (12-17).

$$\left[s_c = \left\{1 - \frac{1}{3}\left(\frac{L/d}{K}\right)^4\right\} s\right] \qquad s_C = \left[1 - \frac{1}{3}\left(\frac{21}{24.26}\right)^4\right] (1300) = 1057 \text{ psi}$$

The $L/d$ ratio in the direction of bending is $184/12 = 14$. Therefore the convenience factor is $J = (14 - 11)/(24.26 - 11) = 0.23$. The maximum bending mo-

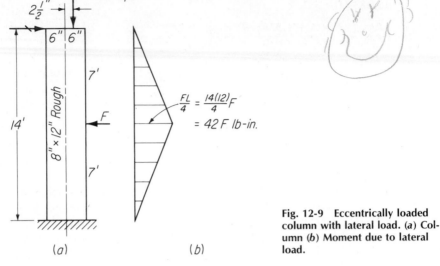

Fig. 12-9  Eccentrically loaded column with lateral load. (a) Column (b) Moment due to lateral load.

ment $M$ is $3.5F$ (see Fig. 12-9b) and $P/A = 40,000/96 = 416.7$ psi. The stress factor is

$$\left[ \frac{P/A}{s_C} + \frac{M/Z + P/A(6 + 1.5J)(e/d)}{s_B - J(P/A)} = 1 \right]$$

$$\frac{416.7}{1057} + \frac{3.5F/192 + 416.7[6 + 1.5(0.23)](2.5/12)}{1900 - 0.23(416.7)} = 1$$

or
$$0.394 + \frac{0.219F + 551}{1804} = 1$$

From which

$$0.219F = 1093 - 551 \quad \text{and} \quad F = 2476 \text{ lb} \qquad Ans.$$

**12-71.** Refer to Fig. 12-8c. A 300- by 400-mm select-structural lodgepole pine column has loads and dimensions as follows: $P_1 = 125$ kN, $P_2 = 100$ kN, $a = 0.6$ m, $L = 6$ m, and $L_1 = 4$ m. Is the column "safe"?

*Solution:* The actual loading can be transformed into the equivalent loading shown in the right-hand sketch of Fig. 12-8c by means of Eq. (12-28).

$$\left[ P' = \frac{3aL'P_2}{L^2} \right] \qquad P' = \frac{(3)(0.5)(4)(100)}{6^2} = 16.67 \text{ kN}$$

The lateral load is, then, 16.67 kN and the axial load is $125 + 100 = 225$ kN. From Table 8-1, the allowable bending stress is 7.6 MPa, the allowable compressive stress parallel to the grain is 5.2 MPa, and the modulus of elasticity is 7600 MPa. The $L/d$ ratio for buckling is $L/d = 6000/300 = 20.0$ and $K = 0.671\sqrt{E/s} = 0.671\sqrt{7600/5.2} = 25.65$. Therefore the column is intermediate in length, and

$$\left[ s_C = \left\{ 1 - \frac{1}{3} \left( \frac{L/d}{K} \right)^4 \right\} s \right] \qquad s_C = \left[ 1 - \frac{1}{3} \left( \frac{20}{25.65} \right)^4 \right] 5.2 = 4.56 \text{ MPa}$$

The $L/d$ ratio in the plane of bending is $6000/400 - 15$ and the convenience factor is then

$$\left[J = \left|\frac{L/d - 11}{K - 11}\right|\right] \qquad J = \frac{15 - 11}{25.65 - 11} = 0.273$$

The bending moment due to the lateral load of 16.67 kN is $M = PL/4 = (16\ 670)(6000)/4 = 25\ 000\ 000$ N·mm. The section modulus $Z = bd^2/6 = [(300)(400^2)]/6 = 8 \times 10^6$ mm³. The equation for the stress factor is (with $e = 0$ and $P/A = 225\ 000/120\ 000 = 1.87$ mPa).

$$\left[\frac{P/A}{s_C} + \frac{M/Z}{s_B - J(P/A)} \leqq 1\right] \qquad \frac{1.87}{4.56} + \frac{(25 \times 10^6)/(8 \times 10^6)}{7.6 - 0.273(1.87)} = 0.41 + 0.44 = 0.85$$

Since the stress factor of 0.85 is less than unity, the design is "safe."    *Ans.*

## PROBLEMS

NOTE: The side dimensions of timber columns subjected to eccentric loads may differ by 2 or 4 in.

**12-72.** Design a timber column 12 ft long of lodgepole pine No. 1 rough to support an eccentric load $P$ of 42,000 lb in a manner similar to that illustrated in Fig. 12-8a. Assume that the eccentricity $e$ is 2 in.

**12-73.** A timber column 5 m long of Douglas fir dense No. 1 rough is subjected to an eccentric load $P$ of 100 kN similar to the column illustrated in Fig. 12-8a. Assume that the column size is 250 by 300 mm. What is the maximum allowable eccentricity?

**12-74.** Design a timber column 12 ft long of Douglas fir select structural rough to support an axial load $P_1$ of 37,500 lb plus an eccentric load $P_2$ of 12,500 lb applied at a point 8 in. from the column axis, in a manner similar to that illustrated in Fig. 12-8c. Length $L_1 = 8$ ft.

*Ans.* 8 by 10 in. rough

**12-75.** A timber column 8 ft long of lodgepole pine No. 1 rough is subjected to an axial load $P$ of 12,000 lb plus lateral load $F$ of 2000 lb applied at midheight of the column in a manner similar to that illustrated in Fig. 12-8b. Select the most economical column. Note that $e = 0$.

**12-76.** A timber column of select-structural Douglas fir is 12 ft long and must support a 40-kip load with an eccentricity $e = 4$ in. Will an 8- by 12-in. rough timber be satisfactory?

**12-77.** A timber column of select-structural Ponderosa pine rough is loaded as illustrated in Fig. 12-8c; $P_1 = 25,000$ lb, $P_2 = 20,000$ lb, and $a = 15$ in. The column is 18 ft long, and the bracket is so placed that $L_1 = 10$ ft. Design the most economical column to use.

**12-78.** Design a timber column 16 ft long of lodgepole pine No. 1 rough to support an axial load $P_1$ of 44 kips and an eccentric load $P_2$ of 6 kips applied at a point 1.5 ft from the column axis in a manner similar to that illustrated in fig. 12-8c. Let $L_1 = 12$ ft.

*Ans.* 10 by 12 in. rough

SUMMARY (By Article Number)_____

**12-1.** Under end loads, concentric or eccentric, columns tend to buckle laterally (sideways). This tendency increases with slenderness.

**12-3.** The **slenderness ratio** of a column, $KL/k$ for metal columns and $L/d$ for timber columns, determines its allowable load $P/A$ per unit (square inch) of cross-sectional area. $P/A$ decreases from a maximum when the slenderness ratio is quite low to approximately 20 percent of the maximum when the slenderness ratio is at the highest permissible in usual practice. Depending on its slenderness ratio, a column is classified as *short, intermediate,* or *long.* Formulas have been devised, nearly all derived from experimental data, giving the allowable unit load $P/A$ for the columns in each of these groups.

**12-5. Euler's formula,** the only one derived entirely mathematically, is

$$\frac{P}{A} = \frac{\pi^2 E}{(L/k)^2} \qquad (12\text{-}2)$$

where $P$ is the critical or *buckling load.* To obtain the safe load, $P$ must be divided by an appropriate factor of safety.

**12-6. End restraint** has great influence on the tendency of a long column to buckle under load. A given long column whose ends are fully restrained against rotation has four times the load-carrying capacity of the same column with hinged ends (free to rotate), *provided, however, that in no case must $P/A$ exceed the value permissible on the simple-column length of the column* in question. Columns in the intermediate range are much less affected by end restraint, and short columns are not affected at all,

**12-11.** For the **design of structural-steel columns and other compression members** in buildings and similar structures, most building codes specify the use of the three AISC formulas, Eqs. (12-10) to (12-12). Values of $P/A$ for these formulas are listed in Table E-8.

**12-15.** The ASCE specifications for aluminum columns give graphically the allowable stress $P/A$ as a function of the slenderness ratio $L/k$.

**12-16.** Axially loaded timber columns are classified as *short, intermediate,* or *long.* Short columns have an $L/d$ ratio of 11 or less and are designed using the allowable compressive stress parallel to the grain. Long columns must have a factor of safety against buckling and are designed using a modified Euler formula.

**12-17. Eccentrically loaded columns** are those in which the action line of the load is parallel to but does not coincide with the axis of the column. Such columns are subjected both to direct axial stress and bending stress. Formulas giving the allowable unit load $P/A$ on such columns become rather complex. A simple rule is that eccentrically loaded columns should be so proportioned that the **stress factor**

$$\frac{f_a}{F_a} + \frac{f_b}{F_b} \text{ will not exceed 1} \qquad (12\text{-}24)$$

REVIEW PROBLEMS_____

**12-79.** A piece of hard-steel shafting having a proportional limit of 60,000 psi is used as a column supporting an axial load. If its diameter is 1 in. and $E =$

30,000,000 psi, calculate (a) the lowest limit of length $L$ for which Euler's formula applies and (b) the buckling load $P$ when $L = 40$ in.

*Ans.* (a) $L = 17.5$ in.; (b) $P = 9080$ lb

**12-80.** A strut in an airplane frame is a 1-in. square, hollow tube of aluminum, 14S-T4, with walls 0.1 in. thick. Using Euler's formula, calculate the buckling load $P$ on this strut if $L = 50$ in.

**12-81.** Using Euler's formula, calculate the safe axial compressive load $P$ on a 2-in.-diameter standard steel-pipe column 8 ft 4 in. long if a factor of safety of $2^1/_2$ is used.

*Ans.* $P = 7680$ lb

**12-82.** The radius of gyration $k$ of a thin-walled square tube, $a$ by $a$, is given by $a/\sqrt{6}$. Develop the equations (a) giving the lower limit of the length $L$ for which Euler's formula applies if the proportional limit is $s_p$, and (b) the corresponding length $L$ for a solid square section with the same outside dimensions.

**12-83.** A 5-m long piece of steel pipe is to be used to withstand a compressive axial load of 40 kN. Its ends are presumed to be hinged. $E$ is 200 000 MPa and a factor of safety of 2.5 is considered appropriate. Determine the required diameter $d$ of pipe, using Euler's formula. Calculate also its slenderness ratio $L/k$ to determine if Euler's formula holds. See Table 12-2.

**12-84.** A 250-mm-long round rod of hard steel having a proportional limit of 275 MPa is to be used as a machine part capable of transmitting an estimated axial compressive load $P$ of 1.3 kN. If $E = 200\,000$ MPa and a factor of safety of 2 is used, what minimum diameter $d$ of rod may be used, using Euler's formula?

**12-85.** A square, hollow tube of structural aluminum, 14S-T6, is to be used as a strut in an airplane frame to withstand a calculated axial compressive load of 9320 lb. Its ends are presumed to be hinged. The proportional limit is 30,000 psi and $E$ is 10,300,000 psi. If the length of the strut is 50 in. and the inside dimension of the tube is 0.9 of the outside dimension $d$, what should be the dimension $d$? Use a factor of safety of 2.

**12-86.** A box column is built up of two 9- by $2^1/_2$-in. standard structural-steel channels, each weighing 13.4 lb/ft, to form a section 5 by 9 in. similar to the one illustrated in Fig. 12-1$k$. If the unsupported length of this column is 24 ft, what is its allowable load $P$? Use AISC specifications.

*Ans.* $P = 57.5$ kips

**12-87.** Select the most economical rolled structural-steel W section to use as a column 10 ft 6 in. long to support an axial compressive load $P$ of 60,000 lb. Use AISC specifications.

**12-88.** A double-angle truss chord member 13 ft long must resist a compressive load of 60,000 lb. Angles are $^3/_8$ in. back to back. Select the most economical angles using AISC specifications.

**12-89.** A structural T section (see Fig. 12 10) is used for the upper (compressive) chord of a steel truss. The tee is formed by shearing the web of an W 8 × 21 into two identical tees with $\bar{y} = 3.309$ in. Calculate the allowable axial compressive load using AISC specifications if the member is braced at 6-ft intervals in the plane of the web and at 12-ft intervals in the plane of the flange.

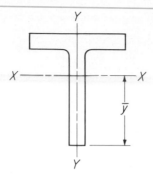

**Fig. 12-10   Structural T section. Prob. 12-89.**

**12-90.** A single-angle structural-steel brace 8 ft long must be capable of with-standing an axial compressive load of 5 kips. Select the most economical equal-leg angle to use. Use AISC specifications.

*Ans.* $2^1/_2$ by $2^1/_2$ by $^3/_{16}$ in.

**12-91.** A latticed column of the type shown in Figs. 12-1*h* and 12-2 is to be 40 ft long and must support an axial load of 172,000 lb. (*a*) Select the most economical standard structural-steel channels to use. (*b*) Calculate the required distance *b* from back to back of channels to give equal moments of inertia with respect to both major axes.

**12-92.** Solve Prob. 12-47 for a length of 25 ft and a load of 265,000 lb.

**12-93.** A round bar of aluminum alloy, 6061-T6, 400 mm long in an aircraft frame is to be subjected to an axial compressive load *P* of 5 kN. Its ends will be free to rotate so that *K* = 1. Determine the required diameter *D* of this bar.

**12-94.** A square tube of aluminum alloy, 6061-T6, 56 in. long is to be subjected to an axial compressive load *P*. Determine the allowable load *P* if the tube is $1^1/_4$- by $1^1/_4$-in. outside dimensions and the wall thickness is $^1/_8$ in. Ends are partially restrained.

**12-95.** Design a timber column 10 ft long of Douglas fir No. 1 S4S to support an axial compressive load of 45,000 lb.

**12-96.** Select the most economical timber column 8 ft 4 in. long of eastern hem-lock select-structural S4S to support an axial load *P* of 8 kips.

*Ans.* 4 by 6 in. S4S

**12-97.** Design a round timber column 4 m long of No. 1 Douglas fir to support an axial load *P* of 200 kN.

**12-98.** Design a round timber column 8 ft long of Ponderosa pine No. 1 to sup-port an axial load *P* of 42,300 lb.

*Ans.* $D = 8.57$ in.

**12-99.** As eastern hemlock select-structural grade column is loaded as shown in Fig. 12-8*c*. The column cross section is 250 by 400 mm, $P_1 = 200$ kN, $P_2 = 50$ kN, $L = 6$ m, and $L' = 4$ m. What is the maximum permissible value of the dimension *a* shown in the figure?

## REVIEW QUESTIONS

**12-1.** Describe a column and its function and the different types of columns.

**12-2.** In what manner does a column usually fail?

**12-3.** What imperfections of material, of manufacture, and of loading would hasten failure?

**12-4.** Which type of column cross section is most efficient?

**12-5.** What is meant by the slenderness ratio of a column?

**12-6.** What slenderness-ratio limits determine (approximately) the group in which a given column belongs?

**12-7.** What is meant by the critical or buckling load on a column?

**12-8.** To what specific group of columns does Euler's formula apply? What load does $P$ represent? Does this formula contain a factor of safety?

**12-9.** In connection with columns, what is meant by end restraint?

**12-10.** Discuss the effect of end restraint on the load-carrying capacity of a column.

**12-11.** In column formulas, why is $P/A$ referred to as the allowable unit load rather than the allowable unit stress?

**12-12.** In structural-steel columns, what distinguishes main members from secondary members?

**12-13.** Do the structural-steel column formulas contain a factor of safety? What load does $P$ represent?

**12-14.** Why must steel columns be designed by a trial-and-error method?

**12-15.** In determining the slenderness ratio of a structural-steel or aluminum column, which of its two radii of gyration must always be used?

**12-16.** When the unsupported column length in one plane differs from that in a perpendicular plane, which $L/k$ governs the allowable load?

**12-17.** Do the curves in Fig. 12-6 for axially loaded aluminum columns include a factor of safety?

**12-18.** Describe the internal stress picture in an eccentrically loaded column.

**12-19.** In what kind of column does the lateral deflection $\Delta$ become an important factor?

**12-20.** When is a timber column considered short? Long?

**12-21.** What is the stress factor and what is its function?

# Chapter 13
# Special Topics

## 13-1 IMPACT OR DYNAMIC LOADING

In the preceding chapters all loads and forces acting on structural members, connections, or parts have been considered to be *static;* that is, they have been *gradually* applied until the maximum was reached and then have remained static.

However, many problems arise, especially in machine design, in which loads are caused by impacts or other dynamic conditions. The solutions of such problems differ radically from those in which only static forces are present, as is shown in the following examples.

## 13-2 STRAIN ENERGY

The work-energy concept is developed in engineering mechanics.[1] A force moving through a distance does work. In the case of a gradually applied load, as in Fig. 13-1, the work done by the varying force is the product of the *average* force and the displacement $\delta$. In symbols,

$$U = \frac{P\delta}{2} = \frac{P}{2}\frac{PL}{AE} \quad \text{or} \quad U = \frac{P^2LA}{2A^2E}$$

Since $P/A$ is the unit stress $s$ and $LA$ is the volume of the member,

$$U = \frac{s^2}{2E} \text{ (volume)} \tag{13-1}$$

The work done on the block is stored as *strain energy* in the material. The block behaves as a very stiff spring, provided the stress $s$ is below the

---

[1] See, for example, the authors' *Applied Engineering Mechanics,* 4th ed., McGraw-Hill Book Company, New York, 1983.

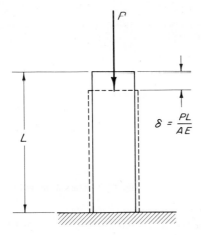

$$\delta = \frac{PL}{AE}$$

Fig. 13-1    Gradually applied axial load *P*.

proportional limit of the material. When the load is removed, the stored energy is recovered.

The strain energy per unit volume is

$$\frac{U}{\text{volume}} = \frac{s^2}{2E} \tag{13-2}$$

The amount of energy transmitted to a structural member under impact is sometimes referred to as an *energy load*. The stresses caused by such energy loads can be computed from Eq. (13-2).

## ILLUSTRATIVE PROBLEMS

**13-1.** The elastic limit of a certain steel is 32,000 psi, and its modulus of elasticity $E$ is 30,000,000 psi. What maximum amount of recoverable energy can be stored in each cubic inch of this material?

*Solution:* If a bar of this material is stressed beyond its elastic limit, it will be permanently deformed, and some of the applied work cannot be recovered. If all the stored energy is to be recovered, $s$ in Eq. (13-2) is the elastic limit, and we then have

$$\left[ \frac{U}{\text{volume}} = \frac{s^2}{2E} \right] \qquad \frac{U}{\text{volume}} = \frac{(32,000)(32,000)}{(2)(30,000,000)} = 17.05 \text{ lb·in./in.}^3 \qquad Ans.$$

This maximum recoverable amount of stored energy for a material is called the *modulus of resilience* of that material. While the recoverable strain energy is determined by the elastic limit of the material, the expression $s^2/2E$ for the strain energy per unit volume is applicable only to the proportional limit. The error in applying $s^2/2E$ to the elastic limit is small.

**13-2.** A 5.099-kg mass $[W = 9.806(5.099) = 50 \text{ N}]$ is dropped from a height of 100 mm onto a spring, as shown in Fig. 13-2. The spring constant $k$ is 20 N/mm; that is, a 20-N force will compress the spring 1 mm, and a 100-N force will compress it 5 mm. Calculate the maximum spring deflection δ.

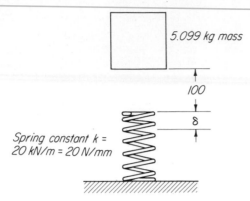

Fig. 13-2    Mass dropped on spring.

*Solution:* The potential energy lost by the falling mass is $U = (50)(100 + \delta)$.

The work done on the spring is stored in it as strain energy and is evaluated from $U = \frac{1}{2}k\delta^2$. Equating the work done on the spring to its strain energy, we obtain

$$(50)(100 + \delta) = \frac{1}{2}k\delta^2$$

Since $k = 20$ N/mm

$$5000 + 50\delta = \frac{20}{2}\delta^2 \quad \text{or} \quad \delta^2 - 5\delta - 500 = 0$$

When this quadratic equation is solved, either by formula or by completing the square, we obtain

$$\delta = 25 \text{ mm} \qquad\qquad Ans.$$

Note that the static spring deflection under the 5.099-kg mass would be

$$[W = k\delta] \qquad\qquad 50 = 20\delta \quad \text{or} \quad \delta = 2.5 \text{ mm}$$

### PROBLEMS

**13-3.** A 1-in.-diameter aluminum rod is 9 ft long. Its yield stress $s$ is 16,000 psi, and its modulus of elasticity $E$ is 10,000,000 psi. What energy load may be applied to the rod without exceeding the yield stress?

$$Ans. \ U = 1086 \text{ lb·in.}$$

**13-4.** The weight $W$ in Fig. 13-3 is 10 lb and drops $h = 3$ in. onto the stop. The rod is 1 in. in diameter and is 20 ft long. Calculate the maximum unit rod stress $s$ if $E = 3,000,000$ psi.

$$Ans. \ s = 990 \text{ psi } (s = 977 \text{ psi if } \delta \text{ neglected})$$

**13-5.** If the weight $W$ in Fig. 13-3 is 1 kN, from what height $h$ may it be dropped if the unit stress $s$ in the steel rod is not to exceed 150 MPa? The rod is 25 mm in diameter, $L$ is 900 mm, and $E$ is 200 000 MPa. Neglect the elongation $\delta$.

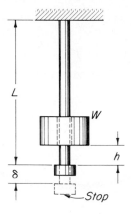

Fig. 13-3   Probs. 13-4 and 13-5.

**13-6.** The cross-sectional area of the upper portion of each of the rods in Fig. 13-4 is 3 in.$^2$ and of the bottom portion is 1 in.$^2$. Determine the maximum unit stress $s$ in each rod if $E = 10,000,000$ psi. Neglect the elongation $\delta$. Note that the unit stress in the lower portion is three times that in the upper portion because $T = S_1A_1 = S_2A_2$. Also note, $U_{\text{total}} = U_{\text{top}} + U_{\text{btm}}$.

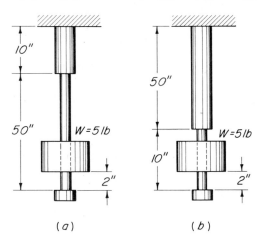

(a)                    (b)                    Fig. 13-4   Probs. 13-6 and 13-22.

## 13-3 IMPACT LOADING

Consider the simple beam shown in Fig. 13-5a. The impact load $W$ is allowed to fall freely onto the midspan of the beam. The beam deflects an amount $\delta$ under this condition, as indicated in Fig. 13-5b. This same deflection $\delta$ could be produced by a gradually applied load of sufficient magnitude. Call this equivalent static load $P_{EQ}$ as indicated in Fig. 13-5d.

Because the deflections and stresses under the equivalent load $P_{EQ}$ are the same as those under the impact load $W$, the internal strain energies are the same. The external work done by the loads must then also be the

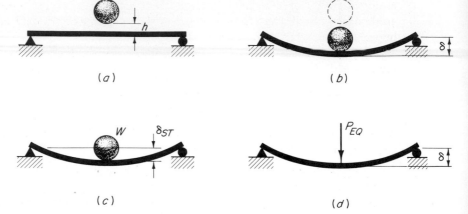

**Fig. 13-5**  Derivation of impact factor. (*a*) Impact load. (*b*) Deflection under impact load. (*c*) Deflection under gradually applied load *W*. (*d*) Equivalent gradually applied load to give some deflection as impact load *W*.

same. Putting this statement in equation form, we obtain

$$W(h + \delta) = \frac{P_{EQ}\delta}{2} \tag{a}$$

Since the deflection of the beam is proportional to the load that produces that deflection,

$$\frac{P_{EQ}}{\delta} = \frac{W}{\delta_{ST}} \quad \text{or} \quad P_{EQ} = W\frac{\delta}{\delta_{ST}} \tag{b}$$

where $\delta_{ST}$ is the static deflection under a gradually applied load $W$, as in Fig. 13-5*c*. Substituting the value of $P_{EQ}$ from Eq. (*b*) in Eq. (*a*) gives

$$W(h + \delta) = \frac{W\delta^2}{2\delta_{ST}} \quad \text{or} \quad h + \delta = \frac{\delta^2}{2\delta_{ST}}$$

Rearranging gives

$$\delta^2 - 2\delta_{ST}\delta = 2h\delta_{ST}$$

Completing the square yields

$$\delta^2 - 2\delta_{ST}\delta + \delta_{ST}^2 = \delta_{ST}^2 + 2h\delta_{ST}$$

or

$$(\delta - \delta_{ST})^2 = \delta_{ST}^2 \left(1 + \frac{2h}{\delta_{ST}}\right)$$

Taking the square root of each side gives

$$\delta - \delta_{ST} = \delta_{ST}\sqrt{1 + \frac{2h}{\delta_{ST}}}$$

or

$$\delta = \delta_{ST}\left(1 + \sqrt{1 + \frac{2h}{\delta_{ST}}}\right) \tag{13-3}$$

We conclude from Eq. (13-3) that the deflection $\delta$ under impact loading is considerably greater than the static deflection $\delta_{ST}$ for the same load gradually applied. The factor $1 + \sqrt{1 + (2h/\delta_{ST})}$ can be thought of as an **impact factor,** which always exceeds unity.

Since the stresses in a beam are proportional to the load that produces them, we may state

$$\frac{s_{ST}}{W} = \frac{s}{P_{EQ}} \qquad\qquad (c)$$

where $s_{ST}$ is the stress due to the gradually applied load $W$ and $s$ is the stress due to the equivalent load $P_{EQ}$.

Recall from Eq. (b) that $P_{EQ} = W(\delta/\delta_{ST})$. Substituting for $P_{EQ}$ in Eq. (c), we obtain

$$\frac{s_{ST}}{W} = \frac{s\delta_{ST}}{W\delta} \qquad \text{or} \qquad s = \frac{\delta}{\delta_{ST}} s_{ST}$$

From Eq. (13-3), $\delta/\delta_{ST} = 1 + \sqrt{1 + (2h/\delta_{ST})}$. Hence

$$s = \left(1 + \sqrt{1 + \frac{2h}{\delta_{ST}}}\right) s_{ST} \qquad\qquad \textbf{(13-4)}$$

The stress $s$ under impact loading is then the impact factor multiplied by the stress $s_{ST}$ under the same loading gradually applied; that is, the static load.

### ILLUSTRATIVE PROBLEMS

**13-7.** The beam shown in Fig. 13-5a is a W 10 × 33 section. Compute the maximum bending stress $s$ if a 1000-lb weight is dropped from a height $h = 4$ in. at midspan. The length of the beam is 12 ft, and $E = 30,000,000$ psi.

*Solution:* The maximum moment under a concentrated static load of 1000 lb is $PL/4$ (see Fig. 9-31). Therefore $M = [(1000)(144)]/4 = 36,000$ lb·in. The maximum bending stress is

$$\left[s = \frac{M}{Z}\right] \qquad\qquad s = \frac{36,000}{35} = 1027 \text{ psi}$$

The maximum deflection under the concentrated static load, from Fig. 9-31, is

$$\left[\delta = \frac{PL^3}{48EI}\right] \qquad \delta_{ST} = \frac{(1000)(144)^3}{(48)(30,000,000)(170.9)} = 0.01214 \text{ in.}$$

The maximum stress under the impact load is

$$\left[s = \left(1 + \sqrt{1 + \frac{2h}{\delta_{ST}}}\right) s_{ST}\right] \qquad s = \left[1 + \sqrt{1 + \frac{(2)(4)}{0.01214}}\right](1,027)$$

or $\qquad\qquad\qquad s = (26.75)(1027) = 27,400 \text{ psi.} \qquad\qquad\qquad Ans.$

**13-8.** The shaft in Fig. 13-6 is subjected to a torsional impact load as shown. The shaft is 6 ft long and its diameter is 2 in. Calculate the maximum torsional shearing stress $s_s$ if $W = 10$ lb, $e = 6$ in., $h = 4$ in., and $G = 10,000,000$ psi.

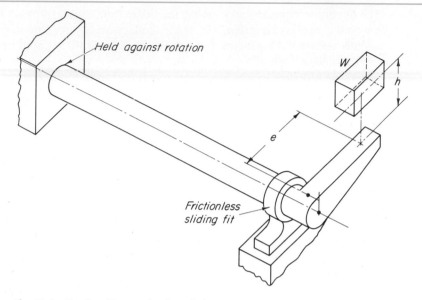

**Fig. 13-6** Torsional impact load on shaft.

*Solution:* The polar moment of inertia $J$ of the shaft is

$$\left[ J = \frac{\pi R^4}{2} \right] \qquad\qquad J = \frac{\pi (1^4)}{2} = 1.57 \text{ in.}^4$$

If the 10-lb load were *gradually* applied, the shearing stress would be

$$\left[ s_s = \frac{TD}{2J} \right] \qquad\qquad s_s = \frac{(10)(6)(2)}{(2)(1.57)} = 38.2 \text{ psi}$$

The angle of twist $\theta$ under this *gradually* applied load would be

$$\left[ \theta = \frac{TL}{GJ} \right] \qquad\qquad \theta = \frac{(10)(6)(72)}{(10,000,000)(1.57)} = 0.000275 \text{ rad}$$

The deflection $\delta$ at the load center would then be

$$[\delta = e\theta] \qquad\qquad \delta = (6)(0.000275) = 0.00165 \text{ in.}$$

The maximum stress $s_s$ under the impact load is given by Eq. (13-4):

$$\left[ s_s = \left( 1 + \sqrt{1 + \frac{2h}{\delta_{ST}}} \right) s_{ST} \right] \quad s_s = \left[ 1 + \sqrt{1 + \frac{(2)(4)}{0.00165}} \right] (38.2)$$

$$s_s = (70.7)(38.2) = 2700 \text{ psi} \qquad\qquad\qquad Ans.$$

---

**PROBLEMS**

**13-9.** From what height $h$ may a 102-kg mass be dropped on a 100 by 250 mm rough timber beam if the bending stress must not exceed 10 MPa? The beam span is 3 m, $E = 12\,000$ MPa, and the load is applied at midspan.

$Ans.\ h = 29.7 \text{ mm}$

**13-10.** Solve Prob. 13-7 if the weight $W = 500$ lb, $h = 3$ in., $L = 10$ ft, and the beam is of dense Douglas fir with a 6- by 10-in. section (S4S).

**13-11.** Calculate the total angle of twist $\theta$ of the shaft shown in Fig. 13-6 under the impact load if the data are as given in Prob. 13-8.

**13-12.** Check the spring deflection of Prob. 13-2 using Eq. (13-3).

**13-13.** Check the rod stress of Prob. 13-4 using Eq. (13-4).

## 13-4 PRESTRESSED BEAMS

A beam material that is relatively weak in tension may be placed in compression by the pull of longitudinal tendons. If this prestressing is done before the application of the usual bending loads, the tendency of the bending moments to place one side of the beam in tension simply results in the reduction of the compressive prestress on that side of the beam. Of course, the bending compressive stress on the other side also adds to any initial compressive stress there. The prestressing tendon may be placed off-center, or may be draped, so that the initial compressive stress will be high on that side of the beam where later bending stresses will tend to be tensile. The off-center tendon may cause small compressive prestress on the opposite side of the beam that will later be compressed still further by the applied bending moments. Prestressed beam problems are solved by superposition, as illustrated below.

**ILLUSTRATIVE PROBLEMS**_____

**13-14.** The beam shown in Fig. 13-7 is composed of concrete blocks, each 6 in. wide and 12 in. deep. A loose-fitting steel rod (or tendon) runs longitudinally

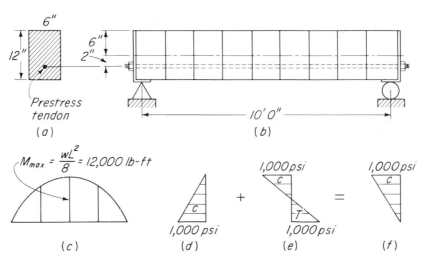

Fig. 13-7   Prestressed beam under uniform load. (a) Beam section. (b) Prestressed beam. (c) Moment diagram. (d) Beam stress due to rod tension. (e) Stress due to bending load. (f) Final stress diagram.

through the beam at a distance of $d/3$ above the bottom. ($a$) Calculate the tendon force required to produce a compressive stress of 1000 psi at the bottom of the beam (see Fig. 13-7$d$). This is accomplished by tightening the nuts on the ends of the rods. ($b$) What is the magnitude of the maximum moment that can be applied to the prestressed beam without causing tensile stresses therein? ($c$) What downward uniform load may be applied to the top of the beam without causing tensile stresses in the prestressed beam?

*Solution:* The tensile tendon force $P$ in the threaded rod is transformed into a compressive force on the ends of the concrete blocks by the bent end plates shown in the illustration. Since the resultant compressive force on the concrete blocks is eccentrically applied ($e = 2$ in.), the compressive stress in the concrete is given by

$$\left[ s = \frac{P}{A} + \frac{Mc}{I} \right] \qquad 1000 = \frac{P}{(6)(12)} + \frac{P(2)(6)}{[(6)(12^3)]/12} = 0.139P + 0.139P$$

or
$$P = \frac{1000}{0.0278} = 36,000 \text{ lb} \qquad\qquad Ans.$$

The stress at the top of the beam due to the tendon force will be

$$\left[ s = \frac{P}{A} + \frac{Pec}{I} \right] \qquad s = \frac{36,000}{72} - \frac{(36,000)(2)(6)}{[(6)(12^3)]/12} = 0$$

Let us now add a bending moment to the beam that will cause a compressive bending stress of 1000 psi at the top of the beam and a tensile bending stress of 1000 psi at the bottom of the beam (see Fig. 13-7$e$). If the stresses due to the tendon force are added to those caused by the later applied bending moment, the result will be the stresses shown in Fig. 13-7$f$. The stresses without the applied bending load are those in Fig. 13-7$d$ and stresses with the applied bending load are shown in Fig. 13-7$f$. In neither case are there any tensile stresses in the beam. Therefore, the beam material need not be capable of withstanding large tensile stresses.

What is the magnitude of the maximum bending moment that may be applied to this prestressed beam? We assume that tension is undesirable. The bending moment that will cause the bending stresses shown in Fig. 13-7$e$ is

$$\left[ M = \frac{sI}{c} \right] \qquad M = \frac{[(1000)(6)(12^3)]/12}{6} = 144,000 \text{ lb·in.} = 12,000 \text{ lb·ft} \qquad Ans.$$

If this moment is the result of a uniformly distributed load, the bending moment diagram will be that shown in Fig. 13-7$c$ and the maximum moment will be $wL^2/8$. The allowable uniform load (dead plus live) is

$$\left[ w = \frac{8 M_{max}}{L^2} \right] \qquad w = \frac{(8)(144,000/12)}{10^2} = 960 \text{ lb/ft} \qquad Ans.$$

**13-15.** A cantilevered prestressed beam is shown in Fig. 13-8. If the allowable compressive stress in the beam is 7 MPa and tension is not allowed, what maximum tendon force may be applied? What uniform load may be applied to the prestressed beam?

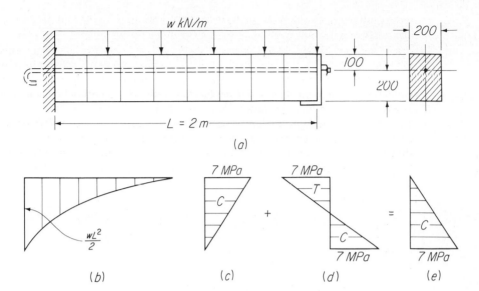

**Fig. 13-8    Prestressed cantilever beam. (a) Beam. (b) Moment diagram. (c) Prestress at wall. (d) Bending stress at wall. (e) Resultant stress at wall.**

*Solution:* The compressive prestress in the concrete due to the tendon force is shown in Fig. 13-8c. The tendon force $P$ can be found from

$$\left[ s = \frac{P}{A} + \frac{Pec}{I} \right] \qquad 7 = \frac{P}{(200)(300)} + \frac{P(50)(150)}{[(200)(300^3)]/12}$$

From this the tendon force is $P = 210\ 000$ N                    *Ans.*

The maximum moment at the wall can be found from

$$\left[ s = \frac{Mc}{I} \right] \qquad 7 = \frac{M(150)}{[(200)(300^3)]/12} \qquad or \qquad M = 21\ 000\ 000 \text{ N·mm}$$

The uniform allowable load is then

$$\left[ M = \frac{wL^2}{2} \right] \qquad 21 \text{ kN·m} = \frac{w(2^2)}{2} \qquad or \qquad w = 10.5 \text{ kN/m} \qquad \textit{Ans.}$$

PROBLEMS_____

**13-16.** The beam shown in Fig. 13-7 is subjected simultaneously to a 36,000-lb tendon pull and a downward uniform load of 960 lb/ft. Calculate the resultant maximum stresses at the top and bottom of the beam at a section 2 ft. 6 in. from the left support.

*Ans.* $s_{top} = 750$ psi; $s_{btm} = 250$ psi

**13-17.** Consider the beam shown in Fig. 13-7, but let the tendon be located at the middepth of the beam. Let $L = 4$ m, $b = 150$ mm, and $d = 400$ mm. (*a*) Calculate the required tendon force to cause a uniform compressive stress of 5 MPa over each section. (*b*) What uniform vertical load, in kN/m, may be placed on the prestressed beam without causing a resultant tensile stress?

*Ans.* (*a*) $F$ = 300 kN; (*b*) $w$ = 10 kN/m

**13-18.** The beam shown in Fig. 13-9 has a rectangular cross section with a width of 6 in. and a depth of 18 in. (*a*) Calculate the maximum flexural bending stress in the beam due to the concentrated load shown. (*b*) Calculate the required tendon force to give sufficient compressive prestress to the beam so that the combined prestress and bending stress at any cross section will be everywhere compressive. Assume that the tendon will be located 6 in. above the bottom of the beam. Neglect the weight of the beam.

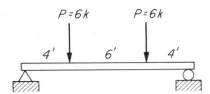

Fig. 13-9   Prob. 13-18.

**13-19.** A beam is composed of concrete blocks held together by a tensioned tendon $4\frac{1}{2}$ in. above the bottom of the beam as shown in Fig. 13-10*a*. The allowable compressive stress in the concrete is 1000 psi, tensile stress in the concrete is to be avoided, and all concrete thicknesses are 1 in. (*a*) Calculate the maximum permissible tendon force and (*b*) the maximum allowable total uniform load on a 10-ft-long simple span.

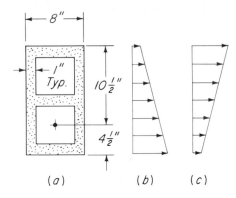

Fig. 13-10   Prob. 13-19. (a) Beam section. (b) Prestress. (c) Final stresses at midspan.

**13-20.** The average weight of the prestressed beam shown in Fig. 13-11 is 2.0 kN/m. The maximum allowable compressive stress in the beam material is 7.0 MPa. and tensile stresses are to be avoided. (*a*) What is the maximum permissible tendon force? [Note that here the critical section for prestress loading is at the outboard end of the beam (see Fig. 13-11*b*).] (*b*) What is the maximum compressive stress in the beam material at the wall due to the tendon force? (*c*) The maximum bending stress that may be superimposed on the prestressed beam without causing tension in the beam material is shown in Fig. 13-11*c*. What is the

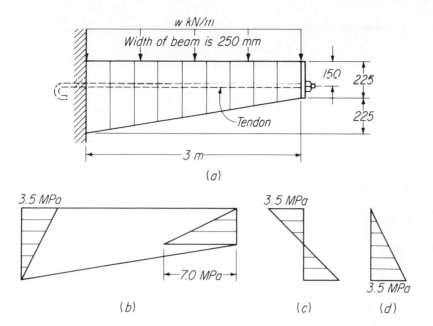

**Fig. 13-11** Probs. 13-20 and 13-26. (*a*) Loaded cantilever beam. (*b*) Prestress at wall and outboard end of beam. (*c*) Bending stress at wall. (*d*) Resultant stress in loaded beam at wall.

numerical value of the corresponding moment? (*d*) What load *w* (in addition to the beam weight) may be placed on the beam?

*Ans.* (*a*) $P = 196.9$ kN; (*b*) $s = 3.5$ MPa; (*c*) $M = 29.5$ kN·m; (*d*) $w = 4.56$ kN/m

**SUMMARY** (By Article Number)

**13-2.** The **strain energy** in a uniformly stressed member is

$$U = \frac{s^2}{2E}(\text{volume}) \qquad (13\text{-}1)$$

**13-3.** The deflection under impact loading is

$$\delta = \delta_{ST}\left(1 + \sqrt{1 + \frac{2h}{\delta_{ST}}}\right) \qquad (13\text{-}3)$$

where $\delta_{ST}$ is the deflection if the load were gradually applied. The parenthetical part of the equation is an impact factor.

The maximum stress under an impact load is

$$s = \left(1 + \sqrt{1 + \frac{2h}{\delta_{ST}}}\right) s_{ST} \qquad (13\text{-}4)$$

## REVIEW PROBLEMS

**13-21.** The weight $W$ falls freely 1 in. onto the spring shown in Fig. 13-12. The spring constant $k$ is 2000 lb/in. and the modulus of elasticity $E$ of the rod is 10,000,000 psi. Calculate the maximum stress in the rod. The elongation of the rod may be neglected, since it is extremely small compared with the deformation of the spring. (HINT: The maximum force in the rod equals the maximum force in the spring.)

*Ans.* $s = 205$ psi

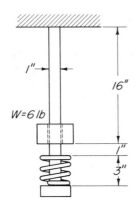

*Fig. 13-12  Prob. 13-21.*

**13-22.** Solve Prob. 13-6 if $W = 10$ lb and $E = 30,000,000$ psi.

*Ans.* $s_a = 4743$ psi; $s_b = 6708$ psi

**13-23.** A weight $W$ is dropped at midspan on a beam as shown in Fig. 13-13. Calculate the maximum bending stress $s$ if $W = 100$ lb, $h = 4$ in., $k = 200$ lb/in., $E = 1,000,000$ psi, $I = 48$ in.$^4$, $L = 48$ in., and the beam is 4 in. deep.

*Ans.* $s = 446$ psi

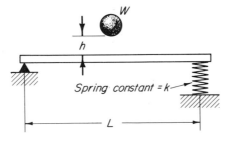

*Fig. 13-13  Probs. 13-23 and 13-24.*

**13-24.** Solve Prob. 13-23 if $W = 500$ N, $h = 75$ mm, $k = 10$ N/mm, $E = 12\,000$ MPa psi, $I = 40 \times 10^6$ mm$^4$, $L = 2$ m, and the depth of the beam is 50 mm.

**13-25.** Solve Prob. 13-18 if the beam carries a uniform load of 240 lb/ft in addition to the two concentrated loads. Let the beam width be 12 in.

**13-26.** Refer to Fig. 13-11. The preload force in the tendon is 196.9 kN. Consider a section 2 m from the wall. (a) What is the maximum compressive prestress at this section due to the tendon force? (b) What superimposed moment can now be applied to this section without causing tension in the beam material here?

## REVIEW QUESTIONS

**13-1.** Distinguish between static and dynamic loads.

**13-2.** Define strain energy.

**13-3.** Define modulus of resilience of a material.

**13-4.** What is meant by the term "impact factor"?

**13-5.** How do stresses due to dynamic loading compare with those caused by static loading?

**13-6.** Show by simple sketches the typical stress distribution in a prestressed beam due to (*a*) the tendon force only, (*b*) bending due to lateral loads only, and (*c*) both causes.

# Appendix A
# Centroids of
# Areas by
# Integration

To locate the centroid of an area, as was stated in Art. 5-1, we need only equate the moment $A\bar{x}$ of the entire area, with respect to some reference axis, to the sum of the moments of the component areas, $\Sigma ax$. When each component area is infinitesimally small, or $dA$, the equations with respect to the $Y$ and $X$ axes become

$$A\bar{x} = \int x \, dA \qquad (a)$$

and
$$A\bar{y} = \int y \, dA \qquad (b)$$

**Rules.** In selecting the differential element $dA$, the following rules must be observed:

1. All points of the element must be the same distance from the moment axis.
2. The position of the centroid of the element must be known, so that the moment of the element with respect to the moment axis is the product of the element and the distance of its centroid from the axis. The locations of the centroids of a triangle, a parabola, and a sector of a circle are shown in Probs. A-1 to A-3.

## ILLUSTRATIVE PROBLEMS

**A-1.** Locate the centroid of the triangle shown in Fig. A-1 with respect to the $Y$ axis.

*Solution:* Let the differential area $dA$ be a vertical strip parallel to the side $a$. This area $dA = y \, dx$. From Eq. $(a)$, we have

$$\left[ A\bar{x} = \int x \, dA \right] \qquad \qquad \frac{1}{2}ba \cdot \bar{x} = \int_0^b xy \, dx \qquad (c)$$

But by similar triangles

$$\frac{y}{x} = \frac{a}{b} \qquad \text{or} \qquad y = \frac{ax}{b} \qquad (d)$$

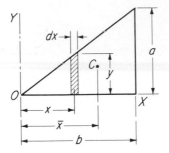

Fig. A-1   Centroid of a triangle.

When this value of $y$ is substituted in Eq. ($c$), we obtain

$$\frac{1}{2}ba \cdot \bar{x} = \frac{a}{b}\int_0^b x^2\,dx \quad \text{and} \quad \bar{x} = \frac{2}{3}b \qquad Ans.$$

**A-2.** Locate the centroid with respect to the $Y$ axis of the parabolic segment shown in Fig. A-2 and bounded by the $X$ axis, the vertical side $b$, and the parabola $y = kx^2$.

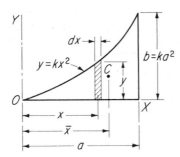

Fig. A-2   Centroid of a parabolic segment.

*Solution:* As the differential area $dA$ we select a vertical strip parallel to side $b$. This area $dA = y\,dx$. But $y = kx^2$, and when $x = a$, $b = ka^2$. Hence

$$\left[A = \int dA\right] \qquad A = \int_0^a y\,dx = \int_0^a kx^2\,dx = \frac{1}{3}ka^3 \qquad (e)$$

To find $\bar{x}$ from Eq. ($a$), we have

$$\left[A\bar{x} = \int x\,dA\right] \qquad \frac{1}{3}ka^3 \cdot \bar{x} = \int_0^a xy\,dx = k\int_0^a x^3\,dx \qquad (f)$$

from which
$$\frac{1}{3}ka^3 \cdot \bar{x} = \frac{1}{4}ka^4 \quad \text{and} \quad \bar{x} = \frac{3}{4}a \qquad Ans.$$

To find $\bar{y}$, we may use the same elemental strip. But our moment axis is now the $X$ axis, and all points of the element are not at the same distance from the axis. Hence rule 2 applies, and the moment of each elemental strip with respect to the $X$ axis is then the product of its area $y\,dx$ and its centroidal distance $1/2y$, or $1/2y \cdot y\,dx$. Then, since $y = kx^2$ and $b = ka^2$,

$$\left[A\bar{y} = \int y\,da\right] \qquad \frac{1}{3}ka^3 \cdot \bar{y} = \int_0^a \frac{1}{2}y \cdot y\,dx = \frac{1}{2}k^2\int_0^a x^4\,dx \qquad (g)$$

or $$\frac{1}{3}ka^3 \cdot \bar{y} = \frac{k^2 a^5}{10}$$ and $$\bar{y} = \frac{3}{10}ka^2 = \frac{3}{10}b$$ *Ans.*

**A-3.** Locate the centroid of the area of the sector of a circle shown in Fig. A-3.

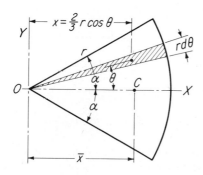

**Fig. A-3    Centroid of a circular sector.**

*Solution:* Let the area of the sector be symmetrical about the $X$ axis. The radius is $r$ and the subtended angle is $2\alpha$. The differential area $dA$ here selected is a triangle whose area is $1/2 r^2 \, d\theta$. The centroid of this triangle lies at a distance $x = 2/3 r \cos \theta$ from the $Y$ axis. The area of the sector is

$$\left[ A = \int dA \right] \qquad A = \int_{-\alpha}^{+\alpha} \frac{1}{2} r^2 \, d\theta = r^2 \alpha \qquad (h)$$

Then from Eq. $(a)$

$$\left[ A\bar{x} = \int x \, dA \right] \qquad r^2 \alpha \cdot \bar{x} = \int_{-\alpha}^{+\alpha} \frac{2}{3} r \cos \theta \frac{1}{2} r^2 \, d\theta \qquad (i)$$

or $$r^2 \alpha \cdot \bar{x} = \frac{1}{3} r^3 \int_{-\alpha}^{+\alpha} \cos \theta \, d\theta = \frac{2}{3} r^3 \sin \alpha \qquad (j)$$

and $$\bar{x} = \frac{2}{3} \frac{r \sin \alpha}{\alpha} \qquad\qquad Ans.$$

When $\alpha = 90°$, as in the semicircle in Fig. A-4, $\bar{x} = 4r/3\pi$.

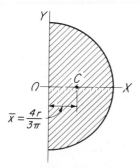

**Fig. A-4    Centroid of a semicircle.**

## PROBLEMS

**A-4.** Find the distance $y$ from the $X$ axis to the centroid of the triangle shown in Fig. A-1 using (a) rule 2 and a strip parallel to the $Y$ axis and (b) rule 1 and a strip parallel to the $X$ axis.

*Ans.* $\bar{y} = {}^1/_3a$

**A-5.** Locate the position of the centroid of the shaded area shown in Fig. A-5.

*Ans.* $\bar{x} = {}^4/_5a; \bar{y} = {}^2/_7b$

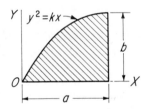

**Fig. A-5   Prob. A-5.**

**A-6.** Determine the position of the centroid of the shaded area shown in Fig. A-6.

*Ans.* $\bar{x} = {}^3/_5a; \bar{y} = {}^3/_8b$

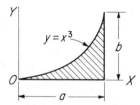

**Fig. A-6   Prob. A-6.**

**A-7.** Locate the centroid of the shaded area shown in Fig. A-7.

*Ans.* $\bar{x} = {}^3/_{10}a; \bar{y} = {}^3/_4b$

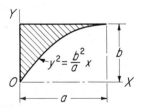

**Fig. A-7   Prob. A-7.**

# Appendix B
# Moments of
# Inertia of Plane
# Areas

**Rectangular Moment of Inertia.** As stated in Art. 5-4, the moment of inertia of an area with respect to an axis in the plane of the area is required in the design of beams and columns. The exact moments of inertia of simple plane areas, such as a rectangle, a triangle, or a circle, must be determined by integration. The basic expression for the moment of inertia $I$ of any plane area with respect to an axis in its plane, as was mentioned in Art. 5-4, is

$$I = \int y^2 \, dA \qquad (a)$$

According to this equation, *the moment of inertia of a plane area with respect to an axis in the plane is the sum of the products of the elemental areas, each multiplied by the square of the distance from the axis to its centroid.*

**Rules.** In selecting the elemental area, the following rules must be observed:

1. All points in the elemental area must be the same distance from the axis.
2. The moment of inertia of the elemental area with respect to the inertia axis must be known. The moment of inertia of the entire area is then the summation of the moments of inertia of all the elements.

## ILLUSTRATIVE PROBLEMS

**B-1.** Find the moment of inertia of the rectangular area shown in Fig. B-1 with respect to ($a$) a centroidal axis parallel to the base $b$ and ($b$) an axis coinciding with the base.

*Solution:* ($a$) *Axis through centroid.* Let the width of the area be $b$ and its depth $h$, and let the elemental area be $b \, dy$, a strip parallel to the inertia axis $X_C$. The moment of inertia of a single strip with respect to this axis is $y^2 \, dA$ and of the entire area between the limits of $h/2$ and $-h/2$ is then

$$\left[ I = \int y^2 \, dA \right] \qquad I_{X_C} = \int_{-h/2}^{+h/2} y^2 b \, dy = b \left[ \frac{y^3}{3} \right]_{-h/2}^{+h/2}$$

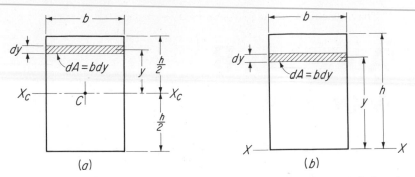

**Fig. B-1   Moment of inertia of rectangular area. (a) Axis through centroid. (b) Axis coinciding with base.**

or

$$I_{X_C} = \frac{b}{3}\left(\frac{h^3}{8} + \frac{h^3}{8}\right) = \frac{bh^3}{12} \qquad\qquad Ans. \quad (b)$$

(*b*) *Axis coinciding with base.* The summation is now made between the limits of 0 and *h*. That is,

$$\left[I = \int y^2\, dA\right] \qquad I_X = \int_0^h y^2 b\, dy = b\left[\frac{y^3}{3}\right]_0^h = \frac{bh^3}{3} \qquad\qquad Ans.$$

**B-2.** Determine the moment of inertia of the triangular area shown in Fig. B-2 with respect to (*a*) a centroidal axis parallel to the base *b* and (*b*) an axis coinciding with the base.

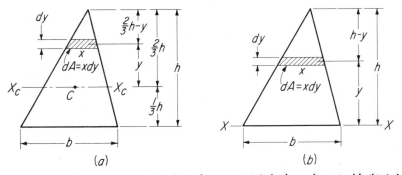

**Fig. B-2   Moment of inertia of a triangular area. (a) Axis through centroid. (b) Axis coinciding with base.**

*Solution:* (*a*) *Axis through centroid.* The elemental area selected is a strip parallel to the centroidal inertia axis $X_C$. The moment of inertia of this strip with respect to axis $X_C$ is $y^2\, dA$, or $y^2 x\, dy$, and the entire triangle is then

$$\left[I = \int y^2\, dA\right] \qquad I_{X_C} = \int_{-h/3}^{+2h/3} y^2 x\, dy \qquad\qquad (c)$$

By similar triangles

$$\frac{x}{\frac{2}{3}h - y} = \frac{b}{h} \qquad or \qquad x = \frac{b}{h}\left(\frac{2}{3}h - y\right) \qquad\qquad (d)$$

The moment of inertia of the entire area is then

$$\left[I = \int y^2 \, dA\right] \quad I_{X_C} = \int_{-h/3}^{+2h/3} y^2 \frac{b}{h}\left(\frac{2}{3}h - y\right) dy$$

$$= \frac{2}{3}b \int_{-h/3}^{+2h/3} y^2 \, dy - \frac{b}{h} \int_{-h/3}^{+2h/3} y^3 \, dy$$

$$= \frac{2}{3}b \left[\frac{y^3}{3}\right]_{-h/3}^{+2h/3} - \frac{b}{h}\left[\frac{y^4}{4}\right]_{-h/3}^{+2h/3}$$

$$= \frac{2}{9}b\left(\frac{8}{27}h^3 + \frac{1}{27}h^3\right) - \frac{b}{4h}\left(\frac{16}{81}h^4 - \frac{1}{81}h^4\right)$$

or
$$I_{X_C} = \frac{bh^3}{36} \qquad\qquad Ans. \quad (e)$$

(b) *Axis coinciding with base.* When the same elemental strip is used,

$$\left[I = \int y^2 \, dA\right] \qquad\qquad I_X = \int_0^h y^2 x \, dy \qquad\qquad (f)$$

But by similar triangles

$$\frac{x}{h - y} = \frac{b}{h} \qquad \text{and} \qquad x = \frac{b}{h}(h - y) \qquad\qquad (g)$$

The moment of inertia of the entire area is then

$$\left[I = \int y^2 \, dA\right] \qquad\qquad I_X = \int_0^h y^2 \cdot \frac{b}{h}(h - y) \, dy$$

$$= b \int_0^h y^2 \, dy - \frac{b}{h}\int_0^h y^3 \, dy$$

$$= b\left[\frac{y^3}{3}\right]_0^h - \frac{b}{h}\left[\frac{y^4}{4}\right]_0^h \qquad\qquad Ans. \quad (h)$$

$$= \frac{bh^3}{3} - \frac{bh^3}{4} = \frac{bh^3}{12}$$

**B-3.** Find the rectangular moment of inertia of the circular area shown in Fig. B-3, with respect to a diameter.

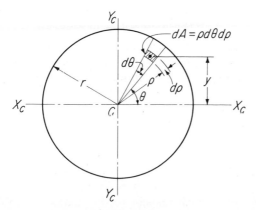

Fig. B-3 Rectangular moment of inertia of a circular area.

*Solution:* The elemental area is selected as shown in Fig. B-3. Its moment of inertia with respect to the $X_C$ axis is $y^2\ dA$, or $\rho^2 \sin^2 \theta\ \rho\ d\theta\ d\rho$, since $y = \rho \sin \theta$. A double integration is now necessary. The moment of inertia of the elemental sector is

$$\int_0^r \rho^2 \sin^2 \theta\ \rho\ d\theta\ d\rho$$

When this quantity is then again integrated between the limits of $\theta = 0$ and $\theta = 2\pi$ radians, we obtain the moment of inertia of the entire circle. That is,

$$\left[ I = \int y^2\ dA \right] \qquad I_{X_C} = \int_0^r \int_0^{2\pi} \rho^3\ d\rho \sin^2 \theta\ d\theta$$

$$= \frac{r^4}{4} \int_0^{2\pi} \sin^2 \theta\ d\theta = \frac{\pi r^4}{4} \qquad \textit{Ans.} \quad (i)$$

**The Transfer Formula for Parallel Axes.** Article 5-7 describes the method of computing the moment of inertia of an area with respect to a noncentroidal axis in the plane of the area. This was accomplished by use of Eq. (5-5), the so-called **transfer formula,** the derivation of which is shown below.

The moment of inertia of the elemental area $dA$ in Fig. B-4 with respect to axis $XX$ is

$$I_X = \int y_1^2\ dA = \int (y + d)^2\ dA \qquad (j)$$

Expanding this, we obtain

$$I_X = \int y^2\ dA + 2d\int y\ dA + d^2 \int dA \qquad (k)$$

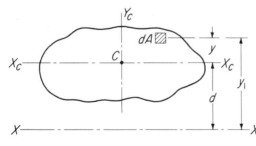

Fig. B-4   Moment of inertia of an area with respect to a noncentroidal axis.

The first term is recognized as the moment of inertia of the area with respect to its own centroidal axis $X_C$. The second term, when integrated, becomes $\bar{y}A$, the first moment of the area with respect to axis $X_C$. When this axis passes through the centroid of the area, as it does, $\bar{y}$ becomes zero, since the first moment of an area with respect to a centroidal axis is zero. Hence the second term becomes zero. The third term, when integrated, becomes $Ad^2$. Accordingly

$$I_X = I_{X_C} + Ad^2 \qquad \textbf{(5-4)}$$

As an example of the use of this equation, let us obtain the moment of inertia of the triangle in Prob. B-2 with respect to the centroidal axis $X_C$, knowing its moment of inertia with respect to the $X$ axis to be $bh^3/12$.

Now $d$, the distance between the two parallel axes, is $1/3 h$. Hence

$$\frac{bh^3}{12} = I_{X_c} + \frac{bh}{2}\left(\frac{h}{3}\right)^2 \tag{l}$$

and

$$I_{X_c} = \frac{bh^3}{12} - \frac{bh^3}{18} = \frac{bh^3}{36} \tag{m}$$

**B-4.** In Prob. B-3 the moment of inertia of the circular area with respect to a diameter was found to be $1/4 \pi r^4$. Consequently the moment of inertia of a semicircle with respect to the same axis would be one-half of that, which is $1/8 \pi r^4$, or $0.3927R^4$, as listed in Table 5-1. By use of the transfer formula, find the moment of inertia of the semicircular area with respect to the centroidal axis $X_C$, which is given as $0.110R^4$ $(r = R)$.

*Solution:* Since $A = \pi R^2/2$ and $d = 0.424R$ (Art. 5-2),

$$[I_X = I_{X_c} + Ad^2] \qquad 0.3927R^4 = I_{X_c} + \frac{1}{2}\pi R^2 (0.424R)^2$$

or

$$I_{X_c} = 0.3927R^4 - 0.2827R^4 = 0.1100R^4 \qquad \text{Ans.} \quad (n)$$

**Polar Moment of Inertia.** In the design of shafts and other members subjected to twisting, we must know the moment of inertia of the cross-sectional area of the member with respect to an axis *perpendicular* to the plane of the area. This is called the *polar moment of inertia* of the area and is denoted by the symbol $J$.

**B-5.** Find the polar moment of inertia of the circular area shown in Fig. B-5 with respect to an axis perpendicular to the plane of the area and passing through its centroid $C$.

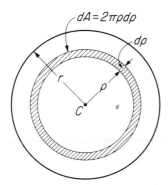

**Fig. B-5** **Polar moment of inertia of a circular area.**

*Solution:* The elemental area selected is circular, as shown in Fig. B-5. Its moment of inertia with respect to an axis through $C$ is $\rho^2\, dA$. Integrating this quantity between the limits of $\rho = 0$ and $\rho = r$ gives the moment of inertia of the entire area. That is, since $dA = 2\pi\rho\, d\rho$,

$$\left[ J = \int \rho^2\, dA \right] \qquad J = \int_0^r \rho^2\, 2\pi\rho\, d\rho$$

$$= 2\pi \int_0^r \rho^3\, d\rho = 2\pi \left[ \frac{\rho^4}{4} \right]_0^r = \frac{\pi r^4}{2} \qquad \text{Ans.} \quad (o)$$

# Appendix C
# Derivation of the
# Laws of Beam
# Diagrams

**Relation between Load and Shear and between Shear and Moment in Beams.** Figure C-1 shows a loaded, simple beam. Below is an enlarged view of a small vertical segment of length $dx$ of this beam. Let the shear and moment on the left side of this segment be $V$ and $M$. If the length of the segment is infinitesimally small and is denoted by $dx$, the resultant of the load on the segment is $w\,dx$ concentrated at a distance $dx/2$ from either vertical face of the segment.

On the right face of the segment the shear has been decreased by an infinitesimally small amount $dV$ equal to the load $w\,dx$ on the segment. We may now write an equation expressing the relation between load and shear at any section in a beam. That is,

$$[\Sigma V = 0] \qquad\qquad V = w\,dx + (V - dV) \qquad\qquad (a)$$

or $\qquad\qquad dV = w\,dx \qquad \text{and} \qquad w = \dfrac{dV}{dx} \qquad\qquad (b)$

Also, the moment of the stresses on the right face of the segment has been increased by an infinitesimally small amount $dM$. Since $\Sigma M = 0$ about any point such as $c$, for example, an equation may now be written expressing the relation between shear and moment at any section in the beam. That is,

$$[\Sigma M_c = 0] \qquad\qquad (M + dM) + w\,dx\,\dfrac{dx}{2} = M + V\,dx$$

The term $w\,dx\,(dx/2)$ may be disregarded, since it becomes $w\,(dx)^2/2$, which is an infinitesimal of the second order. Then,

$$dM = V\,dx \qquad \text{and} \qquad V = \dfrac{dM}{dx} \qquad\qquad (c)$$

Equation $(b)$ is interpreted as follows: At any section along a beam, the

382

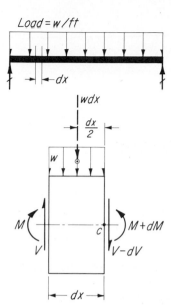

Fig. C-1   Relation between load and shear and between shear and moment.

equation expressing the load $w$ is the first derivative of the equation expressing the shear $V$. Conversely, *the shear equation is the first integral of the load equation.*

In similar manner, from Eq. ($c$), we deduce that *the moment equation is the first integral of the shear equation.*

Expressing these relations mathematically, we have

$$V = \int w \, dx \tag{C-1}$$

and
$$M = \int V \, dx \tag{C-2}$$

**Proof of First and Second Laws of Beam Diagrams.** The **first law of beam diagrams,** as stated in Chap. 6, is

$$\begin{bmatrix} \textbf{Slope of curve at any point} \\ \textbf{in any diagram} \end{bmatrix} = \begin{bmatrix} \textbf{length of ordinate at corresponding} \\ \textbf{point in next higher diagram} \end{bmatrix}$$

*Proof:* In the shear diagram in Fig. C-2, the slope of the shear curve at any point is $dV/dx = w$, the ordinate in the load (next higher) diagram, also verified by Eq. ($b$). Similarly, in the moment diagram the slope of the moment curve at any point is $dM/dx = V$, the ordinate in the shear (next higher) diagram, also verified by Eq. ($c$).

The **second law of beam diagrams,** as stated in Chap. 6, is

$$\begin{bmatrix} \textbf{Difference in length of any two} \\ \textbf{ordinates in any diagram} \end{bmatrix} = \begin{bmatrix} \textbf{area between corresponding ordi-} \\ \textbf{nates in next higher diagram} \end{bmatrix}$$

*Proof:* In the shear diagram in Fig. C-2, the difference in length of any two shear ordinates is $dV$, which equals $w \, dx$ [see Eq. ($b$)], the area

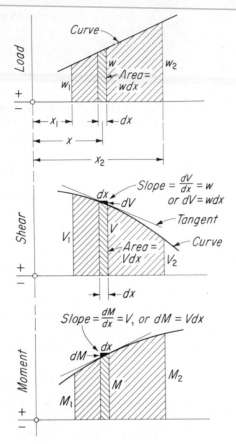

**Fig. C-2   Relation between load and shear diagrams and between shear and moment diagrams.**

between corresponding ordinates in the load (next higher) diagram. Thus the *difference* in length of any two shear ordinates $V_1$ and $V_2$ equals the entire area under the load curve between load ordinates $w_1$ and $w_2$. Similarly, in the moment diagram the difference in length of any two moment ordinates is $dM$, which equals $V\,dx$ [see Eq. $(c)$], the area between corresponding ordinates in the shear (next higher) diagram. Thus the *difference* in length of any two moment ordinates $M_1$ and $M_2$ equals the entire area under the shear curve between shear ordinates $V_1$ and $V_2$.

**Location of Section of Maximum Moment.**   Since the lengths of the shear ordinates are equal to the slope of the moment-diagram curve at corresponding points along the beam, it follows, then, that the shear will be zero or will pass through zero where the bending moment is a maximum. the principle is illustrated in Figs. 6-20, 6-21, and 6-26.

# Appendix D
# Derivation of the
# Flexure Formula.
# Location of the
# Neutral Axis

Figure D-1 shows a side view and a cross-sectional view of a prismatic beam subjected to bending loads. The beam is of irregular cross section but is symmetrical about its vertical axis. Because of the shape of the beam, the neutral axis clearly lies nearer the bottom of the beam. The unit stress on any fiber is proportional to its distance from the neutral axis. Let $s$ be the maximum unit stress in the extreme fiber at distance $c$ from the neutral axis and $f$ the fiber stress at distance $y$ from the neutral axis. Then by similarity of triangles

$$\frac{s}{c} = \frac{f}{y} \qquad \text{and} \qquad f = \frac{s}{c}y \qquad\qquad (a)$$

The total force exerted on the infinitesimal area $dA$ is the product of the unit stress $f$ and the area $dA$. Then, substituting $(s/c)y$ for $f$, we obtain

$$f\, dA = \left(\frac{s}{c}\, y\right) dA \qquad\qquad (b)$$

The internal resisting moment RM of this force about the neutral axis is

$$\text{RM} = \left(\frac{s}{c}\, y\right) dA\ y = \frac{s}{c}\, y^2\, dA \qquad\qquad (c)$$

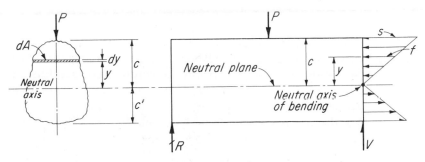

**Fig. D-1** Derivation of the general flexure formula.

The total internal resisting moment RM about the neutral axis is then obtained by a summation of the moments of the forces on all the elemental areas. That is,

$$\text{RM} = \frac{s}{c} \int_{-c'}^{+c} y^2 \, dA \qquad (d)$$

This latter term, $\int y^2 \, dA$, is called the **static moment of inertia,** or second moment of area, of the cross section of the beam and is represented by the symbol $I$. Then, substituting $I$ in Eq. $(d)$, we obtain

$$\text{RM} = M = s\frac{I}{c} \qquad (7\text{-}1)$$

where $M$ is the external bending moment which at any section equals the internal resisting moment RM.

**Location of the Neutral Axis.** In connection with the design of a beam, the required moment of inertia is that with respect to the neutral axis of its cross-sectional area. Consequently it is necessary to locate this neutral axis before a determination of the moment of inertia can be made. The following development is proof that **the neutral axis of the cross-sectional area of a beam passes through the centroid of that area.**

At any section of a bent beam the resultants $C$ and $T$ of the compressive and tensile stresses, respectively, are equal. That is, the total compressive stress on one side of the neutral axis equals the total tensile stress on the other side.

Reference to Fig. D-1 and to Eq. $(b)$ indicates that the force on any elemental area $dA$ on either side of the neutral axis is $(s/c)y \, dA$. The algebraic sum of all forces on the entire cross-sectional area is then

$$(s/c) \int_{-c'}^{+c} y \, dA$$

since the total sum $C$ of the compressive stresses must always equal the total sum $T$ of the tensile stresses. Since $s/c$ obviously cannot equal zero, it follows that the term

$$\int_{-c'}^{+c} y \, dA$$

must equal zero. But the general term $\int y \, dA$ expresses the first moment of the area of the cross section about the neutral axis, and this moment can be zero only if the neutral axis passes through the centroid of the section. The statement above that the neutral axis always passes through the centroid of the cross-sectional area of a beam has thus been proved.

# Appendix E
# Tables

Much of the material in the following tables has been taken from "Steel Construction," the manual of the American Institute of Steel Construction, and is printed here with their permission.

## AVERAGE MECHANICAL PROPERTIES OF SOME ENGINEERING MATERIALS

| Material | Weight lb/ft³ (kN/m³) | Allowable Stresses ksi (MPa) | | | Elastic Limit ksi (MPa) | | Ultimate Strength ksi (MPa) | | | Modulus of Elasticity ksi (MPa) |
|---|---|---|---|---|---|---|---|---|---|---|
| | | Tension | Compression | Shear | Tension | Compression | Tension | Compression | Shear | |
| Steel, carbon ASTM-A36 | 490 (77) | 22 (152) | 22 (152) | 14.5 (100) | 36ᵃ (248)ᵃ | | 58–80 (400–550) | | | 29 000 (200 000) |
| Iron, gray cast | 450 (70) | 5–8 (34–55) | 20–30 (138–210) | 7.5–12.5 (52–86) | | | | | | 16 000 (110 000) |
| malleable | 475 (75) | 6 (40) | 16 (110) | 6 (40) | 18 (124) | 18 (124) | 32 (220) | 46 (320) | 40 (280) | 20 000 (138 000) |
| Aluminum Alloy 14S-T4 | 174 (27.3) | 16 (110) | 16 (110) | 10 (69) | 41 (285) | 41 (285) | 62 (430) | | 38 (262) | 10 000 (69 000) |
| 6061-T6 | 174 (27.3) | 15 (103) | 14 (97) | 10 (69) | 30 (207) | 30 (207) | 42 (290) | | 27 (186) | 10 000 (69 000) |
| Bronze | 535 (84) | 25 (172) | 25 (172) | | 40 (280) | 40 (280) | 75 (520) | 120 (825) | | 14 500 (100 000) |
| Copper, rods, bolts | 555 (87) | 8 (55) | 8 (55) | | 10 (69) | 10 (69) | 32 (220) | 32 (220) | | 16 000 (110 000) |
| Brick, lime mortar | 120 (19) | | 0.25 (1.7) | 0.02 (0.14) | | | | | | 2 000 (13 800) |
| cement mortar | 130 (20) | | 0.3 (2) | 0.025 (0.17) | | | | | | 2 000 (13 800) |
| Concrete, 1:2:4 mix | 150 (23) | | 1 (7) | | | | | | | 2 000 (13 800) |

ᵃ Minimum yield point.

## TIMBER
### PROPERTIES FOR DESIGNING

| Nominal Size in. | American Standard Dressed Size in. | Area of Section in.² | Weight lb/ft | Moment of Inertia in.⁴ | Section Modulus in.³ |
|---|---|---|---|---|---|
| 2 × 4 | 1½ × 3½ | 5.25 | 1.46 | 5.36 | 3.06 |
| 6 | 5½ | 8.25 | 2.29 | 20.8 | 7.56 |
| 8 | 7¼ | 10.88 | 3.02 | 47.6 | 13.14 |
| 10 | 9¼ | 13.88 | 3.85 | 98.9 | 21.4 |
| 12 | 11¼ | 16.88 | 4.69 | 178 | 31.6 |
| 14 | 13¼ | 19.88 | 5.52 | 291 | 43.9 |
| 3 × 4 | 2½ × 3½ | 8.75 | 2.43 | 8.93 | 5.10 |
| 6 | 5½ | 13.75 | 3.82 | 34.7 | 12.6 |
| 8 | 7¼ | 18.12 | 5.04 | 79.4 | 21.9 |
| 10 | 9¼ | 23.12 | 6.42 | 165 | 35.6 |
| 12 | 11¼ | 28.12 | 7.81 | 297 | 52.7 |
| 14 | 13¼ | 33.12 | 9.20 | 485 | 73.2 |
| 4 × 4 | 3½ × 3½ | 12.25 | 3.40 | 12.5 | 7.15 |
| 6 | 5½ | 19.25 | 5.35 | 48.5 | 17.65 |
| 8 | 7¼ | 25.38 | 7.05 | 111.1 | 30.66 |
| 10 | 9¼ | 32.38 | 8.93 | 231 | 49.9 |
| 12 | 11¼ | 39.38 | 10.94 | 415 | 73.8 |
| 14 | 13¼ | 46.38 | 12.88 | 678 | 102.4 |

| Nominal Size in. | American Standard Dressed Size in. | Area of Section in.² | Weight lb/ft | Moment of Inertia in.⁴ | Section Modulus in.³ |
|---|---|---|---|---|---|
| 6 × 6 | 5½ × 5½ | 30.3 | 8.40 | 76.3 | 27.7 |
| 8 | 7½ | 41.3 | 11.4 | 193 | 51.6 |
| 10 | 9½ | 52.3 | 14.5 | 393 | 82.7 |
| 12 | 11½ | 63.3 | 17.5 | 697 | 121 |
| 14 | 13½ | 74.3 | 20.6 | 1128 | 167 |
| 8 × 8 | 7½ × 7½ | 56.3 | 15.6 | 264 | 70.3 |
| 10 | 9½ | 71.3 | 19.8 | 536 | 113 |
| 12 | 11½ | 86.3 | 23.9 | 951 | 165 |
| 14 | 13½ | 101.3 | 28.0 | 1538 | 228 |
| 10 × 10 | 9½ × 9½ | 90.3 | 25.0 | 679 | 143 |
| 12 | 11½ | 109 | 30.3 | 1204 | 209 |
| 14 | 13½ | 128 | 35.6 | 1948 | 289 |
| 16 | 15½ | 147 | 40.9 | 2948 | 380 |
| 12 × 12 | 11½ × 11½ | 132 | 36.7 | 1458 | 253 |
| 14 | 13½ | 155 | 43.1 | 2358 | 349 |
| 16 | 15½ | 178 | 49.5 | 3569 | 460 |
| 18 | 17½ | 201 | 55.9 | 5136 | 587 |

All properties and weights given are for dressed sizes only. Weights are based on 40 lb/ft³ (6.3 kN/m³). This tabulation is a partial listing; other sizes are available.

## TIMBER (continued)
### SI PROPERTIES FOR DESIGNING

| Nominal Size | Dressed Size | Area ÷ $10^3$ | Weight | $I$ ÷ $10^6$ | $Z$ ÷ $10^3$ |
|---|---|---|---|---|---|
| mm | mm | mm$^2$ | N/m | mm$^4$ | mm$^3$ |
| 50 × 100 | 40 × 90 | 3.6 | 22.7 | 2.43 | 54.0 |
| 150 | 140 | 5.6 | 35.3 | 9.15 | 131 |
| 200 | 188 | 7.52 | 47.4 | 22.2 | 236 |
| 250 | 238 | 9.52 | 60.0 | 44.9 | 378 |
| 300 | 288 | 11.52 | 72.6 | 79.6 | 553 |
| 350 | 338 | 13.52 | 85.2 | 129 | 762 |
| 75 × 150 | 65 × 140 | 9.1 | 57.3 | 14.9 | 212 |
| 200 | 188 | 12.2 | 77.0 | 36.0 | 383 |
| 250 | 238 | 15.5 | 97.5 | 73.0 | 614 |
| 300 | 288 | 18.7 | 118 | 129 | 899 |
| 350 | 338 | 22.0 | 138 | 209 | 1238 |
| 100 × 150 | 90 × 140 | 12.6 | 79.4 | 20.6 | 294 |
| 200 | 188 | 16.9 | 107 | 49.8 | 530 |
| 250 | 238 | 21.4 | 135 | 101.1 | 850 |
| 300 | 288 | 25.9 | 163 | 179 | 1244 |
| 350 | 338 | 30.4 | 192 | 290 | 1714 |

| Nominal Size | Dressed Size | Area ÷ $10^3$ | Weight | $I$ ÷ $10^6$ | $Z$ ÷ $10^3$ |
|---|---|---|---|---|---|
| mm | mm | mm$^2$ | N/m | mm$^4$ | mm$^3$ |
| 150 × 150 | 140 × 140 | 19.6 | 123 | 32.0 | 457 |
| 200 | 188 | 26.3 | 166 | 77.5 | 825 |
| 250 | 238 | 33.3 | 210 | 157.3 | 1322 |
| 300 | 288 | 40.3 | 254 | 279 | 1935 |
| 350 | 338 | 47.3 | 298 | 450 | 2666 |
| 200 × 200 | 188 × 188 | 35.3 | 223 | 104 | 1107 |
| 250 | 238 | 44.7 | 282 | 211 | 1775 |
| 300 | 288 | 54.1 | 341 | 374 | 2599 |
| 350 | 338 | 63.5 | 400 | 605 | 3580 |
| 250 × 250 | 238 × 238 | 56.6 | 357 | 267 | 2247 |
| 300 | 288 | 68.5 | 432 | 474 | 3290 |
| 350 | 338 | 80.4 | 507 | 766 | 4531 |
| 300 × 300 | 288 × 288 | 82.9 | 523 | 573 | 3981 |
| 350 | 338 | 97.3 | 613 | 927 | 5484 |
| 400 | 388 | 111.7 | 704 | 1402 | 7226 |

All properties and weights given are for dressed sizes only. Weights are based on 40 lb/ft³ (6.3 kN/m³). This tabulation is a partial listing; other sizes are available.

# SCREW THREADS
## Unified Standard Series—UNC
### ANSI B1.1—1974

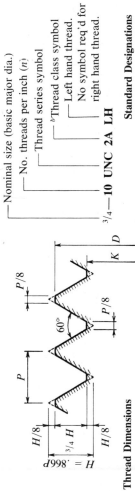

Nominal size (basic major dia.)
No. threads per inch (n)
Thread series symbol
b Thread class symbol
Left hand thread.
No symbol req'd for right hand thread.

**3/4 — 10 UNC 2A LH**

$H = .866P$

### Thread Dimensions

| Diameter | | Area | | | Th'ds per Inch |
|---|---|---|---|---|---|
| Basic Major $D$ | Root $K$ | Gross $A_D$ | Root $A_K$ | [a]Tensile Stress | $n$ |
| in. | in. | in.² | in.² | in.² | |
| 1/4 | 0.185 | 0.049 | 0.027 | 0.032 | 20 |
| 3/8 | 0.294 | 0.110 | 0.068 | 0.078 | 16 |
| 1/2 | 0.400 | 0.196 | 0.126 | 0.142 | 13 |
| 5/8 | 0.507 | 0.307 | 0.202 | 0.226 | 11 |
| 3/4 | 0.620 | 0.442 | 0.302 | 0.334 | 10 |
| 7/8 | 0.731 | 0.601 | 0.419 | 0.462 | 9 |
| 1 | 0.838 | 0.785 | 0.551 | 0.606 | 8 |
| 1 1/8 | 0.939 | 0.994 | 0.693 | 0.763 | 7 |
| 1 1/4 | 1.064 | 1.227 | 0.890 | 0.969 | 7 |

### Standard Designations

| Diameter | | Area | | | Th'ds per Inch |
|---|---|---|---|---|---|
| Basic Major $D$ | Root $K$ | Gross $A_D$ | Root $A_K$ | [a]Tensile Stress | $n$ |
| in. | in. | in.² | in.² | in.² | |
| 1 3/8 | 1.158 | 1.485 | 1.05 | 1.16 | 6 |
| 1 1/2 | 1.283 | 1.767 | 1.29 | 1.41 | 6 |
| 1 3/4 | 1.490 | 2.405 | 1.74 | 1.90 | 5 |
| 2 | 1.711 | 3.142 | 2.30 | 2.50 | 4 1/2 |
| 2 1/4 | 1.961 | 3.976 | 3.02 | 3.25 | 4 1/2 |
| 2 1/2 | 2.175 | 4.909 | 3.72 | 4.00 | 4 |
| 3 | 2.675 | 7.069 | 5.62 | 5.97 | 4 |
| 4 | 3.675 | 12.566 | 10.6 | 11.- | 4 |

[a] Tensile stress area $= 0.7854 \left( D - \dfrac{.9743}{n} \right)^2$.

[b] 2A denotes Class 2 fit external thread. 2B denotes Class 2 fit internal thread.

# SCREW THREADS (continued)

## LENGTH OF THREAD ON BOLTS

ANSI B18.2—1972

| Length of Bolt | Diameter of Bolt, D, in. | | | | | | | | | | | | | | | | |
|---|---|---|---|---|---|---|---|---|---|---|---|---|---|---|---|---|---|
| | 1/4 | 3/8 | 1/2 | 5/8 | 3/4 | 7/8 | 1 | 1 1/8 | 1 1/4 | 1 3/8 | 1 1/2 | 1 3/4 | 2 | 2 1/4 | 2 1/2 | 2 3/4 | 3 |
| To 6 in. incl. | 3/4 | 1 | 1 1/4 | 1 1/2 | 1 3/4 | 2 | 2 1/4 | 2 1/2 | 2 3/4 | 3 | 3 1/4 | 3 3/4 | 4 1/4 | 4 3/4 | 5 1/4 | 5 3/4 | 6 1/4 |
| Over 6 in. | 1 | 1 1/4 | 1 1/2 | 1 3/4 | 2 | 2 1/4 | 2 1/2 | 2 3/4 | 3 | 3 1/4 | 3 1/2 | 4 | 4 1/2 | 5 | 5 1/2 | 6 | 6 1/2 |

Note 1. Thread length for bolts up to 6 in. long is $2D + 1/4$. For bolts over 6 in. long, thread length is $2D + 1/2$. These proportions may be used to compute thread length for diameters not shown in the table. Bolts which are too short for listed or computed thread lengths are threaded as close to the head as possible.

Note 2. For thread lengths for high strength bolts, refer to "Specifications for Structural Joints Using ASTM A325 Bolts."

## METRIC BOLTS

| Nominal bolt size | M5 | M6 | M8 | M10 | M12 | M14 | M16 | M20 | M24 | M30 |
|---|---|---|---|---|---|---|---|---|---|---|
| Thread pitch, mm | 0.8 | 1 | 1.25 | 1.5 | 1.75 | 2 | 2 | 2.5 | 3 | 3.5 |
| Shear area, mm² | 19.6 | 28.3 | 50.3 | 78.5 | 113 | 154 | 201 | 314 | 452 | 706 |
| Tensile area, mm² | 12.7 | 17.9 | 32.8 | 52.3 | 76.2 | 110 | 144 | 225 | 324 | 519 |

| Nominal bolt size | M36 | M42 | M48 | M56 | M64 | M72 | M80 | M90 | M100 |
|---|---|---|---|---|---|---|---|---|---|
| Thread pitch, mm | 4 | 4.5 | 5 | 5.5 | 6 | 6 | 6 | 6 | 6 |
| Shear area, mm² | 1020 | 1380 | 1810 | 2460 | 3220 | 4070 | 5030 | 6360 | 7850 |
| Tensile area, mm² | 759 | 1050 | 1380 | 1910 | 2520 | 3280 | 4140 | 5360 | 6740 |

## W SHAPES     PROPERTIES FOR DESIGNING

| Designation | Flange | | Beam Depth | Web Thickness | Area | Axis X-X | | | Axis Y-Y | | |
|---|---|---|---|---|---|---|---|---|---|---|---|
| | Width | Thickness | | | | $I$ | $Z$ | $k$ | $I$ | $Z$ | $k$ |
| Nom. $D \times$ wt/ft | in. | in. | in. | in. | in.$^2$ | in.$^4$ | in.$^3$ | in. | in.$^4$ | in.$^3$ | in. |
| W21 × 111 | 12³/₈ | ⁷/₈ | 21¹/₂ | 0.550 | 32.7 | 2670 | 249 | 9.05 | 274 | 44.5 | 2.90 |
| 101 | 12¹/₄ | ¹³/₁₆ | 21³/₈ | 0.500 | 29.8 | 2420 | 227 | 9.02 | 248 | 40.3 | 2.89 |
| W21 × 73 | 8¹/₄ | ³/₄ | 21¹/₄ | 0.455 | 21.5 | 1600 | 151 | 8.64 | 70.6 | 17.0 | 1.81 |
| 68 | 8¹/₄ | ¹¹/₁₆ | 21¹/₈ | 0.430 | 20.0 | 1480 | 140 | 8.60 | 64.7 | 15.7 | 1.80 |
| 62 | 8¹/₄ | ⁵/₈ | 21 | 0.400 | 18.3 | 1330 | 127 | 8.54 | 57.5 | 13.9 | 1.79 |
| W21 × 57 | 6¹/₂ | ⁵/₈ | 21 | 0.405 | 16.7 | 1170 | 111 | 8.36 | 30.6 | 9.35 | 1.35 |
| 50 | 6¹/₂ | ⁹/₁₆ | 20⁷/₈ | 0.380 | 14.7 | 984 | 94.5 | 8.18 | 24.9 | 7.64 | 1.30 |
| 44 | 6¹/₂ | ⁷/₁₆ | 20⁵/₈ | 0.350 | 13.0 | 843 | 81.6 | 8.06 | 20.7 | 6.36 | 1.26 |
| W18 × 46 | 6 | ⁵/₈ | 18 | 0.360 | 13.5 | 712 | 78.8 | 7.25 | 22.5 | 7.43 | 1.29 |
| 40 | 6 | ¹/₂ | 17⁷/₈ | 0.315 | 11.8 | 612 | 68.4 | 7.21 | 19.1 | 6.35 | 1.27 |
| 35 | 6 | ⁷/₁₆ | 17³/₄ | 0.300 | 10.3 | 510 | 57.6 | 7.04 | 15.3 | 5.12 | 1.22 |
| W16 × 100 | 10³/₈ | 1 | 17 | 0.585 | 29.4 | 1490 | 175 | 7.10 | 186 | 35.7 | 2.51 |
| 89 | 10³/₈ | ⁷/₈ | 16³/₄ | 0.525 | 26.2 | 1300 | 155 | 7.05 | 163 | 31.4 | 2.49 |
| 77 | 10¹/₄ | ³/₄ | 16¹/₂ | 0.455 | 22.6 | 1110 | 134 | 7.00 | 138 | 26.9 | 2.47 |
| 67 | 10¹/₄ | ¹¹/₁₆ | 16³/₈ | 0.395 | 19.7 | 954 | 117 | 6.96 | 119 | 23.2 | 2.46 |
| W16 × 57 | 7¹/₈ | ¹¹/₁₆ | 16³/₈ | 0.430 | 16.8 | 758 | 92.2 | 6.72 | 43.1 | 12.1 | 1.60 |
| 50 | 7¹/₈ | ⁵/₈ | 16¹/₄ | 0.380 | 14.7 | 659 | 81.0 | 6.68 | 37.2 | 10.5 | 1.59 |
| 45 | 7 | ⁹/₁₆ | 16¹/₈ | 0.345 | 13.3 | 586 | 72.7 | 6.65 | 32.8 | 9.34 | 1.57 |
| 40 | 7 | ¹/₂ | 16 | 0.305 | 11.8 | 518 | 64.7 | 6.63 | 28.9 | 8.25 | 1.57 |
| 36 | 7 | ⁷/₁₆ | 15⁷/₈ | 0.295 | 10.6 | 448 | 56.5 | 6.51 | 24.5 | 7.00 | 1.52 |
| W16 × 31 | 5¹/₂ | ⁷/₁₆ | 15⁷/₈ | 0.275 | 9.12 | 375 | 47.2 | 6.41 | 12.4 | 4.49 | 1.17 |
| 26 | 5¹/₂ | ³/₈ | 15³/₄ | 0.250 | 7.68 | 301 | 38.4 | 6.26 | 9.59 | 3.49 | 1.12 |
| W14 × 99 | 14⁵/₈ | ³/₄ | 14¹/₈ | 0.485 | 29.1 | 1110 | 157 | 6.17 | 402 | 55.2 | 3.71 |
| 90 | 14¹/₂ | ¹¹/₁₆ | 14 | 0.440 | 26.5 | 999 | 143 | 6.14 | 362 | 49.9 | 3.70 |
| W14 × 68 | 10 | ³/₄ | 14 | 0.415 | 20.0 | 723 | 103 | 6.01 | 121 | 24.2 | 2.46 |
| 61 | 10 | ⁵/₈ | 13⁷/₈ | 0.375 | 17.9 | 640 | 92.2 | 5.98 | 107 | 21.5 | 2.45 |
| W14 × 53 | 8 | ¹¹/₁₆ | 13⁷/₈ | 0.370 | 15.6 | 541 | 77.8 | 5.89 | 57.7 | 14.3 | 1.92 |
| 48 | 8 | ⁵/₈ | 13³/₄ | 0.340 | 14.1 | 485 | 70.3 | 5.85 | 51.4 | 12.8 | 1.91 |
| 43 | 8 | ¹/₂ | 13⁵/₈ | 0.305 | 12.6 | 428 | 62.7 | 5.82 | 45.2 | 11.3 | 1.89 |
| W14 × 38 | 6³/₄ | ¹/₂ | 14¹/₈ | 0.310 | 11.2 | 385 | 54.6 | 5.87 | 26.7 | 7.88 | 1.55 |
| 34 | 6³/₄ | ⁷/₁₆ | 14 | 0.285 | 10.0 | 340 | 48.6 | 5.83 | 23.3 | 6.91 | 1.53 |
| 30 | 6³/₄ | ³/₈ | 13⁷/₈ | 0.270 | 8.85 | 291 | 42.0 | 5.73 | 19.6 | 5.82 | 1.49 |

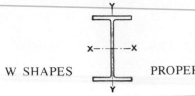

W SHAPES · PROPERTIES FOR DESIGNING
(continued)

| Designation | Flange Width | Flange Thick-ness | Beam Depth | Web Thick-ness | Area | Axis X-X I | Axis X-X Z | Axis X-X k | Axis Y-Y I | Axis Y-Y Z | Axis Y-Y k |
|---|---|---|---|---|---|---|---|---|---|---|---|
| Nom. $D \times$ wt/ft | in. | in. | in. | in. | in.² | in.⁴ | in.³ | in. | in.⁴ | in.³ | in. |
| W14 × 26 | 5 | $7/16$ | $13^7/8$ | 0.255 | 7.69 | 245 | 35.3 | 5.65 | 8.91 | 3.54 | 1.08 |
| 22 | 5 | $5/16$ | $13^3/4$ | 0.230 | 6.49 | 199 | 29.0 | 5.45 | 7.00 | 2.80 | 1.04 |
| W12 × 120 | $12^3/8$ | $1^1/8$ | $13^1/8$ | 0.710 | 35.3 | 1070 | 163 | 5.51 | 345 | 56.0 | 3.13 |
| 106 | $12^1/4$ | 1 | $12^7/8$ | 0.610 | 31.2 | 933 | 145 | 5.47 | 301 | 49.3 | 3.11 |
| 96 | $12^1/8$ | $7/8$ | $12^3/4$ | 0.550 | 28.2 | 833 | 131 | 5.44 | 270 | 44.4 | 3.09 |
| 87 | $12^1/8$ | $13/16$ | $12^1/2$ | 0.515 | 25.6 | 740 | 118 | 5.38 | 241 | 39.7 | 3.07 |
| 79 | $12^1/8$ | $3/4$ | $12^3/8$ | 0.470 | 23.2 | 662 | 107 | 5.34 | 216 | 35.8 | 3.05 |
| 72 | 12 | $11/16$ | $12^1/4$ | 0.430 | 21.1 | 597 | 97.4 | 5.31 | 195 | 32.4 | 3.04 |
| 65 | 12 | $5/8$ | $12^1/8$ | 0.390 | 19.1 | 533 | 87.9 | 5.28 | 174 | 29.1 | 3.02 |
| W12 × 58 | 10 | $5/8$ | $12^1/4$ | 0.360 | 17.0 | 475 | 78.0 | 5.28 | 107 | 21.4 | 2.51 |
| 53 | 10 | $9/16$ | 12 | 0.345 | 15.6 | 425 | 70.6 | 5.23 | 95.8 | 19.2 | 2.48 |
| W12 × 50 | $8^1/8$ | $5/8$ | $12^1/4$ | 0.370 | 14.7 | 394 | 64.7 | 5.18 | 56.3 | 13.9 | 1.96 |
| 45 | 8 | $9/16$ | 12 | 0.335 | 13.2 | 350 | 58.1 | 5.15 | 50.0 | 12.4 | 1.94 |
| 40 | 8 | $1/2$ | 12 | 0.295 | 11.8 | 310 | 51.9 | 5.13 | 44.1 | 11.0 | 1.93 |
| W12 × 35 | $6^1/2$ | $1/2$ | $12^1/2$ | 0.300 | 10.3 | 285 | 45.6 | 5.25 | 24.5 | 7.47 | 1.54 |
| 30 | $6^1/2$ | $7/16$ | $12^3/8$ | 0.260 | 8.79 | 238 | 38.6 | 5.21 | 20.3 | 6.24 | 1.52 |
| 26 | $6^1/2$ | $3/8$ | $12^1/4$ | 0.230 | 7.65 | 204 | 33.4 | 5.17 | 17.3 | 5.34 | 1.51 |
| W12 × 22 | 4 | $7/16$ | $12^1/4$ | 0.260 | 6.48 | 156 | 25.4 | 4.91 | 4.66 | 2.31 | 0.847 |
| 19 | 4 | $3/8$ | $12^1/8$ | 0.235 | 5.57 | 130 | 21.3 | 4.82 | 3.76 | 1.88 | 0.822 |
| 16 | 4 | $1/4$ | 12 | 0.220 | 4.71 | 103 | 17.1 | 4.67 | 2.82 | 1.41 | 0.773 |
| 14 | 4 | $1/4$ | $11^7/8$ | 0.200 | 4.16 | 88.6 | 14.9 | 4.62 | 2.36 | 1.19 | 0.753 |
| W10 × 112 | $10^3/8$ | $1^1/4$ | $11^3/8$ | 0.755 | 32.9 | 716 | 126 | 4.66 | 236 | 45.3 | 2.68 |
| 100 | $10^3/8$ | $1^1/8$ | $11^1/8$ | 0.680 | 29.4 | 623 | 112 | 4.60 | 207 | 40.0 | 2.65 |
| 88 | $10^1/4$ | 1 | $10^7/8$ | 0.605 | 25.9 | 534 | 98.5 | 4.54 | 179 | 34.8 | 2.63 |
| 77 | $10^1/4$ | $7/8$ | $10^5/8$ | 0.530 | 22.6 | 455 | 85.9 | 4.49 | 154 | 30.1 | 2.60 |
| 68 | $10^1/8$ | $3/4$ | $10^3/4$ | 0.470 | 20.0 | 394 | 75.7 | 4.44 | 134 | 26.4 | 2.59 |
| 60 | $10^1/8$ | $11/16$ | $10^1/4$ | 0.420 | 17.6 | 341 | 66.7 | 4.39 | 116 | 23.0 | 2.57 |
| 54 | 10 | $5/8$ | $10^1/8$ | 0.370 | 15.8 | 303 | 60.0 | 4.37 | 103 | 20.6 | 2.56 |
| 49 | 10 | $9/16$ | 10 | 0.340 | 14.4 | 272 | 54.6 | 4.35 | 93.4 | 18.7 | 2.54 |
| W10 × 45 | 8 | $5/8$ | $10^1/8$ | 0.350 | 13.3 | 248 | 49.1 | 4.32 | 53.4 | 13.3 | 2.01 |
| 39 | 8 | $1/2$ | $9^7/8$ | 0.315 | 11.5 | 209 | 42.1 | 4.27 | 45.0 | 11.3 | 1.98 |
| 33 | 8 | $7/16$ | $9^3/4$ | 0.290 | 9.71 | 170 | 35.0 | 4.19 | 36.6 | 9.20 | 1.94 |

# E-4

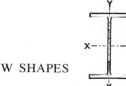

W SHAPES  PROPERTIES FOR DESIGNING
(continued)

| | Flange | | Beam Depth | Web Thickness | Area | Axis X-X | | | Axis Y-Y | | |
|---|---|---|---|---|---|---|---|---|---|---|---|
| Designation | Width | Thickness | | | | $I$ | $Z$ | $k$ | $I$ | $Z$ | $k$ |
| Nom. $D \times$ wt/ft | in. | in. | in. | in. | in.$^2$ | in.$^4$ | in.$^3$ | in. | in.$^4$ | in.$^3$ | in. |
| W10 × 30 | 5³/₄ | ¹/₂ | 10¹/₂ | 0.300 | 8.84 | 170 | 32.4 | 4.38 | 16.7 | 5.75 | 1.37 |
| 26 | 5³/₄ | ⁷/₁₆ | 10³/₈ | 0.260 | 7.61 | 144 | 27.9 | 4.35 | 14.1 | 4.89 | 1.36 |
| 22 | 5³/₄ | ³/₈ | 10¹/₈ | 0.240 | 6.49 | 118 | 23.2 | 4.27 | 11.4 | 3.97 | 1.33 |
| W10 × 19 | 4 | ³/₈ | 10¹/₄ | 0.250 | 5.62 | 96.3 | 18.8 | 4.14 | 4.29 | 2.14 | 0.874 |
| 17 | 4 | ⁵/₁₆ | 10¹/₈ | 0.240 | 4.99 | 81.9 | 16.2 | 4.05 | 3.56 | 1.78 | 0.844 |
| 15 | 4 | ¹/₄ | 10 | 0.230 | 4.41 | 68.9 | 13.8 | 3.95 | 2.89 | 1.45 | 0.810 |
| 12 | 4 | ³/₁₆ | 9⁷/₈ | 0.190 | 3.54 | 53.8 | 10.9 | 3.90 | 2.18 | 1.10 | 0.785 |
| W8 × 67 | 8¹/₄ | ¹⁵/₁₆ | 9 | 0.570 | 19.7 | 272 | 60.4 | 3.72 | 88.6 | 21.4 | 2.12 |
| 58 | 8¹/₄ | ¹³/₁₆ | 8³/₄ | 0.510 | 17.1 | 228 | 52.0 | 3.65 | 75.1 | 18.3 | 2.10 |
| 48 | 8¹/₈ | ¹¹/₁₆ | 8¹/₂ | 0.400 | 14.1 | 184 | 43.3 | 3.61 | 60.9 | 15.0 | 2.08 |
| 40 | 8¹/₈ | ⁹/₁₆ | 8¹/₄ | 0.360 | 11.7 | 146 | 35.5 | 3.53 | 49.1 | 12.2 | 2.04 |
| 35 | 8 | ¹/₂ | 8¹/₈ | 0.310 | 10.3 | 127 | 31.2 | 3.51 | 42.6 | 10.6 | 2.03 |
| 31 | 8 | ⁷/₁₆ | 8 | 0.285 | 9.13 | 110 | 27.5 | 3.47 | 37.1 | 9.27 | 2.02 |
| W8 × 28 | 6¹/₂ | ⁷/₁₆ | 8 | 0.285 | 8.25 | 98.0 | 24.3 | 3.45 | 21.7 | 6.63 | 1.62 |
| 24 | 6¹/₂ | ³/₈ | 7⁷/₈ | 0.245 | 7.08 | 82.8 | 20.9 | 3.42 | 18.3 | 5.63 | 1.61 |
| W8 × 21 | 5¹/₄ | ³/₈ | 8¹/₄ | 0.250 | 6.16 | 75.3 | 18.2 | 3.49 | 9.77 | 3.71 | 1.26 |
| 18 | 5¹/₄ | ⁵/₁₆ | 8¹/₈ | 0.230 | 5.26 | 61.9 | 15.2 | 3.43 | 7.97 | 3.04 | 1.23 |
| W8 × 15 | 4 | ⁵/₁₆ | 8¹/₈ | 0.245 | 4.44 | 48.0 | 11.8 | 3.29 | 3.41 | 1.70 | 0.876 |
| 13 | 4 | ¹/₄ | 8 | 0.230 | 3.84 | 39.6 | 9.91 | 3.21 | 2.73 | 1.37 | 0.843 |
| 10 | 4 | ³/₁₆ | 7⁷/₈ | 0.170 | 2.96 | 30.8 | 7.81 | 3.22 | 2.09 | 1.06 | 0.841 |
| W6 × 25 | 6¹/₈ | ⁷/₁₆ | 6³/₈ | 0.320 | 7.34 | 53.4 | 16.7 | 2.70 | 17.1 | 5.61 | 1.52 |
| 20 | 6 | ³/₈ | 6¹/₄ | 0.260 | 5.87 | 41.4 | 13.4 | 2.66 | 13.3 | 4.41 | 1.50 |
| 15 | 6 | ¹/₄ | 6 | 0.230 | 4.43 | 29.1 | 9.72 | 2.56 | 9.32 | 3.11 | 1.46 |
| W6 × 16 | 4 | ³/₈ | 6¹/₄ | 0.260 | 4.74 | 32.1 | 10.2 | 2.60 | 4.43 | 2.20 | 0.966 |
| 12 | 4 | ¹/₄ | 6 | 0.230 | 3.55 | 22.1 | 7.31 | 2.49 | 2.99 | 1.50 | 0.918 |
| 9 | 4 | ³/₁₆ | 5⁷/₈ | 0.170 | 2.68 | 16.4 | 5.56 | 2.47 | 2.19 | 1.11 | 0.905 |
| W5 × 19 | 5 | ⁷/₁₆ | 5¹/₈ | 0.270 | 5.54 | 26.2 | 10.2 | 2.17 | 9.13 | 3.63 | 1.28 |
| 16 | 5 | ³/₈ | 5 | 0.240 | 4.68 | 21.3 | 8.51 | 2.13 | 7.51 | 3.00 | 1.27 |
| W4 × 13 | 4 | ³/₈ | 4¹/₈ | 0.280 | 3.83 | 11.3 | 5.46 | 1.72 | 3.86 | 1.90 | 1.00 |

W SHAPES  SI PROPERTIES FOR DESIGNING

| Designation Nominal Depth by mass/length | Flange | | Web | | Section Area ÷ 10³ | Axis X-X | | | Axis Y-Y | | |
|---|---|---|---|---|---|---|---|---|---|---|---|
| | Width | Thickness | Depth | Thickness | | $I \div 10^6$ | $Z \div 10^3$ | $k$ | $I \div 10^6$ | $Z \div 10^3$ | $k$ |
| mm × kg/m | mm | mm | mm | mm | mm² | mm⁴ | mm³ | mm | mm⁴ | mm³ | mm |
| W360 × 147 | 370 | 19.8 | 360 | 12.3 | 18.8 | 462 | 2573 | 156.7 | 167 | 905 | 94.2 |
| 134 | 369 | 18.0 | 356 | 11.2 | 17.1 | 416 | 2343 | 156.0 | 151 | 818 | 94.0 |
| W360 × 122 | 257 | 21.7 | 363 | 13.0 | 15.5 | 367 | 2016 | 153.7 | 61.6 | 480 | 63.0 |
| 100 | 256 | 19.9 | 360 | 11.4 | 14.1 | 331 | 1835 | 153.4 | 55.8 | 436 | 63.0 |
| 101 | 255 | 18.3 | 357 | 10.5 | 12.9 | 301 | 1688 | 152.6 | 50.4 | 397 | 62.5 |
| 91 | 254 | 16.4 | 353 | 9.5 | 11.5 | 266 | 1511 | 151.9 | 44.5 | 352 | 62.2 |
| W360 × 79 | 205 | 16.8 | 354 | 9.4 | 10.1 | 225 | 1275 | 149.6 | 24.0 | 234 | 48.8 |
| 72 | 204 | 15.1 | 350 | 8.6 | 9.1 | 202 | 1152 | 148.6 | 21.4 | 210 | 48.5 |
| 64 | 203 | 13.5 | 347 | 7.7 | 8.13 | 178 | 1027 | 147.8 | 18.8 | 185.2 | 48.0 |
| W360 × 57 | 172 | 13.1 | 358 | 7.9 | 7.23 | 160 | 895 | 149.1 | 11.1 | 129.1 | 39.4 |
| 51 | 171 | 11.6 | 355 | 7.2 | 6.45 | 142 | 796 | 148.1 | 9.70 | 113.2 | 38.9 |
| 44.8 | 171 | 9.8 | 352 | 6.9 | 5.71 | 121 | 688 | 145.5 | 8.16 | 95.4 | 37.8 |
| W360 × 39 | 128 | 10.7 | 353 | 6.5 | 4.96 | 102 | 578 | 143.5 | 3.71 | 58.0 | 27.4 |
| 32.9 | 127 | 8.5 | 349 | 5.8 | 4.19 | 82.8 | 475 | 140.7 | 2.91 | 45.9 | 26.4 |

| | | | | | | | | | | | |
|---|---|---|---|---|---|---|---|---|---|---|---|
| W310 × 179 | 313 | 28.1 | 333 | 18.0 | 22.8 | 445 | 2671 | 140.0 | 144 | 918 | 79.5 |
| 158 | 310 | 25.1 | 327 | 15.5 | 20.1 | 388 | 2376 | 138.9 | 125 | 808 | 79.0 |
| 143 | 309 | 22.9 | 323 | 14.0 | 18.2 | 347 | 2147 | 138.2 | 112 | 728 | 78.5 |
| 129 | 308 | 20.6 | 318 | 13.1 | 16.5 | 308 | 1934 | 136.6 | 100 | 651 | 78.0 |
| 117 | 307 | 18.7 | 314 | 11.9 | 15.0 | 276 | 1753 | 135.6 | 89.9 | 587 | 77.5 |
| 107 | 306 | 17.0 | 311 | 10.9 | 13.6 | 248 | 1596 | 134.9 | 81.2 | 531 | 77.2 |
| 97 | 305 | 15.4 | 308 | 9.9 | 12.3 | 222 | 1440 | 134.1 | 72.4 | 477 | 76.7 |
| W310 × 86 | 254 | 16.3 | 310 | 9.1 | 11.0 | 198 | 1278 | 134.1 | 44.5 | 351 | 63.8 |
| 79 | 254 | 14.6 | 306 | 8.8 | 10.1 | 177 | 1157 | 132.8 | 39.9 | 315 | 63.0 |
| W310 × 74 | 205 | 16.3 | 310 | 9.4 | 9.48 | 164 | 1060 | 131.6 | 23.4 | 228 | 49.8 |
| 67 | 204 | 14.6 | 306 | 8.5 | 8.52 | 146 | 952 | 130.8 | 20.8 | 203 | 49.3 |
| 60 | 203 | 13.1 | 303 | 7.5 | 7.61 | 129 | 850 | 130.3 | 18.4 | 180.3 | 49.0 |
| W310 × 52 | 167 | 13.2 | 317 | 7.6 | 6.65 | 119 | 747 | 133.4 | 10.2 | 122.4 | 39.1 |
| 44.5 | 166 | 11.2 | 313 | 6.6 | 5.67 | 99.1 | 633 | 132.3 | 8.45 | 102.3 | 38.5 |
| 38.7 | 165 | 9.7 | 310 | 5.8 | 4.94 | 84.9 | 547 | 131.3 | 7.20 | 87.5 | 38.4 |
| W310 × 32.7 | 102 | 10.8 | 313 | 6.6 | 4.18 | 64.9 | 416 | 124.7 | 1.94 | 37.9 | 21.5 |
| 28.3 | 102 | 8.9 | 309 | 6.0 | 3.59 | 54.1 | 349 | 122.4 | 1.57 | 30.8 | 20.9 |
| 23.8 | 101 | 6.7 | 305 | 5.6 | 3.04 | 42.9 | 280 | 118.6 | 1.17 | 23.1 | 19.6 |
| 21.0 | 101 | 5.7 | 303 | 5.1 | 2.68 | 36.9 | 244 | 117.4 | 0.98 | 15.5 | 19.1 |

# W SHAPES  SI PROPERTIES FOR DESIGNING

| Designation Nominal Depth by mass/length | Flange | | Web | | Section Area | Axis X-X | | | Axis Y-Y | | |
|---|---|---|---|---|---|---|---|---|---|---|---|
| | Width | Thick-ness | Depth | Thick-ness | $\div 10^3$ | $I$ $\div 10^6$ | $Z$ $\div 10^3$ | $k$ | $I$ $\div 10^6$ | $Z$ $\div 10^3$ | $k$ |
| mm × kg/m | mm | mm | mm | mm | mm² | mm⁴ | mm³ | mm | mm⁴ | mm³ | mm |
| W250 × 167 | 265 | 31.8 | 289 | 19.2 | 21.2 | 298 | 2065 | 118.4 | 98.2 | 742 | 68.1 |
| 149 | 263 | 28.4 | 282 | 17.3 | 19.0 | 259 | 1835 | 116.8 | 86.2 | 655 | 67.3 |
| 131 | 261 | 25.1 | 275 | 15.4 | 16.7 | 222 | 1614 | 115.3 | 74.5 | 570 | 66.8 |
| 115 | 259 | 22.1 | 269 | 13.5 | 14.6 | 189 | 1408 | 114.0 | 64.1 | 493 | 66.0 |
| 101 | 257 | 19.6 | 264 | 11.9 | 12.9 | 164 | 1240 | 112.8 | 55.8 | 433 | 65.8 |
| 89 | 256 | 17.3 | 260 | 10.7 | 11.4 | 142 | 1093 | 111.5 | 48.3 | 377 | 65.3 |
| 80 | 255 | 15.6 | 256 | 9.4 | 10.2 | 126 | 983 | 111.0 | 42.9 | 338 | 65.0 |
| 73 | 254 | 14.2 | 253 | 8.6 | 9.29 | 113 | 895 | 110.5 | 38.9 | 306 | 64.5 |
| W250 × 67 | 204 | 15.7 | 257 | 8.9 | 8.58 | 103.2 | 805 | 109.7 | 22.2 | 218 | 51.0 |
| 58 | 203 | 13.5 | 252 | 8.0 | 7.42 | 87.0 | 690 | 108.5 | 18.7 | 185 | 50.3 |
| 49.1 | 202 | 11.0 | 247 | 7.4 | 6.26 | 70.8 | 574 | 106.4 | 15.2 | 151 | 49.3 |
| W250 × 44.8 | 148 | 13.0 | 266 | 7.6 | 5.70 | 70.8 | 531 | 111.2 | 6.95 | 94.2 | 34.8 |
| 38.5 | 147 | 11.2 | 262 | 6.6 | 4.91 | 59.9 | 457 | 110.5 | 5.87 | 80.1 | 34.5 |
| 32.7 | 146 | 9.1 | 258 | 6.1 | 4.19 | 49.1 | 380 | 108.5 | 4.75 | 65.1 | 33.8 |

# E-4

| | | | | | | | | | | | |
|---|---|---|---|---|---|---|---|---|---|---|---|
| W250 × 28.4 | 102 | 10.0 | 260 | 5.4 | 3.63 | 40.1 | 308 | 105.2 | 1.786 | 35.1 | 22.2 |
| 25.3 | 102 | 8.4 | 257 | 6.1 | 3.22 | 34.1 | 265 | 102.9 | 1.482 | 29.2 | 21.4 |
| 22.3 | 102 | 6.9 | 254 | 5.8 | 2.85 | 28.7 | 226 | 100.3 | 1.203 | 23.8 | 20.6 |
| 17.9 | 101 | 5.3 | 251 | 4.8 | 2.28 | 22.4 | 179 | 99.1 | 0.907 | 18.03 | 19.9 |
| W200 × 100 | 210 | 23.7 | 229 | 14.5 | 12.7 | 113.2 | 990 | 94.5 | 36.9 | 351 | 53.8 |
| 86 | 209 | 20.6 | 222 | 13.0 | 11.0 | 94.9 | 852 | 92.7 | 31.3 | 300 | 53.3 |
| 71 | 206 | 17.4 | 216 | 10.2 | 9.1 | 76.6 | 710 | 91.7 | 25.3 | 246 | 52.8 |
| 59 | 205 | 14.2 | 210 | 9.1 | 7.55 | 60.8 | 582 | 90.0 | 20.4 | 199.9 | 51.8 |
| 52 | 204 | 12.6 | 206 | 7.9 | 6.65 | 52.9 | 511 | 89.2 | 17.73 | 173.7 | 51.6 |
| 46.1 | 203 | 11.0 | 203 | 7.2 | 5.89 | 45.8 | 451 | 88.1 | 15.44 | 151.9 | 51.3 |
| W200 × 41.7 | 166 | 11.8 | 205 | 7.2 | 5.32 | 40.8 | 398 | 87.6 | 9.03 | 108.6 | 41.2 |
| 35.7 | 165 | 10.2 | 201 | 6.2 | 4.57 | 34.5 | 342 | 86.9 | 7.62 | 92.3 | 40.9 |
| W200 × 31.3 | 134 | 10.2 | 210 | 6.4 | 3.97 | 31.3 | 298 | 88.6 | 4.07 | 60.8 | 32.0 |
| 26.6 | 133 | 8.4 | 207 | 5.8 | 3.39 | 25.8 | 249 | 87.1 | 3.32 | 49.8 | 31.2 |
| W200 × 22.5 | 102 | 8.0 | 206 | 6.2 | 2.86 | 20.0 | 193 | 83.6 | 1.419 | 27.9 | 22.2 |
| 19.3 | 102 | 6.5 | 203 | 5.8 | 2.48 | 16.5 | 162 | 81.5 | 1.136 | 22.4 | 21.4 |
| 15.0 | 100 | 5.2 | 200 | 4.3 | 1.91 | 12.8 | 128 | 81.8 | 0.870 | 17.37 | 21.4 |

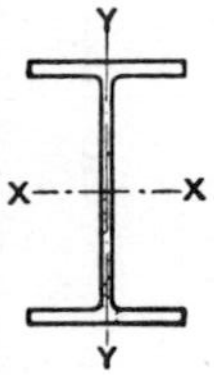

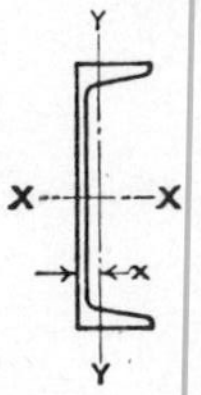

PROPERTIES FOR DESIGNING BEAMS(S)

| Designation Nominal Depth by wt/ft | Flange Width | Flange Thickness | Web Thickness | Section Area | Axis X-X | | | Axis Y-Y | | | x |
|---|---|---|---|---|---|---|---|---|---|---|---|
| | | | | | $I$ | $Z$ | $k$ | $I$ | $Z$ | $k$ | |
| lb/ft | in. | in. | in. | in.² | in.⁴ | in.³ | in. | in.⁴ | in.³ | in. | in. |
| S12 × 50 | 5½ | 11/16 | 0.687 | 14.7 | 305 | 50.8 | 4.55 | 15.7 | 5.74 | 1.03 | |
| 40.8 | 5¼ | 11/16 | 0.462 | 12.0 | 272 | 45.4 | 4.77 | 13.6 | 5.16 | 1.06 | |
| S12 × 35 | 5⅛ | 9/16 | 0.428 | 10.3 | 229 | 38.2 | 4.72 | 9.87 | 3.89 | 0.98 | |
| 31.8 | 5 | 9/16 | 0.350 | 9.35 | 218 | 36.4 | 4.83 | 9.36 | 3.74 | 1.00 | |
| S10 × 35 | 5 | ½ | 0.594 | 10.3 | 147 | 29.4 | 3.78 | 8.36 | 3.38 | 0.901 | |
| 25.4 | 4⅝ | ½ | 0.311 | 7.46 | 124 | 24.7 | 4.07 | 6.79 | 2.91 | 0.954 | |
| S8 × 23 | 4⅛ | 7/16 | 0.441 | 6.77 | 64.9 | 16.2 | 3.10 | 4.31 | 2.07 | 0.798 | |
| 18.4 | 4 | 7/16 | 0.271 | 5.41 | 57.6 | 14.4 | 3.26 | 3.73 | 1.86 | 0.831 | |
| S7 × 20 | 3⅞ | ⅜ | 0.450 | 5.88 | 42.4 | 12.1 | 2.69 | 3.17 | 1.64 | 0.734 | |
| 15.3 | 3⅝ | ⅜ | 0.252 | 4.50 | 36.7 | 10.5 | 2.86 | 2.64 | 1.44 | 0.766 | |
| S6 × 17.25 | 3⅝ | ⅜ | 0.465 | 5.07 | 26.3 | 8.77 | 2.28 | 2.31 | 1.30 | 0.675 | |
| 12.5 | 3⅜ | ⅜ | 0.232 | 3.67 | 22.1 | 7.37 | 2.45 | 1.82 | 1.09 | 0.705 | |

CHANNELS (C)

| | | | | | | | | | | | |
|---|---|---|---|---|---|---|---|---|---|---|---|
| C12 × 30 | 3 1/8 | 1/2 | 0.510 | 8.82 | 162 | 27.0 | 4.29 | 5.14 | 2.06 | 0.763 | 0.674 |
| 25 | 3 | 1/2 | 0.387 | 7.35 | 144 | 24.1 | 4.43 | 4.47 | 1.88 | 0.780 | 0.674 |
| 20.7 | 3 | 1/2 | 0.282 | 6.09 | 129 | 21.5 | 4.61 | 3.88 | 1.73 | 0.799 | 0.698 |
| C10 × 30 | 3 | 7/16 | 0.673 | 8.82 | 103 | 20.7 | 3.42 | 3.94 | 1.65 | 0.669 | 0.649 |
| 25 | 2 7/8 | 7/16 | 0.526 | 7.35 | 91.2 | 18.2 | 3.52 | 3.36 | 1.48 | 0.676 | 0.617 |
| 20 | 2 3/4 | 7/16 | 0.379 | 5.88 | 78.9 | 15.8 | 3.66 | 2.81 | 1.32 | 0.692 | 0.606 |
| 15.3 | 2 5/8 | 7/16 | 0.240 | 4.49 | 67.4 | 13.5 | 3.87 | 2.28 | 1.16 | 0.713 | 0.634 |
| C9 × 20 | 2 5/8 | 7/16 | 0.448 | 5.88 | 60.9 | 13.5 | 3.22 | 2.42 | 1.17 | 0.642 | 0.583 |
| 15 | 2 1/2 | 7/16 | 0.285 | 4.41 | 51.0 | 11.3 | 3.40 | 1.93 | 1.01 | 0.661 | 0.586 |
| 13.4 | 2 3/8 | 7/16 | 0.233 | 3.94 | 47.9 | 10.6 | 3.48 | 1.76 | 0.962 | 0.669 | 0.601 |
| C8 × 18.75 | 2 1/2 | 3/8 | 0.487 | 5.51 | 44.0 | 11.0 | 2.82 | 1.98 | 1.01 | 0.599 | 0.565 |
| 13.75 | 2 3/8 | 3/8 | 0.303 | 4.04 | 36.1 | 9.03 | 2.99 | 1.53 | 0.854 | 0.615 | 0.553 |
| 11.5 | 2 1/4 | 3/8 | 0.202 | 3.38 | 32.6 | 8.14 | 3.11 | 1.32 | 0.781 | 0.625 | 0.571 |
| C7 × 14.75 | 2 1/4 | 3/8 | 0.419 | 4.33 | 27.2 | 7.78 | 2.51 | 1.38 | 0.779 | 0.564 | 0.532 |
| 12.25 | 2 1/4 | 3/8 | 0.314 | 3.60 | 24.2 | 6.93 | 2.60 | 1.17 | 0.703 | 0.571 | 0.525 |
| 9.8 | 2 1/8 | 3/8 | 0.210 | 2.87 | 21.3 | 6.08 | 2.72 | 0.968 | 0.625 | 0.581 | 0.540 |

SI PROPERTIES FOR DESIGNING BEAMS (S)

| Designation Nominal Depth by mass/length | Actual Depth | Flange | | Web Thick-ness | Section Area | Axis X-X | | | Axis Y-Y | | | $\tilde{x}$ |
|---|---|---|---|---|---|---|---|---|---|---|---|---|
| | | Width | Thick-ness | | | $I$ $\div 10^6$ | $Z$ $\div 10^3$ | $k$ | $I$ $\div 10^6$ | $Z$ $\div 10^3$ | $k$ | |
| mm × kg/m | mm | mm | mm | mm | mm² | mm⁴ | mm³ | mm | mm⁴ | mm³ | mm | mm |
| S310 × 74 | 305 | 139 | 16.8 | 17.4 | 9.48 | 127.0 | 832 | 115.6 | 6.53 | 94.1 | 26.2 | |
| 60.7 | 305 | 133 | 16.8 | 11.7 | 7.74 | 113.2 | 744 | 121.2 | 5.66 | 84.6 | 26.9 | |
| S310 × 52 | 305 | 129 | 13.8 | 10.9 | 6.64 | 95.3 | 626 | 120.0 | 4.11 | 63.8 | 24.9 | |
| 47.3 | 305 | 127 | 13.8 | 8.9 | 6.03 | 90.7 | 596 | 122.7 | 3.90 | 61.3 | 25.4 | |
| S250 × 52 | 254 | 126 | 12.5 | 15.1 | 6.64 | 61.2 | 482 | 96.0 | 3.48 | 55.4 | 22.9 | |
| 37.8 | 254 | 118 | 12.5 | 7.9 | 4.81 | 51.6 | 405 | 103.4 | 2.83 | 47.7 | 24.2 | |
| S200 × 34 | 203 | 106 | 10.8 | 11.2 | 4.37 | 27.0 | 265 | 78.7 | 1.794 | 33.9 | 20.3 | |
| 27.4 | 203 | 102 | 10.8 | 6.9 | 3.48 | 24.0 | 236 | 82.8 | 1.553 | 30.5 | 21.1 | |
| S180 × 30 | 178 | 97 | 10.0 | 11.4 | 3.79 | 17.65 | 198.3 | 68.3 | 1.319 | 26.9 | 18.64 | |
| 22.8 | 178 | 92 | 10.0 | 6.4 | 2.89 | 15.28 | 172.1 | 72.6 | 1.099 | 23.6 | 19.46 | |
| S150 × 25.7 | 152 | 90 | 9.1 | 11.8 | 3.27 | 10.95 | 143.7 | 57.9 | 0.961 | 21.3 | 17.15 | |
| 18.6 | 152 | 84 | 9.1 | 5.8 | 2.36 | 9.2 | 120.8 | 62.2 | 0.758 | 17.86 | 17.91 | |

# E-5

CHANNELS (C)

| Designation | | | | | | | | | | | | |
|---|---|---|---|---|---|---|---|---|---|---|---|---|
| C310 × 45 | 305 | 80 | 12.7 | 13.0 | 5.69 | 67.4 | 442 | 109.0 | 2.14 | 33.8 | 19.38 | 17.12 |
| 37 | 305 | 77 | 12.7 | 9.8 | 4.74 | 59.9 | 395 | 112.5 | 1.861 | 30.8 | 19.81 | 17.12 |
| 30.8 | 305 | 74 | 12.7 | 7.2 | 3.93 | 53.7 | 352 | 117.1 | 1.615 | 28.4 | 20.3 | 17.73 |
| C250 × 45 | 254 | 76 | 11.1 | 17.1 | 5.69 | 42.9 | 339 | 86.9 | 1.640 | 27.0 | 16.99 | 16.48 |
| 37 | 254 | 73 | 11.1 | 13.4 | 4.74 | 38.0 | 298 | 89.4 | 1.399 | 24.2 | 17.17 | 15.67 |
| 30 | 254 | 69 | 11.1 | 9.6 | 3.79 | 32.8 | 259 | 93.0 | 1.170 | 21.6 | 17.58 | 15.39 |
| 22.8 | 254 | 65 | 11.1 | 6.1 | 2.90 | 28.1 | 221 | 98.3 | 0.949 | 19.01 | 18.11 | 16.10 |
| C230 × 30 | 229 | 67 | 10.5 | 11.4 | 3.79 | 25.3 | 221 | 81.8 | 1.007 | 19.17 | 16.31 | 14.81 |
| 22 | 229 | 63 | 10.5 | 7.2 | 2.84 | 21.2 | 185.2 | 86.4 | 0.803 | 16.55 | 16.79 | 14.88 |
| 19.9 | 229 | 61 | 10.5 | 5.9 | 2.54 | 19.9 | 173.7 | 88.4 | 0.733 | 15.76 | 16.99 | 15.27 |
| C200 × 27.9 | 203 | 64 | 9.9 | 12.4 | 3.56 | 18.31 | 180.3 | 71.6 | 0.824 | 16.55 | 15.21 | 14.35 |
| 20.5 | 203 | 59 | 9.9 | 7.7 | 2.61 | 15.03 | 148.0 | 76.0 | 0.637 | 13.99 | 15.62 | 14.05 |
| 17.1 | 203 | 57 | 9.9 | 5.6 | 2.18 | 13.57 | 133.4 | 79.0 | 0.549 | 12.80 | 15.88 | 14.50 |
| C180 × 22.0 | 179 | 58 | 9.3 | 10.6 | 2.79 | 11.32 | 127.5 | 63.8 | 0.574 | 12.77 | 14.33 | 13.51 |
| 18.2 | 178 | 55 | 9.3 | 8.0 | 2.32 | 10.07 | 113.6 | 66.0 | 0.487 | 11.52 | 14.50 | 13.34 |
| 14.6 | 178 | 53 | 9.3 | 5.3 | 1.85 | 8.87 | 99.6 | 69.1 | 0.403 | 10.24 | 14.76 | 13.72 |

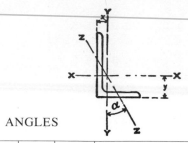

ANGLES

SI PROPERTIES
FOR DESIGNING

| Designation Size and Thickness | Mass per Meter | Section Area | Axis X-X | | | | Axis Y-Y | | | | Axis Z-Z |
|---|---|---|---|---|---|---|---|---|---|---|---|
| | | | $I \div 10^6$ | $Z \div 10^3$ | $k$ | $y$ | $I \div 10^6$ | $Z \div 10^3$ | $k$ | $z$ | $k_{min}$ |
| mm | kg | mm² | mm⁴ | mm³ | mm | mm | mm⁴ | mm³ | mm | mm | mm |
| L127 × 76 × 12.7 | 19.0 | 2419 | 3.93 | 47.7 | 40.4 | 44.5 | 1.074 | 18.8 | 21.1 | 19.1 | 16.5 |
| 9.5 | 14.5 | 1845 | 3.07 | 36.7 | 40.9 | 43.2 | 0.849 | 14.6 | 21.5 | 17.9 | 16.6 |
| 7.9 | 12.1 | 1548 | 2.61 | 31.0 | 40.9 | 42.7 | 0.728 | 12.3 | 21.7 | 17.3 | 16.7 |
| 6.4 | 9.8 | 1252 | 2.13 | 25.1 | 41.1 | 42.2 | 0.599 | 10.1 | 21.9 | 16.7 | 16.8 |
| L102 × 102 × 15.9 | 23.4 | 2974 | 2.77 | 39.3 | 30.5 | 31.2 | 2.77 | 39.3 | 30.5 | 31.2 | 19.8 |
| 12.7 | 19.0 | 2419 | 2.31 | 32.3 | 31.0 | 30.0 | 2.31 | 32.3 | 31.0 | 30.0 | 19.9 |
| 9.5 | 14.6 | 1845 | 1.81 | 24.9 | 31.2 | 29.0 | 1.81 | 24.9 | 31.2 | 29.0 | 20.0 |
| 7.9 | 12.2 | 1548 | 1.54 | 21.1 | 31.5 | 28.4 | 1.54 | 21.1 | 31.5 | 28.4 | 20.1 |
| 6.4 | 9.8 | 1252 | 1.27 | 17.2 | 31.8 | 27.7 | 1.27 | 17.2 | 31.8 | 27.7 | 20.2 |
| L102 × 89 × 12.7 | 17.6 | 2258 | 2.21 | 31.8 | 31.2 | 31.8 | 1.578 | 24.9 | 26.4 | 25.4 | 18.3 |
| 9.5 | 13.5 | 1723 | 1.74 | 24.4 | 31.8 | 30.7 | 1.228 | 19.2 | 26.9 | 24.3 | 18.5 |
| 7.9 | 11.4 | 1452 | 1.48 | 20.6 | 32.0 | 30.0 | 1.061 | 16.3 | 27.2 | 23.7 | 18.5 |
| 6.4 | 9.2 | 1168 | 1.21 | 16.9 | 32.3 | 29.5 | 0.870 | 13.2 | 27.2 | 23.1 | 18.6 |
| L102 × 76 × 12.7 | 16.4 | 2097 | 2.10 | 31.0 | 31.8 | 33.8 | 1.007 | 18.4 | 21.9 | 21.0 | 16.2 |
| 9.5 | 12.6 | 1600 | 1.65 | 23.9 | 32.0 | 32.5 | 0.799 | 14.2 | 22.3 | 19.9 | 16.4 |
| 7.9 | 10.7 | 1348 | 1.41 | 20.2 | 32.3 | 32.0 | 0.687 | 12.0 | 22.5 | 19.3 | 16.4 |
| 6.4 | 8.6 | 1090 | 1.15 | 16.4 | 32.5 | 31.5 | 0.566 | 9.82 | 22.8 | 18.7 | 16.5 |
| L89 × 89 × 9.5 | 12.6 | 1600 | 1.19 | 18.8 | 27.2 | 25.7 | 1.195 | 18.8 | 27.2 | 25.7 | 17.4 |
| 7.9 | 10.7 | 1348 | 1.02 | 16.0 | 27.4 | 25.1 | 1.020 | 16.0 | 27.4 | 25.1 | 17.5 |
| 6.4 | 8.6 | 1090 | 0.84 | 13.0 | 27.7 | 24.6 | 0.837 | 13.0 | 27.7 | 24.6 | 17.6 |
| L89 × 76 × 9.5 | 11.7 | 1484 | 1.13 | 18.5 | 27.7 | 27.4 | 0.770 | 13.95 | 22.8 | 21.1 | 15.9 |
| 7.9 | 9.8 | 1245 | 0.970 | 15.6 | 27.9 | 26.9 | 0.658 | 11.83 | 23.0 | 20.5 | 15.9 |
| 6.4 | 8.0 | 1006 | 0.795 | 12.7 | 28.2 | 26.4 | 0.541 | 9.65 | 23.2 | 19.9 | 16.0 |
| L89 × 64 × 9.5 | 10.7 | 1361 | 1.066 | 17.9 | 27.9 | 29.5 | 0.454 | 9.70 | 18.3 | 16.8 | 13.6 |
| 7.9 | 9.0 | 1148 | 0.912 | 15.2 | 28.2 | 29.0 | 0.391 | 8.26 | 18.5 | 16.2 | 13.7 |
| 6.4 | 7.3 | 929 | 0.749 | 12.4 | 28.4 | 28.2 | 0.323 | 6.75 | 18.7 | 15.6 | 13.8 |
| L76 × 76 × 12.7 | 14.0 | 1774 | 0.924 | 17.5 | 22.8 | 23.7 | 0.924 | 17.5 | 22.8 | 23.7 | 14.8 |
| 9.5 | 10.7 | 1361 | 0.733 | 13.7 | 23.2 | 22.6 | 0.733 | 13.7 | 23.2 | 22.6 | 14.9 |
| 7.9 | 9.1 | 1148 | 0.629 | 11.6 | 23.4 | 22.0 | 0.629 | 11.6 | 23.4 | 22.0 | 15.0 |
| 6.4 | 7.3 | 929 | 0.516 | 9.46 | 23.6 | 21.4 | 0.516 | 9.46 | 23.6 | 21.4 | 15.0 |

# E-6

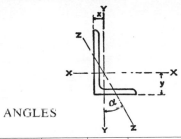

ANGLES                                    PROPERTIES FOR DESIGNING

| Designation Size and Thickness | Weight per Foot | Section Area | Axis X-X | | | | Axis Y-Y | | | | Axis Z-Z $k_{min}$ |
|---|---|---|---|---|---|---|---|---|---|---|---|
| | | | $I$ | $Z$ | $k$ | $y$ | $I$ | $Z$ | $k$ | $x$ | |
| in. | lb | in.² | in.⁴ | in.³ | in. | in. | in.⁴ | in.³ | in. | in. | in. |
| L6 × 6 × ³/₄ | 28.7 | 8.44 | 28.2 | 6.66 | 1.83 | 1.78 | 28.2 | 6.66 | 1.83 | 1.78 | 1.17 |
| ⁵/₈ | 24.2 | 7.11 | 24.2 | 5.66 | 1.84 | 1.73 | 24.2 | 5.66 | 1.84 | 1.73 | 1.18 |
| ¹/₂ | 19.6 | 5.75 | 19.9 | 4.61 | 1.86 | 1.68 | 19.9 | 4.61 | 1.86 | 1.68 | 1.18 |
| ³/₈ | 14.9 | 4.36 | 15.4 | 3.53 | 1.88 | 1.64 | 15.4 | 3.53 | 1.88 | 1.64 | 1.19 |
| L6 × 4 × ³/₄ | 23.6 | 6.94 | 24.5 | 6.25 | 1.88 | 2.08 | 8.68 | 2.97 | 1.12 | 1.08 | 0.860 |
| ⁵/₈ | 20.0 | 5.86 | 21.1 | 5.31 | 1.90 | 2.03 | 7.52 | 2.54 | 1.13 | 1.03 | 0.864 |
| ¹/₂ | 16.2 | 4.75 | 17.4 | 4.33 | 1.91 | 1.99 | 6.27 | 2.08 | 1.15 | 0.987 | 0.870 |
| ³/₈ | 12.3 | 3.61 | 13.5 | 3.32 | 1.93 | 1.94 | 4.90 | 1.60 | 1.17 | 0.941 | 0.877 |
| L6 × 3¹/₂ × ³/₈ | 11.7 | 3.42 | 12.9 | 3.24 | 1.94 | 2.04 | 3.34 | 1.23 | 0.988 | 0.797 | 0.767 |
| ⁵/₁₆ | 9.8 | 2.87 | 10.9 | 2.73 | 1.95 | 2.01 | 2.85 | 1.04 | 0.996 | 0.763 | 0.772 |
| L5 × 5 × ³/₄ | 23.6 | 6.94 | 15.7 | 4.53 | 1.51 | 1.52 | 15.7 | 4.53 | 1.51 | 1.52 | 0.975 |
| ¹/₂ | 16.2 | 4.75 | 11.3 | 3.16 | 1.54 | 1.43 | 11.3 | 3.16 | 1.54 | 1.43 | 0.983 |
| ³/₈ | 12.3 | 3.61 | 8.74 | 2.42 | 1.56 | 1.39 | 8.74 | 2.42 | 1.56 | 1.39 | 0.990 |
| ⁵/₁₆ | 10.3 | 3.03 | 7.42 | 2.04 | 1.57 | 1.37 | 7.42 | 2.04 | 1.57 | 1.37 | 0.994 |
| L5 × 3¹/₂ × ¹/₂ | 13.6 | 4.00 | 9.99 | 2.99 | 1.58 | 1.66 | 4.05 | 1.56 | 1.01 | 0.906 | 0.755 |
| ³/₈ | 10.4 | 3.05 | 7.78 | 2.29 | 1.60 | 1.61 | 3.18 | 1.21 | 1.02 | 0.861 | 0.762 |
| ⁵/₁₆ | 8.7 | 2.56 | 6.60 | 1.94 | 1.61 | 1.59 | 2.72 | 1.02 | 1.03 | 0.838 | 0.766 |
| L5 × 3 × ¹/₂ | 12.8 | 3.75 | 9.45 | 2.91 | 1.59 | 1.75 | 2.58 | 1.15 | 0.829 | 0.750 | 0.648 |
| ³/₈ | 9.8 | 2.86 | 7.37 | 2.24 | 1.61 | 1.70 | 2.04 | 0.888 | 0.845 | 0.704 | 0.654 |
| ⁵/₁₆ | 8.2 | 2.40 | 6.26 | 1.89 | 1.61 | 1.68 | 1.75 | 0.753 | 0.853 | 0.681 | 0.658 |
| ¹/₄ | 6.6 | 1.94 | 5.11 | 1.53 | 1.62 | 1.66 | 1.44 | 0.614 | 0.861 | 0.657 | 0.663 |
| L4 × 4 × ³/₈ | 12.8 | 3.75 | 5.56 | 1.97 | 1.22 | 1.18 | 5.56 | 1.97 | 1.22 | 1.18 | 0.782 |
| ⁵/₁₆ | 9.8 | 2.86 | 4.36 | 1.52 | 1.23 | 1.14 | 4.36 | 1.52 | 1.23 | 1.14 | 0.788 |
| ⁵/₁₆ | 8.2 | 2.40 | 3.71 | 1.29 | 1.24 | 1.12 | 3.71 | 1.29 | 1.24 | 1.12 | 0.791 |
| ¹/₄ | 6.6 | 1.94 | 3.04 | 1.05 | 1.25 | 1.09 | 3.04 | 1.05 | 1.25 | 1.09 | 0.795 |
| L4 × 3¹/₂ × ¹/₂ | 11.9 | 3.50 | 5.32 | 1.94 | 1.23 | 1.25 | 3.79 | 1.52 | 1.04 | 1.00 | 0.722 |
| ³/₈ | 9.1 | 2.67 | 4.18 | 1.49 | 1.25 | 1.21 | 2.95 | 1.17 | 1.06 | 0.955 | 0.727 |
| ⁵/₁₆ | 7.7 | 2.25 | 3.56 | 1.26 | 1.26 | 1.18 | 2.55 | 0.994 | 1.07 | 0.932 | 0.730 |
| ¹/₄ | 6.2 | 1.81 | 2.91 | 1.03 | 1.27 | 1.16 | 2.09 | 0.808 | 1.07 | 0.909 | 0.734 |
| L4 × 3 × ¹/₂ | 11.1 | 3.25 | 5.05 | 1.89 | 1.25 | 1.33 | 2.42 | 1.12 | 0.864 | 0.827 | 0.639 |
| ³/₈ | 8.5 | 2.48 | 3.96 | 1.46 | 1.26 | 1.28 | 1.92 | 0.866 | 0.879 | 0.782 | 0.644 |
| ⁵/₁₆ | 7.2 | 2.09 | 3.38 | 1.23 | 1.27 | 1.26 | 1.65 | 0.734 | 0.887 | 0.759 | 0.647 |
| ¹/₄ | 5.8 | 1.69 | 2.77 | 1.00 | 1.28 | 1.24 | 1.36 | 0.599 | 0.896 | 0.736 | 0.651 |

ANGLES

PROPERTIES FOR DESIGNING

| Designation Size and Thickness | Weight per Foot | Section Area | Axis X-X | | | | Axis Y-Y | | | | Axis Z-Z |
|---|---|---|---|---|---|---|---|---|---|---|---|
| | | | I | Z | k | y | I | Z | k | x | $k_{min}$ |
| in. | lb | in.² | in.⁴ | in.³ | in. | in. | in.⁴ | in.³ | in. | in. | in. |
| L3½ × 3½ × 3/8 | 8.5 | 2.48 | 2.87 | 1.15 | 1.07 | 1.01 | 2.87 | 1.15 | 1.07 | 1.01 | 0.687 |
| 5/16 | 7.2 | 2.09 | 2.45 | 0.976 | 1.08 | 0.990 | 2.45 | 0.976 | 1.08 | 0.990 | 0.690 |
| 1/4 | 5.8 | 1.69 | 2.01 | 0.794 | 1.09 | 0.968 | 2.01 | 0.794 | 1.09 | 0.968 | 0.694 |
| L3½ × 3 × 3/8 | 7.9 | 2.30 | 2.72 | 1.13 | 1.09 | 1.08 | 1.85 | 0.851 | 0.897 | 0.830 | 0.625 |
| 5/16 | 6.6 | 1.93 | 2.33 | 0.954 | 1.10 | 1.06 | 1.58 | 0.722 | 0.905 | 0.808 | 0.627 |
| 1/4 | 5.4 | 1.56 | 1.91 | 0.776 | 1.11 | 1.04 | 1.30 | 0.589 | 0.914 | 0.785 | 0.631 |
| L3½ × 2½ × 3/8 | 7.2 | 2.11 | 2.56 | 1.09 | 1.10 | 1.16 | 1.09 | 0.592 | 0.719 | 0.660 | 0.537 |
| 5/16 | 6.1 | 1.78 | 2.19 | 0.927 | 1.11 | 1.14 | 0.939 | 0.504 | 0.727 | 0.637 | 0.540 |
| 1/4 | 4.9 | 1.44 | 1.80 | 0.755 | 1.12 | 1.11 | 0.777 | 0.412 | 0.735 | 0.614 | 0.544 |
| L3 × 3 × 1/2 | 9.4 | 2.75 | 2.22 | 1.07 | 0.898 | 0.932 | 2.22 | 1.07 | 0.898 | 0.932 | 0.584 |
| 3/8 | 7.2 | 2.11 | 1.76 | 0.833 | 0.913 | 0.888 | 1.76 | 0.833 | 0.913 | 0.883 | 0.587 |
| 5/16 | 6.1 | 1.78 | 1.51 | 0.707 | 0.922 | 0.865 | 1.51 | 0.707 | 0.922 | 0.865 | 0.589 |
| 1/4 | 4.9 | 1.44 | 1.24 | 0.577 | 0.930 | 0.842 | 1.24 | 0.577 | 0.930 | 0.842 | 0.592 |
| 3/16 | 3.71 | 1.09 | 0.962 | 0.441 | 0.939 | 0.820 | 0.962 | 0.441 | 0.939 | 0.820 | 0.596 |

| Designation | | | | | | | | | | | |
|---|---|---|---|---|---|---|---|---|---|---|---|
| L3 × 2½ × 3/8 | 6.6 | 1.92 | 1.66 | 0.810 | 0.928 | 0.956 | 1.04 | 0.581 | 0.736 | 0.706 | 0.522 |
| 1/4 | 4.5 | 1.31 | 1.17 | 0.561 | 0.945 | 0.911 | 0.743 | 0.404 | 0.753 | 0.661 | 0.528 |
| 3/16 | 3.39 | 0.996 | 0.907 | 0.430 | 0.954 | 0.888 | 0.577 | 0.310 | 0.761 | 0.638 | 0.533 |
| L3 × 2 × 3/8 | 5.9 | 1.73 | 1.53 | 0.781 | 0.940 | 1.04 | 0.543 | 0.371 | 0.559 | 0.539 | 0.430 |
| 5/16 | 5.0 | 1.46 | 1.32 | 0.664 | 0.948 | 1.02 | 0.470 | 0.317 | 0.567 | 0.516 | 0.432 |
| 1/4 | 4.1 | 1.19 | 1.09 | 0.542 | 0.957 | 0.993 | 0.392 | 0.260 | 0.574 | 0.493 | 0.435 |
| 3/16 | 3.07 | 0.902 | 0.842 | 0.415 | 0.966 | 0.970 | 0.307 | 0.200 | 0.583 | 0.470 | 0.439 |
| L2½ × 2½ × 3/8 | 5.9 | 1.73 | 0.984 | 0.566 | 0.753 | 0.762 | 0.984 | 0.566 | 0.753 | 0.762 | 0.487 |
| 5/16 | 5.0 | 1.46 | 0.849 | 0.482 | 0.761 | 0.740 | 0.849 | 0.482 | 0.761 | 0.740 | 0.489 |
| 1/4 | 4.1 | 1.19 | 0.703 | 0.394 | 0.769 | 0.717 | 0.703 | 0.394 | 0.769 | 0.717 | 0.491 |
| 3/16 | 3.07 | 0.902 | 0.547 | 0.303 | 0.778 | 0.694 | 0.547 | 0.303 | 0.778 | 0.694 | 0.495 |
| L2½ × 2 × 3/8 | 5.3 | 1.55 | 0.912 | 0.547 | 0.768 | 0.831 | 0.514 | 0.363 | 0.577 | 0.581 | 0.420 |
| 5/16 | 4.5 | 1.31 | 0.788 | 0.466 | 0.766 | 0.809 | 0.446 | 0.310 | 0.584 | 0.559 | 0.422 |
| 1/4 | 3.62 | 1.06 | 0.654 | 0.381 | 0.784 | 0.787 | 0.372 | 0.254 | 0.592 | 0.537 | 0.424 |
| 3/16 | 2.75 | 0.809 | 0.509 | 0.293 | 0.793 | 0.764 | 0.291 | 0.196 | 0.600 | 0.514 | 0.427 |
| L2 × 2 × 3/8 | 4.7 | 1.36 | 0.479 | 0.351 | 0.594 | 0.636 | 0.479 | 0.351 | 0.594 | 0.636 | 0.389 |
| 5/16 | 3.92 | 1.15 | 0.416 | 0.300 | 0.601 | 0.614 | 0.416 | 0.300 | 0.601 | 0.614 | 0.390 |
| 1/4 | 3.19 | 0.938 | 0.348 | 0.247 | 0.609 | 0.592 | 0.348 | 0.247 | 0.609 | 0.592 | 0.391 |
| 3/16 | 2.44 | 0.715 | 0.272 | 0.190 | 0.617 | 0.569 | 0.272 | 0.190 | 0.617 | 0.569 | 0.394 |
| 1/8 | 1.65 | 0.484 | 0.190 | 0.131 | 0.626 | 0.546 | 0.190 | 0.131 | 0.626 | 0.546 | 0.398 |

TWO UNEQUAL ANGLES

PROPERTIES OF SECTIONS
SHORT LEGS BACK TO BACK

| Size | Thickness | Weight per Foot 2 Angles | Area of 2 Angles | Axis X-X | | | | Radii of Gyration about Axis Y-Y Back to Back of Angles, in. | | | | | |
|---|---|---|---|---|---|---|---|---|---|---|---|---|---|
| | | | | I | Z | k | y | 0 | 1/4 | 3/8 | 1/2 | 5/8 | 3/4 |
| in. | in. | lb | in.² | in.⁴ | in.³ | in. | in. | | | | | | |
| 6 × 4 | 5/8 | 40.0 | 11.72 | 15.0 | 5.1 | 1.13 | 1.03 | 2.78 | 2.87 | 2.92 | 2.97 | 3.01 | 3.06 |
| | 9/16 | 36.2 | 10.62 | 13.8 | 4.6 | 1.14 | 1.01 | 2.77 | 2.86 | 2.91 | 2.96 | 3.00 | 3.05 |
| | 1/2 | 32.4 | 9.50 | 12.5 | 4.2 | 1.15 | .99 | 2.76 | 2.85 | 2.90 | 2.95 | 2.99 | 3.04 |
| | 7/16 | 28.6 | 8.36 | 11.2 | 3.7 | 1.16 | .96 | 2.75 | 2.84 | 2.88 | 2.93 | 2.98 | 3.03 |
| | 3/8 | 24.6 | 7.22 | 9.8 | 3.2 | 1.17 | .94 | 2.74 | 2.83 | 2.87 | 2.92 | 2.97 | 3.02 |
| 6 × 3½ | 1/2 | 30.6 | 9.00 | 8.5 | 3.2 | .97 | .83 | 2.83 | 2.92 | 2.97 | 3.02 | 3.07 | 3.12 |
| | 3/8 | 23.4 | 6.84 | 6.7 | 2.5 | .99 | .79 | 2.81 | 2.90 | 2.95 | 3.00 | 3.05 | 3.09 |
| | 5/16 | 19.6 | 5.74 | 5.7 | 2.1 | 1.00 | .76 | 2.80 | 2.89 | 2.94 | 2.99 | 3.03 | 3.08 |
| 5 × 3½ | 3/4 | 39.6 | 11.62 | 11.1 | 4.4 | .98 | 1.00 | 2.34 | 2.43 | 2.48 | 2.53 | 2.58 | 2.63 |
| | 5/8 | 33.6 | 9.84 | 9.7 | 3.8 | .99 | .95 | 2.31 | 2.40 | 2.45 | 2.50 | 2.55 | 2.60 |
| | 1/2 | 27.2 | 8.00 | 8.1 | 3.1 | 1.01 | .91 | 2.29 | 2.38 | 2.43 | 2.48 | 2.53 | 2.58 |
| | 7/16 | 24.0 | 7.06 | 7.3 | 2.8 | 1.01 | .88 | 2.28 | 2.37 | 2.41 | 2.46 | 2.51 | 2.56 |
| | 3/8 | 20.8 | 6.10 | 6.4 | 2.4 | 1.02 | .86 | 2.27 | 2.36 | 2.40 | 2.45 | 2.50 | 2.55 |
| | 5/16 | 17.4 | 5.12 | 5.4 | 2.0 | 1.03 | .84 | 2.26 | 2.35 | 2.38 | 2.43 | 2.48 | 2.53 |
| 5 × 3 | 1/2 | 25.6 | 7.50 | 5.2 | 2.3 | .83 | .75 | 2.36 | 2.46 | 2.50 | 2.55 | 2.60 | 2.65 |
| | 3/8 | 19.6 | 5.72 | 4.1 | 1.8 | .84 | .70 | 2.34 | 2.43 | 2.48 | 2.53 | 2.58 | 2.63 |
| | 5/16 | 16.4 | 4.80 | 3.5 | 1.5 | .85 | .68 | 2.33 | 2.42 | 2.47 | 2.52 | 2.57 | 2.62 |

| Size | Thick. | | | | | | | | | | | | |
|---|---|---|---|---|---|---|---|---|---|---|---|---|---|
| 4 × 3½ | ½ | 23.8 | 7.00 | 7.6 | 3.0 | 1.04 | 1.00 | 1.76 | 1.85 | 1.89 | 1.94 | 1.99 | 2.04 |
| | 7/16 | 21.2 | 6.18 | 6.8 | 2.7 | 1.05 | .98 | 1.75 | 1.84 | 1.89 | 1.94 | 1.98 | 2.03 |
| | 3/8 | 18.2 | 5.34 | 6.0 | 2.3 | 1.06 | .96 | 1.74 | 1.83 | 1.88 | 1.92 | 1.97 | 2.02 |
| | 5/16 | 15.4 | 4.50 | 5.1 | 2.0 | 1.07 | .93 | 1.73 | 1.81 | 1.86 | 1.91 | 1.96 | 2.00 |
| | ¼ | 12.4 | 3.62 | 4.2 | 1.6 | 1.07 | .91 | 1.72 | 1.80 | 1.85 | 1.90 | 1.94 | 1.99 |
| 4 × 3 | 5/8 | 27.2 | 7.96 | 5.7 | 2.7 | .85 | .87 | 1.84 | 1.94 | 1.99 | 2.03 | 2.08 | 2.14 |
| | ½ | 22.2 | 6.50 | 4.8 | 2.2 | .86 | .83 | 1.82 | 1.92 | 1.96 | 2.01 | 2.06 | 2.11 |
| | 7/16 | 19.6 | 5.74 | 4.4 | 2.0 | .87 | .80 | 1.81 | 1.90 | 1.95 | 1.99 | 2.04 | 2.09 |
| | 3/8 | 17.0 | 4.96 | 3.8 | 1.7 | .88 | .78 | 1.80 | 1.89 | 1.94 | 1.98 | 2.03 | 2.08 |
| | 5/16 | 14.4 | 4.18 | 3.3 | 1.5 | .89 | .76 | 1.79 | 1.88 | 1.93 | 1.97 | 2.02 | 2.07 |
| | ¼ | 11.6 | 3.38 | 2.7 | 1.2 | .90 | .74 | 1.78 | 1.87 | 1.92 | 1.96 | 2.01 | 2.06 |
| 3½ × 3 | ½ | 20.4 | 6.00 | 4.7 | 2.2 | .88 | .88 | 1.56 | 1.65 | 1.70 | 1.75 | 1.80 | 1.85 |
| | 7/16 | 18.2 | 5.30 | 4.2 | 2.0 | .89 | .85 | 1.54 | 1.63 | 1.68 | 1.73 | 1.78 | 1.83 |
| | 3/8 | 15.8 | 4.60 | 3.7 | 1.7 | .90 | .83 | 1.53 | 1.62 | 1.67 | 1.72 | 1.77 | 1.82 |
| | 5/16 | 13.2 | 3.86 | 3.2 | 1.4 | .90 | .81 | 1.52 | 1.61 | 1.66 | 1.71 | 1.76 | 1.81 |
| | ¼ | 10.8 | 3.12 | 2.6 | 1.2 | .91 | .79 | 1.52 | 1.61 | 1.65 | 1.70 | 1.75 | 1.80 |
| 3½ × 2½ | ½ | 18.8 | 5.50 | 2.7 | 1.5 | .70 | .70 | 1.62 | 1.71 | 1.76 | 1.81 | 1.86 | 1.91 |
| | 7/16 | 16.6 | 4.86 | 2.5 | 1.4 | .71 | .68 | 1.61 | 1.70 | 1.75 | 1.80 | 1.85 | 1.90 |
| | 3/8 | 14.4 | 4.22 | 2.2 | 1.2 | .72 | .66 | 1.61 | 1.69 | 1.74 | 1.79 | 1.84 | 1.89 |
| | 5/16 | 12.2 | 3.56 | 1.9 | 1.0 | .73 | .64 | 1.60 | 1.68 | 1.73 | 1.77 | 1.82 | 1.88 |
| | ¼ | 9.8 | 2.88 | 1.6 | 0.8 | .74 | .61 | 1.58 | 1.67 | 1.71 | 1.76 | 1.81 | 1.86 |
| 3 × 2½ | ½ | 17.0 | 5.00 | 2.6 | 1.5 | .72 | .75 | 1.35 | 1.45 | 1.50 | 1.55 | 1.60 | 1.65 |
| | 7/16 | 15.2 | 4.42 | 2.4 | 1.3 | .73 | .73 | 1.34 | 1.44 | 1.49 | 1.54 | 1.59 | 1.64 |
| | 3/8 | 13.2 | 3.84 | 2.1 | 1.2 | .74 | .71 | 1.34 | 1.43 | 1.48 | 1.53 | 1.58 | 1.63 |
| | 5/16 | 11.2 | 3.24 | 1.8 | 1.0 | .74 | .68 | 1.32 | 1.41 | 1.46 | 1.51 | 1.56 | 1.50 |
| | ¼ | 9.0 | 2.62 | 1.5 | 0.8 | .75 | .66 | 1.31 | 1.40 | 1.45 | 1.50 | 1.55 | 1.50 |

PROPERTIES OF SECTIONS

LONG LEGS BACK TO BACK

(continued)

TWO UNEQUAL ANGLES

| Size | Thick-ness | Weight per Foot 2 Angles | Area of 2 Angles | Axis X-X | | | | Radii of Gyration about Axis Y-Y | | | | | |
|---|---|---|---|---|---|---|---|---|---|---|---|---|---|
| | | | | | | | | Back to Back of Angles, in. | | | | | |
| | | | | $I$ | $Z$ | $k$ | $y$ | 0 | $1/4$ | $3/8$ | $1/2$ | $5/8$ | $3/4$ |
| in. | in. | lb | in.² | in.⁴ | in.³ | in. | in. | | | | | | |
| 6 × 4 | $5/8$ | 40.0 | 11.72 | 42.1 | 10.6 | 1.90 | 2.03 | 1.53 | 1.62 | 1.66 | 1.71 | 1.76 | 1.80 |
| | $9/16$ | 36.2 | 10.62 | 38.5 | 9.7 | 1.90 | 2.01 | 1.52 | 1.61 | 1.66 | 1.70 | 1.75 | 1.79 |
| | $1/2$ | 32.4 | 9.50 | 34.8 | 8.7 | 1.91 | 1.99 | 1.52 | 1.60 | 1.65 | 1.69 | 1.74 | 1.78 |
| | $7/16$ | 28.6 | 8.36 | 31.0 | 7.7 | 1.92 | 1.96 | 1.50 | 1.59 | 1.63 | 1.68 | 1.72 | 1.77 |
| | $3/8$ | 24.6 | 7.22 | 26.9 | 6.6 | 1.93 | 1.94 | 1.50 | 1.58 | 1.62 | 1.67 | 1.71 | 1.76 |
| 6 × 3½ | $1/2$ | 30.6 | 9.00 | 33.2 | 8.5 | 1.92 | 2.08 | 1.27 | 1.36 | 1.40 | 1.45 | 1.49 | 1.55 |
| | $3/8$ | 23.4 | 6.84 | 25.7 | 6.5 | 1.94 | 2.04 | 1.26 | 1.34 | 1.39 | 1.43 | 1.48 | 1.53 |
| | $5/16$ | 19.6 | 5.74 | 21.8 | 5.5 | 1.95 | 2.01 | 1.26 | 1.33 | 1.38 | 1.42 | 1.46 | 1.51 |
| 5 × 3½ | $3/4$ | 39.6 | 11.62 | 27.8 | 8.6 | 1.55 | 1.75 | 1.40 | 1.49 | 1.54 | 1.59 | 1.64 | 1.69 |
| | $5/8$ | 33.6 | 9.84 | 24.1 | 7.3 | 1.56 | 1.70 | 1.37 | 1.46 | 1.51 | 1.56 | 1.60 | 1.65 |
| | $1/2$ | 27.2 | 8.00 | 20.0 | 6.0 | 1.58 | 1.66 | 1.36 | 1.44 | 1.49 | 1.54 | 1.58 | 1.63 |
| | $7/16$ | 24.0 | 7.06 | 17.8 | 5.3 | 1.59 | 1.63 | 1.35 | 1.43 | 1.47 | 1.52 | 1.57 | 1.62 |
| | $3/8$ | 20.8 | 6.10 | 15.6 | 4.6 | 1.60 | 1.61 | 1.34 | 1.42 | 1.46 | 1.51 | 1.55 | 1.60 |
| | $5/16$ | 17.4 | 5.12 | 13.2 | 3.9 | 1.61 | 1.59 | 1.33 | 1.41 | 1.45 | 1.50 | 1.54 | 1.59 |
| 5 × 3 | $1/2$ | 25.6 | 7.50 | 18.9 | 5.8 | 1.59 | 1.75 | 1.11 | 1.21 | 1.25 | 1.30 | 1.35 | 1.40 |
| | $3/8$ | 19.6 | 5.72 | 14.7 | 4.5 | 1.61 | 1.70 | 1.09 | 1.18 | 1.23 | 1.27 | 1.32 | 1.37 |
| | $5/16$ | 16.4 | 4.80 | 12.5 | 3.8 | 1.61 | 1.68 | 1.09 | 1.17 | 1.22 | 1.26 | 1.31 | 1.36 |

| Size | Thick. | | | | | | | | | | | | |
|---|---|---|---|---|---|---|---|---|---|---|---|---|---|
| 4 × 3½ | 5/8 | 29.4 | 8.60 | 12.7 | 4.7 | 1.22 | 1.29 | 1.46 | 1.55 | 1.60 | 1.65 | 1.70 | 1.75 |
| | 1/2 | 23.8 | 7.00 | 10.6 | 3.9 | 1.23 | 1.25 | 1.44 | 1.53 | 1.58 | 1.63 | 1.67 | 1.72 |
| | 7/16 | 21.2 | 6.18 | 9.5 | 3.4 | 1.24 | 1.23 | 1.44 | 1.52 | 1.57 | 1.62 | 1.66 | 1.71 |
| | 3/8 | 18.2 | 5.34 | 8.4 | 3.0 | 1.25 | 1.21 | 1.43 | 1.52 | 1.56 | 1.61 | 1.66 | 1.70 |
| | 5/16 | 15.4 | 4.50 | 7.1 | 2.5 | 1.26 | 1.18 | 1.42 | 1.50 | 1.55 | 1.59 | 1.64 | 1.69 |
| | 1/4 | 12.4 | 3.62 | 5.8 | 2.1 | 1.27 | 1.16 | 1.41 | 1.49 | 1.54 | 1.58 | 1.63 | 1.67 |
| 4 × 3 | 5/8 | 27.2 | 7.96 | 12.1 | 4.6 | 1.23 | 1.37 | 1.22 | 1.31 | 1.36 | 1.41 | 1.46 | 1.51 |
| | 1/2 | 22.2 | 6.50 | 10.1 | 3.8 | 1.25 | 1.33 | 1.20 | 1.29 | 1.33 | 1.38 | 1.43 | 1.48 |
| | 7/16 | 19.6 | 5.74 | 9.0 | 3.4 | 1.25 | 1.30 | 1.18 | 1.27 | 1.32 | 1.36 | 1.41 | 1.46 |
| | 3/8 | 17.0 | 4.96 | 7.9 | 2.9 | 1.26 | 1.28 | 1.18 | 1.26 | 1.31 | 1.35 | 1.40 | 1.45 |
| | 5/16 | 14.4 | 4.18 | 6.8 | 2.5 | 1.27 | 1.26 | 1.17 | 1.25 | 1.30 | 1.35 | 1.39 | 1.44 |
| | 1/4 | 11.6 | 3.38 | 5.5 | 2.0 | 1.28 | 1.24 | 1.16 | 1.25 | 1.29 | 1.34 | 1.38 | 1.43 |
| 3½ × 3 | 1/2 | 20.4 | 6.00 | 6.9 | 2.9 | 1.07 | 1.13 | 1.25 | 1.34 | 1.38 | 1.43 | 1.48 | 1.53 |
| | 7/16 | 18.2 | 5.30 | 6.2 | 2.6 | 1.08 | 1.10 | 1.23 | 1.32 | 1.37 | 1.41 | 1.46 | 1.51 |
| | 3/8 | 15.8 | 4.60 | 5.4 | 2.3 | 1.09 | 1.08 | 1.22 | 1.31 | 1.36 | 1.40 | 1.45 | 1.50 |
| | 5/16 | 13.2 | 3.86 | 4.7 | 1.9 | 1.10 | 1.06 | 1.22 | 1.30 | 1.35 | 1.39 | 1.44 | 1.49 |
| | 1/4 | 10.8 | 3.12 | 3.8 | 1.6 | 1.11 | 1.04 | 1.21 | 1.29 | 1.34 | 1.38 | 1.43 | 1.48 |
| 3½ × 2½ | 1/2 | 18.8 | 5.50 | 6.5 | 2.8 | 1.09 | 1.20 | .99 | 1.08 | 1.13 | 1.18 | 1.23 | 1.29 |
| | 7/16 | 16.6 | 4.86 | 5.8 | 2.5 | 1.09 | 1.18 | .98 | 1.07 | 1.12 | 1.17 | 1.22 | 1.27 |
| | 3/8 | 14.4 | 4.22 | 5.1 | 2.2 | 1.10 | 1.16 | .97 | 1.07 | 1.11 | 1.16 | 1.21 | 1.26 |
| | 5/16 | 12.2 | 3.56 | 4.4 | 1.9 | 1.11 | 1.14 | .96 | 1.05 | 1.10 | 1.15 | 1.20 | 1.24 |
| | 1/4 | 9.8 | 2.88 | 3.6 | 1.5 | 1.12 | 1.11 | .95 | 1.04 | 1.09 | 1.13 | 1.18 | 1.23 |

## TWO UNEQUAL ANGLES

## SI PROPERTIES OF SECTIONS

LONG LEGS BACK TO BACK

| Size | Thick-ness | Mass 2 Angles | Area 2 Angles | Axis X-X | | | | Radii of Gyration about Axis $Y$-$Y$, mm | | | | | |
| | | | | $I \div 10^6$ | $Z \div 10^3$ | $k$ | $y$ | Back to Back of Angles, mm | | | | | |
| mm × mm | mm | kg/m | mm² | mm⁴ | mm³ | mm | mm | 0 | 8 | 10 | 12 | 16 | 20 |
|---|---|---|---|---|---|---|---|---|---|---|---|---|---|
| 152 × 89 | 9.5 | 34.6 | 4412 | 10.7 | 106 | 49.3 | 51.8 | 32.0 | 34.7 | 35.4 | 36.1 | 37.6 | 39.1 |
| | 7.9 | 29.0 | 3704 | 9.07 | 89.6 | 49.5 | 51.0 | 32.0 | 34.4 | 35.1 | 35.8 | 37.3 | 38.8 |
| 127 × 89 | 19.0 | 58.6 | 7496 | 11.6 | 140 | 39.4 | 44.4 | 35.6 | 38.4 | 39.2 | 39.9 | 41.5 | 43.2 |
| | 12.7 | 40.4 | 5162 | 8.32 | 97.8 | 40.1 | 42.2 | 34.3 | 37.2 | 38.0 | 38.7 | 40.2 | 41.8 |
| | 9.5 | 30.8 | 3936 | 6.49 | 75.2 | 40.6 | 40.9 | 34.0 | 36.6 | 37.3 | 38.0 | 39.5 | 41.1 |
| | 7.9 | 25.8 | 3304 | 5.49 | 63.4 | 40.9 | 40.4 | 33.8 | 36.4 | 37.1 | 37.8 | 39.3 | 40.8 |
| 127 × 76 | 12.7 | 38.0 | 4838 | 7.87 | 95.4 | 40.4 | 44.4 | 28.4 | 31.2 | 32.0 | 32.7 | 34.3 | 35.9 |
| | 9.5 | 29.0 | 3690 | 6.12 | 73.2 | 40.9 | 43.2 | 27.9 | 30.6 | 31.4 | 32.1 | 33.6 | 35.2 |
| | 7.9 | 24.2 | 3098 | 5.20 | 61.8 | 40.9 | 42.7 | 27.7 | 30.4 | 31.1 | 31.8 | 33.3 | 34.8 |
| | 6.4 | 19.6 | 2504 | 4.25 | 50.1 | 41.2 | 42.2 | 27.5 | 30.1 | 30.8 | 31.5 | 33.0 | 34.5 |
| 102 × 89 | 12.7 | 35.2 | 4516 | 4.41 | 63.4 | 31.2 | 31.8 | 36.6 | 39.5 | 40.3 | 41.0 | 42.6 | 44.2 |
| | 9.5 | 27.0 | 3446 | 3.48 | 49.0 | 31.8 | 30.7 | 36.1 | 39.0 | 39.7 | 40.5 | 42.0 | 43.5 |
| | 7.9 | 22.8 | 2904 | 2.96 | 41.5 | 32.0 | 30.0 | 36.1 | 38.6 | 39.3 | 40.1 | 41.6 | 43.1 |
| | 6.4 | 18.4 | 2336 | 2.43 | 33.6 | 32.3 | 29.5 | 35.8 | 38.4 | 39.1 | 39.8 | 41.3 | 42.8 |
| 102 × 76 | 12.7 | 32.8 | 4194 | 4.20 | 61.9 | 31.8 | 33.8 | 30.5 | 33.3 | 34.0 | 34.8 | 36.4 | 38.0 |
| | 9.5 | 25.2 | 3200 | 3.30 | 47.8 | 32.0 | 32.5 | 30.0 | 32.7 | 33.4 | 34.2 | 35.7 | 37.3 |
| | 7.9 | 21.4 | 2696 | 2.81 | 40.5 | 32.3 | 32.0 | 29.7 | 32.4 | 33.1 | 33.9 | 35.4 | 36.9 |
| | 6.4 | 17.2 | 2180 | 2.31 | 32.8 | 32.5 | 31.5 | 29.5 | 32.1 | 32.8 | 33.6 | 35.1 | 36.6 |

SHORT LEGS BACK TO BACK

| | | | | | | | | | | | | | |
|---|---|---|---|---|---|---|---|---|---|---|---|---|---|
| 152 × 89 | 9.5 | 34.6 | 4412 | 2.78 | 40.3 | 25.1 | 20.0 | 71.4 | 74.5 | 75.2 | 76.0 | 77.5 | 79.0 |
|  | 7.9 | 29.0 | 3704 | 2.37 | 34.1 | 25.3 | 19.4 | 71.1 | 74.1 | 74.8 | 75.6 | 77.1 | 78.6 |
| 127 × 89 | 19.0 | 58.6 | 7496 | 4.62 | 72.6 | 24.8 | 25.3 | 59.2 | 62.4 | 63.2 | 64.0 | 65.6 | 67.2 |
|  | 12.7 | 40.4 | 5162 | 3.37 | 51.1 | 25.6 | 23.0 | 58.2 | 61.2 | 61.9 | 62.7 | 64.2 | 65.8 |
|  | 9.5 | 30.8 | 3936 | 2.65 | 39.5 | 25.9 | 21.9 | 57.7 | 60.6 | 61.3 | 62.0 | 63.6 | 65.1 |
|  | 7.9 | 25.8 | 3304 | 2.26 | 33.4 | 26.2 | 21.3 | 57.4 | 60.4 | 61.1 | 61.8 | 63.4 | 64.9 |
| 127 × 76 | 12.7 | 38.0 | 4838 | 2.15 | 37.5 | 21.1 | 19.0 | 59.9 | 63.1 | 63.8 | 64.6 | 66.2 | 67.8 |
|  | 9.5 | 29.0 | 3690 | 1.70 | 29.2 | 21.5 | 17.9 | 59.4 | 62.4 | 63.2 | 64.0 | 65.5 | 67.1 |
|  | 7.9 | 24.2 | 3098 | 1.45 | 24.7 | 21.7 | 17.3 | 59.2 | 62.0 | 62.8 | 63.6 | 65.1 | 66.7 |
|  | 6.4 | 19.6 | 2504 | 1.20 | 20.2 | 21.9 | 16.7 | 59.0 | 61.9 | 62.6 | 63.4 | 64.9 | 66.5 |
| 102 × 89 | 12.7 | 35.2 | 4516 | 3.16 | 49.6 | 26.4 | 25.4 | 44.7 | 47.5 | 48.2 | 49.0 | 50.6 | 52.2 |
|  | 9.5 | 27.0 | 3446 | 2.48 | 38.5 | 26.9 | 24.3 | 44.2 | 47.1 | 47.8 | 48.6 | 50.1 | 51.6 |
|  | 7.9 | 22.8 | 2904 | 2.12 | 32.6 | 27.2 | 23.7 | 43.9 | 46.7 | 47.4 | 48.2 | 49.7 | 51.2 |
|  | 6.4 | 18.4 | 2336 | 1.74 | 26.6 | 27.2 | 23.1 | 43.7 | 46.5 | 47.2 | 47.9 | 49.4 | 51.0 |
| 102 × 76 | 12.7 | 32.8 | 4194 | 2.02 | 36.5 | 22.0 | 21.0 | 46.2 | 49.4 | 50.1 | 50.9 | 52.5 | 54.1 |
|  | 9.5 | 25.2 | 3200 | 1.60 | 28.4 | 22.3 | 19.9 | 45.7 | 48.6 | 49.3 | 50.1 | 51.6 | 53.2 |
|  | 7.9 | 21.4 | 2696 | 1.37 | 24.1 | 22.5 | 19.3 | 45.5 | 48.3 | 49.1 | 49.8 | 51.4 | 53.0 |
|  | 6.4 | 17.2 | 2180 | 1.13 | 19.7 | 22.8 | 18.7 | 45.2 | 48.1 | 48.9 | 49.6 | 51.2 | 52.7 |

ALLOWABLE STRESS, ksi, FOR COMPRESSION
MEMBERS OF 36 ksi (248 MPa) YIELD POINT STEEL

| Main and Secondary Members KL/k not over 120 | | | | | | Main Members KL/k 121 to 200 | | | | Secondary Members L/k 121 to 200 | | | |
|---|---|---|---|---|---|---|---|---|---|---|---|---|---|
| $\frac{KL}{k}$ | $\frac{P}{A}$ | $\frac{KL}{k}$ | $\frac{P}{A}$ | $\frac{KL}{k}$ | $\frac{P}{A}$ | $\frac{KL}{k}$ | $\frac{P}{A}$ | $\frac{KL}{k}$ | $\frac{P}{A}$ | $\frac{L}{k}$ | $\frac{P}{A}$ | $\frac{L}{k}$ | $\frac{P}{A}$ |
| 1 | 21.56 | 41 | 19.11 | 81 | 15.24 | 121 | 10.14 | 161 | 5.76 | 121 | 10.19 | 161 | 7.25 |
| 2 | 21.52 | 42 | 19.03 | 82 | 15.13 | 122 | 9.99 | 162 | 5.69 | 122 | 10.09 | 162 | 7.20 |
| 3 | 21.48 | 43 | 18.95 | 83 | 15.02 | 123 | 9.85 | 163 | 5.62 | 123 | 10.00 | 163 | 7.16 |
| 4 | 21.44 | 44 | 18.86 | 84 | 14.90 | 124 | 9.70 | 164 | 5.55 | 124 | 9.90 | 164 | 7.12 |
| 5 | 21.39 | 45 | 18.78 | 85 | 14.79 | 125 | 9.55 | 165 | 5.49 | 125 | 9.80 | 165 | 7.08 |
| 6 | 21.35 | 46 | 18.70 | 86 | 14.67 | 126 | 9.41 | 166 | 5.42 | 126 | 9.70 | 166 | 7.04 |
| 7 | 21.30 | 47 | 18.61 | 87 | 14.56 | 127 | 9.26 | 167 | 5.35 | 127 | 9.59 | 167 | 7.00 |
| 8 | 21.25 | 48 | 18.53 | 88 | 14.44 | 128 | 9.11 | 168 | 5.29 | 128 | 9.49 | 168 | 6.96 |
| 9 | 21.21 | 49 | 18.44 | 89 | 14.32 | 129 | 8.97 | 169 | 5.23 | 129 | 9.40 | 169 | 6.93 |
| 10 | 21.16 | 50 | 18.35 | 90 | 14.20 | 130 | 8.84 | 170 | 5.17 | 130 | 9.30 | 170 | 6.89 |
| 11 | 21.10 | 51 | 18.26 | 91 | 14.09 | 131 | 8.70 | 171 | 5.11 | 131 | 9.21 | 171 | 6.85 |
| 12 | 21.05 | 52 | 18.17 | 92 | 13.97 | 132 | 8.57 | 172 | 5.05 | 132 | 9.12 | 172 | 6.82 |
| 13 | 21.00 | 53 | 18.08 | 93 | 13.84 | 133 | 8.44 | 173 | 4.99 | 133 | 9.03 | 173 | 6.79 |
| 14 | 20.95 | 54 | 17.99 | 94 | 13.72 | 134 | 8.32 | 174 | 4.93 | 134 | 8.94 | 174 | 6.76 |
| 15 | 20.89 | 55 | 17.90 | 95 | 13.60 | 135 | 8.19 | 175 | 4.88 | 135 | 8.86 | 175 | 6.73 |
| 16 | 20.83 | 56 | 17.81 | 96 | 13.48 | 136 | 8.07 | 176 | 4.82 | 136 | 8.78 | 176 | 6.70 |
| 17 | 20.78 | 57 | 17.71 | 97 | 13.35 | 137 | 7.96 | 177 | 4.77 | 137 | 8.70 | 177 | 6.67 |
| 18 | 20.72 | 58 | 17.62 | 98 | 13.23 | 138 | 7.84 | 178 | 4.71 | 138 | 8.62 | 178 | 6.64 |
| 19 | 20.66 | 59 | 17.53 | 99 | 13.10 | 139 | 7.73 | 179 | 4.66 | 139 | 8.54 | 179 | 6.61 |
| 20 | 20.60 | 60 | 17.43 | 100 | 12.98 | 140 | 7.62 | 180 | 4.61 | 140 | 8.47 | 180 | 6.58 |
| 21 | 20.54 | 61 | 17.33 | 101 | 12.85 | 141 | 7.51 | 181 | 4.56 | 141 | 8.39 | 181 | 6.56 |
| 22 | 20.48 | 62 | 17.24 | 102 | 12.72 | 142 | 7.41 | 182 | 4.51 | 142 | 8.32 | 182 | 6.53 |
| 23 | 20.41 | 63 | 17.14 | 103 | 12.59 | 143 | 7.30 | 183 | 4.46 | 143 | 8.25 | 183 | 6.51 |
| 24 | 20.35 | 64 | 17.04 | 104 | 12.47 | 144 | 7.20 | 184 | 4.41 | 144 | 8.18 | 184 | 6.49 |
| 25 | 20.28 | 65 | 16.94 | 105 | 12.33 | 145 | 7.10 | 185 | 4.36 | 145 | 8.12 | 185 | 6.46 |
| 26 | 20.22 | 66 | 16.84 | 106 | 12.20 | 146 | 7.01 | 186 | 4.32 | 146 | 8.05 | 186 | 6.44 |
| 27 | 20.15 | 67 | 16.74 | 107 | 12.07 | 147 | 6.91 | 187 | 4.27 | 147 | 7.99 | 187 | 6.42 |
| 28 | 20.08 | 68 | 16.64 | 108 | 11.94 | 148 | 6.82 | 188 | 4.23 | 148 | 7.93 | 188 | 6.40 |
| 29 | 20.01 | 69 | 16.53 | 109 | 11.81 | 149 | 6.73 | 189 | 4.18 | 149 | 7.87 | 189 | 6.38 |
| 30 | 19.94 | 70 | 16.43 | 110 | 11.67 | 150 | 6.64 | 190 | 4.14 | 150 | 7.81 | 190 | 6.36 |
| 31 | 19.87 | 71 | 16.33 | 111 | 11.54 | 151 | 6.55 | 191 | 4.09 | 151 | 7.75 | 191 | 6.35 |
| 32 | 19.80 | 72 | 16.22 | 112 | 11.40 | 152 | 6.46 | 192 | 4.05 | 152 | 7.69 | 192 | 6.33 |
| 33 | 19.73 | 73 | 16.12 | 113 | 11.26 | 153 | 6.38 | 193 | 4.01 | 153 | 7.64 | 193 | 6.31 |
| 34 | 19.65 | 74 | 16.01 | 114 | 11.13 | 154 | 6.30 | 194 | 3.97 | 154 | 7.59 | 194 | 6.30 |
| 35 | 19.58 | 75 | 15.90 | 115 | 10.99 | 155 | 6.22 | 195 | 3.93 | 155 | 7.53 | 195 | 6.28 |
| 36 | 19.50 | 76 | 15.79 | 116 | 10.85 | 156 | 6.14 | 196 | 3.89 | 156 | 7.48 | 196 | 6.27 |
| 37 | 19.42 | 77 | 15.69 | 117 | 10.71 | 157 | 6.06 | 197 | 3.85 | 157 | 7.43 | 197 | 6.26 |
| 38 | 19.35 | 78 | 15.58 | 118 | 10.57 | 158 | 5.98 | 198 | 3.81 | 158 | 7.39 | 198 | 6.24 |
| 39 | 19.27 | 79 | 15.47 | 119 | 10.43 | 159 | 5.91 | 199 | 3.77 | 159 | 7.34 | 199 | 6.23 |
| 40 | 19.19 | 80 | 15.36 | 120 | 10.28 | 160 | 5.83 | 200 | 3.73 | 160 | 7.29 | 200 | 6.22 |

Note: MPa = 6.895 × ksi.

# E-9

PIPE DIMENSIONS AND PROPERTIES

| Nominal Diameter | Outside Diameter | Inside Diameter | Thick-ness | Weight | Moment of Inertia | Sectional Area | Radius of Gyration |
|---|---|---|---|---|---|---|---|
| in. | in. | in. | in. | lb/ft | in.⁴ | in.² | in. |

STANDARD WEIGHT—ANSI SCHEDULE 40

| Nominal Diameter | Outside Diameter | Inside Diameter | Thick-ness | Weight | Moment of Inertia | Sectional Area | Radius of Gyration |
|---|---|---|---|---|---|---|---|
| $^1/_2$ | 0.84 | 0.622 | 0.109 | 0.85 | 0.017 | 0.250 | 0.26 |
| $^3/_4$ | 1.050 | 0.824 | 0.113 | 1.13 | 0.037 | 0.333 | 0.33 |
| 1 | 1.315 | 1.049 | 0.133 | 1.68 | 0.087 | 0.494 | 0.42 |
| $1^1/_4$ | 1.660 | 1.380 | 0.140 | 2.27 | 0.195 | 0.669 | 0.54 |
| $1^1/_2$ | 1.900 | 1.610 | 0.145 | 2.72 | 0.310 | 0.799 | 0.62 |
| 2 | 2.375 | 2.067 | 0.154 | 3.65 | 0.666 | 1.075 | 0.79 |
| $2^1/_2$ | 2.875 | 2.469 | 0.203 | 5.79 | 1.530 | 1.704 | 0.95 |
| 3 | 3.500 | 3.068 | 0.216 | 7.58 | 3.017 | 2.228 | 1.16 |
| 4 | 4.500 | 4.026 | 0.237 | 10.79 | 7.233 | 3.174 | 1.51 |
| 6 | 6.625 | 6.065 | 0.280 | 18.97 | 28.14 | 5.581 | 2.25 |

EXTRA STRONG—ANSI SCHEDULE 80

| Nominal Diameter | Outside Diameter | Inside Diameter | Thick-ness | Weight | Moment of Inertia | Sectional Area | Radius of Gyration |
|---|---|---|---|---|---|---|---|
| $1^1/_2$ | 1.900 | 1.500 | 0.200 | 3.63 | 0.391 | 1.068 | 0.61 |
| 2 | 2.375 | 1.939 | 0.218 | 5.02 | 0.868 | 1.477 | 0.77 |
| $2^1/_2$ | 2.875 | 2.323 | 0.276 | 7.66 | 1.924 | 2.254 | 0.92 |
| 3 | 3.500 | 2.900 | 0.300 | 10.25 | 3.894 | 3.016 | 1.14 |
| 4 | 4.500 | 3.826 | 0.337 | 14.98 | 9.610 | 4.407 | 1.48 |
| 6 | 6.625 | 5.761 | 0.432 | 28.57 | 40.49 | 8.405 | 2.20 |

DOUBLE EXTRA STRONG

| Nominal Diameter | Outside Diameter | Inside Diameter | Thick-ness | Weight | Moment of Inertia | Sectional Area | Radius of Gyration |
|---|---|---|---|---|---|---|---|
| $1^1/_2$ | 1.900 | 1.100 | 0.400 | 6.41 | 0.568 | 1.88 | 0.55 |
| 2 | 2.375 | 1.503 | 0.436 | 9.03 | 1.31 | 2.66 | 0.703 |
| $2^1/_2$ | 2.875 | 1.771 | 0.552 | 13.69 | 2.87 | 4.03 | 0.844 |
| 3 | 3.500 | 2.300 | 0.600 | 18.58 | 5.99 | 5.47 | 1.05 |
| 4 | 4.500 | 3.152 | 0.674 | 27.54 | 15.3 | 8.10 | 1.37 |
| 6 | 6.625 | 4,897 | 0.864 | 53.16 | 66.33 | 15.64 | 2.06 |
| 8 | 8.625 | 6.875 | 0.875 | 72.42 | 162.0 | 21.3 | 2.76 |

PIPE    DIMENSIONS AND PROPERTIES

## SI DIMENSIONS AND PROPERTIES

| O.D. | I.D. | $t$ | Mass | $I \div 10^6$ | $A$ | $k$ |
|---|---|---|---|---|---|---|
| mm | mm | mm | kg/m | mm$^4$ | mm$^2$ | mm |
| 88.9 | 77.7 | 5.6 | 11.5 | 1.28 | 1470 | 29.5 |
| | 72.9 | 8.0 | 16.0 | 1.68 | 2030 | 28.7 |
| | 56.9 | 16.0 | 28.8 | 2.55 | 3660 | 26.4 |
| 101.6 | 90.4 | 5.6 | 13.3 | 1.95 | 1690 | 34.0 |
| | 85.6 | 8.0 | 18.5 | 2.60 | 2350 | 33.2 |
| 114.3 | 102.5 | 5.9 | 15.8 | 2.96 | 2010 | 38.4 |
| | 96.7 | 8.8 | 22.9 | 4.09 | 2920 | 37.4 |
| | 80.1 | 17.1 | 41.0 | 6.36 | 5220 | 34.9 |
| 141.3 | 128.1 | 6.6 | 21.9 | 6.35 | 2790 | 47.7 |
| | 122.3 | 9.5 | 30.9 | 8.59 | 3930 | 46.7 |
| | 103.3 | 19.0 | 57.3 | 14.0 | 7300 | 43.8 |
| 168.3 | 154.1 | 7.1 | 28.2 | 11.7 | 3600 | 57.0 |
| | 146.3 | 11.0 | 42.7 | 16.9 | 5440 | 55.8 |
| 219.1 | 202.7 | 8.2 | 42.7 | 30.3 | 5430 | 74.6 |
| | 193.7 | 12.7 | 64.6 | 44.0 | 8230 | 73.1 |

# Index
## Part 1

# Index
## Part 2